Respiration in Aquatic Ecosystems

Respiration in Aquatic Ecosystems

EDITED BY

Paul A. del Giorgio
Université du Québec à Montréal, Canada

Peter J. le B. Williams
University of Wales, Bangor, UK

OXFORD
UNIVERSITY PRESS

OXFORD

UNIVERSITY PRESS

Great Clarendon Street, Oxford OX2 6DP

Oxford University Press is a department of the University of Oxford.
It furthers the University's objective of excellence in research, scholarship,
and education by publishing worldwide in

Oxford New York

Auckland Bangkok Buenos Aires Cape Town Chennai
Dar es Salaam Delhi Hong Kong Istanbul Karachi Kolkata
Kuala Lumpur Madrid Melbourne Mexico City Mumbai Nairobi
São Paulo Shanghai Taipei Tokyo Toronto

Oxford is a registered trade mark of Oxford University Press
in the UK and in certain other countries

Published in the United States
by Oxford University Press Inc., New York

A catalogue record for this title is available from the British Library

Library of Congress Cataloging in Publication Data
(Data available)
ISBN 0 19 852709 8 (hbk)
ISBN 0 19 852708 X (pbk)

10 9 8 7 6 5 4 3 2 1

Typeset by Newgen Imaging Systems (P) Ltd., Chennai, India

Printed in Great Britain
on acid-free paper by Antony Rowe Ltd., Chippenham

Preface

Respiration, in its various biochemical manifestations, is the process by which all organisms obtain vital energy from a variety of reduced compounds, and represents the largest sink of organic matter in the biosphere. Although respiration is at the center of the functioning of ecosystems, much of contemporary ecology has chosen to focus attention and research on the productive processes rather than on catabolism. Almost certainly, respiration represents a major area of ignorance in out understanding the global carbon cycle. We decided to embark in this project during the 2000 ASLO meeting in Copenhagen, right after a session on aquatic respiration organized by Erik Smith and one of us. To our knowledge, it was among the first sessions entirely devoted to aquatic respiration at an international meeting, and although we had hoped to attract some interest, we never expected to fill the largest room in the conference center! It become clear that there was a tremendous interest in the topic within the aquatic scientific community, and that there was a need for synthesis and direction in this area of research. No textbook that had reviewed and synthesized the extant information on respiration in natural aquatic systems, independent of production or other traditional areas of focus. In spite of its obvious ecological and biogeochemical importance, most oceanographic and limnological textbooks usually deal with respiration superficially and only as an extension of production. We set out to fill this gap and provide the first comprehensive review of respiration in aquatic systems, with the help of an outstanding and diverse group of researchers who laid down a biochemical basis, examined the patterns and scales of respiration in diverse aquatic ecosystems. The result is a synthesis that spans the major aquatic ecosystems of the biosphere, and an in depth analysis of the current state of understanding of respiration in aquatic systems. It is our hope that this collective effort will help establish the main scientific questions and challenges that face this particular field, and more important, help bring respiration into focus as a priority area of future research.

The book is mainly directed to respiration occurring in the water column of aquatic systems, because there are other excellent textbooks that extensively discuss sediment metabolism. Sediment respiration is discussed in the broader context in individual chapters, as part of the specific aim of integrating water column and benthic metabolic processes. We first (Chapters 1-5) lay down a basis to understand respiration within major categories of organisms, the photoautotrophs, the bacteria, the protozoa and the planktonic metazoa. We then (Chapters 6-12) consider the process in a series of aquatic ecosystems and the current status of attempts to model respiration in various plankton groups (Chapter 13). As the book progresses we shall seek to assess the current state of knowledge concerning respiration in aquatic ecosystems, and to address a series of questions that are key to understanding this process at the ecosystem and biospheric level.

Finally, we wish to thank Hugh Ducklow, Jonathan Cole, Mike Pace, Ian Joint and Morten Søndergaard for reviewing various portion of this book. We also thank Ian Sherman and Anita Petrie of Oxford University Press for their patience and encouragement for this project, Ann Mitchel for providing a cover design and Paula Conde for help in editing manuscripts. Paul del Giorgio acknowledges the financial support of the National Science and Engineering Council of Canada. Finally, we would like to dedicate this book to Lawrence Pomeroy and to the late Robert Peters, two pioneers and visionaries who understood early on the importance of respiration in aquatic ecosystems.

Contents

List of contributors   ix

1 Respiration in aquatic ecosystems: history and background   1
Peter J. le B. Williams and Paul A. del Giorgio

2 Ecophysiology of microbial respiration   18
Gary M. King

3 Respiration in aquatic photolithotrophs   36
John A. Raven and John Beardall

4 Respiration in aquatic protists   47
Tom Fenchel

5 Zooplankton respiration   57
Santiago Hernández-León and Tsutomu Ikeda

6 Respiration in wetland ecosystems   83
Charlotte L. Roehm

7 Respiration in lakes   103
Michael L. Pace and Yves T. Prairie

8 Estuarine respiration: an overview of benthic, pelagic, and whole system respiration   122
Charles S. Hopkinson, Jr and Erik M. Smith

9 Respiration and its measurement in surface marine waters   147
Carol Robinson and Peter J. le B. Williams

10 Respiration in the mesopelagic and bathypelagic zones of the oceans   181
Javier Arístegui, Susana Agustí, Jack J. Middelburg, and Carlos M. Duarte

11 Respiration in coastal benthic communities   206
Jack J. Middelburg, Carlos M. Duarte, and Jean-Pierre Gattuso

12 Suboxic respiration in the oceanic water column   225
Louis A. Codispoti, Tadashi Yoshinari, and Allan H. Devol

13 Incorporating plankton respiration in models of aquatic ecosystem function   248
Kevin J. Flynn

**14 The global significance of respiration in aquatic ecosystems: from single cells
to the biosphere**   267
Paul A. del Giorgio and Peter J. le B. Williams

List of contributors

Susana Agustí
IMEDEA (CSIC-UIB)
Grupo de Oceanografía Interdisciplinar (GOI)
Instituto Mediterráneo de Estudios
Avanzados, C
Miquel Marqués, 21
07190 Esporles (Mallorca)
Spain

Javier Arístegui
Facultad de Ciencias del Mar
Campus Universitario de Tafira
Universidad de las Palmas de
Gran Canaria
35017 Las Palmas de Gran Canaria
Spain

John Beardall
School of Biological Sciences
Monash University
Clayton
VIC 3800
Australia

Louis A. Codispoti
University of Maryland Center for
Environmental Science
Horn Point Laboratory
P.O. Box 775
Cambridge
MD 21613
USA

Paul A. del Giorgio
Département des sciences biologiques
Université du Québec à Montréal
CP 8888, Succ. Centre Ville
Montréal, Québec
Canada H3C3P8

Allan H. Devol
School of Oceanography
University of Washington
Seattle, WA 98195
USA

Carlos M. Duarte
IMEDEA (CSIC-UIB)
Grupo de Oceanografía Interdisciplinar (GOI)
Instituto Mediterráneo de Estudios Avanzados, C
Miquel Marqués, 21
07190 Esporles (Mallorca)
Spain

Tom Fenchel
Marine Biological Laboratory
University of Copenhagen
DK-3000 Helsingør
Denmark

Kevin J. Flynn
Ecology Research Unit
University of Wales Swansea
Swansea SA2 8PP
UK

Jean-Pierre Gattuso
Laboratoire d'Océanographie de Villefranche
CNRS and University of Paris 6
B. P. 28
06234 Villefranche-sur-Mer Cedex
France

Santiago Hernández-León
Biological Oceanography Laboratory
Facultad de Ciencias del Mar
Universidad de Las Palmas de Gran Canaria
Campus Universitario de Tafira
35017 Las Palmas de GC
Canary Islands
Spain

Charles S. Hopkinson, Jr
Ecosystems Center
Marine Biological Laboratory
Woods Hole, MA 02543
USA

Tsutomu Ikeda
Marine Biodiversity Laboratory
Graduate School of Fisheries Sciences
Hokkaido University
3-1-1 Minato-cho
Hakodate 041-0821
Japan

Gary M. King
Darling Marine Center
University of Maine
Walpole
USA 04573

Jack J. Middelburg
Netherlands Institute of Ecology
(NIOO-KNAW)
P.O. Box 140
4400 AC Yerseke
The Netherlands

Michael L. Pace
Institute of Ecosystem Studies
Millbrook, NY, 12545
USA

Yves T. Prairie
Département des sciences biologiques
Université du Québec à Montréal
CP 8888, Succ. Centre Ville
Montréal
Québec
Canada H3C3P8

John A Raven
Division of Environmental and Applied Biology
University of Dundee at SCRI
Invergowrie
Dundee DD2 5DA
UK

Carol Robinson
Plymouth Marine Laboratory
Prospect Place
West Hoe
Plymouth, PL1 3DH
UK

Charlotte Roehm
Département des sciences biologiques
Université du Québec à Montréal
CP 8888, Succ. Centre Ville
Montréal, Québec
Canada H3C3P8

Erik M. Smith
Department of Biological Sciences
University of South Carolina
Columbia, SC 29208
USA

Peter J. le B. Williams
School of Ocean Sciences
University of Wales, Bangor
Menai Bridge
Anglesey, LL59 5EY
UK

Tadashi Yoshinari
Wadsworth Center
New York State Department of Health
Albany, NY 12201
USA

CHAPTER 1

Respiration in aquatic ecosystems: history and background

Peter J. le B. Williams[1] and Paul A. del Giorgio[2]

[1] *School of Ocean Sciences, University of Wales, Bangor, UK*
[2] *Département des sciences biologiques, Université du Québec à Montréal, Canada*

Outline

This chapter sets out a broad conceptual basis of the biochemistry and ecology of respiration, designed to provide a framework for the subsequent chapters. The various forms of respiration are identified and categorized, as well as the fundamental basis of the trophic connections through the passage of the products of photosynthesis (the acquired protons and electrons) through the food web. The historical development of the understanding of respiration in aquatic ecosystems and communities is mapped out. The chapter concludes by raising a set of key questions, which are addressed in subsequent chapters.

1.1 Introduction

Respiration occurs in all organisms, save obligate fermenters. In its different biochemical manifestations, respiration is the process whereby organisms obtain vital energy from a variety of reduced compounds. At the cellular and organism level, respiration is recognized as a key function and has been extensively studied, with excellent textbooks available on the subject. At the ecosystem level, respiration represents the largest sink for organic matter in the biosphere. Further, the reactants and products of respiration, such as oxygen, carbon dioxide, methane, and low molecular weight compounds, are key players in the function of the biosphere. Although respiration is at the center of the functioning of ecosystems, much of contemporary ecology has chosen to focus attention and research on the productive processes rather than on catabolism. There is some justification for this focus, as without production of new organic matter, respiration would cease. However, the coupling between production and decomposition processes

in natural aquatic ecosystems is much weaker than has traditionally been assumed. There are separations in time—as the products of photosynthesis flow and cycle through the food web, and there are separations in space—as organic material diffuses, advects, or settles out of the area in which it was originally produced. No ecosystem is closed. All aquatic systems receive organic material from, and export material to, adjacent ecosystems and more so between areas within single ecosystems. Production alone thus does not fully explain the magnitude, patterns, and regulation of catabolism nor the accumulation of biomass and other organic material in aquatic systems. The various forms of catabolism may not only be regulated by factors different from those affecting production, but also may have very different consequences at the ecosystem level. Almost certainly respiration represents a major, if not the major, area of ignorance in our understanding of the global carbon cycle. Repeatedly, we find that a lack of understanding of respiration invariably limits our interpretation of photosynthetic measurements (Williams 1993).

In spite of its obvious ecological and biogeochemical importance, most oceanographic and limnological textbooks usually deal with respiration superficially and only as an extension of production. There is at present no textbook that reviews and synthesizes the extant information on respiration *per se* in natural aquatic systems, independent of production or other traditional areas of focus. The exception is perhaps sediment metabolism, which has received considerable attention in past decades, and there are excellent textbooks that extensively discuss sediment respiration (i.e. Fenchel *et al.* 1998). The objective of our book is to fill this gap and provide the first comprehensive review of respiration in aquatic ecosystems. The book emphasizes respiration occurring in the water column of aquatic systems, because this is the aspect that has received the least attention, but benthic and sediment respiration will also be covered in the various chapters, as we explain below.

The book has a group of introductory chapters that deal with the general biochemistry and biology of respiration in the main planktonic organisms: bacteria (King, Chapter 2), phytoplankton (Raven and Beardall, Chapter 3), protists (Fenchel, Chapter 4), and zooplankton (Hernández-León and Ikeda, Chapter 5). The book does not explicitly include respiration of larger metazoans, such as fish and aquatic mammals. Although respiration is key to understanding the function and performance of these organisms in nature, from an ecosystem standpoint the contribution of larger organisms to total respiration is generally small. Exceptions to this are coral reefs and benthic meiofauna, which are treated in Chapter 11 by Middelburg *et al.* Likewise, there is no chapter that deals exclusively with the respiration of macrophytes (i.e. sea grasses, macroalgae, and various freshwater vascular plants), but the metabolism of these groups is discussed in separate chapters.

These introductory sections are followed by a group of chapters devoted to exploring the magnitude, variation, and regulation of respiration at the ecosystem level in the major aquatic systems of the world: freshwater bogs and swamps (Roehm, Chapter 6), lakes (Pace and Prairie, Chapter 7), estuaries (Hopkinson and Smith, Chapter 8), surface layers of coastal and open oceans (Robinson and

Williams, Chapter 9), dark open ocean layers (Arístegui *et al.*, Chapter 10), and coastal benthic communities (Middelburg *et al.*, Chapter 11). Middelburg *et al.* (Chapter 11) deal exclusively with sediment and benthic respiration in coastal marine ecosystems, including that of macrophyte beds and coral reefs. Also explored are various aspects of sediments respiration in lakes (Pace and Prairie, Chapter 7), estuaries (Hopkinson and Smith, Chapter 8), and deep ocean sediments (Arístegui *et al.*, Chapter 10). Codispoti *et al.* (Chapter 12) discuss the magnitude, regulation, and biogeochemical importance of nitrogen respiration in suboxic layers in the column of the world's oceans, and Flynn in Chapter 13 explores conceptual issues concerning the modeling of respiration at different levels of organization, and discusses the inclusion of respiration in existing physiological and carbon flow models. The final chapter (del Giorgio and Williams, Chapter 14) synthesizes the ecological and biogeochemical importance of respiration across aquatic systems and in the biosphere and integrates the information and the main conclusions of the different chapters of the book.

The aim of this introductory chapter is to provide a general historical, biochemical, and ecological background to the topics covered in this book, and to set the following chapters in some broad context.

1.2   Respiration in a biochemical context

Barber and Hilting (2002), when considering the history of the measurement of plankton productivity, observed that "It is difficult to study something without a concept of what is to be studied." This would seem to apply equally to the history of the measurement of respiration and to some degree contrives to bedevil its study.

The term respiration is used to cover a variety of very different processes: for example, Raven and Beardall (Chapter 3) identify 6 oxygen-consuming reactions in algae; King in Chapter 2 lists 14 electron acceptors other than oxygen (NO_3^-, NO_2^-, Fe^{3+}, Mn^{4+}, SO_4^{2-}, S^0, CO_2, quinone function groups of humic acids, arsenate (As^{6+}), perchlorate, uranium (U^{6+}), chromium (Cr^{6+}), selinate (Se^{6+}), halogen-C bonds) that organisms exploit for respiratory purposes.

At its most fundamental, respiration involves the transfer of protons and electrons from an internal donor to a receptor—the latter almost invariably drawn into the cell by diffusion or facilitated diffusion from the external environment. Oxygen in the contemporary environment is the principal, and in most cases, the ultimate electron acceptor. Again in the contemporary environment, for heterotrophic organisms, organic material is the principal donor in the majority of, but not all, cases.

We may divide the many processes described as respiration into two broad categories:

1. Reactions that occur in the light and involve oxygen cycling and energy dissipation.
2. Reactions that occur both in the light and dark and effect the acquisition of energy.

1.2.1 Reactions involved in oxygen cycling and energy dissipation

The first class of reactions seems to be restricted to the photoautotrophs. They occur in the light and are closely linked with either the electron transport or the carbon assimilation processes of the photosynthetic mechanism. Importantly these involve energy dissipation and oxygen cycling. They take two forms, which are intimately associated with photosynthesis and both associated with light. The Mehler reaction (see Raven and Beardall, Chapter 3, Section 3.3.3 and Fig. 3.2) takes the reductant and electrons from photosystem I, through a series of reactions and disposes them by the conversion of oxygen, ultimately to water. It gives rise to little or no provision of energy and most probably its function is to dispose of oxygen or reductant in excess of those required to reduce carbon dioxide, nitrate, and sulfate for organic production (Falkowski and Raven 1997). There is no concomitant release of carbon dioxide associated with the Mehler reaction.

The second reaction, or complex of reactions, in this class is the ribulose-*bis*-phosphate carboxylase oxygenase (RUBISCO)/photorespiration pathway. These reactions are a consequence of the remarkable dual oxygenase–carboxylase function of enzyme RUBISCO. The balance between these two functions is controlled by the P_{O2}/P_{CO2} ratio: high ratios facilitate the expression of the oxygenase function, low ratios the carboxylase function. At high P_{O2}/P_{CO2} ratios, RUBISCO effects a substrate level oxidation of ribulose-1,5-*bis*-phosphate with the immediate formation of phosphoglycolate, one molecule of oxygen is then consumed in the process, and the phosphoglycolate converted to glycolate. Then follows a series of reactions (see Raven and Beardall, Chapter 3, Fig. 3.3) that involve the release of a single molecule of carbon dioxide and the eventual reformation of ribulose-1,5-*bis*-phosphate. In some respects, the term photo*respiration* is a misnomer when applied to this cycle of reactions, as it is a substrate level oxidation and the term *respiration* is not normally extended to this form of oxygen consumption. The reaction sequence does generate some energy in the form of reduced nicotinamide-adenine dinucleotide (NADH). Raven and Beardall regard the reaction sequence to be primarily a means of scavenging the unavoidable phosphoglycolate that arises at high P_{O2}/P_{CO2} ratios. However, many algae, and all cyanobacteria possess active CO_2-concentrating mechanisms, based on active transport of inorganic carbon species that elevate CO_2 concentrations, and hence lower P_{O2}/P_{CO2} ratios, at the active site of RUBISCO and thereby suppress photorespiration.

The two reactions, the Mehler reaction and the RUBISCO oxygenase, would seem to serve as safety valves to release the excess reductants associated with the photoreactions. They would seem to have little, in the case of the Mehler reaction nothing, directly to do with organic production or decomposition, although they often and wrongly are perceived to have a bearing on the measurement of rates of organic respiration and production in aquatic ecosystems.

1.2.2 Reactions involved in the provision of energy

Reactions involved in the provision of energy may be regarded as respiration in the true sense. They incur the transfer of protons and electrons from reduced substrates (internally derived in the case of the pure autotrophs, externally in the case of the pure heterotrophs) to a proton and electron acceptor. The proton and electron donors are commonly organic compounds, but may also comprise a range of reduced inorganic substrates (see King, Chapter 2). The predominant proton acceptor is oxygen but again a range of other oxidized inorganic and organic (e.g. the quinine part of humic acids) substrates may also serve as acceptors. If the thermodynamic circumstances are favorable, then microorganisms seem to have evolved the facility to extract the energy from almost any coupled redox reaction. Ecologically,

reactions associated with this form of respiration are *significant*—in that they provide energy for all but a small fraction of heterotrophic organisms; *evident*—in that they help shape the organic and many parts of the inorganic environment; and *important*—in that they represent the largest sink of organic matter in the biosphere. They are a major factor determining net ecosystem production and also the degradation (from the human perspective) of the environment by giving rise to anoxia.

The biochemical processes involved, at their simplest, are summarized in Fig. 1.1. The overall respiration process may be reduced to a small

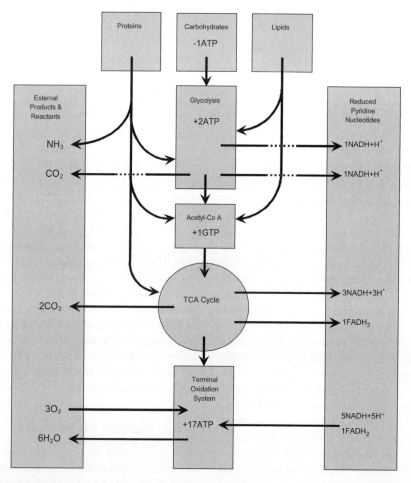

Figure 1.1 The stoichiometry for the metabolism of a three carbon compound entering glycolysis. ATP yield (19ATP/3C, that is, 38/hexose) assumes 1ATP/GTP, 2ATP/FADH$_2$, and 3ATP/NADH. Details from Mathews and Holde (1990). (TCA = tricarboxylic acid cycle; NADH = reduced nicotinamide-adenine dinucleotide; FADH$_2$ = reduced flavin-adenine dinucleotide).

number of stages. The core may be taken to be the Krebs or tricarboxylic acid (TCA) cycle and the terminal oxidation system (TOS). The former generates two molecules of carbon dioxide and a number of molecules of reduced nucleotides (three molecules of NADH and one of FADH$_2$ (reduced flavin adenine dinucleotide) per cycle). The latter are then fed into the terminal oxidation system. At this, the final stage, the electrons and protons associated with the reduced nucleotides (arising from the TCA cycle and other reactions) are used to reduce oxygen to water. The resultant electron flow along a voltage gradient gives rise to the formation of metabolic energy in the form of adenosine triphosphate (ATP). The TCA cycle in the case of carbohydrates and other classes of compounds is fed by glycolysis. This results in the formation of small quantities of NADH$^+$ (two molecules per three carbon atoms) and the eventual production of a molecule of acetyl coenzyme A, a molecule of guanosine triphosphate (GTP) and a molecule of carbon dioxide. The combined stoichiometry of this sequence of reactions (glycolysis, the TCA cycle, and the TOS) results in the production of three molecules of CO$_2$ and the consumption of an equal number of O$_2$ molecules. This stoichiometry is the basis of the respiratory quotient (RQ $= \Delta$CO$_2/-\Delta$O$_2$), which in the simple case of carbohydrates is unity, that is, 3CO$_2$/3O$_2$. Other classes of molecules enter the glycolysis/TCA/TOS cluster of reactions at a number of points and this, as well as substrate level oxidations *en route*, give rise to a range of RQ values. Substrate level oxidations in themselves yield little energy as compared with respiration in the "true sense." The pathway for fats (triglycerides) involves both glycolysis (the glycerol part) and the β-oxidation pathway (the fatty acid part), the latter giving rise to the production of acetyl coenzyme A, a variable amount of reduced pyridine nucleotides, but no carbon dioxide. This results in the RQ for lipids being reduced to a value nearer to 2/3. The metabolism of amino acids is more complex, as there are multiple point entry into both glycolysis and the TCA cycle.

It is important to note that there is no catabolic generation of ammonia or phosphate within the glycolysis/TCA/TOS complex of reactions; the formation of these molecules occurs elsewhere

during metabolism—outside the reactions that may be seen to constitute respiration. This has both ecological and practical implications, in part because respiration has often been estimated indirectly on the basis of nutrient cycling, and conversely, nutrient regeneration has been deduced from respiration, despite the fact that the relationship between the two processes is complex and highly variable.

1.2.3 The respiratory quotient

The common preference to use oxygen as a determinant for the measurement of respiration and the customary use of carbon as currency in ecosystem models always brings to the fore the question of the appropriate conversion factor that is, what value to use for the RQ. This matter is raised in a number of chapters. This has been dealt with briefly from a fundamental biochemical aspect in the previous section. It may also be derived from simple stoichiometric calculations for potential organic substrates. The general solution for the oxidation of a compound with an elemental ratio C$_\alpha$H$_\beta$O$_\chi$N$_\delta$P$_\varepsilon$S$_\phi$ is:

$$C_\alpha H_\beta O_\chi N_\delta P_\varepsilon S_\phi + \gamma O_2$$
$$= \alpha\, CO_2 + \delta NH_3 + \varepsilon H_3PO_4$$
$$+ 0.5[\beta - 3(\delta + \varepsilon) + 2\phi]H_2O + \phi H_2SO_4$$
$$\gamma = \alpha + 0.25\beta - 0.5\chi - 0.75\delta + 1.25\varepsilon + 1.5\phi$$
$$\text{Thus RQ} = \alpha/\gamma$$

The extreme range of RQ values based upon the oxidation of organic material varies from 0.5 (methane) to 1.33 (glycollic acid), the more conventional biochemical compounds and food stuffs fall in the much narrower range of 0.67–1.24, values for planktonic material fall in an even narrower range (see Rodrigues and Williams 2001 and Table 1.1).

This analysis only applies to conventional organic respiration, with oxygen as the proton acceptor. As one moves away from this to the events in Box II in Fig. 1.3, where no carbon dioxide is produced, to Box VII where there in no consumption of oxygen and to Box VIII (the chemoautotrophs) where there is autotrophic assimilation of CO$_2$ based on the energy obtained by respiration, any

Table 1.1 Stoichiometric calculations of the respiratory coefficient.

Category	Respiratory substrate	Elemental composition or proportions	RQ
Upper boundary	Glycolic acid	$C_2H_4O_3$	1.33
Biochemical groupings	Nucleic acid	$C_1H_{1.3}O_{0.7}N_{0.4}P_{0.11}$	1.24
	Polysaccharide	$(C_6H_{10}O_5)n$	1.0
	Protein	$C_1H_{1.58}O_{0.33}N_{0.26}$	0.97
	Saturated fatty acid	$(-CH_2-)n$	0.67
Planktonic materials	Algal cell material (Williams and Robertson 1991)	40% protein, 40% carbohydrate, 15% lipid, 5% nucleic acid $C_1H_{1.7}O_{0.43}N_{0.12}P_{0.0046}$	0.89
	Planktonic material (Hedges *et al.* 2002)	65% protein, 19% lipid, 16% carbohydrate $C_1H_{1.7}O_{0.35}N_{0.16}S_{0.004}$	0.89
Lower boundary	Methane	CH_4	0.5

Note: the calculations assume that the respiratory product is ammonia, i.e. there is no nitrification.

close biochemical constraints on the $\Delta CO_2/\Delta O_2$ ratio are lost. The RQ has no useful meaning in these situations. Experimental observations of RQ may exceed the theoretical boundaries of aerobic heterotrophic respiration (see Hernández-León and Ikeda, Chapter 5; Roehm, Chapter 6; and Arístegui *et al.*, Chapter 10). This may be in part a consequence of chemoautotrophy, but more likely is a problem associated with the difficulties of obtaining sufficiently accurate measurements of carbon dioxide production. There are still relatively few direct measurements of RQ for the various components of aquatic ecosystems, and the vast majority of studies of respiration are based on assumed values, generally between 0.8 and 1.2. However, it is unlikely that this uncertainty in RQ will result in fundamentally biased data in carbon currency, but this is clearly one more element that adds uncertainty to current estimates of aquatic respiration.

1.3 History of the measurement of respiration in aquatic ecosystems

Much of the work on the study of respiration in aquatic ecosystems, and certainly the early work, employed the Winkler technique. The technique was developed in 1888 by Winkler, applied quite soon afterward in 1892 by Natterer to seawater, and then apparently first used in 1917 by Gran

to measure biological oxygen flux. This paper was largely missed and it awaited the subsequent publication by Gaarder and Gran (1927) before the approach received any recognition. These workers' studies were directed to the understanding of photosynthesis and the dark bottle (from which the measurement of respiration derives) merely served as a means of correcting the measurement of photosynthesis. This relegation of oxygen consumption to a correction rather that a process in its own right, persisted through the work of Steemann Nielsen in the 1930s and 1950s, and it would seem beyond. Riley, no doubt as a result of his schooling under Hutchinson at Yale, was probably the first to clearly treat respiration as a process in its own right. The near exclusive focus on photosynthesis remained in the marine field arguably until the latter part of 1960s when specific studies began of the respiratory consumption of oxygen (Pomeroy and Johannes 1966, 1968) and the production of carbon dioxide from added labeled organic substrates (Williams and Askew 1968; Andrews and Williams 1971).

It is not entirely clear why there should have been this single-minded attention to photosynthesis. It is of course the principal mechanism of formation of organic material in the biosphere, but the scope of modern ecology lies well beyond this single focus. It may have arisen from the expectation of early marine scientists, such as Hensen, who felt that if they could develop a simple method to measure marine productivity then they would be able to predict the potential of fisheries (Barber and Hilting 2002). If this were so, then it is ironic to note that within a year of the publication of what appears to have been a remarkably accurate map of the distribution of global primary productivity (Koblenz-Mishke *et al.* 1968) it was shown by Ryther (1969) that primary production itself was a minor factor determining fisheries potential, which he demonstrated to be overwhelmingly controlled by food chain structure and trophic efficiency.

Perhaps one of the reasons why respiration was not regarded as ecologically important during the first half of the twentieth-century was that the prevailing view at the time did not consider microbes as quantitatively important in aquatic ecosystems. This view was the direct result of the massive underestimation of microbial biomass by the methods employed at the time, which led to the conclusion that the respiration of aquatic heterotrophs was probably very modest and of little or no biogeochemical significance (Zobell 1976). The respiration of the algae, on the other hand, was considered worth measuring if only to derive an estimate of net primary production.

The principal exception to this line of thinking occurred in eastern Europe, and particularly in the Russian school of aquatic ecology, which early on placed a strong focus on mineralization of organic matter in lakes, rivers, and the oceans, as opposed to the almost exclusive focus on the production of organic matter that has long dominated most in Western countries. One of the reasons the Russians may have been more preoccupied with respiration may be precisely that they had developed techniques to directly assess aquatic microbes, and early on had realized the quantitative importance of bacteria in aquatic ecosystems. The Russian researchers had hypothesized that bacteria played a much larger role in carbon cycling in both lakes and oceans (Romenenko 1964) long before this became an accepted paradigm in the West, and the obvious consequence of the realization that these high rates of organic matter "destruction" (as was often termed in the Russian literature, e.g. Gambaryan 1962), had to be accompanied by high rates of heterotrophic metabolism, particularly respiration. For example, Kusnetzow and Karsinkin (1938) had estimated that bacterial abundance in the mesotrophic lake, Glubokoye, was in the range of 2×10^6 cells cm^{-3}, many orders of magnitude higher than what researchers in other countries were obtaining for similar systems based on culturing techniques (c.f. figure for bacterial numbers from viable plate counts is given in Section 1.4.4), and concluded that much of the oxygen consumption in the lake was due to these very abundant bacteria. There is a stark contrast between the messages that come out of ZoBell's (ZoBell 1946) and Kriss's (Kriss 1963) books on marine microbiology. There are differences that can be attributed to the passage of time between the publication of the two books, but fundamental underpinning philosophies did not change that rapidly. A modern microbial ecologist would recognize the system Kriss was describing, not so Zobell's description of marine microbiology.

This early view of respiration as an index of the degradation of organic matter did not only have an ecological interest, but took an important practical twist. During the first half of the twentieth-century, Europe, and to a lesser extent North America, was facing critical problems of water pollution, particularly organic pollution from untreated sewage affecting rivers. Researchers searching for tools that would allow them to quantify and characterize the extent of the organic matter pollution in rivers (Kolkwitz and Marsson 1908) realized early on that it was often more effective to determine the organic matter degradation from either changes in *in situ* oxygen concentration, or from the rates of oxygen consumption in bottle incubations, than from changes in the organic matter pool itself (Sladécek 1973). In addition, researchers began to develop the concept of labile organic matter, as opposed to total organic matter load, and further realized that respiration provided a more realistic measure of the labile pool than any chemical measure that could be made (Streeter and Phelps 1925). Thus, measurements of respiration (or biochemical oxygen demand as referred to in this context, Dobbins *et al.* 1964) remained much more widely used in sanitary engineering than in aquatic ecology studies, in spite of the obvious ecological value of these measurements.

The effect of the introduction of the ^{14}C technique in 1952 was to focus work to the photosynthetic process. Up until then, a dark bottle measurement would have been made along with the light bottle measurement of photosynthesis and so the databases of photosynthesis and respiration progressed at more or less the same rate. The arrival of the ^{14}C technique gave rise to an unprecedented step change in the measurement of oceanic photosynthesis (see Fig. 1.2). Between the publication of the ^{14}C technique in 1952 and the time the two Pomeroy and Johannes papers on respiration were published (in 1966 and 1968), we had accrued some 7000 profiles of ^{14}C-determined production (probably of the order of 35 000 or so individual measurements) and we were able to produce credible maps of the distribution of oceanic photosynthesis (Koblenz-Mishke et al. 1968). Over the same period, it is hard to find a single reported measurement of respiration in the oceans. Without doubt, the introduction of the ^{14}C technique seriously held back the development of our understanding of respiration and net community production in the oceans and still does. It is interesting to pose the question, if in the long run the enthusiastic adoption of the ^{14}C-technique has been detrimental to our broader understanding of the carbon cycle and the ecology and trophodynamics of aquatic systems. Without the advent of the ^{14}C technique we would almost certainly know much more about net community production in lakes and oceans. The question is very topical as, with the development of the fast repetition rate fluorometer, we have what many biological oceanographers may see as the holy grail—an instrument that will provide real time in situ measurements of photosynthesis. This could produce a second step change in the measurement of planktonic productivity. It will be interesting to see if history repeats itself as there is a real danger that the introduction of the instrument could for the second time detract from the study of respiration which has evolved, not without difficulty or opposition, over the past decade. In our judgment that would be tragic and a major setback to our understanding of the carbon cycle.

The two papers of Pomeroy and Johannes (1966, 1968) kindled interest in oceanic respiration. Pomeroy and Johannes (1966) worked with concentrates (up to 4000-fold), which allowed rates to be measured using oxygen electrodes over periods of an hour. Previous work by Riley, and Gaarder and Gran had used several day incubations. Riley's observations were criticized for the long incubation times and it was largely because of this that he lost the debate with Steemann Nielsen over the productivity of oligotrophic waters—despite the fact that 50 years later his observations, crude though they were, are probably no less at variance to contemporary values than Steemann Nielsen's ^{14}C measurements. There was a long and unnecessary delay before techniques were developed to allow the measurement of respiration (and photosynthesis) with what may be regarded acceptable incubation times (12–24 h). Such measurements awaited the development of high precision oxygen titrations in the mid- to late 1970s (e.g. Bryan et al. 1976), although in principle the enabling techniques had been present since the early 1960s, if not before. Clearly although the paper by Pomeroy and Johannes had awakened interest in respiration, by no means did it send a shock wave through biological oceanography!

The sparsity of direct respiration measurements is similarly characteristic of lakes and rivers, in spite of the early recognition that respiration plays a major role in determining oxygen concentration in these systems, of its link to organic matter supply and lability, and to whole ecosystem metabolism (Odum 1956; Hutchinson 1957). One of the most evident signs of respiration in lakes is the systematic decline in oxygen concentration in the hypolimnetic layers, especially during summer stratification, which often leads to anoxia and its associated chemical and biological consequences (Hutchinson 1957). Much of this oxygen consumption was thought to occur in the sediments, not in the water column (Hayes and Macaulay 1959), as the former was known to contain a high abundance of bacteria. The consequence was that there were comparatively few direct measurements made of plankton respiration in the decades that followed. The few measurements reported during these decades were carried out in the context of lake primary production studies and were mostly "dark oxygen bottle" values, used for the calculation of gross primary production (i.e. Talling 1957). As a result, when del Giorgio and Peters (1993) reviewed

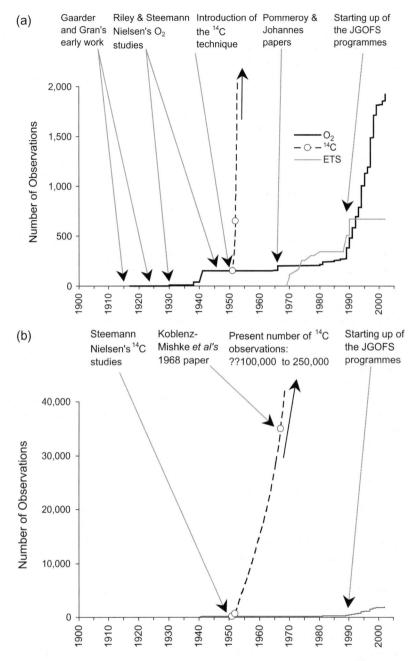

Figure 1.2 The time development of respiration (both electron transport system (ETS) and O_2-determined and ^{14}C-determined photosynthesis in the study of oceanic productivity. The 1968 figure for the number of ^{14}C-observations is taken from the paper of Koblenz-Mishske et al. (1968), assuming five depths per profile. (a) Figure scaled to show the development of the number of respiration measurement up to 2002. (b) Figure scaled to the estimated number of ^{14}C observation reported in the 1968 paper of Koblenz-Mishske et al. Note that up until the 1980s the O_2-based observation was almost contained within the thickness of the line! The range given for the number of accumulated ^{14}C observations is a best guess, derived from estimates of numbers of biological oceanographic cruises and the productivity measurements made as part of them, the preceeding question marks draw attention the uncertainty in the estimate.

the published literature on plankton respiration and production in lakes, they were able to obtain less than 100 individual published measurements of respiration worldwide, performed in the previous three decades, in contrast to the many hundreds or possibly thousands of measurements of primary production published during the same period. The situation has improved somewhat since 1993, and Pace and Prairie (Chapter 7) were able to extract roughly the same number of measurements of lake plankton respiration published during the past decade alone, suggesting that more researchers may be focusing on lake respiration. In any event, the total available dataset for lake respiration remains extremely small and largely inadequate, as Pace and Prairie discuss in Chapter 7.

Thus, the histories of the development of the study of respiration in fresh and marine waters would appear to show distinct parallels, the immediate cause for the lack of attention to pelagic respiration may have been different in the two ecosystems, but very likely the ultimate villain—the viable plate count technique for enumerating bacteria—was the same.

An expansion of the study of respiration in biological oceanography could have been expected in the mid-1980s as interest in the carbon cycle in the ocean broadened and the transfer of organic and inorganic carbon into the deeper parts of the ocean was recognized as an important biogeochemical step in the overall cycle (Berger *et al.* 1989). A major stimulus to the study of the export of carbon to deeper water came from an awareness that the deep water in the oceans was a sink for the accumulating carbon dioxide in the atmosphere. Part of the export to the deeper water is driven by biological processes, collectively termed the biological pump. In considering the export of organic material, the rate of photosynthesis is of secondary interest, the major determinant being the net productivity of surface waters, that is, the difference between photosynthesis and respiration. The interest in the biological pump might have been expected to have been a major stimulus to studies of respiration in the oceans as net community production (i.e. photosynthesis minus respiration) would provide a measure of the potential for export.

Instead, much of the emphasis was directed to the study of the flux of organic material to the deeper parts of the oceans using sediment traps, and the prediction of export as new production from mass-balance considerations based on the input of inorganic nitrogen from deeper water. No measurement of either net community production or respiration was included in the core set of measurements for major programs such as JGOFS. Despite this, groups of workers on both sides of the Atlantic used these major programs as an opportunity to make measurements of respiration and net community production. As a consequence, the measurements of respiration increased exponentially from about 1989 onward (see Fig. 1.2(a)), lagging some 40–50 years behind the growth in photosynthetic measurements and still on a much smaller scale.

In the 1970s, there was a development parallel to the classical Winkler dark bottle measurement of respiration. Packard (1971) developed a procedure to estimate respiration based on the assay of the terminal oxidation system (see Fig. 1.1)—the so-called ETS (electron transport system) technique. The technique had the capability to record respiration at all depths and all times in the oceans. For reasons that remain something of a mystery, the technique attracted attention, but few followers. In part, Packard was probably ahead of his time and suffered as others have done for it. His timing was also perhaps unfortunate in that he probably hit a perigee of interest in the study of respiration in aquatic ecosystems. There was no objective reason to reject the technique, its uncertainties were certainly no less than, for example, the thymidine method for measuring bacterial growth, which was introduced a few years later and was widely and enthusiastically adopted. For whatever reasons, despite heroic efforts by Packard and his colleagues, the technique was largely put to one side for plankton community respiration measurements in the latter part of the 1980s, only to be revived again in the mid-1990s mainly on the European side of the Atlantic by Arístegui, Lefèvre, Dennis, and others (see Hernández-León and Ikeda, Chapter 5, and Arístegui *et al.*, Chapter 10). At about the same time the technique was taken up by freshwater scientists.

1.4 Respiration in a community context

1.4.1 Ecological terminology

The biochemical breakdown of extracellular organic material is the prerogative of heterotrophic organisms. A variety of terms is associated with the activity of these organisms: oxidation, decomposition, degradation, mineralization, remineralization, and respiration. Although these terms are frequently used as synonyms in the ecological literature, in practice they are not, and Section 1.2.2 allows a basis to distinguish between them. The terms mineralization and remineralization are associated with the return of nitrogen and phosphorus to their inorganic state, although it could also be taken to include the return of carbon to carbon dioxide. Whereas the last process could be regarded to be respiration, as pointed out in Section 1.2.2, the release of inorganic nitrogen and phosphate is biochemically distinct from respiration in the strict sense, thus logically mineralization (or remineralization) should be seen to be a process separate from, but in a loose way connected to, respiration. The terms degradation and decomposition probably describe the overall sequence of transformations, mediated by enzymes produced by heterotrophic microbes, that lead from particulate to dissolved, from large to small molecules, and eventually to either inorganic carbon or organic compounds that are resistant to further degradation or decomposition. Decomposition of organic matter may begin outside the bacterial cells, involving abiotic and/or exoenzymatic breakdown of large molecules to produce smaller units that can be taken up by the cells. Not all the organic material that results from the extracellular decomposition or degradation will be taken up by bacteria and of that taken up, a portion will eventually fuel respiration and thus be completely oxidized. Another portion will be used for biosynthesis. In this context, respiration is not equal to organic matter decomposition or degradation, but rather quantifies one of the possible outcomes of the degradation process. The products of organic matter degradation eventually fuel respiration, and the energy obtained through respiration supports further degradation of organic matter, the link between the two processes however is not simple. At the ecosystem or biosphere level and over relatively long timescales, most of the organic matter synthesized by plants, as well as most of the secondary production supported by it, is not only degraded but also eventually respired. So on these scales, respiration is an index of total organic matter degradation, total oxidation, and therefore, of the total organic carbon flux through the biota. But exactly over what temporal and spatial scales of respiration and organic matter degradation are linked remains a major gap in our current knowledge of aquatic metabolism.

These arguments would lead to the view that although all these processes are ultimately linked together biochemically, it is useful to separate the processes, respiration, decomposition, and (re)mineralization, when we consider their measurement and interpretation.

1.4.2 Conceptual basis

Respiration in some respects is a more complex process to study than photosynthesis as it occurs in a number of forms (see Raven and Beardall, Chapter 3), involves a much greater variety of reactants (King, Chapter 2), is a feature of all organisms, and occurs in all parts of aquatic environments (Williams 1984). We commonly treat it mathematically and conceptually as a process linked to photosynthesis, the linkages occurring at different scales (from physiological to ecological) and having characteristic timescales from milliseconds to millennia. The relationship between respiration and primary production in aquatic systems is complex, and certainly not well understood, and modeling this relationship is still a major conceptual challenge, as is explored in various chapters of the book, notably by Flynn in Chapter 13.

At its simplest, the ecosystem may be seen as a production line with protons and electrons being abstracted by photosynthesis from water at one end (oxygen being the primary by-product), and at the other the protons and electrons being returned by respiration to form water from the oxygen by-product. Falkowski (2002) observed that the ecosystem was a market place for the protons and electrons generated by photosynthesis. Figure 1.3 is an attempt to portray the essential elements of this sequence of events and flow of biological currency within an aquatic community. It depicts the major traders—autotrophs, aerobic and anaerobic heterotrophs, and chemoautotrophs, and the currency—electrons and photons, the last transported characteristically, but not invariably, within the food web associated with carbon in the form of organic material. The figure also shows the inputs and outputs of the primary reactants and products: oxygen, carbon dioxide, water, and energy (photons and heat). The diagram makes clear that when all the processes (Boxes I–VIII) are considered, the fluxes of oxygen and carbon dioxide are not tightly coupled and should not be seen as simple alternative measurements. As they may be associated with different processes, they provide different information.

1.4.3 Measurement of respiration

As is depicted in Fig. 1.3 and discussed in the section of the biochemistry, the term respiration includes a wide and heterogeneous set of reactions. The aerobic water column and aerated sediments primarily support the reactions I–VI and VIII. Anoxic sediments, and waters of depleted and exhausted oxygen content, mainly sustain reactions shown in Box VII.

Oxic environments
When studying respiration (and in the case of the oxygen technique also photosynthesis) with the conventional dark/light bottle technique it is clear that respiration can only be measured in the dark and so the Mehler and the RUBISCO oxidase/-phosphoglycollate reactions (Boxes II and III) will not be included in the measurement. Thus, total oxygen-based respiration is underestimated. If the respiration measurement is used in conjunction with a measurement of net oxygen production to calculate photosynthesis, then gross production will be similarly underestimated (this would also be the case for the ETS technique). It is possible to circumvent this problem and derive a measurement

of "total respiration," by working backwards by using an $H_2^{18}O$ tracer to enable the measurement of "true" gross photosynthetic oxygen production (it can be achieved also by isotope dilution of added $^{18}O_2$). Then, from measurements of net oxygen production, total O_2-based respiration can be calculated (Bender *et al.* 1987). This of course would include all substrate-level oxidations, not just the RUBISCO oxidase/phosphoglycollate reaction. The step up in technology and particularly expense is significant and it is not given that measurements of "total" respiration and "true" gross oxygen production are more valuable to the ecologist than the rates provided by the classical light and dark bottle. The standard productivity method—the ^{14}C technique—likewise would fail to measure photosynthesis that feeds the Mehler and possibly also the RUBISCO oxidase/phosphoglycollate reactions. Whether one needs to go to the additional effort and expense rests on the question to be answered. The Mehler and RUBISCO oxidase reactions are intimately linked to the light reaction and can be viewed as tightly coupled cycles within the photoautotroph and by themselves not a matter of great significance to the ecologist. If however, one wants a fundamental understanding of quantum yield with the intention of developing ecosystem models that give an accurate and fundamental account of the transfer of photons to electrons and protons, then there is no question that the $^{18}O_2$ approach must be used, as rates determined by either the classical oxygen, or the ^{14}C technique, cannot be used to parameterize such models. Models of this type are rare, if they exist at all, in the field of aquatic ecology. If one is ultimately interested in organic production, then with such models one would face the subsequent dilemma to build in and parameterize a theoretical basis for disposal of the electrons and protons that are not used for organic production, that is, one would need submodels of the Mehler and the RUBISCO oxidase complex of reactions. Such models would need to be extremely accurate as the flux of protons/electrons through the Mehler reaction alone is substantial and may exceed those passing onto carbon fixation by a factor of five (Luz *et al.* 2002). It is doubtful that in the foreseeable future that such models could

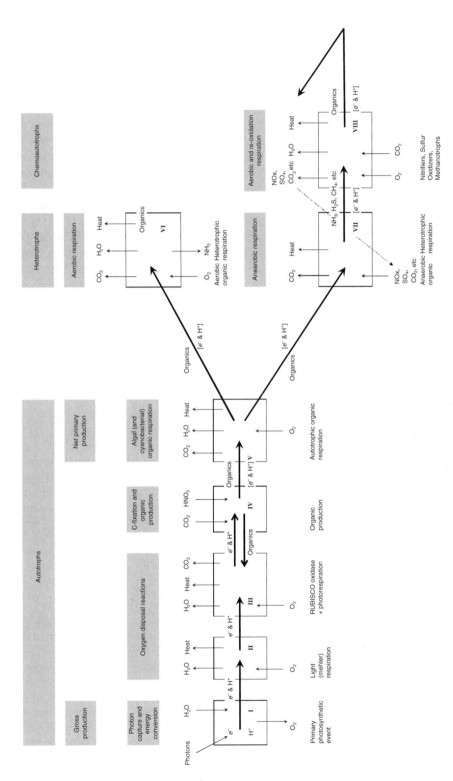

Figure 1.3 Conceptual diagram of the flux of protons (H^+) and electrons (e^-) through an ecosystem. The diagram shows photosynthesis (Box I) and the four basic categories of aerobic respiration: oxygen disposal (Mehler, Box II), substrate level RUBISCO oxidase (Box III), organic (Boxes V and VI), and inorganic (Box VIII). The various forms of anaerobic respiration are given in Box VII. Note that CO_2 is only a product of the middle four forms of respiration (Boxes III, V, VI, and VII). The chemoautotrophic reactions (Box VIII) involves the consumption of both CO_2 and O_2. The NO_x associated with boxes VII and VIII refer to nitrate and nitrite.

be written, certainly with the accuracy required (better than 95% if one wanted organic production measurements to have an accuracy of 25% or better). Whether or not one chooses to measure the light-associated forms of respiration is really a matter of "horses for courses," the crux being whether one is driven by physiological or ecological questions. The physiologist will want to have accurate models of quantum yield, whereas the ecologist, who is mainly interested in organic production, will be principally concerned in the events from Box IV onward in Fig. 1.3, and presently seems prepared to settle for empirical rather than fundamental models of quantum efficiency. This being the case, it may be regarded as no great loss, perhaps even fortuitous, that neither the ^{14}C nor the conventional light/dark O_2 technique takes account of these light-associated energy disposing reactions.

Anoxic and low oxygen environments

The problems associated with the measurement of respiration in oxic environments pale into insignificance when compared with those associated with anoxic and suboxic systems, where there are fourteen or so alternative proton acceptors, some of which can act as proton donors as well as acceptors. In some cases (Fe^{2+}, H_2S) there is the additional complication of comparatively fast competing chemical reactions. Two broad approaches exist, either to treat environments containing the complex of mix of oxic and anoxic subsystems, as a black box and simply determine the exchange between adjacent fully oxic environments of a terminal reactant, carbon dioxide, sometimes nitrate but more often oxygen. Alternatively one may attempt to disentangle the processes going on within samples taken from suboxic/anoxic systems. The former approach is predominantly used with sediments; the rates being determined by rates of concentration change in restricted enclosures placed above sediments or from flux measurements derived from concentration profiles and diffusion coefficients. In the water column where the inhomogeneities and gradients are less marked than in sediments, the discrete sample technique have more application. Conventional chemical, as well as enzymatic and stable isotope procedures, have been used to identify processes and to determine rates.

1.4.4 Distribution of respiration within the community

Once an understanding of the rates of respiration is gained then follows the question of the organisms responsible. In the case of sediments the answer has always been very clear—there were evident large numbers of bacteria and associate protozoa and so these organisms were seen to be the major players. With pelagic systems early workers found themselves facing a dilemma. It is clear that researchers, such as Krogh (1934), knew intuitively that in environments such as the oceans, bacteria must play an important quantitative role in the decomposition process, however this could not be reconciled with the then estimates of bacterial numbers in the water column (<10 cells cm^{-3}, i.e., about 0.05 µg dm^{-3}) obtained by plate counts on agar media. ZoBell, whose work had a widespread effect on the thinking of marine scientists at the time, and even more unfortunately subsequently, was of the view that bacteria were responsible for less than 1% of overall respiration of organic material (ZoBell 1976). The breakthrough to understanding the true scale of microbial processes in the sea followed from the development of the acridine orange technique for estimating bacterial numbers (Daley and Hobbie 1975). Despite the resultant 100 000-fold or so upward revision in bacterial numbers, the view that heterotrophic respiration and mineralization was dominated by the zooplankton persisted for another 15 years— notably in the modeling community. For example, the first widely accepted marine planktonic ecosystem model to incorporate bacteria as a specific component did not appear until 1990 (Fasham *et al.* 1990). Even now the information we have available to us on the distribution of respiration within the broad sectors of the planktonic community is pitifully small.

1.5 Respiration in a whole ecosystem context

As one moves from the biochemistry and physiology of respiration to processes on a whole ecosystem scale, the focus moves to the sources and the nature of organic matter which cycles in the system and supports respiration, and to the difference between production and consumption processes and the linkages between them. The linkage may be on a variety of timescales so the balance will shift with time, with short or extended periods of positive and negative imbalance.

One of the problems that complicates our understanding of the links between respiration and local production in aquatic ecosystems is that there are often large movements of organic matter from one ecosystem to another, or between various parts of any given ecosystem. These movements of material occur because no ecosystem is closed and all export and import material to varying degrees. The subsidy that inland waters may receive from allochthonous sources, both natural and anthropogenic, may be substantial, in acute pollution circumstances overwhelming the internal inputs. All water bodies export organic material. The control of the rate of export has as its basis the excess between the rates of the internal and external supply, the rates of internal removal (by respiration and net sedimentation), and the flushing time of the water body. In the case of lakes, streams, and rivers, the geographical exit from the water bodies is generally reasonably well defined. As one moves seaward, through estuaries, to the coastal ocean and to the open ocean the boundaries become large and frequently difficult to define. Very large bodies of water, such as the open oceans, contain geographical, biogeochemical, and hydrographic subregions (Longhurst 1998). Between these regions there is the possibility of transfer by advection and a resultant imbalance of their individual local budgets (e.g. Duarte and Agusti 1998). The transfer between sectors of the oceans may be massive in their own right, yet remain a very small fraction (a few percentage) of the overall balance sheet of a particular part of the ocean (Hansell *et al.* 2004). With the lack of clear boundaries and the lack of geographical confines, auditing the carbon balance of major sectors of the oceans is a formidable challenge. As a simple illustration, our budgets of the delivery of dissolved organic material from the river to the oceans would appear to be reasonably well constrained, estimates range over a factor of 2 (10–21 Tmol C a^{-1}; Cauwet 2002), whereas the span of estimates for the export of carbon from the coastal ocean to the open ocean has an uncertainty of an order of magnitude greater (17–390 Tmol C a^{-1}; Liu *et al.* 2000).

One of the theses that we attempt to develop in the book is that respiration is perhaps the best index of the flow of organic matter in aquatic ecosystems, because it integrates all the different sources of organic carbon, as well as the temporal variability in its rate of supply. In this regard, respiration is an effective tracer of the movement of organic matter in the environment, and as such should be an extremely useful tool in constraining biogeochemical models of organic matter flux in aquatic ecosystems. In practice, however, respiration has most often been derived from these models rather than used to independently constrain them. This may be in part due to the scarcity of direct respiration data, and in part to the traditional view of respiration as a process that is completely dependent on production.

The advancement of our understanding of respiration in aquatic ecosystems will require the acquisition of more and better data, but perhaps first and more importantly a profound adjustment of the conceptual framework in which we place respiration in aquatic ecosystems. Before this can be effectively attempted, it is important to assess the contemporary theoretical and empirical understanding of aquatic respiration and this is the prime objective of the book. We first (Chapters 1–5) lay down a basis to understand respiration within major categories of organisms, the photoautotrophs, the bacteria, the protozoa and the planktonic metazoa. We then (Chapters 6–12) consider the process in a series of aquatic ecosystems and the current status of attempts to model respiration in various plankton groups (Chapter 13). As the book progresses we shall seek to assess the current state of knowledge concerning respiration in aquatic ecosystems, and to address a series of questions that are key to understanding this process at the ecosystem and biospheric level:

1. What data do we have on respiration for the different aquatic ecosystems of the world?

2. What is the magnitude of respiration in the major aquatic ecosystems of the biosphere?

3. How is this total respiration partitioned between the benthos and the pelagia, and among the major biological components of aquatic food webs?

4. How does respiration vary seasonally and along spatial environmental gradients?

5. What are the main factors controlling respiration in aquatic ecosystems?

6. What are the major connections between primary production and respiration, what is the balance between these two processes, and what are the primary factors controlling this balance?

7. How do we incorporate respiration to models of ecosystem function and carbon and energy dynamics—specifically, by difference, or as a function of photosynthesis?

8. What the main technical and conceptual obstacles to furthering our understanding of aquatic respiration?

9. How will global climate change and other major environmental perturbations influence respiration in aquatic ecosystems?

We return to these matters in our synthesis in Chapter 14. As the readers will find as they cover the different sections of the book, we still do not have satisfactory answers for most of these questions, and the general conclusion of most chapters is that at present, we can barely begin to address even the most basic problems, such as the actual magnitude of respiration in the major aquatic ecosystems. But this exercise allows us to define the major gaps and weaknesses in our understanding, and to suggest avenues that might guide future research efforts in this critical area of ecology and biogeochemistry.

Acknowledgments

The authors are very grateful for thoughtful comments on the chapter from Hugh Ducklow, Tony Fogg, Ian Joint, John Raven, John Beardall and Morten Søndegaard.

References

Andrews, P. and Williams, P. J. le B. 1971. Heterotrophic utilisation of dissolved compounds in the sea. III: measurements of the oxidation rates and concentrations of glucose and amino acids in sea water. *J Mar. Biol. Assoc. UK*, **51**: 111–125.

Barber, R. T. and Hilting, A. K. 2002. History of the study of plankton productivity. In P. J. le B. Williams, D. N. Thomas, and C. S. Reynolds (eds) *Phytoplankton Productivity: Carbon Assimilation in Marine and Freshwater Ecosystems*. Blackwell Science Ltd, Oxford, UK, pp. 16–43.

Bender, M., Grande, K., Johnson, K., Marra, J., Williams, P. J. le B., Sieburth, J., Pilson, M., Langdon, C., Hitchcock, G., Orchardo, J., Hunt, C., Donaghay, P., and Heinemann, K. 1987. A comparison of four methods for determining planktonic community production. *Limnol. Oceanogr.*, **32**: 1085–1098.

Berger, W. H., Smetacek, V. S., and Wefer, G. 1989. Ocean productivity and paleoproductivity—an overview. In W. S. Berger, V. S. Smetacek, and G. Wefer (eds) *Productivity of the Ocean: Present and Past*. John Wiley and Sons, Chichester, pp. 1–34.

Bryan, J. R., Riley, J. P., and Williams, P. J. le B. 1976. A procedure for making precise measurements of oxygen concentration for productivity and related studies *J. Exp. Mar. Biol., Ecol.*, **21**: 191–197.

Cauwet, G. 2002. DOM in the coastal zone. In D. A. Hansell and A. Craig Carlson (eds) *Biogeochemistry of Marine Dissolved Matter*. Academic Press, San Diego, pp. 579–609.

Daley, R. J. and Hobbie, J. E. 1975. Direct counts of aquatic bacteria in a modified epifluoresence technique. *Limnol. Oceanogr.*, **31**: 875–882.

del Giorgio, P. A. and Peters, R. H. 1993. Balance between phytoplankton production and plankton respiration in lakes. *Can. J. Fish. Aquat. Sci.*, **50**: 282–289.

Dobbins, W. E. 1964. BOD and oxygen relationships in streams. *J. San. Eng. Div. Am. Soc. Civil. Eng.*, **90**(5A3): 53–78.

Duarte, C. M. and Agusti, S. 1998. The CO_2 balance of unproductive aquatic ecosystems. *Science*, **281**: 234–236.

Falkowski, P. G. 2002. On the evolution of the carbon cycle. In P. J. le B. Williams, D. N. Thomas, and C. S. Reynolds (eds) *Phytoplankton Productivity: Carbon Assimilation in Marine and Freshwater Ecosystems*. Blackwell Science Ltd, Oxford, UK, pp. 318–349.

Falkowski, P. G. and Raven, J. A. 1997. *Aquatic Photosynthesis*. Blackwell Science, Malden, Massachussets.

Fasham, M. J. R., Ducklow, H. W., and McKelvie, S. M. 1990. A nitrogen based model of plankton dynamics in the ocean mixed layer. *J. Mar. Res.*, **48**: 591–639.

Fenchel, T., King, G. M. and Blackburn. T. H. 1998. *Bacterial Biogeochemistry: The Ecophysiology of Mineral Cycling*, 2nd edition. Academic Press, San Diego.

Gaarder, T. and Gran, H. H. 1927. Investigations of the production of plankton in the Oslo Fjord. *Rapp. et Proc. Verb. Cons. Int. Explor. Mer.*, **42**: 1–48.

Gambaryan, S. 1962. On the method of estimation of the intensity of destruction of organic matter in silts of deep water bodies. *Microbiologia*, **31**: 895–899.

Hayes, F. R. and Macaulay, M. A. 1959. Lake water and sediment. V. Oxygen consumption in water over sediment cores. *Limnol. Oceanogr.*, **4**: 291–298.

Hedges, J. I., Baldock, J. A., Gélinas, Y., Lee, C., Peterson, M. L., and Wakeham, S. G. 2002. The biochemical and elemental composition of marine plankton: a NMR perspective. *Mar. Chem.*, **78**: 47–63.

Hutchinson, G. H. 1957. *A Treatise on Limnology*, Vol. I. John Wiley and Sons, New York.

Koblentz-Mishke, O. J., Volkovinsky, V. V., and Kabanova, J. G. 1968. Noviie dannie o velichine pervichnoi produktsii mirovogo okeana. *Dokl. Akad. Nauk SSSR*, **183**: 1186–1192.

Kolkwitz, R. and Marsson, M. 1908. Olologie des pflanzenlichen. Saprobien. *Ber. Deutsch. Bot. Ges.*, **26**: 505–510.

Kriss, A. E. 1963. *Marine Microbiology: Deep Sea*. Oliver & Boyd, London.

Krogh A. 1934. Conditions of life in the sea. *Ecol. Monogr.*, **4**: 421–429.

Kusnetzow, S. I. and Karsinkin, G. S. 1938. Direct method for the quantitative study of bacteria in water, and some considerations on the causes which produce a zone of oxygen-minimum in lake Glubokoye. *Zbl. Bakt. II* **83**: 169–174.

Liu K.-K., Iseki, K., and Chao, S.-Y. 2000. Continental margin carbon fluxes. In R. B. Hanson, H. W. Ducklow, and J. G. Field (eds) *The Changing Ocean Carbon Cycle: A Midterm Synthesis of the Joint Global Ocean Flux Study* IGBP Series 3, Cambridge University press, Cambridge, pp. 187–239.

Longhurst, A. 1998. *Ecological Geography of the Sea*. Academic Press, New York.

Luz, B., Barkan, E., Sagi, Y., and Yacobi, Y. Z. 2002. Evaluation of community respiration mechanisms with oxygen isotopes: a case study in Lake Kinneret. *Limnol. Oceanogr.*, **47**: 33–42.

Mathews, C. K. and Holde, van, K. E. 1990. *Biochemistry*. Benjamin/Cummings Publishers, Redwood, CA.

Odum, H. T. 1956. Primary production in flowing waters. *Limnol. Oceanogr.*, **1**: 102–117.

Packard, T. T. 1971. The measurement of respiratory electron transport activity in marine phytoplankton. *J Mar. Res.*, **29**: 235–244.

Pomeroy, L. R. and Johannes, R. E. 1966. Total plankton respiration. *Deep-Sea Res.*, **13**: 971–973.

Pomeroy, L. R. and Johannes, R. E. 1968. Occurance and respiration of ultraplankton in the upper 500 meters of the ocean. *Deep-Sea Res.*, **15**: 381–391.

Rodrigues, R. M. N. V. and Williams, P. J. le B. 2001. Bacterial utilisation of nitrogenous and non-nitrogenous substrates determined from ammonia and oxygen fluxes. *Limnol. Oceanogr.*, **46**: 1675–1683.

Romanenko, W. I. 1964. The dependence between the amount of the O_2 and CO_2 consumed by heterotrophic bacteria. *Dokl. Acad. Sci. USSR*, **157**: 178–179.

Ryther, J. H. 1969. Potential productivity of the sea. *Science*, **130**: 602–608.

Slàdecek, V. 1973. System of water quality from a biological point of view. *Ergeb. Limnol.*, **7**: 1–218.

Streeter, H. W. and Phelps, E. B. 1925. A study of the pollution and natural purification of the Ohio River. *III. Public Health Bulletin*. Ohio University Press, Athens, USA.

Talling, J. F. 1957. Diurnal changes of stratification and photosynthesis in some tropical African waters. *Proc. R. Soc. B.*, **147**: 57–83.

Williams, P. J. le B. 1984. A review of measurements of respiration rates of marine plankton populations. In J. E. Hobbie and P. J. le B. Williams (eds) *Heterotrophy in the Sea* Plenum Press, New York, pp. 357–389.

Williams, P. J. le B. 1993. Chemical and tracer methods of measuring plankton production: what do they in fact measure? The ^{14}C technique considered. *ICES Mar. Sci. Symp.*, **197**: 20–36.

Williams, P. J. le B. and Askew, C. 1968. A method of measuring the mineralization by micro-organisms of organic compounds in sea water. *Deep-Sea Res.*, **15**: 365–375.

Williams, P. J. le B. and Robertson, J. E. 1991. Overall plankton oxygen and carbon dioxide metabolism: the problem of reconciling observations and calculations of photosynthetic quotients. *J. Plankton Res.*, **13**: 153–169.

ZoBell, C. E. 1946. *Marine Microbiology*. Chronica Botanica Press, Waltham, Mass.

ZoBell, C. E. 1976. Discussion, In C. Litchfield (ed.) *Marine Microbiology*. Academic Press, London, p. 198.

CHAPTER 2

Ecophysiology of microbial respiration

Gary M. King

Darling Maine Center, University of Maine, USA

Outline

This chapter summarizes the physiology of bacterial respiration with an emphasis on topics relevant for understanding processes at a cellular level, as well as for understanding the larger-scale implications of respiration within aquatic ecosystems. The text describes biochemical systems involved in respiration and reviews the various electron acceptors used by bacteria in the absence of oxygen. A summary of the literature reveals the extraordinary capacity of bacteria to couple organic matter mineralization efficiently to the reduction of a very wide range of inorganic oxidants, including various metals, metalloids, nitrogen, and sulfur species. Thus, bacterial respiration determines the dynamics of numerous important elements in addition to playing a major role in carbon dynamics. Although much is known about the physiology of bacterial respiration, novel processes continue to be discovered and a great deal remains to be learned about the relationship between microbial diversity and rates of specific respiratory pathways.

2.1 Introduction

Respiration is a fundamental component of contemporary biospheric metabolism. This also appears true throughout virtually all of Earth's history. The temporal appearance of various forms of metabolism remains uncertain, but several considerations support an early and central role for respiration (Gest and Schopf 1983; Jones 1985; Wächtershäuser 1988). There is also little doubt that respiration evolved as an aquatic process, perhaps in hydrothermal vents, but certainly not in a terrestrial context.

Given the ubiquity of simple to relatively complex organics, along with several potential inorganic oxidants at early stages of planetary evolution, some form of respiratory metabolism alone may be sufficient for the development and even maintenance of primitive biotic systems. This applies not only to Earth, but also to extraterrestrial systems such as Mars and Europa.

Of course, the evolution of oxygenic photosynthesis promoted the accumulation of various electron acceptors (e.g. nitrate, metal oxides), and accounts at least in part for the diversity of microbial respiratory systems (e.g. Rao *et al.* 1985). Obviously, photosynthesis also contributes to the dynamics of virtually all ecosystems, including hydrothermal vents. Nonetheless, many benthic systems and the water column of deep or stratified environments function effectively as heterotrophic systems, since they are often temporally and spatially remote from photosynthetic organic matter sources. The structure and dynamics of such systems are determined by the modes of respiration available for metabolizing organic inputs, as well as the quantity and quality of inputs. Oxygen is often considered the most important respiratory electron acceptor, but aerobic respiration does not necessarily support the fastest rates of organic matter mineralization, nor even dominate organic turnover (e.g. Kristensen *et al.* 1995; Fenchel *et al.* 1998).

Basic ecophysiological aspects of the various modes of respiration in these functionally heterotrophic systems are summarized here. Microbial respiration forms the focus of the summary. The analysis begins with definitions and examples of respiratory systems, and reviews essential aspects of their biochemistry. Modes of respiration and aspects of their dynamics are surveyed next, with a brief consideration of the organisms involved, their distribution and constraints on their activity.

2.2   Characteristics of respiration

Respiration can be defined from two perspectives, physiological and ecological. From a physiological perspective, respiration encompasses electron (hydrogen) flow through membrane-associated transport systems from donors to acceptors while generating a proton gradient (Fig. 2.1). The chemical potential energy in proton (and electrochemical) gradients can be coupled to ATP synthesis according to chemiosmotic theory, or can otherwise drive flagellar motors, substrate transport, and similar phenomena (White 2000). Molecular oxygen supports aerobic respiration; all remaining electron acceptors support anaerobic respiration, a process largely confined to bacteria, but that can be carried out by some protists (see Fenchel, Chapter 4). Other than oxygen, common electron acceptors in aquatic systems include nitrate, manganic ion (Mn^{4+}), ferric iron (Fe^{3+}), sulfate, elemental sulfur (S^0), CO_2, and the quinone functional groups of humic acids. In some aquatic systems, non-oxygen electron acceptors may be much more abundant than molecular oxygen. This is especially true in marine systems, which contain 28 mM sulfate in contrast

to saturation oxygen concentrations ≤ 0.3 mM. Less common electron acceptors include various metals (e.g. uranium, U^{6+}; chromium, Cr^{6+}); metalloids (e.g. arsenate, As^{6+}; selenate, Se^{6+}); perchlorate, and halogen–carbon bonds. Utilization of oxygen and other available electron acceptors follows a well-understood pattern that reflects a combination of thermodynamic and kinetic constraints. Specifically, electron acceptors are utilized in order of decreasing free energy yields, with oxygen followed by nitrate, manganic and ferric oxides, sulfate, and CO_2. This pattern is manifest in marine sediments by vertical zonation of anaerobic metabolism (Fig. 2.2) as described classically by Froelich *et al.* (1979) and Middelburg *et al.* (Chapter 11).

Electron flow in association with membranes and generation of a proton gradient distinguish respiration from other cellular oxidation–reduction (redox) processes coupled to energy conservation (White 2000). For example, fermentation pathways (see below) include redox transformations and support substrate-level ATP synthesis, but typically do not depend on cell membranes or result in direct proton transport across membranes. In addition, fermentative reactions are less efficient at conserving chemical energy than are respiratory reactions and only partially oxidize electron donors. Both of these constraints have substantial physiological and ecological consequences.

At a cellular level, respiration consists of two tightly linked but distinct systems (e.g. Perry and Staley 1997; White 2000) one for substrate oxidation

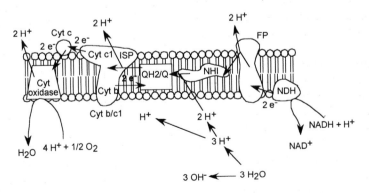

Figure 2.1 Schematic diagram of membrane-associated electron transport proteins, and flow of electrons and protons. NDH, NADH + H$^+$ dehydrogenase; FP, flavoprotein; NHI, non-heme iron; QH2/Q, reduced/oxidized quinone; ISP, iron–sulfur protein; Cyt, cytochrome.

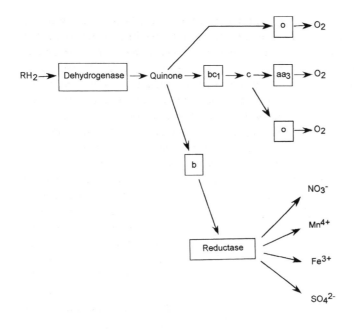

Figure 2.2 Structure of bacterial respiratory systems, showing branched pathways for electron flow to terminal oxidases (cytochromes) for aerobic respiration, and various reductases for anaerobic respiration.

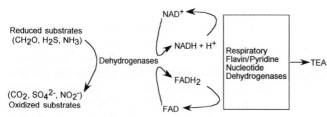

Figure 2.3 Schematic diagram indicating coupling of distinct cellular systems for substrate oxidation via dehydrogenases to systems for terminal electron acceptor reduction (TEA).

and the second for electron acceptor reduction and energy conservation (Fig. 2.3). Substrate oxidation typically involves diverse soluble (or cytoplasmic) enzymes regulated by substrate availability and cell energy status. Cofactors, such as nicotinic adenine dinucleotide (NAD) or flavin adenine dinucleotide (FAD), function as intermediate hydrogen or electron transporters that link substrate oxidation systems with electron acceptor reduction. In aerobic organisms, including bacteria and micro and macroeukaryotes, catabolism often involves coupling between a set of fermentative pathways (e.g. glycolysis) that incompletely oxidize substrates, and a distinct set of pathways that completely oxidize intermediates produced by the first (e.g. the tricarboxylic acid cycle, TCA). This is particularly true for sugar oxidation, but other compounds, for example, fatty acids, amino acids, and aromatic compounds, are also typically degraded to the level of acetate, pyruvate, or TCA intermediates before subsequent complete oxidation.

Electron acceptor reduction and energy conservation involve membrane-bound dehydrogenases that oxidize reduced cofactors (e.g. $NADH + H^+$). Specific electron transport proteins, the various cytochromes, serve as intermediates in electron flow from reduced cofactors to terminal electron acceptors, for example, molecular oxygen (see also Fig. 2.1). The number and identity of cytochromes involved in electron transport depends on the nature of the terminal electron acceptor, the reduced cofactor, growth and environmental conditions, and the organism in question. For example, eukaryotes tend to express a limited set of cytochromes (b, c_1, c) with cytochrome aa_3 as a terminal oxidase (White 2000).

In contrast, cytochrome architecture in bacteria varies considerably, reflecting the multiplicity of growth conditions and electron acceptors used in bacterial metabolism (White 2000). Branched cytochrome chains are typical, with branches occurring at the level of dehydrogenases and terminal oxidases (Fig. 2.4). Terminal oxidases include five basic groups that are distinguished by the nature of their heme moiety: aa_3, a_1, o, d and cd_1. Terminal oxidases have also been categorized according to their electron donors. Class I oxidases are coupled to cytochrome c, and class II oxidases are coupled to quinols, for example, ubiquinone and menaquinone.

The diversity of terminal oxidases and options for coupling them to upstream components of electron transport systems provide bacteria with mechanisms for fine-tuning responses to types and amounts of substrates and oxidants (White 2000). Oxygen respiration typically depends on cytochromes o and d, while c-type cytochromes function with alternate electron acceptors, for example, nitrate, trimethylamine oxide, and dimethyl sulfoxide (White 2000). Different cytochromes are also expressed in response to different electron acceptor concentrations. In *Escherichia coli* (and many other aerobes), cytochrome o functions as a high V_{max}, high K_m system for elevated oxygen concentrations. Cells express a low V_{max}, low K_m

cytochrome d at low oxygen tensions. In addition, some aerobic bacteria produce heme-containing proteins that bind oxygen. These proteins function both as oxygen sensors, regulating expression of various catabolic systems, and in oxygen transport, thereby playing a role similar to that of hemoglobin (Rodgers 1999; Hou *et al.* 2000).

Irrespective of electron transport system architecture, aerobic respiration is virtually all encompassing with respect to potential electron donors, since almost any reduced compound or element may be coupled to oxygen as an electron acceptor. This is because the electrochemical potential for the oxygen reduction "half-reaction" is typically more positive than the oxidation potentials for the half-reactions of physiologically and ecologically relevant substrates. Half-reactions for oxidation and reduction take the schematic form below; each has a characteristic electrical potential (E_h) measured in volts relative to a hydrogen electrode.

Oxidation: $CH_4 + 2H_2O \rightarrow CO_2 + 8H^+ + 8e^-$

Reduction: $8H^+ + 8e^- + 2O_2 \rightarrow 4H_2O$

Sum: $CH_4 + 2O_2 \rightarrow CO_2 + 2H_2O$

In principle, the sum of the potentials for any given pair of oxidation and reduction reactions determines whether or not the hypothetical reaction is

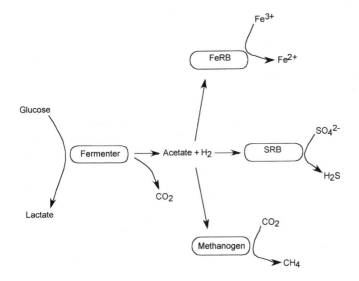

Figure 2.4 Illustration of central role of acetate and hydrogen as substrates coupling organic matter fermentation to major anaerobic respiratory processes in aquatic systems.

Table 2.1 Standard electrical potentials (E_0', pH 7) for selected physiologically and biogeochemically important redox pairs

Redox pair	Redox potential (V)
Cytochrome a (oxidation/reduction)	+0.38
Cytochrome c_1 (oxidation/reduction)	+0.23
Ubiquinone (oxidation/reduction)	+0.11
Cytochrome b (oxidation/reduction)	+0.03
FAD/FADH	−0.22
NAD+/NADH + H+	−0.32
Flavodoxin (oxidation/reduction)	−0.37
Ferridoxin (oxidation/reduction)	−0.39
Fumarate/succinate	+0.03
CO_2/pyruvate	−0.31
O_2/H_2O	+0.82
Fe^{3+}/Fe^{2+}	+0.77
NO_3^-/N_2	+0.75
$MnO_2/MnCO_3$	+0.52
NO_3^-/NO_2^-	+0.43
NO_3^-/NH_4^+	+0.38
SO_4^{2-}/HS^-	−0.22
CO_2/CH_4	−0.24
S^0/HS^-	−0.27
H_2O/H_2	−0.41

Source: Data from Stumm and Morgan (1981) and Thauer *et al.* (1977).

thermodynamically feasible, since the Gibb's free energy change is given by $\Delta G = -nEF$, where F is the Faraday constant and E is the net potential. Positive E values indicate thermodynamically favorable redox reactions, while negative values indicate prohibited reactions.

Using this formalism and tables of relevant half-reactions (Table 2.1), oxygen reduction provides the greatest free energy yields for most biologically relevant substrates, with decreasing energy yields for alternate electron acceptors. Calculated energy yields for a given redox couple provide an indication of the extent to which energy might be conserved as ATP, although biochemical and physiological constraints can lead to significantly lower levels of conservation than might be predicted from thermodynamics alone. For example, the electron transport system associated with nitrate reduction to dinitrogen provides fewer sites for ATP synthesis than does the system for oxygen reduction (Zumft 1997). Thus, even though redox potentials

for nitrate and oxygen reduction are similar (0.82 V versus 0.77 V, respectively), ATP yields and growth yields are significantly less for nitrate reduction. ATP yields for sulfate reduction are also lower than might be expected from the sulfate/sulfide redox potential. This is because sulfate must be activated by ATP hydrolysis before reduction (LeGall and Fauque 1988).

Decreasing energy yields in the electron acceptor series from oxygen to carbon dioxide are also associated with metabolism of fewer electron donors. As noted previously, oxygen serves as an electron acceptor for essentially all reduced or partially reduced compounds and elements. Suitable electron donors are fewer for anaerobic than aerobic respiration for two fundamental reasons. First, oxidation of certain substrates may not be thermodynamically feasible for certain electron acceptors. For example, neither sulfate nor carbon dioxide can oxidize ammonium due to thermodynamic constraints. Second, certain substrates, especially saturated hydrocarbons, usually require molecular oxygen for activation before further metabolism. Once activated, a variety of electron acceptors can be utilized for oxidation of such compounds. In addition, sulfate is not typically associated with oxidation of some substrates, for example, sugars, although such substrates appear highly suitable from a thermodynamic perspective. Reasons for substrate exclusion in this case are unclear, but may reflect tradeoffs between energy yields and maintenance costs for the genetic and enzyme systems required for sugar utilization.

Although it yields relatively little energy, the process of fermentation uses a broad range of electron donors, including most biogeochemically relevant organics with the notable exception of saturated aliphatics and some aromatics. Aside from distinctions previously enumerated, fermentation differs fundamentally from respiration in that partially oxidized organic intermediates generated by transformations of the starting substrate serve as electron acceptors in the final steps (White 2000). Accordingly, fermenting organisms excrete reduced end-products into their growth medium, that is, the water column or porewater in aquatic ecosystems. Thus, pyruvate produced during glucose fermentation serves as an electron acceptor in reactions leading to lactate, an excreted end-product. Anaerobic fermentations produce numerous additional end-products, including acetate, propionate, butyrate, succinate, ethanol, acetone, and butanol among others. Of these, acetate is typically the most important, and dominates carbon and reduces equivalent flow *in situ* (e.g. Sansone and Martens 1981; Sørensen *et al.* 1981; Balba and Nedwell 1982).

In addition to organic end-products, fermenting bacteria can excrete molecular hydrogen (White 2000). Hydrogen production occurs via two systems, one linked to ferridoxin, the other to $NADH + H^+$. The former produces small amounts of hydrogen and is relatively insensitive to ambient concentrations. The latter functions in reduced pyridine nucleotide recycling and is capable of

producing considerable amounts of hydrogen, but is inhibited by ambient levels $>10^{-4}$–10^{-3} atm (about 100–1000 Pa). Uptake by iron-, sulfate-, and CO_2-reducing bacteria maintains hydrogen concentrations at levels less than these (Lovley and Phillips 1987), and promotes more efficient fermentation pathways leading to more oxidized fermentation end-products (e.g. acetate instead of lactate). The symbiotic relationship between hydrogen-producing fermenters and various anaerobic respiring bacteria (Fig. 2.5) has been termed "interspecies hydrogen transfer," and shown to play an important role in aquatic carbon dynamics (e.g. Lovley and Klug 1982).

In aquatic ecosystems, anaerobic metabolism, whether respiratory or fermentative, occurs primarily in sediments due to the limited distribution of oxygen within them (e.g. Revsbech *et al.* 1980; Fenchel *et al.* 1998; and see Pace and Praire,

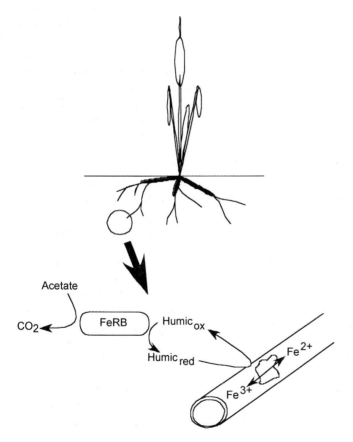

Figure 2.5 Schematic illustration of interactions between plant roots and iron-reducing microbes involving cyclic oxidation and reduction of iron oxide plaques on root surfaces with humic acids as electron shuttles.

Chapter 7 and Middelburg *et al.*, Chapter 11). Exceptions include seasonally stratified oxygen-depleted systems such as eutrophic lakes and basins (e.g. the Baltic Sea), fjords (e.g. Saanich Inlet), permanently stratified systems such as the Cariaco Trench and Black Sea, and the oxygen minimum zone of the marine water column, which support relatively high levels of denitrification (e.g. Codispoti and Christensen 1985; Lipschultz *et al.* 1990; Brettar and Rheinheimer 1991; and see Codispoti *et al.*, Chapter 12). More typically, water column anaerobiosis is limited to the interior of organic aggregates and animal guts.

Irrespective of the locus of activity, the end-products of anaerobic metabolism can affect the structure and dynamics of aerobic systems. Methane and sulfide transport from anoxic sediments or water masses to oxic zones provide well-known examples, since both support sometimes extensive populations of methane- and sulfide-oxidizing bacteria, respectively (e.g. Rudd and Taylor 1980; Howarth 1984). Indeed, sulfide oxidation can account for a very large fraction of total benthic oxygen uptake and lead to the development of microbial mats that contribute measurably to food webs (Howarth 1984). When transported to the atmosphere, anaerobic end-products such as methane and nitrous oxide can also have global impacts through their effects on radiative forcing and climate.

2.3 Modes of anaerobic respiration

2.3.1 N-oxide respiration

N-oxides, dominated by nitrate and nitrite, occur commonly in aquatic ecosystems, albeit at relatively low concentrations. Accordingly, N-oxide respiration usually accounts for only a small fraction of total aquatic organic matter turnover. However, N-oxides assume somewhat greater significance in systems affected by anthropogenic nitrogen inputs, and can vary seasonally in importance with organic loading, water column depth, sediment bioturbation, and numerous other parameters. Thus, elevated denitrification rates have been reported for animal burrow sediments, relatively

oxidized sediments during winter, continental shelf sediments, and hypoxic regions of the water column (e.g. Codispoti and Christensen 1985; Jørgensen and Sørensen 1985; Kristensen *et al.* 1985; Kristensen and Blackburn 1987; Capone and Kiene 1988; Lipschultz *et al.* 1990; Brettar and Rhineheimer 1991; Devol 1991; Binnerup *et al.* 1992; and see Codispoti *et al.*, Chapter 12).

N-oxide respiration consists of three distinct processes: denitrification, dissimilatory nitrate reduction, and ammonifying nitrate (or nitrite) respiration (Stouthamer 1988; Tiedje 1988; Zumft 1997). During denitrification *sensu strictu* nitrate is reduced to nitrous oxide, or more commonly, dinitrogen, while generating a proton motive force that promotes ATP synthesis. Dissimilatory nitrate reduction also promotes ATP synthesis, but differs in that nitrite is the end-product of a single step process that is functionally equivalent to the first step in denitrification. In contrast, relatively low energy yields accompany ammonifying nitrate respiration, which results in the formation of ammonium from either nitrate or nitrite. In many cases, nitrite reduction to ammonium represents a mechanism for oxidizing reduced pyridine nucleotides or disposing of excess reducing equivalents. This process does not pump protons, and does not produce ATP via oxidative phosphorylation (i.e. via membrane bound ATPases).

Numerous microbial taxa respire N-oxides (Shapleigh 2002). Many of these occur in aquatic systems as transients or as components of an autochthonous microbiota. Representatives of the α-, β-, and γ-Proteobacteria commonly denitrify; so to do various members of the Firmicutes (high and low G+C bacteria) and Archaea. The Enterobacteriaceae commonly respire nitrate to nitrite (Perry and Staley 1997), while a range of anaerobes ammonify nitrate or nitrite. N-oxide respiration occurs in numerous biogeochemically important functional groups, for example, heterotrophs, sulfide oxidizers, ammonia oxidizers, methylotrophs, hydrogenotrophs, nitrogen fixers, etc. (Perry and Staley 1997). With few exceptions, if thermodynamics permit coupling of a given electron donor with nitrate, there exists a microbial representative for the reaction.

Denitrification

Energy yields for denitrification rival those for oxygen respiration (see Table 2.1), but the two processes seldom fully coexist *in situ*, although distributions of activity may overlap, as has been well documented for the water column and sediments (Codispoti and Christensen 1985; Jensen *et al.* 1988; Kristensen *et al.* 1985; Blackburn and Blackburn 1992; and see Codispoti *et al.*, Chapter 12). Physical separation between aerobic respiration and denitrification arises because anoxia is usually essential for induction of denitrifying enzymes. However, in sediments denitrification is often dependent on nitrates produced by aerobic ammonia oxidation (or in some cases by abiological processes involving manganic oxides; Hulth *et al.* 1999.) Thus, the denitrification and nitrification are usually closely coupled spatially and temporally (e.g. Blackburn and Blackburn 1992).

Denitrification occurs within numerous proteobacterial and archaeal lineages, and encompasses a wide range of physiological characteristics and ecological functions. However, in spite of this diversity, denitrifiers are unified as facultative anaerobes, that is, bacteria that respire aerobically in the presence of oxygen and anaerobically in the presence of suitable alternative electron acceptors, for example, nitrate, nitrite, or in some cases, nitrous oxide. Anoxia, along with the presence of nitrate, induces expression of a membrane-bound, molybdenum-containing dissimilatory (or respiratory) nitrate reductase (NR), the activity of which is coupled to ATP synthesis (Stouthamer 1988; Zumft 1997). Respiratory NRs exhibit a modestly high affinity for nitrate *in vitro*, with K_ms ranging from 0.3–3.8 mM. However, denitrifiers *in situ* likely exhibit much higher affinities for nitrate, since ambient concentrations in most aquatic systems seldom exceed a few hundred micromolar, and are often considerably lower.

Denitrifiers often express a second, molybdenum-containing NR that is periplasmic in location (Zumft 1997). Periplasmic NR is formed and is active in the presence of oxygen. Though its function remains uncertain, it appears to be involved in transitions from oxic to anoxic conditions, and may function in aerobic denitrification (see below).

The remaining denitrifying enzymes—nitrite reductase (NiR), NO reductase, and nitrous oxide reductase—are also induced by their respective substrates and anoxia, and are expressed sequentially in cultures (Zumft 1997). These reductases are each membrane bound, contain metal cofactors, and are associated with distinct cytochromes expressed only under anoxic conditions. NiR exists in two forms, a copper protein and a cd_1 cytochrome. However, unlike NR, only one form occurs in a given organism. Cu–NiR is the most common form, but presently there is no clear phylogenetic pattern for the distribution of the two types, since they occur throughout the Proteobacteria and other denitrifying lineages. Reported K_m values for Cu–NiR range from 30–700 µM nitrite and from 6–53 for cd_1-NiR. These values likely reflect conditions within actively denitrifying cells rather than in the external environment, since most of the nitrite consumed during denitrification presumably originates from NR.

The product of NiR, the radical gas NO, is significant for a number of reasons. First, NO is an extremely important reactant in the troposphere, and denitrification is a globally significant NO source (Fenchel *et al.* 1998). Second, NO is potentially toxic. Accordingly, at least some denitrifiers possess specific mechanisms for mitigating toxicity. These include synthesis of cytochrome *c*, which has a high affinity for NO, but a low affinity for oxygen (Lawson *et al.* 2000). In effect, it acts as an NO "sponge." The amount of NO formed during denitrification varies among organisms and with culture conditions, but low nanometer levels have been commonly reported for cultures in steady state (Zumft 1997).

Nitric oxide (NO) is reduced to nitrous oxide by the product of the *nor* operon, nitric oxide reductase, which contains a cytochrome *bc* (Zumft 1997). NO reductase is periplasmic, with a very low apparent K_m (<10 nM). Exogenous NO has been shown to support growth and carbon catabolism in cultures, but it is unlikely that it plays such a role *in situ*, since ambient NO concentrations are typically extremely low (<1 nM).

In most denitrifiers, N_2O is reduced to dinitrogen by nitrous oxide reductase, the product of the *nos* operon (Stouthamer 1988; Zumft 1997). N_2O reductase is localized in the cytoplasmic membrane, contains Cu, and a cytochrome bc. N_2O reductase is coupled to proton pumping and ATP synthesis, and can support growth with N_2O as a sole electron acceptor. Some denitrifiers lack nitrous oxide reductase, and for these N_2O is the terminal product. N_2O excretion contributes to pools of dissolved N_2O. Dissolved N_2O also arises from losses by leakage through the outer cell (periplasmic) membrane due to inefficiencies in the coupling of N_2O production and consumption.

Substrates that support denitrification are as diverse as the organisms themselves, and include a wide range of simple and complex organics, hydrogen, carbon monoxide, sulfide, iron, and manganese. Accordingly, when nitrate is available denitrifiers can play an important role in virtually all aspects of carbon mineralization (Tiedje 1988). Although methane is not oxidized anaerobically by denitrifiers, in principle a methanogen–denitrifier consortium should even be feasible, with the methanogen oxidizing methane to CO_2 and hydrogen, and the denitrifier consuming the hydrogen. The major constraint on such a consortium may be inhibition of the methanogenic partner by nitrate or other N-oxides. Nonetheless, denitrifier–methanogen consortia might be active in certain nitrate-limited freshwater systems.

Ammonifying nitrate (nitrite) reduction
Nitrate can be reduced to nitrite by a periplasmic enzyme, which is the product of the *nap* operon (Zumft 1997). This reductase does not pump protons, and does not contribute to generation of membrane proton potentials or proton-linked ATP synthesis. Nitrite may be excreted as an endproduct, or further reduced to ammonia. In the latter case, three two-electron reductions are required, no free intermediates are released and no gaseous products (e.g. NO) are formed. By definition, ammonifying nitrate reduction is not a respiratory process. Instead, it appears to function as a mechanism for some anaerobes to recycle NAD^+ reduced during fermentations. Though probably not a major sink for nitrate in general, in some organic-rich, highly reducing systems, ammonifying nitrate reduction

accounts for a significant fraction of total nitrate reduction (Sørensen 1978; Buresh and Patrick 1981; Binnerup *et al.* 1992).

2.3.2 Metal reduction

Metal oxide solubility and residence times substantially limit the role of metal reduction in the water column of most aquatic ecosystems. In contrast, significant metal oxide (e.g. iron and manganese) concentrations and reduction rates can occur in sediments (e.g. Sørensen 1982; Canfield *et al.* 1993; Lovley 2000). This is particularly true in sedimentary environments characterized by low to moderate rates of organic carbon input and clay-rich inorganic phases. Particle reworking by macrobenthos can further enhance metal oxide availability by promoting porewater oxygenation and reoxidation of reduced mineral phases (Canfield *et al.* 1993).

Although iron and manganese reduction have been known for many years, the nature and importance of dissimilatory iron reduction in particular have been appreciated for only a decade or so (Lovley 2000). The importance of iron reduction derives in part from the ubiquity of ferric oxides, the ability of iron-reducing bacteria to use multiple electron acceptors, and the fact that energy yields from iron reduction exceed yields from sulfate and CO_2 reduction. The ability of anaerobes that respire different electron acceptors to compete for the two major intermediates of anaerobic organic matter mineralization, acetate and hydrogen, correlates with energy yields (Löffler *et al.* 1999; Lovley 2000). Accordingly, in sediments where a dynamic iron cycle maintains pools of amorphous ferric iron oxides, iron reduction can account for a significant fraction of carbon oxidation and limit the activity of sulfate-reducing and methane-producing bacteria. Contributions of iron reduction to total respiration of about 10% to >50% have been reported for several freshwater and marine sediments (e.g. Kostka *et al.* 1999; Lovley 2000).

Dissimilatory iron-reducing bacteria have been the subjects of extensive research. Two physiologically and ecologically distinct groups have been described in some detail (Lovley 2000). *Shewanella putrifaciens* and related taxa belonging to the γ-Proteobacteria respire oxygen and a variety of other electron acceptors, including ferric iron. Typically, these organisms do not oxidize acetate or hydrogen, but utilize more complex organics instead. In contrast, taxa within the *Geobacteriaceae* (δ-Proteobacteria) respire anaerobically only, using iron, a number of other metals and N-oxides. A wide range of organic substrates, including acetate and hydrogen serve as electron donors. Consequently, members of the *Geobacteriaceae* can function at several trophic levels in aquatic anaerobic food webs. Additional iron-reducing taxa belonging to the α- and β-Proteobacteria have been reported, but less is known about their ecological significance.

Shewanella, Geobacter, and other iron-reducing bacteria have been obtained from a wide range of marine and freshwater habitats, including thermophilic and hyperthermophilic systems (Coates *et al.* 1996; Lovley 2000). In addition, these organisms are active in the anoxic zones of aquifers and in the water-saturated soils. It should also be noted that many iron-reducing bacteria reduce manganic oxides. As a result, they play important roles in manganese cycling. Indeed, in some sedimentary systems, manganic oxides may account for a greater fraction of carbon mineralization than iron oxides (Aller 1990, 1994; Canfield *et al.* 1993; Aller *et al.* 1998).

The mechanisms for metal respiration remain somewhat speculative (Lovley 2000). A membrane-bound, *c*-type cytochrome iron reductase has been described for *Geobacter sulfurreducens*, but a growing body of information indicates that organisms such as *Geobacter* produce a variety of cytochrome systems depending on the availability of specific metal oxides. The architecture of these systems supports chemiosmotic ATP synthesis and provides for considerable flexibility in the nature of the electron acceptors used. In addition, different strategies exist for accessing insoluble metal oxides. Some organisms, such as *Shewanella* and *Geothrix*, produce low molecular weight quinones that act as chelators and solubilizers, thereby minimizing the need for direct contact with metal oxides. Others, such as *Geobacter metallireducens*, possess adaptations that optimize direct contact with metal oxides.

Many dissimilatory metal-reducing bacteria also reduce the quinone moieties of humic acids as well as simple humic acid analogs (e.g. anthraquinone-2,6-disulfate, AQDS; Lovley *et al.* 1996). Humic acids may serve as extracellular electron shuttles that promote metal oxide reduction. Such a system might be particularly effective at maintaining the metabolic activity of metal reducers when either the humic acids or reduced metals can be rapidly reoxidized. The rhizosphere of aquatic plants might provide a model system (Fig. 2.6). Iron reducers in an anoxic region around a root (e.g. King and Garey 1999) could transfer electrons to amorphous iron oxide plaques on the root surface with humic acids as an intermediate. The resulting ferrous iron could then be reoxidized chemically or microbiologically using oxygen diffusing for the root aerenchyma.

2.3.3 Sulfate (sulfur) respiration

Several oxidized sulfur species occur in aquatic environments, with sulfate particularly important. Indeed, sulfate arguably dominates the pool of available electron acceptors in many aquatic environments, since it occurs at concentrations approximately 100-fold greater than those of molecular oxygen in marine systems, and at levels similar to those of oxygen in numerous freshwater systems. In some sediments, sulfate may also rival or exceed metal oxide concentrations. Of course, since sulfate respiration occurs primarily under anoxic conditions, only a small fraction of the total aquatic sulfate pool is typically reduced. Nonetheless, a substantial fraction of available sulfate can be reduced in seasonally or permanently stratified systems such as lakes, fjords, and certain basins with restricted water flow. Rapid sulfate depletion to near-zero concentrations also occurs in sediments subjected to high rates of organic loading.

Although oxygen serves as the primary electron acceptor in most contemporary aquatic ecosystems, the significance of sulfate respiration is difficult to overestimate for several reasons. First, molecular biological and geochemical evidence indicate that sulfate respiration arose early in microbial evolution, on the order of 3.1 Gya or earlier (Wagner *et al.* 1998), possibly pre-dating oxygenic

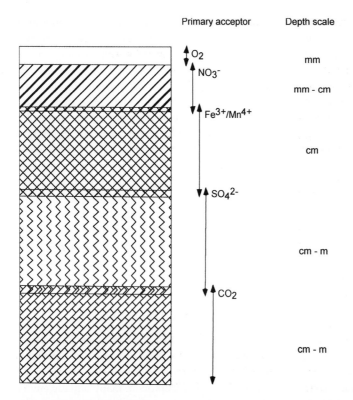

Primary acceptor   Depth scale

O_2

NO_3^- mm

                  mm - cm

Fe^{3+}/Mn^{4+}

                  cm

SO_4^{2-}

                  cm - m

CO_2

Figure 2.6 Approximate depth ranges dominated by specific terminal electron acceptors in sedimentary systems. Substantial variations occur among systems depending on organic loading and levels of sulfate in the water column.

cm - m

photosynthesis. Sulfate respiration did not become globally significant until much later though, about 2.3 Gya, when the advent of an oxygenic atmosphere resulted in high sulfate concentrations (Habicht and Canfield 1996). Still, some form of sulfate respiration may have been an ancestral trait for organisms at or near the base of the phylogenetic tree.

Second, sulfate respiration plays a major role in marine benthic biogeochemistry. In many coastal and continental environments sulfate reduction accounts for a major fraction (up to 90%) of carbon mineralization (Sørensen *et al.* 1979; Howes *et al.* 1984; Henrichs and Reeburgh 1987; Capone and Kiene 1988; King 1988; Canfield 1989, 1991; Sampou and Oviatt 1991; Middelburg 1992; Kostka *et al.* 1999; and see Middelburg *et al.*, Chapter 11). In such systems, the end-product of sulfate respiration, sulfide, impacts benthic floral and faunal distributions and activity, transformations and cycling of nitrogen, phosphorous, and metals, and trophic exchanges between benthos and the overlying water column. In addition, benthic biological and chemical

sulfide oxidation account for a significant fraction of oxygen utilization, in some cases contributing to water column hypoxia or anoxia.

Sulfate-respiring bacteria (sulfidogens) are active even when sulfate is effectively absent (Widdel 1988). In these situations, sulfate respirers oxidize organic substrates and transfer reducing equivalents as hydrogen to methanogenic bacteria, which use them to reduce CO_2 to methane. In sediments with low sulfate and high methane concentrations, a flow of electrons in the opposite direction, that is, from methanogens to sulfate respirers, promotes anaerobic methane oxidation, a process with considerable biogeochemical significance on a global scale (Henrichs and Reeburgh 1987; Hoehler *et al.* 1994).

Third, the dynamics of sulfate respiration and sulfide oxidation are sensitive to anthropogenic disturbances, especially in coastal systems subject to eutrophication. Elevated rates of sulfate respiration and sulfide oxidation due to increased organic matter inputs have contributed to wide-spread hypoxic conditions in Chesapeake Bay and elsewhere, with major consequences for the structure and function of these ecosystems (e.g. Kemp *et al.* 1992). Thus, mobilization and transport of organic matter through land use change and stimulation of photosynthesis by increased nitrogen availability become manifest in marine systems through increased rates of sulfur cycling. This may be as pervasive and serious in marine systems as the impacts of climate change, the direct effects of which might well be minimal on processes such as sulfate respiration.

As might be expected for an ancient and globally significant group of organisms, sulfate- and sulfur-respiring bacteria exhibit a wide range of characteristics (Widdel 1988). This includes diverse morphologies, flagellation and motility, responses to temperature and salinity, and ability to grow autotrophically. Sulfate- and sulfur-respiring bacteria occur primarily in the δ-subgroup of the Proteobacteria, and to a limited extent within the Archaea. Two Gram-positive sulfate-respiring genera have been described (*Desulfotomaculum* and *Desulfosporosinus*) and sulfidogens also occur in two deeply branching thermophilic genera, *Thermodesulfobacterium* and *Thermodesulfovibrio*. Based on physiological attributes, these varied organisms are commonly divided into four major groups: (i) sporeformers that respire sulfate and completely or incompletely oxidize organic substrates; (ii) nonsporeformers that respire sulfur or both sulfate and sulfur and incompletely oxidize organic substrates; (iii) nonsporeformers that respire sulfate but not sulfur and that completely oxidize organic substrates; (iv) sulfur respirers that do not reduce sulfate and that completely oxidize organic substrates. Of these, organisms in groups (ii) and (iii) are perhaps the best known and most significant ecologically. Representatives of group (ii) include genera such as *Desulfovibrio*, *Desulfobulbus*, and *Desulfomicrobium*; representatives of group (iii) include *Desulfobacter*, *Desulfobacterium*, and *Desulfonema*.

Sulfate respiration occurs through the action of three key enzymes (LeGall and Fauque 1988; Widdel 1988): an ATP sulfurylase, an adenosinephosphosulfate reductase (APS reductase) and a bisulfite reductase (Fig. 2.7). ATP sulfurylase catalyzes sulfate activation by adenosylation and pyrophosphate

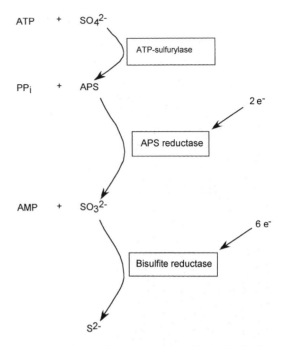

Figure 2.7 Schematic diagram of the sequential reduction of sulfate by sulfate-reducing bacteria indicating primary enzymes and metabolites. APS, adenophosphosulfate.

formation. This reaction represents a significant energy cost that can only be minimized to the extent that the chemical energy in pyrophosphate is conserved through reactions with AMP or ADP. Recovery of pyrophosphate appears only partial at best. Several lines of evidence suggest that most of the pyrophosphate formed during sulfate activation is simply hydrolyzed to phosphate.

Adenosinephospho sulfate reductase, a membrane-bound, proton pumping system, forms sulfite and AMP while consuming two reducing equivalents. APS reductase contains a nonheme iron–sulfur complex (or complexes) and a flavin moiety essential for activity. Because APS reductase pumps protons, its activity is coupled to ATP synthesis.

Bisulfite reductase reduces the product of APS reductase (or exogenous sulfite) to sulfide while consuming six reducing equivalents. Though two intermediates, trithionate and thiosulfate, occur during sulfide generation, they normally remain bound. Bisulfite reductase also pumps protons and contributes to the bulk of ATP synthesis during sulfate reduction. At least four distinct types of bisulfate reductase have been described, with differences occurring in absorption spectra, siroheme and iron content, and subunit structure.

In addition, to sulfate and other sulfur oxyanions, a limited number of sulfidogens reduce other electron acceptors (Widdel 1988), including nitrate and nitrite, ferric iron, fumarate, acrylate, carbon–halogen bonds, arsenate, chromate, and uranate (U^{6+}). Reduction of these various species occurs under specialized conditions that typically have little ecological impact, but may have considerable significance for mobilization of toxic elements and bioremediation.

Some sulfidogens also reduce molecular oxygen, a process that may be ecologically significant in biofilms, sediments, or other systems that experience transiently oxic conditions (Cypionika 2000). Although once considered oxygen-insensitive strict anaerobes, a variety of sulfidogens clearly respire oxygen at low concentrations ($<1\,\mu M$) and couple this activity to energy conservation (e.g. Finster et al. 1997; Cypionika 2000). Aerotaxis has also been demonstrated for Desulfovibrio species. These and other results suggest that sulfidogens have adapted to limited oxygen exposure (e.g. Canfield and DesMarais 1991; Fründ and Cohen 1992) in spite of the fact that oxygen reduction apparently does not promote growth.

Regardless of the electron acceptors they reduce and their ubiquity in aquatic systems, sulfidogens, use only a relatively modest array of simple substrates as electron donors (Widdel 1988). These include organic acids and alcohols, some amino acids, certain methylated amines and sulfur compounds, monoaromatics, and n-alkanes, but generally not sugars. Substrate utilization has been characterized as "complete," resulting in CO_2, or "incomplete," resulting in organic end-products. Partial oxidation of lactate with acetate and CO_2 as products represents a classic example of the latter. In an ecological context, the products of incomplete substrate oxidation are typically consumed by sulfidogens with complete oxidation pathways.

A number of studies in marine systems have shown that molecular hydrogen and acetate account for a large fraction of the substrates used for sulfate reduction (Sansone and Martens 1981; Sørensen et al. 1981; Balba and Nedwell 1982). Sulfidogens consuming these substrates exist in consortia with hydrogen-producing and acetogenic fermenters, which degrade complex organics, especially polymeric species that constitute the bulk of aquatic organic matter. Hydrogen consumption in such consortia is especially important, since hydrogen partial pressures regulate fermentation pathways, and in some cases even determine whether certain substrates (e.g. propionate) can be degraded at all. By maintaining low hydrogen partial pressures, sulfidogens promote more efficient substrate utilization patterns, and thus play a key role in the trophic structure and dynamics of anaerobic food webs.

In certain circumstances, sulfidogens also apparently exist in consortia with methanogenic bacteria (e.g. Hoehler et al. 1994; Orphan et al. 2002). These consortia occur at specific depths in anoxic sediments that contain low sulfate and elevated methane concentrations. Within the consortia, methane is oxidized anaerobically by what is presumably a respiratory process in which the methanogenic partner converts methane to CO_2 and hydrogen (reverse methanogenesis). The sulfidogenic partner plays a critical role in promoting reverse methanogenesis by maintaining hydrogen partial pressures at levels sufficiently low to provide for a negative ΔG for the overall process:

$$CH_4 + SO_4^{2-} \rightarrow S^{2-} + CO_2 \quad \Delta G = -25\,\text{kJ}\,\text{mol}^{-1}$$

While anaerobic methane oxidation remains poorly understood, it contributes significantly to the total global methane budget.

2.3.4   Carbon dioxide reduction (hydrogenotrophic methanogenesis)

Methanogenesis seldom occurs within the water column per se, since conditions that promote methane production, particularly low redox potentials and complete absence of molecular oxygen, usually do not occur. Nonetheless, respiratory CO_2 reduction to methane occurs commonly in aquatic sediments, especially those that lack nitrate, sulfate, or significant metal oxide concentrations. Although methanogenesis is the least favorable mode of respiration from a thermodynamic perspective, it is widespread in sedimentary systems since the demand for electron acceptors, such as metal oxides, nitrate, and sulfate, often exceeds the supply, leaving CO_2 as the most abundant final option.

Though least in terms of energy yields, methanogenesis is hardly least in general significance. Methane production via both CO_2 reduction and disproportionation of certain methyl compounds and acetate promotes carbon flow through more efficient fermentation pathways, and facilitates degradation of fatty acids that might otherwise accumulate (Fenchel *et al.* 1998). Methane transport supports populations of methanotrophic bacteria that can account for a large fraction of microbial biomass in the oxic–anoxic transition zone of some lakes (Rudd and Taylor 1980). Obviously, methane release to the atmosphere impacts both tropospheric chemistry and radiative forcing at global scales (Fenchel *et al.* 1998).

Respiratory methanogenesis involves CO_2 reduction by molecular hydrogen through seven sequential steps (White 2000). The mechanism includes an initial activation step with formation of formyl-methanofuran. Subsequent steps reduce the formyl group to a methyl group. Finally, the methyl group is transferred to a unique methanogenic cofactor, coenzyme M, and reduced to methane. Methanogens also contain a number of other unique cofactors, for example, F_{420} and methanopterin, which are critical for CO_2 reduction. The process is coupled to chemiosmotic ATP synthesis, but ATP yields are very low and methanogens appear to operate near the lower limits predicted from thermodynamic considerations.

A second form of methanogenesis involves substrates such as methylamines and acetate (Whitman *et al.* 1999; White 2000). Methylamines are essentially disproportionated, with methyl groups serving as both a source of reductant (from oxidation to CO_2 and hydrogen) and an electron acceptor. An aceticlastic reaction cleaves acetate, producing a CoM bound methyl group and a carboxyl group bound to CO dehydrogenase. The latter is oxidized to CO_2 and hydrogen, with the hydrogen serving to reduce methyl-CoM to methane. The mechanism of ATP synthesis is not clear, but the process overall functions more as a fermentation than respiration.

In spite of the thermodynamic constraints under which they operate, the methanogenic bacteria are relatively diverse and widespread. They are restricted to the Archaea, but include more than 19 genera and 50 species representing thermophiles, extreme thermophiles, halophiles, and acidophiles along with taxa that require more moderate conditions (Whitman *et al.* 1999). Methanogens represent an ancient but highly successful lineage that has managed to maintain a prominent position in the global carbon dynamics of a planet commonly viewed as dominated by oxygen.

2.3.5 Other electron acceptors

In addition to the electron acceptors discussed above, numerous other oxidized inorganics support anaerobic respirations. These include arsenate, selenate, C–halogen bonds, and perchlorate. It has become axiomatic that if thermodynamics permit reduction by hydrogen of a given potential electron acceptor, there is a microbe somewhere that uses it for metabolic gain. In some cases, it appears that there are even organisms that specialize on unusual electron acceptors. For example, several arsenate- and selenate-respiring bacteria are now known (Oremland *et al.* 1999; Oremland and Stolz 2000), and perchlorate reduction has been associated with both conventional and novel taxa (Logan 1998; Coates *et al.* 1999; Zhang *et al.* 2002).

The ecological significance of these less abundant electron acceptors in aquatic systems remains uncertain. Various organohalides occur naturally at low concentrations in aquatic (especially marine) environments (e.g. Neidleman and Geigert 1986), and may serve as electron acceptors for a group of halo-respiring anaerobes related to sulfate reducers. Arsenate and selenate reduction clearly contribute to the biogeochemical cycling of these elements (e.g. Oremland *et al.* 1989; Oremland *et al.* 1990; Steinberg and Oremland 1990; Oremland and Stolz 2000; Hoeft *et al.* 2002), and even if the magnitude of reduction is relatively small, the end-products, for example, arsenite and selenite can pose serious environmental problems (e.g. Oldendorf and Santolo 1994; Nickson *et al.* 1998; Zobrist *et al.* 2000). In some unusual circumstances, certain electron acceptors may also contribute measurably to total anaerobic respiration. For example, arsenate reduction may account for as much as 14% of photosynthetic carbon fixation during meromixis in Mono Lake, California (Oremland *et al.* 2000). More typically, it appears that many of the various minor electron acceptors are reduced incidentally along with other processes, or assume significance due to the enhanced toxicity of the reduced products (e.g. arsenite).

The respiratory mechanisms for most of these minor electron acceptors are only modestly understood, and much remains to be learned about the relevant microbial populations that use them (Oremland and Stoltz 2000). Observations of growth at the expense of arsenate and selenate reduction and characterization of a few terminal oxidases strongly suggest that unconventional electron acceptor reduction proceeds with conventional respiratory system architecture and chemiosmotic energy conservation. This reflects the fact that most of the bacterial isolates capable of respiring oxidants such as arsenate, selenate, and perchlorate use other more typical electron acceptors (e.g. sulfate, nitrate) as well (Oremland *et al.* 1999; Oremland and Stolz 2000). Thus, respiration of "exotic" oxidants likely involves adaptations extant biochemical systems that might initially have played a role in detoxification, but that were ultimately coupled to growth in certain systems. It is noteworthy that bacterial capacities for utilization of exotic oxidants may be undergoing a new round of selection due to large-scale anthropogenic environmental disturbances.

2.4 Concluding comments

The advent of lipid membranes marked a seminal event in cellular evolution, but also set the stage for exploitation by the earliest life forms of a variety of common abiological redox processes. These various processes provided the basis for primitive respiratory systems that rapidly assumed control over many elemental cycles and largely determined Earth's subsequent geochemical history. Respiration

in its multiple forms remains a major determinant of the structure and dynamics of aquatic ecosystems, the yin to the yang of photosynthesis.

References

Aller, R. C. 1990. Bioturbation and manganese cycling in hemipelagic sediments. *Phil. Trans. R. Soc. Lond.*, **331**: 51–68.

Aller, R. C. 1994. The sedimentary Mn cycle in Long Island Sound: its role as intermediate oxidant and the influence of bioturbation, O_2, and C_{org} flux on diagenetic reaction balances. *J. Mar. Res.*, **52**: 259–295.

Aller, R. C., Hall, P. O., Rude, P. D., and Aller, J. Y. 1998. Biogeochemical heterogeneity and suboxic diagenesis in hemipelagic sediments of the Panama Basin. *Deep-Sea Res.*, **45**: 133–165.

Balba, M. T. and Nedwell, D. B. 1982. Microbial metabolism of acetate, propionate, and butyrate in anoxic sediment from the Colne Point saltmarsh, Essex, U. K. *Appl. Environ. Microbiol.*, **42**: 985–992.

Binnerup, S. J., Jensen, K., Revsbech, N. P., Jensen, M. H., and Sørensen, J. 1992. Denitrification, dissimilatory reduction of nitrate to ammonium, and nitrification in a bioturbated estuarine sediment as measured with ^{15}N and microsensor techniques. *Appl. Environ. Microbiol.*, **58**: 303–313.

Blackburn, T. H. and Blackburn, N. D. 1992. Model of nitrificaton and denitrification in marine sediments. *FEMS Microbiol. Lett.*, **100**: 517–522.

Brettar, I. and Rheinheimer, G. 1991. Denitrification in the Central Baltic—evidence for H_2S-oxidation as motor of denitrification at the oxic–anoxic interface. *Mar. Ecol. Prog. Ser.*, **77**: 157–169.

Buresh, R. J. and Patrick, J. W. H. 1981. Nitrate reduction to ammonium and organic nitrogen in an estuarine sediment. *Soil Biol. Biochem.*, **13**: 279–283.

Canfield, D. and DesMarais, D. J. 1991. Aerobic sulfate reduction in microbial mats. *Science*, **251**: 1471–1473.

Canfield, D. E. 1989. Sulfate reduction and oxic respiration in marine sediments: implications for organic carbon preservation in euxinic environments. *Deep-Sea Res.*, **36**: 121–138.

Canfield, D. E. 1991. Sulfate reduction in deep-sea sediments. *Am. J. Sci.*, **291**: 177–188.

Canfield, D. E., Thamdrup, B., and Hansen, J. W. 1993. The anaerobic degradation of organic matter in Danish coastal sediments: iron reduction, manganese reduction and sulfate reduction. *Geochim. Cosmochim. Acta*, **57**: 3867–3883.

Capone, D. G. and Kiene, R. P. 1988. Comparison of microbial dynamics in marine and freshwater sediments: contrasts in anaerobic carbon metabolism. *Limnol. Oceanogr.*, **33**: 725–749.

Coates, J. D., Phillips, E. J. P., Lonergan, D. J., Jenter, H., and Lovley, D. R. 1996. Isolation of *Geobacter* species from diverse sedimentary environments. *Appl. Environ. Microbiol.*, **62**: 1531–1536.

Coates, J. D., Michaelidou, U., Bruce, R. A., O'Conner, S. M., Crespi, J. N., and Achenbach, L. A. 1999. Ubiquity and diversity of dissimilatory (per)chlorate-reducing bacteria. *Appl. Environ. Microbiol.*, **65**: 5234–5241.

Codispoti, L. A. and Christensen, J. P. 1985. Nitrification, denitrification and nitrous oxide cycling in the eastern tropical south Pacific ocean. *Mar. Chem.*, **16**: 277–300.

Cypionika, H. 2000. Oxygen respiration by *Desulfovibrio* species. *Annu. Rev. Microbiol.*, **54**: 827–848.

Devol, A. H. 1991. Direct measurement of nitrogen gas fluxes from continental shelf sediments. *Nature (Lond.)*, **349**: 319–321.

Fenchel, T., King, G. M., and Blackburn, T. H. 1998. *Bacterial Biogeochemistry: An Ecophysiological Analysis of Mineral Cycling.* Academic Press, New York, 307 pp.

Finster, K., Liesack, W., and Tindall, B. J. 1997. *Sulfospirillum arachonense* sp. nov., a new microaerophilic sulfur-reducing bacterium. *Int. J. Syst. Bacteriol.*, **47**: 1212–1217.

Froelich, P. N., Klinkhammer, G. P., Bender, M. L., Luedtke, N. A., Heath, G. R., Cullen, D., Dauphin, P., Hammond, D., Hartman, B., and Maynard, V. 1979. Early oxidation of organic matter in pelagic sediments of the eastern equatorial Atlantic: suboxic diagenesis. *Geochim. Cosmochim. Acta*, **43**: 1075–1090.

Fründ, C. and Cohen, Y. 1992. Diurnal cycles of sulfate reduction under oxic conditions in cyanobacterial mats. *Appl. Environ. Microbiol.*, **58**: 70–77.

Gest, H. and Schopf, J. W. 1983. Biochemical evolution of anaerobic energy conversion: the transition from fermentation to anoxygenic photosynthesis. In J. W. Schopf, (ed.) *Earth's Earliest Biosphere: Its Origin and Evolution.* Princeton University Press, Princeton, NJ, pp. 135–148.

Habicht, K. S. and Canfield, D. E. 1996. Sulphur isotope fractionation in modern microbial mats and the evolution of the sulphur cycle. *Nature (Lond.)*, **382**: 342–343.

Henrichs, S. M. and Reeburgh, W. S. 1987. Anaerobic mineralization of marine sediment organic matter: rates and the role of anaerobic processes in the oceanic carbon economy. *Geomicrobiol. J.*, **5**: 191–237.

Hoeft, S. E., Lucas, F., Hollibaugh, J. T., and Oremland, R. S. 2002. Characterization of microbial arsenate reduction in the anoxic bottom waters of Mono Lake, California. *Geomicrobiol. J.*, **19**: 1–18.

Hoehler, T. M., Alperin, M. J., Albert, D. B., and Martens, C. S. 1994. Field and laboratory studies of methane oxidation in anoxic marine sediment: evidence for methanogen-sulfate reducer consortium. *Glob. Biogeochemistry Cyc.*, **8**: 451–463.

Hou, S., Larsen, R. W., Boudko, D., Riley, C. W., Karatan, E., Zimmer, M., Ordal, G. W., and Alam, M. 2000. Myoglobin-like aerotaxis transducers in Archaea and Bacteria. *Nature (Lond.)*, **403**: 540–544.

Howarth, R. W. 1984. The ecological significance of sulfur in the energy dynamics of salt marsh and coastal marine sediments. *Biogeochem.*, **1**: 5–27.

Howes, B. L., Dacey, J. W. H., and King, G. M. 1984. Carbon flow through oxygen and sulfate reduction pathways in salt marsh sediments. *Limnol. Oceanogr.*, **29**: 1037–1051.

Hulth, S., Aller, R. C., and Gilbert, F. 1999. Coupled anoxic nitrification/manganese reduction in marine sediments. *Geochim. Cosmochim. Acta*, **63**: 49–66.

Jensen, M. H., Andersen, T. K., and Sørensen, J. 1988. Denitrification in coastal bay sediment: regional and seasonal variations in Aarhus Bight, DK. *Mar. Ecol.*, **48**: 155–162.

Jones, C. W. 1985. The evolution of bacterial respiration. In K. H. Schliefer and E. Stackenbrandt (eds) *Evolution of Prokaryotes.* Academic Press, London, pp. 175–204.

Jørgensen, B. B. and Sørensen, J. 1985. Seasonal cycles of O_2, NO_3^-, and SO_4^{2-} reduction in estuarine sediments: the significance of an NO_3^- reduction maximum in spring. *Mar. Ecol. Prog. Ser.*, **24**: 65–74.

Kemp, W. M., Sampou, P. A., Garber, J., Tuttle, J., and Boynton, W. R. 1992. Seasonal depletion of oxygen from bottom waters of Chesapeake Bay: roles of benthic and planktonic respiration and physical exchange processes. *Mar. Ecol. Prog. Ser.*, **85**: 137–152.

King, G. M. 1988. The dynamics of sulfur and sulfate reduction in a South Carolina salt marsh. *Limnol. Oceanogr.*, **33**: 376–390.

King, G. M. and Garey, M. A. 1999. Ferric iron reduction by bacteria associated with the roots of freshwater and marine macrophytes. *Appl. Environ. Microbiol.*, **65**: 4393–4398.

Kostka, J. E., Thamdrup, B., Glud, R. N., and Canfield, D. E. 1999. Rates and pathways of carbon oxidation in permanently cold Arctic sediments. *Mar. Ecol. Prog. Ser.*, **180**: 7–21.

Kristensen, E. and Blackburn, T. H. 1987. The fate of organic carbon and nitrogen in experimental marine sediments: influence of bioturbation and anoxia. *J. Mar. Res.*, **45**: 231–257.

Kristensen, E., Jensen, M. H., and Andersen, T. K. 1985. The impact of polychaete (*Nereis virens* sars) burrows on nitrification and nitrate reduction in estuarine sediments. *J. Exp. Mar. Biol. Ecol.*, **85**: 75–91.

Kristensen, E., Ahmed, S. I., and Devol, A. H. 1995. Aerobic and anaerobic decomposition of organic matter in marine sediment: which is fastest? *Limnol. Oceanogr.*, **40**: 1430–1437.

Lawson, D. M., Stevenson, C. E. M., Andrew, C. R., and Eady, R. R. 2000. Unprecedented proximal binding of nitric oxide to heme: implications for guanylate cyclase. *EMBO J.*, **19**: 5661–5671.

LeGall, J. and Fauque, G. 1988. Dissimilatory reduction of sulfur compounds. In A. J. B. Zehnder, (ed.) *Biology of Anaerobic Microorganisms*. Wiley Interscience, New York, pp. 587–640.

Lipschultz, F., Wofsy, S. C., Ward, B. B., Codispoti, L. A., Friedrich, G., and Elkins, J. W. 1990. Bacterial transformations of inorganic nitrogen in the oxygen-deficient waters of the Eastern Tropical Pacific Ocean. *Deep-Sea Res.*, **37**: 1513–1542.

Löffler, F. E., Tiedje, J. M., and Sanford, R. A. 1999. Fraction of electrons consumed in electron acceptor reduction and hydrogen thresholds as indicators of halorespiratory physiology. *Appl. Environ. Microbiol.*, **65**: 4049–4056.

Logan, B. E. 1998. A review of chlorate and perchlorate-respiring microorganisms. *Biomed. J.*, **2**: 69–79.

Lovley, D. R. 2000. Fe(III) and Mn(IV) reduction. In D. R. Lovley (ed.) *Environmental Microbe-metal Interactions*. ASM Press Washington DC, pp. 3–30.

Lovley, D. R. and Klug, M. J. 1982. Intermediary metabolism of organic matter in the sediments of a eutrophic lake. *Appl. Environ. Microbiol.*, **43**: 552–560.

Lovley, D. R. and Phillips, E. J. P. 1987. Competitive mechanisms for inhibition of sulfate reduction and methane production in the zone of ferric iron reduction in sediments. *Appl. Environ. Microbiol.*, **53**: 2636–2641.

Lovley, D. R., Coates, J. R., Blunt-Harris, E. L., Phillips, E. J. P., and Woodward, J. C. 1996. Humic substances as electron acceptors for microbial respiration. *Nature (Lond.)*, **382**: 445–448.

Middelburg, J. J. 1992. Organic matter decomposition in the marine environment. *Encycloped. Earth Syst. Sci.*, **3**: 493–499.

Neidleman, S. L. and Geigert, J. 1986. *Biohalogenation: Principles, Basic Roles and Applications*. Ellis Horwood, Ltd, Chichester, 203 pp.

Nickson, R., McArthur, J., Burgess, W., Ahmed, K. M., Ravenscroft, P., and Rahman, M. 1998. Arsenic poisoning of Bangaladesh groundwater. *Nature (Lond.)*, **395**: 338.

Oldendorf, H. M. and Santolo, G. M. 1994. Kesterton Reservoir—past, present and future: an ecological risk assessment. In W. T. Frankenberger *et al.* (eds) *Selenium in the Environment*. Marcel Dekker, New York, pp. 69–117.

Oremland, R. S. and Stolz, J. 2000. Dissimilatory reduction of selenate and arsenate in nature. In D. R. Lovley (ed.) *Environmental Microbe–Metal Interactions*. ASM Press, Washington DC, pp. 199–224.

Oremland, R. S., Hollibaugh, J. T., Maest, A. S., Presser, T. S., Miller, L. G., and Culbertson, C. W. 1989. Selenate reduction to elemental selenium by anaerobic bacteria in sediments and culture: biogeochemical significance of a novel, sulfate-independent respiration. *Appl. Environ. Microbiol.*, **55**: 2333–2343.

Oremland, R. S., Steinberg, N. A., Maest, A. S., Miller, L. G., and Hollibaugh, J. T. 1990. Measurement of *in situ* rates of selenate removal by dissimilatory bacterial reduction in sediments. *Environ. Sci. Technol.*, **24**: 1157–1164.

Oremland, R. S., Blum, J. S., Bindi, A. B., Dowdle, P. R., Herbel, M., and Stolz, J. F. 1999. Simultaneous reduction of nitrate and selenate by cell suspensions of selenium-respiring bacteria. *Appl. Environ. Microbiol.*, **65**: 4385–4392.

Oremland, R. S., Dowdle, P. R., Hoeft, S., Sharp, J. O., Schaefer, J. K., Miller, L. G., Blum, J. S., Smith, R. L., Bloom, N. S., and Wallschlaeger, D. 2000. Bacterial dissimilatory reduction of arsenate and sulfate in meromictic Mono Lake, California. *Geochim. Cosmochim. Acta*, **64**: 3073–3084.

Orphan, V. J., House, C. H., Hinrichs, K. U., McKeegan, K. D., and DeLong, E. F. 2002. Multiple archaeal groups mediate methane oxidation in anoxic cold seep sediments. *Proc. Natl. Acad. Sci. USA*, **99**: 7663–7668.

Perry, J. J. and Staley, J. T. 1997. *Microbiology: Dynamics and Diversity*. Saunders College Publishing, Fort Worth, TX, 911 pp.

Rao, K. K., Cammack, R., and Hall, D. O. 1985. Evolution of light energy conservation. In K. H. Schliefer and E. Stackenbrandt (eds) *Evolution of Prokaryotes*. Academic Press, London, pp. 143–173.

Revsbech, N. P., Sørensen, J., Blackburn, T. H., and Lomholt, J. P. 1980. Distribution of oxygen in marine sediments measured with microelectrodes. *Limnol. Oceanogr.*, **25**: 403–411.

Rodgers, K. R. 1999. Heme-based sensors in biological systems. *Curr. Opin. Chem. Biol.*, **3**: 158–167.

Rudd, J. W. M. and Taylor, C. D. 1980. Methane cycling in aquatic environments. *Adv. Aquat. Microbiol.*, **1**: 77–150.

Sampou, P. and Oviatt, C. A. 1991. A carbon budget for a eutrophic marine ecosystem and the role of sulfur metabolism in sedimentary carbon, oxygen and energy dynamics. *J. Mar. Res.*, **49**: 825–844.

Sansone, F. J. and Martens, C. S. 1981. Determination of volatile fatty acid turnover rates in organic-rich sediments. *Mar. Chem.*, **10**: 233–247.

Shapleigh J. P. 2002. The denitrifying prokaryotes. In M. Dworkin (ed.) *The Prokaryotes: An Evolving Electronic Resource for the Microbiological Community*. Springer-Verlag, New York. http://141.150.157.117:8080/prokPUB/index.htm

Sørensen, J. 1978. Capacity for denitrification and reduction of nitrate to ammonia in a coastal marine sediment. *Appl. Environ. Microbiol.*, **35**: 301–305.

Sørensen, J. 1982. Reduction of ferric iron in anaerobic, marine sediment and interaction with reduction of nitrate and sulfate. *Appl. Environ. Microbiol.*, **43**: 319–324.

Sørensen, J., Jørgensen, B. B., and Revsbech, N. P. 1979. A comparison of oxygen, nitrate, and sulfate respiration in coastal marine sediments. *Microb. Ecol.*, **5**: 105–115.

Sørensen, J., Christensen, D., and Jørgensen, B. B. 1981. Volatile fatty acids and hydrogen as substrates for sulfate-reducing bacteria in anaerobic marine sediments. *Appl. Environ. Microbiol.*, **42**: 5–11.

Steinberg, N. A. and Oremland, R. S. 1990. Dissimilatory selenate reduction potentials in a diversity of sediment types. *Appl. Environ. Microbiol.*, **56**: 3550–3557.

Stumm, W. and Morgan, J. J. 1981. *Aquatic Chemistry*, 2nd edition. Wiley-Interscience, New York, 780 pp.

Stouthamer, A. H. 1988. Dissimilatory reduction of oxidized nitrogen compounds. In A. J. B. Zehnder (ed.) *Biology of Anaerobic Microorganisms*. Wiley Interscience, New York, pp. 245–304.

Thauer, R. K., Jungerman, K., and Decker, K. 1977. Energy conservation in chemotrophic anaerobic bacteria. *Bacteriol. Rev.*, **41**: 100–108.

Tiedje, J. M. 1988. Ecology of denitrification and dissimilatory nitrate reduction to ammonium. In A. J. B. Zehnder (ed.) *Biology of Anaerobic Microorganisms*. Wiley Interscience, New York, pp. 179–248.

Wächtershäuser, G. 1988. Before enzymes and templates: theory of surface metabolism. *Microbiol. Rev.*, **52**: 452–484.

Wagner, M., Roger, A. J., Flax, J. L., Brusseau, G. A., and Stahl, D. A. 1998. Phylogeny of dissimilatory sulfite reductase supports an early origin of sulfate reduction. *J. Bacteriol.*, **180**: 2975–2982.

White, D. 2000. *The Physiology and Biochemistry of Prokaryotes*, 2nd edition. Oxford University Press, Oxford, 565 pp.

Whitman, W. B., Bowen, T. L., and Boone, D. R. 1999. The methanogenic bacteria. In M. Dworkin *et al.* (eds) *The prokaryotes: An Evolving Electronic Resource for the Microbiological Community*, 3rd edition, release 3.0, May 21, 1999. Springer-Verlag, New York.

Widdel, F. 1988. Microbiology and ecology of sulfate- and sulfur-reducing bacteria. In A. J. B. Zehnder (ed.) *Biology of Anaerobic Microorganisms*. Wiley Interscience, New York, pp. 469–586.

Zhang, H., Bruns, M. A., and Logan, B. E. 2002. Perchlorate reduction by a novel chemolithoautorophic hydrogen-oxidizing bacterium. *Environ. Microbiol.*, **4**: 570–576.

Zobrist, J., Dowdle, P. R., Davis, J. A., and Oremland, R. S. 2000. Mobilization of arsenite by dissimilatory reduction of adsorbed arsenate. *Environ. Sci. Technol.*, **34**: 4747–4753.

Zumft, W. G. 1997. Cell biology and molecular basis of denitrification. *Microbiol. Mol. Biol. Rev.*, **61**: 533–616.

CHAPTER 3

Respiration in aquatic photolithotrophs

John A. Raven[1] and John Beardall[2]

[1] *Division of Environmental and Applied Biology, University of Dundee, UK*
[2] *School of Biological Sciences, Monash University, Australia*

Outline

This chapter discusses respiratory processes in phytoplankton, the objectives being to demonstrate the range of respiratory processes found in phytoplankton, to describe the functions of these individual pathways, and to explore their interactions. Phytoplankton organisms have the core respiratory processes common to most planktonic organisms; these reactions occur in the dark and, to varying extents, in the light. These organisms also possess oxygen uptake and carbon dioxide production reactions related to their photosynthetic apparatus. Two of these, the Mehler-peroxidase reaction and the oxidase function of ribulose biphosphate carboxylase-oxygenase (RUBISCO), are light dependent; a third, chlororespiration, occurs mainly in the dark. Determining the contribution of these processes to overall respiratory gas exchange of phytoplankton is beset with problems. While significant advances in the understanding of phytoplankton respiration have recently come from molecular genetic, biochemical, and physiological investigations, further methodological advances are needed if the role of the different pathways under natural conditions is to be understood.

3.1 Introduction

Using a broad definition of respiration, that is, oxygen uptake and/or carbon dioxide production, the photolithotrophs have the greatest range of respiratory mechanisms of any aquatic organisms. This is because the photolithotrophs have the normal respiratory mechanisms of aerobic chemoorganotrophs and, in addition, oxygen uptake and carbon dioxide release mechanisms associated with the photosynthetic apparatus. A further complication in eukaryotes is the existence of another major compartment, the plastid, which can supplement, or largely replace the cytosol as the site for some normal respiratory enzymes. By considering all aquatic photolithotrophs, in this chapter we are dealing with non-oxygen evolving photolithotrophs (e.g. green sulfur bacteria) and oxygen evolving photolithotrophs (e.g. the cyanobacteria *sensu lato* as well as the polyphyletic eukaryotic algae and the higher plants). The combination of genomes in the primary and secondary endosymbioses which gave rise to the plastids of the various algae contributed not only photosynthesis and other plastid genes but also other genes including some respiratory genes to the chemoorganotrophic host.

3.2 The diversity of aquatic photolithotrophs and their respiratory pathways

Table 3.1 shows the range of photolithotrophs in relation to their status as gram-negative bacteria and, for the eukaryotes, as the products of endosymbioses of cyanobacterial or eukaryotic photolithotrophs with a range of chemoorganotrophs. The green photosynthetic sulfur bacteria do not evolve, consume, or tolerate oxygen, that is, never have

Table 3.1 The range of aquatic photolithotrophs and photo-organotrophs. (Falkowski and Raven 1997; Kolber *et al*. 2000; Oliviera and Bhattacharya 2000)

Taxon	Cell structure type	O_2 evolution capability	Respiratory O_2 uptake
Green S bacteria	Gram negative bacterium	−	−
Aerobic purple bacteria	Gram negative bacterium	−	+
Cyanobacteria	Gram negative bacterium	+	+
Red algae, green algae, and embryophytes	Eukaryotic resulting from primary endosymbiosis of a cyanobacterium with a phagotrophic chemoorganotroph	+	+
Chlororachniophyte and euglenoid algae	Eukaryote resulting from secondary endosymbiosis of green algae with two different phagotrophic chemoorganotrophs	+	+
Cryptophyte, haptophyte, and heterokont (and dinophyte?) algae	Eukaryotes resulting from secondary endosymbiosis of red algae with three (four?) phagotrophic chemoorganotrophs	+	+

had oxygen-based respiration. Most other photosynthetic bacteria have aerobic respiration, but only the aerobic purple bacteria photosynthesize under oxic conditions (Kolber *et al*. 2000). Cyanobacteria and the eukaryotes evolve oxygen, and assimilate carbon dioxide in the light, and evolve carbon dioxide and reduce oxygen in the dark. The oxygen uptake processes in the dark are usually dominated by cytochrome oxidase or a prokaryotic analog, with a contribution from some alternative oxidase and, in eukaryotes, of chlororespiration. These oxygen uptake processes are coupled to evolution of carbon dioxide (Table 3.2). In the light, these two processes can also occur, usually at a lower rate than in the dark, but the predominant oxygen uptake processes in the light are usually the Mehler-peroxidase reaction and the RUBISCO oxygenase and the reactions involved in metabolizing the glycolate (Table 3.2). The Mehler-peroxidase and RUBISCO oxygenase activities are not coupled to carbon dioxide evolution, while glycolate metabolism does involve carbon dioxide evolution (Table 3.2).

The various major oxygen uptake processes have a wide range of affinities for oxygen (Table 3.2), with the processes which can occur in the dark

having much higher affinities than those which are obligately dependent on photosynthetically active radiation. Only RUBISCO oxygenase requires more than air-equilibrium oxygen concentrations in solution to give half of the oxygen saturated rate. The various major oxygen uptake processes also have a wide range of oxygen isotope discrimination values (Table 3.2); the potential for *in situ* estimation of the fraction of respiration involving these different pathways from $^{18}O/^{16}O$ ratio measurements (Nagel *et al*. 2001) is rather small, especially in the light with so many possible oxygen uptake pathways (Raven 1990).

3.3 Functions of the various respiratory pathways

3.3.1 Dark respiration: generalizations

Glycolysis, the oxidative pentose phosphate cycle, the tricarboxylic acid (TCA) cycle and the oxidative phosphorylation pathway have the roles that they have in other aerobic organisms (Buchanan *et al*. 2000, and see Fenchel, Chapter 4). Glycolysis, the (oxidative) pentose pathway, and the TCA

Table 3.2 Comparison of properties of the major oxygen-consuming processes in aquatic photolithotrophs. Data from Raven and Beardall (1981), Beardall and Raven (1990), Guy et al. (1993), Asada (2000), Badger et al. (2000), Nixon (2000), and Nagel et al. (2001)

Property	Oxygen-consuming reactions					
	Mehler-peroxidase reaction	RUBISCO oxygenase	Glycolate oxidase	Chloro-respiration	Cytochrome oxidase	Alternate oxidase
V_{max} (catalytic capacity at O_2 saturation)	$\le V_{max}$ for gross O_2 evolution	≈ 0.15 of V_{max} of RUBISCO carboxylase or gross O_2 evolution	0.08 of V_{max} of gross O_2 evolution	~ 0.1 of V_{max} of gross O_2	~ 0.1 of V_{max} of gross O_2 evolution (range 0.02–0.3)	As for cytochrome oxidase, but not additive with it
$K_{1/2}(O_2)$ mmol m^{-3}	50–190	500–1000	170	?6	≤ 2	≤ 6
In vivo effect of full non-cyclic chain	Absolute requirement	Absolute requirement	Absolute requirement	Inhibition	Variable (inhibits or stimulates)	Variable (inhibits or stimulates)
In vivo coupling to CO_2 evolution	No	Via the pathways of glycolate metabolism	Via the pathway of glyoxylate metabolism	Via the pathways of supply of NAD(P)H	Via the pathways that supply NAD(P)H	Via the pathways that supply NAD(P)H
Rate of $^{16}O_2$ uptake relative to rate of $^{18}O_2$ uptake	1.015	1.021	1.023	?	1.018	1.025
Inhibition by cyanide	—	+	—	(−)?	+	—
Inhibition by salicylhydroxamic acids	(−)?	(−)?	(−)?	(+)?	—	+

cycle are the pathways from carbohydrates to all carbon skeletons needed in biosynthesis. The oxidative part of the pentose phosphate pathway is not essential to carbon skeleton production. The oxidative mode of the pentose phosphate pathway also reduces nicotinamide adenine dinucleotide phosphate (NADP$^+$) to produce NADPH, used in reductive biosyntheses and reduction of SO_4^{2-} and, via ferredoxin, reduction of NO_2^- and SO_3^{2-}. Glycolysis reduces nicotinamide adenine dinucleotide (NAD$^+$) to produce NADH, which is used in NO_3^- reduction. The TCA cycle generates NADH and (via succinic dehydrogenase) reduced flavoproteins; this NADH can possibly be used in NO_3^- reduction, but is (with some supplementation from glycolytic NADH) mainly used in oxidative phosphorylation. Glycolysis and oxidative phosphorylation generate the ATP used in maintenance processes in the dark and, with NAD(P)H for reductive syntheses and

the C skeletons mentioned above, in such growth processes as occur in the dark in photolithotrophs (and in the light in non-green cells of multicellular photolithotrophs).

These generalizations, with the exception of the "dark" and "non-green cells" provisos, could apply to almost any aerobic organism, albeit with restricted biosynthetic capacities (e.g. of amino acids) in metazoa. Specifically photolithotrophic, bacterial, and protist aspects of these dark respiratory processes require various modifications to the generalizations (Buchanan et al. 2000; Raven and Beardall 2003a).

Bacteria

Cyanobacterial (and other bacterial) cells have the water-soluble respiratory enzymes in the cytosol with the photosynthetic carbon reduction cycle and the oxidative phosphorylation sequence located in thylakoid membranes and the plasma membrane (see King, Chapter 2). In photosynthetic bacteria, the sole nicotinamide adenine dinucleotide cofactor is NAD^+, while in cyanobacteria it is $NADP^+$. The oxidative phosphorylation mechanism is mainly in the same membrane as photosynthetic electron transport and phosphorylation reactions, and shares intermediates such as NAD(P)H dehydrogenase ubiquinone/menaquinone (UQ/MQ) (photosynthetic bacteria) or plastoquinone (PQ) (cyanobacteria), cytochrome b/c_1 (photosynthetic bacteria) or cytochrome b_6/f (cyanobacteria), and cytochrome c (photosynthetic bacteria) or cytochrome c_6/plastocyanin (cyanobacteria). These spatial arrangements mean that oxidative phosphorylation is suppressed in the light in cyanobacteria, and that respiratory fluxes from carbohydrate stores to 3-phosphoglycerate are much restricted in the light. Cyanobacteria also generally seem to lack the 2-oxyglutarate dehydrogenase component of the TCA cycle, although the gene for this enzyme is present in *Synechocystis* sp. PCC 6803 (Kaneko *et al.* 1996; Kotani and Tabata 1998). Biosynthesis of glutamate-derived amino acids and tetrapyrrols is provided for by the operation of the citrate to 2-oxoglutarate part of the cycle. Finally, cyanobacteria have not only cytochrome a/a_3 but also other cytochrome-containing oxidases, one of which may be analogous to the alternative oxidase of many eukaryotes (excluding metazoa).

Eukaryotes

In aerobic eukaryotes there is the mitochondrial compartment with the TCA cycle and electron transport and oxidative phosphorylation machinery. In algae and higher plants, like other non-metazoan eukaryotes, there is also an alternative oxidase which oxidizes UQ by a pathway which does not pump H^+ and hence does not phosphorylate ADP (Buchanan *et al.* 2000). The eukaryotic photolithotrophs may also have a non H^+-pumping NADH–UQ oxidoreductase, and also have a NADH dehydrogenase that uses cytosolic (e.g. glycolytic) NADH, again without H^+ pumping. Diatoms have triose phosphate isomerase and glyceraldehyde-3-phosphate dehydrogenase in their mitochondria in addition to those in their cytosol and their plastids;

the function of these enzymes in the mitochondrion is unclear (Lund *et al.* 2000), but has been used as evidence that the eukaryotic glycolytic pathway had a mitochondrial (proteobacterial) origin (Stibitz *et al.* 2000; Henzer and Martin 2001; Rujan and Martin 2001; Martin *et al.* 2003). As will be seen later when considering photorespiration, eukaryotic photolithotroph mitochondria have some of the enzymes of glycolate metabolism.

The energetics of oxidative phosphorylation in eukaryotic aquatic photolithotrophs is probably more similar to that of protista than of metazoa with the possibility of oxidation of external NAD(P)H feeding electrons into UQ as well as the pathway from internal NADH to UQ without pumping H^+, and the alternative oxidase which oxidizes UQH_2 without pumping H^+ mentioned above. Raven and Beardall (2003*b*) consider the energetics of algal oxidative phosphorylation. Endogenous NADH (from the tricarboxylic cycle or glycine decarboxylase) pumps $5H^+$ from matrix to cytosol for each electron transferred to oxygen. The mitochondrial ATP synthetase probably transports $3H^+$ from cytosol to matrix in phosphorylating 1 ADP to ATP, and a further $1H^+$ is used in moving ATP (in exchange for 1 ADP and 1 phosphate) to the cytosol where most of the mitochondrial ATP is used. This means that the maximum number of moles of ATP produced per mole of oxygen taken up or carbon dioxide released with carbohydrate as the organic C source is $5 \times 4/(3+1)$ or 5, that is, 30 ATP per hexose. Despite substrate level phosphorylation in glycolysis and the TCA cycle adding to this ATP yield, the occurrence of NADH generation in the cytosol and the succinate dehydogenase step of the TCA cycle only pump $3H^+$ per electron transferred to oxygen, that is, yielding an ATP/O_2 ratio of $3 \times 4/(3 + 1)$ or 3. Furthermore, some energy (H^+ gradient) is used in moving pyruvate into the mitochondria. For cyanobacteria the (chloroplast-type) ATP synthetase probably has an H^+/ATP of 4 rather than 3 in mitochondria, so that the absence of an energy requirement for ATP export from the mitochondria is exactly balanced and thus the conclusions are similar to those for eukaryotes (except for the absence of energy costs of pyruvate transport and a higher H^+ per electron yield (5) for glycolytic NAD(P)H.

Based on the structural biology of the ATP synthetases rather than on measured H^+/ATP stoichiometries, the H^+/ATP ratio of the mitochondrial ATP synthetase may be 4 and that of the plastid and cyanobacterial ATP synthetase may be 4.67 (Allen 2003; Beardall *et al.* 2003; Kramer *et al.* 2003; Raven and Beardall 2003*a*,*b*). This would give not more than 24 ATP per hexose completely oxidized in mitochondria, and not more than 26 ATP per hexose completely oxidized in cyanobacteria. Further measurements of H^+/ATP ratios for the cyanobacterial and mitochondrial ATP synthetases are needed to decide whether the structural biological data suggesting a higher H^+/ATP give the more correct view of the energetics of respiratory ADP phosphorylation.

Peculiarities of green algae
The plastids of green algae and higher plants have, in addition to the photosynthetic machinery, the two dehydrogenases of the oxidative pentose phosphate pathway and one or more of the enzymes of glycolysis which are not common to the photosynthetic carbon reduction cycle. In some green algae, there is a complete glycolytic sequence in the plastids but not, apparently, in the cytosol (Falkowski and Raven 1997). Evidence for the occurrence of enzymes specific to the oxidative pentose phosphate pathway and glycolysis in algae other than chlorophytes is not abundant, although enolase (a glycolytic enzyme not common to the photosynthetic carbon reduction cycle) occurs in the plastids of *Euglena* (Hannaert *et al.* 2000). This partial or complete duplication of glycolytic and oxidative pentose phosphate pathways presumably gives more flexibility of control, especially in the light, than is the case in cyanobacteria. The dehydrogenases of the oxidative phosphorylation pathway and the ATP production sites of glycolysis can generate NADPH and ATP in the plastid in the dark, supplementing any reductant and ATP entry via shuttles and adenylate transporters, thus supporting maintenance and growth-related (e.g. NO_2^- reduction; net protein synthesis) processes in the plastid in the dark.

3.3.2 Chlororespiration

The photosynthetically competent plastid also has the possibility of generating ATP in the dark by oxidation of organic substrates via NAD(P)H using components common to the cyclic electron transport pathway around photosystem I and, in part, the non-cyclic electron transport chain, and a terminal oxidase related to the alternate oxidase of mitochondria (Beardall *et al.* 2003). The occurrence of this pathway, known as chlororespiration (Fig. 3.1), in all aquatic oxygen evolvers is still in doubt, and its quantitative importance is also not well characterized. In addition to a role in carbohydrate breakdown and ATP synthesis in plastids in the dark, chlororespiration might also be involved in photoprotection and, more generally, in regulation of photosynthetic electron flow, as well as in carotenoid desaturation (Bennoun 2001; Beardall *et al.* 2003). The ratio of ADP phosphorylation to carbohydrate consumed in chlororespiration is not yet established (see Allen 2003; Beardall *et al.* 2003; Kramer *et al.* 2003).

3.3.3 Reactions that require light

The two main light-dependent oxygen uptake processes which occur in plastids (Table 3.2) are the Mehler-peroxidase reaction and RUBISCO oxygenase activity, and associated oxygen uptake and carbon dioxide release reactions.

Mehler reaction
The Mehler-peroxidase reaction (Asada 2000; Badger *et al.* 2000; Allen 2003; Kramer *et al.* 2003; Fig. 3.2) involves electrons from the reducing end of PSI passing to oxygen to produce superoxide, which is then dismuted to H_2O_2 and oxygen using superoxide dismutase. Finally, the H_2O_2 is reduced to H_2O using a peroxidase which consumes more electrons from the reducing end of PSI. The overall result is that for every two molecules of H_2O oxidized to one molecule of oxygen at the oxidizing end of PSII, one molecule of oxygen is reduced to H_2O at the reducing end of PSI. This reaction, involving as it does electron flow through PSII, the inter-system electron transport components and PSI, would be expected to be coupled to H^+ pumping from the stroma into the thylakoid lumen. The occurrence of a Q cycle in electron flow between PSII and PSI means that $3H^+$ are translocated per electron moving from H_2O to H_2O in the water–water cycle.

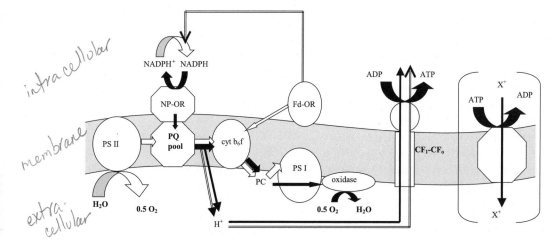

intracellular

membrane

extra-cellular

Figure 3.1 A model for chlororespiratory electron flow, shown by the black arrows. NAPDH$_2$ is oxidized via NADP-dependent plastoquinone oxidoreductase (NP-OR) and passes electrons and protons to the plastoquinone pool. Electron transport to cytochromes leads to proton re-translocation to the intrathylakoid space and maintenance of a proton gradient in the dark. Electrons from cytochromes are passed on via plastocyanin (PC) to an oxidase where molecular oxygen is reduced to water. In cyanobacteria, with shared respiratory and photosynthetic electron transport chains, this oxidase may be cytochrome c oxidase. An analogous enzyme presumably operates in eucaryotes, but has yet to be identified. Protons can cross the thylakoid membrane through the CF$_1$–CF$_0$ ATP synthetase. Electron transport during photosynthesis, in the light, is shown by the white arrows. In brackets on the right-hand side is shown a putative cation transporting ATPase according to Bennoun (1994).

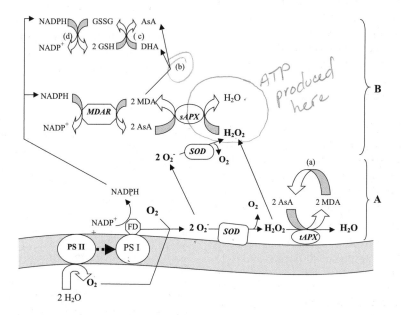

ATP produced here

Figure 3.2 The Mehler-peroxidase reaction. (A) shows the use of superoxide dismutase (SOD) and the thylakoid-bound ascorbate peroxidase (tAPX), with ascorbic acid (AsA) dehydrogenated to monodehydroascorbate (MDA) by the tAPX (reaction a). The MDA is reduced back to AsA as shown under (B). (B) shows the use of SOD and a soluble (stromal) ascorbate peroxidase (sAPX). The MDA is reduced back to AsA by monodehydroascorbate reductase (MDAR) using NADPH generated by the thylakoid photochemical apparatus. Alternatively, the MDA disproportionates to AsA and dehydroascorbate (DHA) (reaction b). The DHA is reduced to (another) AsA by reduced glutathione (GSH)–DHA oxidoreductase (reaction c), and the oxidized glutathione (GSSG) is reduced back to GSH by NADPH generated by the thylakoid photochemical apparatus (reaction d).

With a thylakoid membrane H^+/ATP of 4, the ATP/electron ratio is 0.75 and the ATP/absorbed photon ratio is 0.375 (Raven *et al*. 1999). As a means of photogeneration of ATP, cyclic electron transport around PSI could have an ATP/absorbed photon ratio as high as 1.0 (Raven *et al*. 1999). Despite the lower ATP yield per photon (even under optimal photon allocation conditions) of the Mehler-peroxidase reaction, cyclic electron transport is very sensitive to the redox state of the electron transport components. Consequently ATP generation by the Mehler-peroxidase reaction is perhaps a more reliable, if less energetically efficient, source of ATP in the light in addition to that generated in ferredoxin and $NADP^+$ reduction with subsequent consumption of the reduced ferredoxin and NADPH coupled to carbon dioxide, NO_3^- and SO_4^{2-} reduction to, respectively, $[CH_2O]$, NH_4^+, and HS^-. There is evidence that the ATP used in some manifestations of inorganic carbon concentrating mechanisms comes from the Mehler-peroxidase reaction (Raven and Beardall 2003*a*). It may also be relevant that if only 8 absorbed photons are needed to convert $1CO_2$ to $1[CH_2O]$ and ATP is regenerated by glycolysis and oxidative phosphorylation, then up to 5 ATP can be produced from 8 absorbed photons, that is, an ATP/absorbed photon ratio of 0.625 (see above, and Raven and Beardall 2003*b*). However, as with the computations in 3.3.1. and 3.3.2 of the ATP yield of respiratory processes per unit carbohydrate consumed, these estimates of the photon yield of the Mehler-peroxidase reaction and of other photophosphorylation reactions are subject to reconsideration in the context of the structural biological data indicating a higher H^+/ATP ratio for the chloroplast ATP synthetase (Allen 2003; Kramer *et al*. 2003).

Other (and perhaps more widespread) roles of the Mehler-peroxidase reaction are in removing active oxygen species (in addition to the oxygen and H_2O_2 generated in the Mehler-peroxidase reaction itself), in restricting photoinactivation of PSII by limiting over-reduction of intersystem electron carriers and by generating a large proton gradient across the thylakoid membrane which leads to downregulation of PSII (Asada 2000). While all of these roles of the Mehler-peroxidase reaction are plausible, some of them are mutually incompatible. Thus, restriction

of the rate of electron transport by the rates at which ADP is regenerated through the use of ATP in inorganic C accumulation is not compatible with increased electron transport rate in energy dissipation, unless there is facultative uncoupling of electron transport from H^+ pumping, or an uncoupling protein is functioning in the thylakoid membrane (Beardall *et al*. 2003).

RUBISCO oxygenase and photorespiration

RUBISCO oxygenase is the other major light-dependent oxygen uptake process in photosynthetic organisms, which use RUBISCO as their core carboxylase and photosynthesize in the presence of oxygen, with metabolism of glycolate involving further oxygen uptake as well as carbon dioxide production. The carboxylase and oxygenase activities are competitive and hence are a function of the carbon dioxide and oxygen concentrations around RUBISCO during steady-state photosynthesis. Intracellular carbon dioxide and oxygen during steady-state photosynthesis range from carbon dioxide: O_2 lower than that in the medium when carbon dioxide supply to RUBISCO is solely by diffusion, to ratios greatly in excess of those in the medium, with carbon dioxide concentrations also very significantly in excess of those in the medium, in organisms with a well-developed carbon dioxide-concentrating mechanism (CCM). The selectivity for carbon dioxide over oxygen shows significant phylogenetic variability, with almost an order of magnitude variation in the carboxylase: oxygenase ratio for a given carbon dioxide: O_2 between the high-oxygenase cases (many purple bacteria; cyanobacteria; peridinin-containing dinoflagellates) and the extreme low-oxygenase case of thermophilic red algae (Badger *et al*. 1998; Badger and Spalding 2000). Aquatic organisms with diffusive carbon dioxide entry generally live in habitats with high carbon dioxide and/or low temperatures, thereby minimizing oxygenase activity relative to carboxylase activity (Raven and Beardall 2003*a*). The majority of aquatic photolithotrophs have CCMs and maintain carbon dioxide and oxygen levels around RUBISCO which minimize, but generally do not eliminate, the oxygenase function of RUBISCO (Badger and

Spalding 2000; Badger *et al*. 2000; Franklin and Badger 2001; Raven and Beardall 2003*a*; Beardall *et al*. 2003).

Fig 3.3 The oxygenase function of RUBISCO generates phosphoglycolate and thence glycolate. Some of this glycolate is lost to the aquatic medium but some (usually most) is further metabolized. As in terrestrial higher plants, the charophycean green algae and aquatic higher plants metabolize glycolate via a peroxisomal H_2O_2-producing glycolate oxidase, with the resulting glyoxylate metabolized in peroxisomes, mitochondria, and plastids to produce phosphoglycerate. This phosphoglycerate is treated as is the phosphoglycerate resulting from the carboxylase (and oxygenase) function of RUBISCO. This photorespiratory carbon oxidation cycle involves oxygen uptake by glycolate oxidase and, in mitochondria, as a result of the glycine decarboxylase reaction which produces carbon dioxide equal to 0.25 of the C in glycolate. A similar pathway probably occurs in brown and red algae (Raven *et al*. 2000). Figure 3.3 shows this pathway, but with the glycolate oxidase replaced by mitochondrial glycolate dehydrogenase as occurs in *Chlamydomonas* (see below).

Other algae (including cyanobacteria and non-charophycean green algae) have mitochondrial (in cyanobacteria, thylakoidal) glycolate dehydrogenase coupled to the UQ(PQ) to cytochrome oxidase segment of respiratory electron transport, with various, and often poorly known, pathways of glyoxylate metabolism (Raven *et al*. 2000). While the energetics of the photorespiratory carbon oxidation cycle with glycolate oxidase or glycolate dehydrogenase is quite well understood, the energetics of other modes of glycolate metabolism are not well defined (Raven *et al*. 2000; Wingler *et al*. 2000). Since the energetics of the carbon-concentrating mechanisms are also poorly constrained, the relative energetics of CCMs and the alternative of glycolate synthesis and metabolism are similarly not well defined (Raven *et al*. 2000).

The function of the pathways of phosphoglycolate oxidation is primarily a means of scavenging the unavoidable production of phosphoglycolate (unavoidable granted the CO_2/O_2 ratio around RUBISCO, the temperature

and RUBISCO kinetics: Wingler *et al*. 2000). All other functions, such as the supply of biosynthetic intermediates and stress protection, are subsidiary to this, and clearly other pathways can substitute for phosphoglycolate metabolism when RUBISCO oxygenase activity is minimal (e.g. in many cyanobacteria).

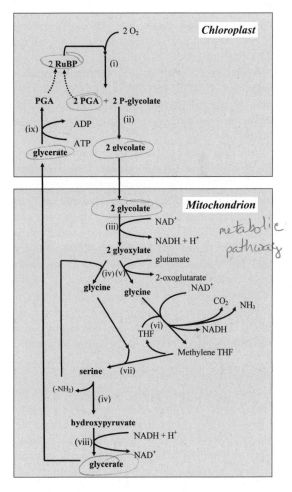

Figure 3.3 The photorespiratory carbon oxidation cycle in a non-charophycean green alga such as *Chlamydomonas*. PGA = phosphoglycerate, RuBP = ribulose-1,5-bisphosphate, THF = tetrahydrofolic acid. Enzymes involved: (i) Ribulose-1,5-bisphosphate carboxylase–oxygenase, (ii) phosphoglycolate phosphatase, (iii) glycerate dehydrogenase, (iv) serine-glyoxylate amino-transferase, (v) glutamate-glyoxylate amino-transferase, (vi) glycine decarboxylase, (vii) serine hydroxymethyl transferase, (viii) hydroxypyruvate reductase, and (ix) glycerate kinase.

3.3.4 Relative importance of the various respiration processes

These tentative conclusions on the energetics of these respiratory processes depend on an ability to estimate the rate of operation of RUBISCO oxygenase under natural, or even experimental, conditions. Such estimates are difficult if not impossible, especially when the other oxygen uptake mechanisms in the light (Mehler-peroxidase reaction; cytochrome oxidase, and alternative oxidase "dark" respiration) are considered. Even in the dark there are problems with distinguishing among the cytochrome oxidase, and alternative oxidase pathways of "dark" respiration and chlororespiration. In general, pathway-specific carbon dioxide fluxes are more difficult to estimate than are oxygen fluxes.

We deal first with "dark" respiration and chlororespiration. In pure cultures, the cytochrome oxidase and alternate oxidase pathway are potentially distinguished in a non-invasive manner from natural abundance stable O isotope measurements (Nagel et al. 2001). However, work on microalgae and other marine planktonic organisms (Kiddon et al. 1993) shows that the variability at a community level, including bacteria and zooplankton, render this method relatively insensitive as an indicator of phytoplankton respiratory function (Bender et al. 2000). Discrimination between chlororespiration and the other two pathways is, of course, impossible if only a single value $^{18}O/^{16}O$ is available when there are three pathways involved.

In the light, the use of pulse-amplitude modulated fluorescence and measurements of incident photon flux density permit estimation (with several assumptions) of gross oxygen evolution; comparison with net oxygen evolution gives an estimate of oxygen uptake in the light. Monitoring $^{16}O_2$ and $^{18}O_2$ after supplying $^{18}O_2$ or $H_2^{18}O$ also give estimates of gross oxygen evolution and oxygen uptake in cultures and (with measurements of respiration at the community level) in situ (Grande et al. 1991; Badger et al. 2000; Franklin and Badger 2001). Natural abundance triple isotope studies (^{16}O, ^{17}O, ^{18}O; Luz and Barkan 2000) give estimates of gross oxygen evolution in situ and, with estimates of net oxygen changes, permit estimation

of oxygen uptake in the light, again at a community level. Even in pure cultures the use of changes in the $^{18}O/^{16}O$ natural abundance ratio to estimate the mean discrimination against ^{18}O in oxygen uptake in the light (bearing in mind that oxygen evolution in gross photosynthesis has the same $^{18}O/^{16}O$ as the substrate water) as an indicator of the quantitative importance of the different oxygen uptake processes is complicated by the possibility of four (or, if glycolate is oxidized by glycolate oxidase, five) oxygen uptake processes in the light, each with a characteristic $^{18}O/^{16}O$ discrimination (Table 3.2; Raven 1990; Grande et al. 1991; Badger et al. 2000; Franklin and Badger 2001).

Techniques which analyze the contribution of components of the oxygen uptake/CO_2 release pathways in the light which do not involve isotopes of C or O, or fluorescence methodology are interventionist in that they must involve changes in oxygen or inorganic C concentrations or photon flux density, or the use of inhibitors (Table 3.2; Badger et al. 2000; Franklin and Badger 2001).

What information is available on the relative significance of the Mehler-peroxidase reaction and RUBISCO oxygenase (and glycolate oxidase) in oxygen uptake in illuminated algal cells suggests that Mehler-peroxidase activity exceeds that of RUBISCO oxygenase activity in green, red, brown, haptophyte and dinophyte algae, and greatly exceeds RUBISCO oxygenase activity in cyanobacteria (Badger et al. 2000; Franklin and Badger 2001). Estimates of the flux through glycolate using $^{14}CO_2$ or $^{13}CO_2$ labeling is another method for estimating oxygen uptake in RUBISCO oxygenase (and downstream reactions) as well as carbon dioxide release from the downstream reactions. However, all of the available data relate to algae with carbon-concentrating mechanisms, so that relative independence of oxygen uptake from carbon dioxide concentration (except in cyanobacteria where oxygen uptake is *decreased* at low carbon dioxide) is to be expected. The contribution of chlororespiration, cytochrome oxidase, and alternative oxidase to oxygen uptake in the light is not readily discerned from the $^{18}O_2$ data. A minimal estimate of carbon dioxide release in the tricarboxylic acid cycle comes from estimates of the rate of synthesis of compounds involving C skeleton supplied by the TCA cycle.

3.4 Summary

In summarizing respiration *sensu lato* in aquatic photolithotrophs, we reiterate the multiplicity of oxygen uptake and carbon dioxide release mechanisms which could contribute to gas exchange in the dark and, especially, the light. Attempts to quantitate the processes contributing to gas exchange in the light in aquatic photolithotrophs fall foul of problems with the invasive nature of the techniques which are perforce employed. However, we can conclude that the dominant oxygen uptake mechanism in illuminated eukaryotes is the Mehler-peroxidase reaction, followed by RUBISCO oxygenase with, even in organisms employing a carbon concentrating mechanism, a rather lower rate of oxygen uptake via the cytochrome oxidase and alternative oxidase. For carbon dioxide evolution, the rate of decarboxylation in glycolate metabolism probably does not exceed the rate of carbon dioxide production in C skeleton synthesis by the tricarboxylic acid cycle. For cyanobacteria the extent of oxygen uptake by the Mehler-peroxidase reaction is higher, and that by RUBISCO oxygenase and downstream reactions of glycolate metabolism, is lower than in eukaryotes.

References

Allen, J. F. 2003. Cyclic, pseudocyclic and non-cyclic photophosphorylation: new links the chain. *Trends Plant Sci.*, **8**: 15–19.

Asada, K. 2000. The water-water cycle as alternative photon and electron sinks. *Phil. Trans. R. Soc. Lond. B*, **355**: 1419–1431.

Badger, M. R. and Spalding, M. H. 2000. CO_2 acquisition, concentration and fixation in cyanobacteria and algae. In R. C. Leegood, T. D. Sharkey, and S. von Caemmerer (eds) *Photosynthesis: Physiology and Metabolism*. Kluwer, Dordrecht, pp. 369–397.

Badger, M. R., Andrews, T. J., Whitney, S. M., Ludwig, M., Yellowlees, D. C., Leggat, W., and Price, G. D. 1998. The diversity and co-evolution of Rubisco plastids, pyrenoids and chloroplast-based CO_2-concentrating mechanisms in the algae. *Can. J. Bot.*, **117**: 1052–1071.

Badger, M. R., von Caemmerer, S., Ruuska, S., and Nakano, H. 2000. Electron transport to oxygen in higher plants and algae: rates and control of direct photoreduction (Mehler reaction) and rubisco oxygenase. *Phil. Trans. R. Soc. Lond. B*, **355**: 1433–1446.

Beardall, J. and Raven, J. A. 1990. Pathways and mechanisms of respiration in microalgae. *Mar. Microb. Food Webs*, **4**: 7–30.

Beardall, J., Quigg, A., and Raven, J. A. 2003. Oxygen consumption: Photorespiration and chlororespiration. In A. W. D. Larkum, S. Douglas, and J. A. Raven (eds) *Algal Photosynthesis and Respiration*. Kluwer, Dordrecht, pp. 157–181.

Bender, M. L., Dickson, M. L., and Orchards, J. 2000. Net and gross production in the Ross Sea as determined by incubation experiments and dissolved O_2 studies. *Deep-Sea Res. II*, **47**: 3141–3158.

Bennoun, P. 2001. Chlororespiration and the process of carotenoid biosynthesis. *Biochim. Biophys. Acta—Bioenerg.*, **1506**: 133–142.

Buchanan, B. B., Gruissem, W., and Jones, R. L. 2000. *Biochemistry and Molecular Biology of Plants*. American Society of Plant Physiologists, Rockville, Maryland, Baltimore, MD.

Falkowski, P. G. and Raven, J. A. 1997. *Aquatic Photosynthesis*. Blackwell Science, Malden, MA.

Franklin, L. A. and Badger, M. R. 2001. A comparison of photosynthetic electron transport in macroalgae measured by pulse amplitude modulated chlorophyll fluorometry and mass spectrometry. *J. Phycol.*, **37**: 756–767.

Grande, K. D., Bender, M. L., Irwin, B., and Platt, T. 1991. A comparison of net and gross rates of oxygen production as a function of light-intensity in some natural plankton populations and in a *Synechococcus* culture. *J. Plankton Res.*, **13**: 1–16.

Guy, R. D., Fogel, M. L., and Berry, J. A. 1993. Photosynthetic fractionation of the stable isotopes of oxygen and carbon. *Plant Physiol.*, **101**: 37–47.

Hannaert, V., Brinkmann, H., Norwitzki, U., Lee, J. A., Albert, M. A., Senser, C. W., Goasterland, T., Muller, M., Michels, P., and Martin, W. 2000. Enolase from *Trypanosomena brucei*, from the amitochondrial protist *Mastigamoeba balamuthi*, and from the chloroplast and cytosol of *Euglena gracilis*: pieces in the evolutionary puzzle of the eukaryotic glycolytic pathway. *Mol. Biol. and Evol.*, **17**: 989–1000.

Henzer, K. and Martin, W. 2001. How do mitochondrial genes get into the nucleus? *Trends Genet.*, **17**: 383–387.

Kaneko, T., *et al.* 1996. Sequence analysis of the genome of the unicellular cyanobacterium *Synechocystis* sp. strain PCC 6803. II. Sequence determination of the entire genome and assignment of the entire genome and assignment of potential protein-coding regions. *DNA Res.*, **3**: 109–136.

Kiddon, J., Bender, M. L., Orchards, J., Caven, D. A., Goldman, J. C., and Dennett, M. 1993. Isotopic fractionation of oxygen by respiring marine organisms. *Glob. Biogeochem. Cyc.*, **7**: 679–694.

Kolber, Z. S., Van Dover, C. L., Niedeman, P. A., and Falkowski, P. G. 2000. Bacterial photosynthesis in surface waters of the open ocean. *Nature*, **407**: 177–180.

Kotani, H. and Tabata, S. 1998. Lessons from sequencing of the genome of a unicellular cyanobacterium, *Synechocystis* sp. PCC 6803. *Annu. Revi. Plant Physiol. Plant Mol. Biol.*, **49**: 151–171.

Kramer, D. M., Cruz, J. A., and Kanazawa, A. 2003. Balancing the central roles of the thylakoid proton gradient. *Trends Plant Sci.*, **8**: 27–32.

Lund, M. F., Lichtle, C., Apt, K., Martin, W., and Ceriff, R. 2000. Compartment-specific isoforms of TPI and GAPDH are imported into diatom mitochondria as a fusion protein: evidence in favour of a mitochondrial origin of the eukaryotic glycolytic pathway. *Mol. Biol. Evol.*, **17**: 213–223.

Luz, B. and Barkan, E. 2000. Assessment of oceanic productivity with triple-isotope composition of dissolved oxygen. *Science*, **288**: 2028–2031.

Martin, W., Rujan, T., Richly, E., Hansen, A., Cornelson, S., Lins, T., Leister, D., Stoebe, B., Hasegawa, M., and Penny, D. 2003. Evolutionary analysis of *Arabidopsis*, cyanobacterial and chloroplast genomes reveals plastid phylogeny and thousands of cyanobacterial genes in the nucleus. *Proc. Nat. Acad. Sci. USA*, **99**: 12246–12251.

Nagel, O. W., Waldron, S., and Jones, H. G. 2001. An off-line implementation of the stable isotope technique for measurements of alternative respiratory pathway activities. *Plant Physiol.*, **127**: 1279–1286.

Nixon, P. J. 2000. Chlororespiration. *Phil. Trans. R. Soc. Lond. B*, **355**: 1541–1547.

Oliveira, M. C. and Bhattacharya, D. 2000. Phylogeny of the Bangiophycidae (Rhodophyta) and the secondary endosymbiotic origin of algal plastids. *Am. J. Bot.*, **87**: 482–492.

Raven, J. A. 1990. Use of isotopes in estimating respiration and photorespiration in microalgae. *Mar. Microb. Food Webs*, **4**: 59–86.

Raven, J. A. and Beardall, J. 1981. Respiration and photorespiration. In T. Platt (ed) *Physiological Bases of Phytoplankton Ecology. Can. Bull. Fish. Aquat. Sci.*, No. 210, pp. 55–82.

Raven, J. A. and Beardall, J. 2003a. Carbon acquisition mechanisms of algae: carbon dioxide diffusion and carbon concentrating mechanisms. In A. W. D. Larkum, S. Douglas, and J. A. Raven (eds) "Algal photosynthesis and photorespiration", Kluwer, Dordrecht, pp. 225–244.

Raven, J. A. and Beardall, J. 2003b. Carbohydrate metabolism and respiration. In A. W. D. Larkum, S. Douglas, and J. A. Raven, (eds) *Algal Photosynthesis and Respiration*. Kluwer, Dordrecht, pp. 205–224.

Raven, J. A., Evans, M. C. W., and Korb, R. E. 1999. The role of trace metals in photosynthetic electron transport in O_2-evolving organisms. *Photosynth. Res.*, **760**: 111–149.

Raven, J. A., Kübler, J. I., and Beardall, J. 2000. Put out the light, and then put out the light. *J. Mar. Biol. Assoc. UK*, **80**: 1–25.

Rujan, T. and Martin, W. 2001. How many genes in *Arabidopsis* come from cyanobacteria? An estimate from 386 protein phylogenies. *Trends Genet.*, **17**: 113–120.

Stibitz, T. B., Keeling, P. J., and Bhattacharya, D. 2000. Symbiotic origin of a novel actin gene in the cryptophyte. *Pyrenomonas helgolandii. Mol. Biol. Evol.*, **17**: 1731–1738.

Wingler, A., Lea, P. J., Quick, W. P., and Leegood, R. C. 2000. Photorespiration: metabolic pathways and their role in stress protection. *Phil. Trans. R. Soc. Lond. B*, **355**: 1517–1529.

CHAPTER 4

Respiration in aquatic protists

Tom Fenchel

Marine Biological Laboratory, University of Copenhagen, Denmark

Outline

This chapter considers aspects of oxygen uptake in protists, emphasizing the coupling to energy metabolism. Microorganisms potentially grow very fast and so for growing cells the largest part of energy metabolism is spent on macromolecular synthesis and respiration is almost proportional to the growth rate. For a given species, oxygen uptake may vary by a factor of up to 50; measurements of respiration are therefore only meaningful in the light of the physiological state of cells. For rapidly growing cells, weight specific respiration scales as (weight)$^{-0.25}$ when different species are compared. For equally sized species, some taxonomic groups show lower respiratory rates than others, but in general interspecific differences in respiratory rates remain to be studied in detail.

4.1 Introduction

Respiration is a manifestation of energy metabolism. In many contexts, respiration is considered synonymous with the uptake of oxygen. Oxygen uptake can be considered as a measure of energy generation in organisms assuming ordinary aerobic metabolism based on organic matter, and this applies to almost all aerobic, phagotrophic eukaryotes. "Anaerobic respiration" implies the use of external electron receptors other than oxygen (see King, Chapter 2); among protists there is, so far, only one known example (nitrate respiration); anaerobic respiration always implies a lower energy yield per unit substrate. Anaerobic protists with a fermentative energy metabolism occur in several taxonomic groups; they do not use external electron acceptors and anaerobic energy generation cannot be measured in terms of respiration. Some protists are facultative anaerobes: they are capable of oxidative phosphorylation in the presence of oxygen, but they can grow and divide in the absence of oxygen on the basis of fermentation. All nongreen protists are heterotrophs depending on organic matter

for food or substrate, and carbon dioxide generation would be an alternative measure of respiration. Since the measurement of oxygen uptake is methodologically much simpler, almost all existing data on respiration are based on oxygen uptake. Some anaerobes respire oxygen, but this is not coupled directly to energy conservation.

By far the greatest energy expenditure of microorganisms is spent for the synthesis of macromolecules (nucleic acids, proteins) and to a lesser degree for active transport across the cell membrane, whereas energy consumption for maintenance and cell motility is relatively small. Consequently, respiratory rates are closely coupled to growth and reproduction; in fact, respiration increases in almost linear proportion to the growth rate constant during balanced growth. Respiratory rates are therefore, to a much larger extent than is the case for animals, a function of the physiological state of the cell and the concept of basal metabolism has no meaning for microbes. For respiratory rates to have a meaning, they must be related to a physiological state such as balanced growth with a given growth rate constant or starvation for a given period of time. The way in which the energy metabolism of individual species responds to different ambient conditions varies, however, and this reflects different adaptive "strategies" for coping with heterogeneous habitats.

Respiration is also stoichiometrically coupled to other aspects of metabolism. For example, nitrogen excretion is proportional to respiration for a given growth efficiency and for given C/N-ratios of food particles and of the cell itself.

There are inherent variations in respiration rates among different species of similar sizes and physiological conditions. Overlying this, energy metabolism scales with cell size: everything else being equal, smaller cells have a higher rate of living. Generally it has been found that the weight specific metabolic rate increases proportionally to $W^{-0.25}$, so that increase in cell volume by a factor of 10 000 means a decrease in weight specific metabolic rate by a factor of 10.

Most data on respiratory rates of protozoa are found scattered in a large number of publications. Most recent reviews of this literature are Fenchel and Finlay (1983) and Caron et al. (1990), these papers compile all available data on protist respiration rates.

4.2   Energy metabolism of protists

4.2.1   Aerobic metabolism

The vast majority of protists are aerobic heterotrophs with a glycolytic pathway and a citric acid cycle and an electron transport chain residing in mitochondria. Since growth is closely coupled to respiration, this provides a way to calculate the maximum growth efficiency from respiratory rates. From bacteria it has been generalized that 1 mol of ATP is required to produce 4 g of cell carbon (Bauchop and Elsden 1960). In an aerobic organism, the complete oxidation of 1 mol of glucose ($= 72\,g\,C$), corresponding to the consumption of 6 mol of oxygen, yields 32 mol ATP. To synthesize 4×32 g of cell carbon the cells must also assimilate this amount of organic carbon in addition to that necessary for energy metabolism. The ideal net growth efficiency is therefore expected to be $(32 \times 4)/[(32 \times 4) + 72]$ or 0.64. Actual data on growth efficiencies of protists (Calow

1977; Fenchel and Finlay 1983; Caron and Goldman 1990) are largely consistent with this consideration. Estimates of growth efficiency vary substantially (cf. compilation of data in the literature in Caron and Goldman 1990). This foremost reflects different methodologies applied and it may not be meaningful to take a mean value. Most estimates have been gross growth efficiencies and they are therefore somewhat lower than 60%—in most cases between 30% and 60%. Attempts to distinguish between net and gross growth efficiencies have resulted in values to close to 60% (Fenchel 1982a). Low values for growth efficiency may also reflect that the food offered had deficiencies with respect to N-content or to certain micronutrients. Some organisms have been shown to be capable of mitochondrial sulfide oxidation and in some metazoans this is coupled to energy conservation (e.g. Oeschger and Vismann 1994). Sulfide oxidation has so far been found only in one protozoan species (the soil amoeba Acanthamoeba). Terminal oxidases other than cytochrome a has been demonstrated in some protozoa (Lloyd et al. 1980; 1981); these are characterized by insensitivity to certain respiratory inhibitors (CN^-, S^{2-}). The role of this is probably oxygen detoxification and the oxygen reduction is not coupled to energy conservation.

4.2.2   Anaerobic metabolism

So far only one species, the ciliated protozoon Loxodes, is known to carry out anaerobic respiration (Finlay 1985; Finlay et al. 1983). Loxodes (with a couple of species) is a limnic form that lives in the microaerobic layers of sediments and in the hypoxic zone of the water column in stratified lakes. When exposed to anoxia it can perform nitrate reduction (to nitrite) in its mitochondria and the process is coupled to energy conservation. It is possible that denitrifying forms are more widespread, especially among microaerophilic or facultatively anaerobic protists, but this has not been studied.

All other anaerobic protists have a fermentative metabolism. Anaerobic lifestyle has evolved independently within many free-living protists including several flagellate groups, some amoebae, and ciliates (Fenchel and Finlay, 1995; Smirnov and Fenchel 1998; Bernard *et al.* 2000). Free-living, anaerobic protists occur in the anaerobic and sulfidic zone of aquatic sediments, in the anoxic part of stratified water columns, and in sewage digesters. The diplomonad flagellates and the pelobiont amoebae do not have mitochondria (or mitochondria-derived organelles). Fermentation is cytosolic, producing various low molecular weight organic compounds (acetate, ethanol, lactate, etc) as end products. Other groups of anaerobic protists have retained organelles with a mitochondrial structure. In many cases these have been shown to be hydrogenosomes. Hydrogenosomes are now known to derive from mitochondria and this evolution has taken place independently in many protist taxa (Biagini *et al.* 1997*a*). Hydrogenosomes are capable of fermenting pyruvate (produced by the glycolytic pathway in the cytosol) into acetate and H_2. This mechanism is the energetically most effective type of fermentation altogether yielding 4 mol of ATP per mole of dissimilated glucose whereas fermentation without H_2-formation yields only 2–3 mol of ATP (Müller 1988; Fenchel and Finlay 1995). Complete fermentation of carbohydrates into acetate $+ H_2$ requires a low ambient H_2-tension. In many anaerobic protists this is accomplished by endosymbiotic methanogenic bacteria, by ectosymbiotic sulfate reducing bacteria, or, in one case, by endosymbiotic purple non-sulfur bacteria (Fenchel and Finlay 1995). Artificial removal of endosymbiotic methanogens results in a decrease in H_2-production and a decrease in growth rate and growth yield by about 25% (Fenchel and Finlay 1995).

Applying considerations similar to those for aerobes, it can be calculated that the growth efficiency of fermenting protists should be 14% (i.e. one-fourth of that of aerobes), and under given food concentrations it should lead to growth rate constants that are one-fourth of that of similarly sized aerobes; this prediction has been shown to hold (Fenchel and Finlay 1990*a*).

Many obligate anaerobes have been shown to take up oxygen under microaerobic conditions (<5–10% atmospheric saturation; higher concentrations are usually lethal) at rates that are comparable to those of aerobic cells. This is not connected to electron transport phosphorylation. The adaptive significance seems to be twofold. It is in part a means of oxygen detoxification and such that at oxygen tensions below about 2% cells can thus maintain a completely anaerobic intracellular environment. It also enhances the energetic efficiency of fermentation in that oxygen may act as a sink for reduction equivalents as an alternative to H_2-production. Microaerobic conditions (1–2% atmospheric oxygen tension) has been shown to enhance growth in some obligate anaerobic protists (Fenchel and Finlay 1990*b*; Biagini *et al.* 1997*b*).

4.2.3 Microaerophiles and facultative anaerobes

Many free-living protists have proven to be microaerophiles with distinct preferences for certain low oxygen tensions (typically within the range of 2–10% atmospheric oxygen tension). This can be shown through their chemosensory behavior in oxygen gradients and from suboptimal growth or pathological responses at higher oxygen tensions (Fenchel *et al.* 1989; Bernard and Fenchel 1996). The adaptive significance is probably that these organisms can to some extent dispense from energetically costly defenses against oxygen toxicity. In addition, the microaerobic part of stratified water columns and sediments represent zones of enhanced bacterial activity, for example, by sulfur bacteria; chemosensory attraction to areas with low oxygen tension is therefore also a mechanism for congregating at high food particle concentrations. Many microaerophiles have also been shown to be facultative anaerobes: they are capable of growing in anoxia, albeit with lower growth rates than under microaerobic conditions. Their anaerobic growth is unaffected by CN^- and HS^-, suggesting fermentative energy generation (Bernard and Fenchel 1996).

4.3 Methodology

Methodology of respiratory rates in protists is largely a question of measuring oxygen concentration or tension with a sufficient sensitivity within a small volume of cell suspension. Earlier, the most common method was the Warburg technique.

It is based on the decreasing headspace volume, resulting from oxygen consumption, under constant pressure and with concomitant absorption of CO_2 produced during respiration. The Cartesian diver technique is a more elegant version of this principle. It allows for measuring oxygen uptake by single cells (Zeuthen 1943; Scholander *et al.* 1952). The somewhat more advanced gradient-diver technique is based on the vertical migration of the divers in a density gradient (Hamburger and Zeuthen 1973). A simple method is based on the chemosensory motile behavior of microaerophilic protists (Fenchel *et al.* 1989). Cell suspensions are added to one end of a glass capillary with atmospheric oxygen saturation. The cells will form a well-defined band since their own oxygen consumption will lead to the optimal low oxygen tension. As the oxygen tension falls below this value the band will slowly migrate along the capillary while consuming oxygen in front of the band. Knowing the diameter of the capillary and the number of added cells, the migration velocity is a direct measure of the per capita oxygen consumption.

The advent of amperometric oxygen electrodes has simplified respiration measurements substantially. Oxygen microelectrodes, that are (almost) insensitive to stirring and with an oxygen consumption that is so low that it can be ignored in practice, are especially useful (Revsbech and Jørgensen 1986). Microelectrodes can be inserted through a capillary opening into a temperature-controlled respiration chamber (with or without stirring) containing cell suspensions and the oxygen tension can be monitored continuously with some sort of recording device (Fig. 4.1). Under some circumstances, oxygen microelectrodes can also be used to measure the oxygen consumption (or production) of individual cells (Jørgensen *et al.* 1985). It is based on the oxygen concentration gradient surrounding the respiring cell and equations describing the diffusion gradients surrounding a spherical absorber. The method is, however, limited to nonmotile, and rather large cells.

For all methods it applies that the most important aspect is to know and maintain a particular physiological state of the cells during measurements. For example, even a brief period of starvation may

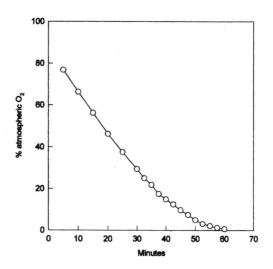

Figure 4.1 Decrease in oxygen tension in a respiration chamber with a suspension of 5.54×10^4 *Euplotes* cells cm^{-3} measured with an O_2-electrode. Data from Fenchel *et al.* 1989.

reduce respiratory rates substantially (see below). On the other hand, inclusion of living bacteria during measurements must be avoided.

All the methods that have been described so far are incapable of measuring oxygen consumption at very low oxygen tensions (in the case of oxygen microelectrodes the detection limit is about 0.5% atmospheric saturation or about 1 μM O_2). At such low oxygen tensions, the luminescence of the bacterium *Photobacterium* can be used as a sensitive oxygen sensor (Lloyd *et al.* 1982). Another very sensitive method is a membrane inlet mass spectrometer (e.g. Yarlett *et al.* 1983).

4.4 Factors affecting oxygen uptake

4.4.1 Oxygen tension

The K_m of cytochrome *a* (cytochrome oxidase) is extremely low. Larger aquatic animals have an absolute requirement for oxygen tensions that is typically considerably higher than 10% atmospheric saturation. The reason for this is that the oxygen transport takes place via a number of steps based on molecular diffusion: from the water into the blood stream via diffusion through gills, from the blood and into tissue, from tissue into cells, and from the cytosol into the mitochondria. Each such step requires concentration gradients to drive the diffusive fluxes. Therefore, in order to provide the mitochondria with an oxygen tension around 1% atmospheric saturation, a much higher oxygen tension in the environment is necessary. The oxygen supply of spherical organisms measuring less than 1–2 mm is solely based on molecular diffusion (the absence of internal advective oxygen transport generally sets the upper size limit for unicellular organisms). Oxygen uptake is therefore limited by diffusion in the water film surrounding the cells and by diffusion within the cells; this means that access to oxygen is a decreasing function of cell size.

Using spherical diffusion equations and the fact that oxygen uptake of cells increases with cell size proportional to volume$^{0.75}$, it is readily shown that the minimum necessary ambient oxygen tension that can sustain aerobic metabolism increases proportional to $R^{1.25}$, where R is the radius of a spherical cell (Fenchel and Finlay 1995). Figure 4.2 shows the half-saturation constant for oxygen uptake for some protists (see also Fig. 4.1). The slope of the regression line is somewhat lower than 1.25; this may perhaps be explained by the fact that the larger protists deviate more from a spherical shape than do smaller ones. In general, large protists have a K_m that is 1–2% atmospheric saturation and smaller ones (like most flagellates and amoebae have a K_m

that is 0.1–0.2% . For mitochondria and for aerobic bacteria, the K_m is even lower. For protists, aerobic metabolism is thus possible at oxygen tensions that are at or below the detection limit of conventional methods for oxygen quantification (O_2-electrodes, chemical methods).

4.4.2 Temperature

All biological process rates are affected by temperature. Different species have different ranges in temperature tolerance. Within the largest part of this range (and other physiological conditions being equal) respiration (and other physiological rate constants such as growth) has a Q_{10}-value of 2–3 (meaning that an increase by 10°C means an increase in respiration rate by a factor between 2 and 3 (Caron *et al.* 1990). Close to the maximum tolerable temperatures, Q_{10} decreases and eventually becomes negative (respiration decreases with increasing temperature); close to the lower temperature range for growth Q_{10} increases with decreasing temperature.

4.4.3 Variation in oxygen uptake during the cell cycle

Even for cells that experience balanced growth, the respiration of individual cells is not constant during

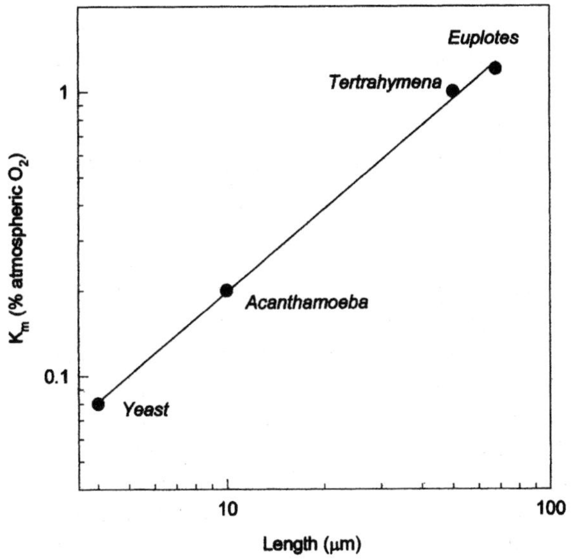

Figure 4.2 The half-saturation constant for oxygen uptake for four protists as function of cell size. Data from Fenchel and Finlay (1995).

the cell cycle. Studies on the amoeba *Acanthamoeba* and the ciliate *Tetrahymena* have shown that oxygen uptake increases linearly (by a factor of about 2) from the moment of cell division and until shortly prior to a new cell division, during which it remains constant (Løvlie 1963; Hamburger 1975).

4.4.4 Food access, coupling to growth

Cells need energy to grow, foremost reflecting the energetic cost of the synthesis of DNA, RNA, and proteins. Cells also have a requirement for energy to maintain their integrity (maintenance energy) and some other cell functions such as motility. Small organisms potentially have such high growth rates that maintenance energy constitutes only a small fraction of energy generation during rapid growth. In resting cysts, the oxygen uptake almost completely vanishes. Energy generation, and thus respiration, is therefore almost linearly proportional to the growth rate constant during balanced growth. This can be demonstrated directly and it is also evident from the fact that growth yield is almost invariant with growth rate (Fenchel 1982a; Fenchel and Finlay 1983).

In most species starvation quickly induces a drop in oxygen uptake. The decrease in respiratory rates induced by starvation differs among species. In the flagellate *Ochromonas* oxygen uptake eventually drops to <5% of that of growing cells following starvation; in the ciliate *Tetrahymena* respiration rate falls only to 30% (Hamburger and Zeuthen 1971; Fenchel 1982a; Fenchel and Finlay 1983; Finlay *et al.* 1983; Fig. 4.3). It has been speculated that this effect is more pronounced in smaller than in larger species (as Fig. 4.3 would indicate), but there is also considerable variation among different, similarly sized organisms. Decrease in oxygen uptake, following the onset of starvation, primarily reflects that transcription and translation is slowed down or almost stops; after a longer period of starvation the number of mitochondria also decreases due to autophagy (Nilsson 1970; Trinci and Thurston 1976; Fenchel 1982b).

The adaptive significance of this is, of course, that it prolongs the survival time during starvation. Assuming an organism that divides, for example, every 4 h with a growth efficiency of 60% will have a complete carbon turnover every 6 h; deprivation of food would therefore mean that cells dissimilated

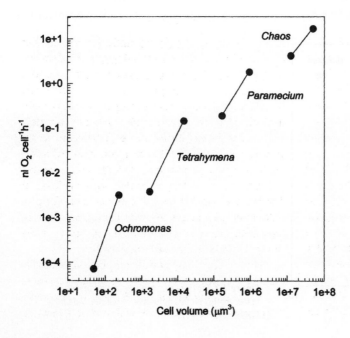

Figure 4.3 Oxygen uptake and cell volume for growing and for starving cells belonging to four protist species. For each species the high values correspond to cells showing maximum growth and the lower values for starving cells. Redrawn from Fenchel and Finlay (1983).

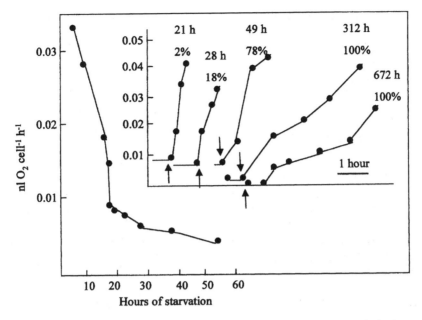

Figure 4.4 Oxygen uptake as a function of time after an exponentially growing culture of the ciliate *Pseudocohnilembus* was deprived of food. Insert: oxygen uptake of starved cells at different times between 21 and 672 h following starvation and respiratory response to addition of a bacterial suspension. The percentage of encysted individuals is also indicated. Data from Fenchel (1990).

their own carbon content within 6 h unless they reduced the rate of respiration. Some protists may thus survive starvation for hundred or more hours without producing resting cysts. The cost of this is that there is a long lag time before cell divisions can be resumed when food is again available, because much of the enzymatic machinery must first be rebuilt. In some protists, starvation only induces a moderate reduction in the rate of respiration; they can only survive for a short time, but during this period they can resume cell divisions after a shorter lag. The ciliate *Pseudocohnilembus* uses both strategies. Starvation first leads to a phenotype in which respiration rates decrease slowly to about 30% of that of growing cells and these can proliferate quickly if food becomes available. If this fails, most cells will produce resting cysts; their respiration rates fall gradually to become undetectable after about 300 h. The resting cysts remain viable for months; addition of food bacteria (or dissolved organic matter) induces increasing oxygen uptake again and eventually excystment and growth. But the lag time before cell divisions can take place

increases with increasing age of the cysts (Fenchel 1990; Fig. 4.4).

4.5 Scaling with size

Respiration scales according to the power function $R = a \times W^b$, where R is respiration rate of individual cells, W is body weight (or volume) and a and b are constants. This relation holds when a sufficiently large range of body sizes are compared. It has empirically been found that the value of b is approximately 0.75 and so the weight specific respiration rate scales according to $W^{-0.25}$ (Hemmingsen 1960; Schmidt-Nielsen 1984). Hemmingsen (1960), who compiled all data on respiratory rates that were available at the time, found that the value of a is lower for protists than it is for invertebrate organisms. This is, however, an artifact that is inherent in data from the literature if these are used uncritically.

Fenchel and Finlay (1983) compiled all available data on measurements of protozoa. These data show a cell size scaling with $b \approx 0.75$, but they also show a considerable range in the value of a exceeding

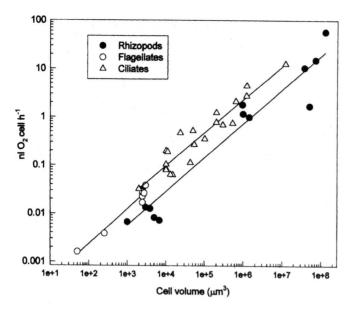

Figure 4.5 Oxygen uptake for different protist species during growth as function of cell volume. Data from Fenchel and Finlay (1983).

a factor of 50, and sometimes this even applies for a given species. The average value of a then corresponded to that of Hemmingsen (1960). Most reports on protozoan respiration rates in the literature refer to cells in an unknown physiological state and in other cases sources mention that cells came from cultures in a "stationary phase" or that the cells were not fed. Selecting only data referring to growing cells (Fig. 4.5) reduced the scatter substantially and the regression line for ciliates (Fig. 4.5) represents an extrapolation of the (metazoan) heterotherm line in Hemmingsen's graph.

It is difficult to determine to what extent the residual variation reflects something real (i.e. interspecies variations in the values of the constant a and to what extent it reflects experimental artifacts. Exact volume determinations are difficult to make and reported values may be based on rather simplifying assumptions on the geometry of the cells. Furthermore, for a given species, the cell volume may vary considerably according to culture conditions. Measurements have been made at different temperatures and data on respiration rates (Fig. 4.5) were therefore corrected to 20°C assuming $Q_{10} = 2$ and this value may not generally apply. Finally, the fact that cells

were reported to be "growing" does not necessarily reflect maximum growth rates, but could well be a function of different culture conditions and this would be reflected by respiratory rates.

Data on respiratory rates for rhizopods (largely naked amoebae) do seem consistently to be lower than for similarly sized ciliates (Fig. 4.5). This is consistent with reported data for growth rate constants of amoebae that also seem to be systematically lower than that of similarly sized ciliates. Similarly, the maximum growth rate constants recorded for heterotrophic flagellates appear to be somewhat lower than those of similarly sized ciliates (Fenchel 1991; Hansen *et al.* 1997).

The simple scaling laws for respiration as function of cell size is real. But they apply only to cells undergoing the maximum rate of balanced growth. As already emphasized in this chapter, *per capita* rates of natural populations depend on the physiological state of the cells and the respiration rate may respond rapidly to changes in environmental conditions. Estimates of metabolic activity and growth of natural populations are therefore unreliable if based solely on cell sizes and numbers.

References

Bauchop, T. and Elsden, S.R. 1960. The growth of micro-organisms in relation to their energy supply. *J. Gen. Microbiol.*, **23**: 457–469.

Bernard, C. and Fenchel, T. 1996. Some microaerobic ciliates are faculative anaerobes. *Eur. J. Protistol.*, **32**: 293–247.

Bernard, C., Simpson, A. G. B., and Patterson, D. J. 2000. Some free-living flagellates (Protista) from anoxic habitats. *Ophelia*, **52**: 113–142.

Biagini, G. A., Finlay, B. J., and Lloyd, D. 1997*a*. Evolution of the hydrogenosome. *FEMS Microbiol. Lett.*, **155**: 133–140.

Biagini, G. A., Suller, M. T. E., Finlay, B. J., and Lloyd, D. 1997*b*. Oxygen uptake and antioxidant responses of the free-living diplomonad *Hexamita* sp. *J. Eukar. Microbiol.*, **44**: 447–453.

Calow, P. 1977. Conversion efficiencies in heterotrophic organisms. *Biol. Rev.*, **52**: 385–409.

Caron, A. D. and Goldman, J. C. 1990. Protozoan mutrient regeneration. In G. M. Capriulo, (ed.) *Ecology of Marine Protozoa*, Oxford University Press, pp. 283–306.

* Caron, A. D., Goldman, J. C., and T. Fenchel. 1990. Protozoan respiration and metabolism. In G. M. Capriulo (ed) *Ecology of Marine Protozoa*. Oxford University Press, New York, pp. 307–322.

Fenchel, T. 1982*a*. Ecology of heterotrophic microflagellates. II. Bioenergetics and growth. *Mar. Ecol. Prog. Ser.*, **8**: 225–231.

Fenchel, T. 1982*b*. Ecology of heterotrophic microflagellates. III. Adaptations to heterogeneous environments. *Mar. Ecol. Prog. Ser.*, **9**: 25–33.

Fenchel, T. 1990. Adaptive significance of polymorphic life cycles in protozoa: responses to starvation and refeeding in two species of marine ciliates. *J. Exp. Mar. Biol. Ecol.*, **136**: 159–177.

* Fenchel, T. 1991. Flagellate design and function. pp. 7–19. In Patterson and J. Larsen, (eds) *The Biology of Free-living Heterotrophic Flagellates*. Oxford University Press, Oxford, pp. 7–19.

Fenchel, T. and Finlay, B. J. 1983. Respiration rates in heterotrophic free-living protozoa. *Microb. Ecol.*, **9**: 99–122.

Fenchel, T. and Finlay, B. J. 1990*a*. Anaerobic free-living protozoa: growth efficiencies and the structure of anaerobic communities. *FEMS Microb. Ecol.*, **74**: 269–276.

Fenchel, T. and Finlay, B. J. 1990*b*. Oxygen toxicity, respiration and behavioural responses to oxygen in free-living anaerobic ciliates. *J. Gen. Microbiol.*, **136**: 1953–1959.

* Fenchel, T., and Finlay, B. J. 1995. Ecology and Evolution in Anoxic Worlds. Oxford University Press, Oxford, UK.

Fenchel, T., Finlay, B. J., and A. Giannì. 1989. Microaerophily in ciliates: responses of an *Euplotes* species (Hypotrichida) to oxygen tension. *Arch. Protist.*, **137**: 317–330.

Finlay, B. J. 1985. Nitrate respiration by protozoa (*Loxodes* spp.) in the hypolimnetic nitrite maximum of a productive freshwater pond. *Freshwater Biol.*, **15**: 333–346.

Finlay, B. J., Span, A. S. W., and Harman, J. M. P. 1983. Nitrate respration in primitive eukaryotes. *Nature*, **303**: 333–336.

Finlay, B. J., Span, A., and Ochsenbein-Gatlen, C. 1983. Influence of physiological state on indices of respiration rate in protozoa. *Comp. Biochem. Physiol.*, **74A**: 211–219.

Hamburger, K. 1975. Respiratory rate through the growth-division cycle of *Acanthamoeba* sp. *C. R. Trav. Lab. Carlsberg*, **40**: 175–185.

Hamburger, K. and E. Zeuthen. 1971. Respiratory responses to dissolved food of starved, normal and division synchronized Tetrahymena cells. *C. R. Trav. Lab. Carlsberg*, **38**: 145–161.

Hamburger, K. and E. Zeuthen. 1973. Recording mittic cycles in single cleaving frog eggs. Gasometric studies with the gradient diver. *C. R. Trav. Lab. Carlsberg*, **39**: 415–432.

Hansen, P. J., Bjørnsen, P. K., and Hansen, B. W. 1997. Zooplankton grazing and growth: scaling within the 2–2,000-μm body size range. *Limnol. Ocenogr.*, **42**: 687–704.

Hemmingsen, A. M. 1960. Energy metabolism as related to body size and respiratory surfaces and its evolution. *Rep. Steno. Mem. Hosp. Copenhagen*, **9**: 1–110.

Jørgensen, B. B., Erez, J., Revsbech, N. P., and Cohen, Y. 1985. Symbiotic photosynthesis in planktonic foraminifera, *Globigerinoides sacculifer* (Brady), studied with microelectrodes. *Limnol. Oceanogr.*, **30**: 1253–1267.

Lloyd, D., Kristensen, B., and Degn, H. 1980. The effect of inhibitors on the oxygen kinetics of terminal oxidases of *Tetrahymena pyriformis* ST. *J. Gen. Microbiol.*, **121**: 117–125.

Lloyd, D., Kristensen, B., and Degn, H. 1981. Oxidative detoxification of hydrogen sulphide detected by mass spectrometry in the soil amoeba *Acanthamoeba castellanii*. *J. Gen. Microbiol.*, **126**: 167–170.

Lloyd, D., Williams, J., Yarlett, N., Williams, A. G. 1982. Oxygen affinities of the hydrogenosome-carrying protozoa *Tritrichomonas foetus* and *Dasytricha ruminantium*, and two aerobic protozoa determined by bacterial bioluminescence. *J. Gen. Microbiol.*, **128**: 1019–1022.

Løvlie, A. 1963. Growth in mass and respiration during the cell cycle of *Tetrahymena pyriformis. C. R. Trav. Lab. Carlsberg*, **33**: 372–411.

Müller, M. 1988. Energy metabolism of protozoa without mitochondria. *Ann. Rev. Microbiol.*, **42**: 465–488.

Nilsson, J. R. 1970. Cytolysomes in *Tetrahymena pyriformis. G. L. C. R. Trav. Lab. Carlsberg*, **38**: 87–121.

Oeschger, R., and Vismann, B. 1994. Sulphide tolerance in heteromastus filiformis (Polychaeta): mitochondrial adaptations. *Ophelia*, **40**: 147–158.

Revsbech, N. P., and Jørgensen, B. B. 1986. Microelectrodes: their use in microbial ecology. *Adv. Microb. Ecol.*, **9**: 293–353.

* Schmidt-Nielsen, K. 1984. *Scaling. Why is Animal Size So Important*. Cambridge University Press, New York.

Scholander, P. F., Claff, C. L., and Sveinsson, S. L. 1932. Respiratory studies in single cells. II. Observations on the oxygen consumption in single protozoans. *Biol. Bull.*, **102**: 178–184.

Smirnov, A. V., and Fenchel, T. 1996. *Vahlkampfia anaerobica* n.sp. and *Vanella peregrina* n. sp. (Rhizopoda)—anaerobic amoebae from a marine sediment. *Arch. Protist.*, **147**: 189–198.

*Trinci, A. O. J., and Thurston, C. F. 1976. Transition to the non-growing state in eukaryotic micro-organisms. In T. R. G. Gray and J. R. Postgate (eds) *The Survival of Vegetative Microbes*. Cambridge University Press, New York, pp. 55–79.

Yarlett, N., Scott, R. I., Williams, A. G., and Lloyd, D. 1983. A note on the effects of oxygen on hydrogen production by the rumen protozoon *Dasytricha ruminantium* Schuberg. *J. Appl. Microbiol.*, **55**: 359–361.

Zeuthen, E. 1943. A cartesian diver micro-respirometer with a gas volume of 0.1 μl. Respiration measurements with an experimental error of 2×10^{-5} μl. *C. R. Trav. Carlsberg*, **24**: 479–518.

CHAPTER 5

Zooplankton respiration

Santiago Hernández-León[1] and Tsutomu Ikeda[2]

[1] *Biological Oceanography Laboratory, Universidad de Las Palmas de Gran Canaria, Spain*
[2] *Marine Biodiversity Laboratory, Hokkaido University, Japan*

Outline

In this chapter metazooplankton ($> 200\,\mu m$) respiration in aquatic systems is reviewed, including methodology, modifying agents (physical, chemical, and biological factors), definitions, and carbon equivalents. The major determinants of zooplankton respiration rate are argued to be body mass and temperature. The causes underlying temporal and spatial (vertical and horizontal) variations in metazooplankton community respiration (as a product of specific respiration rate and biomass) in nature are discussed, with special reference to the mesoscale variations linked with frontal systems, eddies, and filaments of upwellings. Evidence is given that respiration of vertically migrating zooplankton contributes to downward carbon flux in the oceans. Metazooplankton respiration is reviewed in terms of latitudinal patterns and variation with depth (epi, meso, and bathypelagic zones). We emphasize the need to fill the gap in our knowledge regarding metazooplankton biomass data, in order to assess better the role of those organisms in the oceanic carbon cycle.

5.1 Introduction

The oceans and seas occupy a total of 71% of the entire hydrosphere on the earth. As heterotrophs, planktonic animals, including diverse taxa and size groups, are distributed throughout the entire water column of the world's oceans and seas and lakes. Marine planktonic animals may be divided broadly into protozoans (protozooplankton) of nano-($2–20\,\mu m$) and micro-size ($20–200\,\mu m$) and metazoans (metazooplankton) of micro, meso, or larger size. Our knowledge of protozooplankton, a key component of microbial food webs in the oceans, has increased in recent years (cf. Gifford and Caron 2000). However, information about biomass and physiological rates of protozooplankton are still sparse compared to that of metazooplankton. Metazooplankton biomass in the world ocean has been estimated to be 21.5×10^9 tons, which is much greater than that of consumers such as

zoobenthos (10×10^9 tons) and nekton (1×10^9 tons) (cf. Conover 1978).

As heterotrophs, metazooplankton (referred to as "zooplankton" hereafter) require organic carbon to fuel their metabolism. Part of the ingested organic carbon is assimilated and used for metabolism, growth, and reproduction. That part of organic carbon not assimilated is defecated. Thus, information about assimilation efficiency and growth efficiency of ingested carbon is of prime importance to quantify energy transfer and matter cycling through zooplankton in pelagic ecosystems. Among the physiological processes in zooplankton, metabolism (in terms of "oxygen consumption" or "respiration") has been most extensively studied (cf. Omori and Ikeda 1984). This is largely because metabolism accounts for a major component of energy/material budgets in all zooplankton species (42–72%, cf. Ikeda 1985), and is closely coupled to growth ("physiological" method, cf. Ikeda and Motoda 1978). Information on metabolism is thus useful to define the minimum food demand of zooplankton.

The measurement of oxygen consumption or respiration in freshwater or marine zooplankton is relatively simple in design and inexpensive in practice (see below), but direct extrapolation of laboratory results to field populations requires caution. Studies have shown that respiration rates of zooplankton may vary to a great extent under different experimental conditions (Lampert 1984; Omori and Ikeda 1984; Ikeda *et al.* 2000). In addition, respiration rates per unit body mass of zooplankton change with their body size. Our understanding of this variability, inherent with respiration rates of zooplankton, combined with variations in their community structure and biomass at various spatial/temporal scales in the oceans, is of critical importance to elucidate the role of zooplankton respiration in carbon cycles in aquatic systems of our planet.

In this chapter, we first briefly review conventional and new methodologies for measuring respiration rates of zooplankton and note the advantages and shortcomings of each method. More detail of the methods can be found in the comprehensive review by Ikeda *et al.* (2000). Finally, temporal and spatial variability of respiration on various scales in the ocean are reviewed, and a tentative account of zooplankton community respiration in the ocean is given, as a first step toward assessing their role in the carbon cycle in the ocean.

5.2 Methodology

5.2.1 The sealed-chamber method

This method (the "water-bottle method") has long been used by zooplanktologists (cf. Omori and Ikeda 1984). Live zooplankton specimens are confined in air-tight containers (glass bottle, bell-jar, etc.) filled with filtered seawater for several hours to a day, and the decrease in dissolved oxygen content in seawater is measured at the end of incubation (i.e. single end-point method). The difference in oxygen concentration between experimental and control containers (without animals) is assumed

to be due to respiration by the zooplankton incubated. Typically, oxygen concentration of seawater from experimental and control containers is determined precisely by the Winkler method. The use of syringe-like incubators allows time-series determinations of respiration rates although, the volume of the incubator reduces progressively during experiment (Torres 2000). Alternatively, an oxygen sensor may be placed inside each experimental container to monitor the decrease in oxygen concentrations in the course of experiments. There is no change in container volumes in this variation, but may require gentle agitation to avoid polarization of the oxygen sensor (Ikeda *et al.* 2000). High respiration rates at the beginning of the experiment may be due to handling of specimens, reduction in the rates toward the end of experiments may also occur due to starvation (cf. Skjoldal *et al.* 1984; Ikeda *et al.* 2000).

5.2.2 The flow-through method — H₂O exchange?

The flow-through method is superior to sealed-chamber method in that the short-term changes in zooplankton respiration rates may be determined at near constant oxygen concentrations and without accumulation of metabolic excreta. Despite these advantages, the method is not commonly used largely because of complexities in construction, operation, and calibration of the system and requirement of a system-specific calculation algorithm taking into account flow rate, chamber size, and metabolic rate of zooplankton to obtain reliable results (Ikeda *et al.* 2000). It is also noted that the results from flow-through systems may be biased because of the unavoidable delay in detecting the metabolic response of the specimens, as well as the problem of unstable rates. Another disadvantage is the need to use very small chamber volumes (see Kiørboe *et al.* 1985; Thor 2000) and/or high concentration of specimens, often well beyond the conditions found in nature.

5.2.3 Enzymatic approaches

Biochemical methods for oxygen consumption rates of marine zooplankton involve the determination of the activity of various enzymatic systems, including succinate dehydrogenase (Curl and Sandberg

1961), electron transfer system (ETS) (Packard 1971), lactate dehydrogenase (LDH), pyruvate kinase (PK) and citrate synthase (CS) (Berges *et al.* 1990), and malate dehydrogenase (MDH) (Thuesen and Childress 1993). Of these indices, ETS activity has been employed most extensively among biological oceanographers (see also Robinson and Williams, Chapter 9, Arístegui *et al.*, Chapter 10).

Enzymatic methods provide good estimates of respiration rates provided the enzyme studied is not limited by intracellular substrates (Hernández-León and Gómez 1996; Packard *et al.* 1996; Hernández-León and Torres 1997; Roy and Packard 1998; Roy *et al.* 1999). The *in vitro* measurements of enzyme activity as presently being practised represent maximum activities under saturated substrate concentrations, thereby equivalent to the measurement of enzyme concentrations. In nature, the cells of organisms may be substrate limited and therefore the extrapolation of enzyme activities *in vitro* to those *in vivo* is not straightforward. The relationship between enzymatic activity and respiration rate is known to vary with body size, even within the same species (Berges *et al.* 1990). The influence of different isozymes within species with different K_m and V_{max} values also introduces uncertainty to the interpretation of these assays (Runge and Roff 2000).

According to Hernández-León and Gómez (1996), the *in vivo* respiration to ETS activity ratios of marine zooplankton ranged from 0.5 to 1.0, the upper range was from specimens from seasons characterized by higher chlorophyll concentrations and primary production, suggesting that well-fed specimens are generally not substrate limited. Hernández-León and Torres (1997) also observed the effect of different substrates on the relationships between glutamate dehydrogenase (GDH) activity and ammonia excretion rates in zooplankton. In general, the results from optimized enzyme assays result in an overestimation of metabolic activity. Packard *et al.* (1996), Roy and Packard (1998), and Roy *et al.* (1999) showed the effect of substrate limitation on enzyme activities and the subsequent respiration rates of planktonic organisms. Packard *et al.* (1996) estimated *in vivo* ETS activity in the marine bacterium *Pseudomonas nautica* from a model incorporating a bisubstrate

enzyme kinetics algorithm to account for the limitation of the intracellular substrates. From the model, respiration rates of the bacterium were predicted with high precision (correlation coefficient between observed and predicted rates, $r^2 > 0.92$). Båmstedt (2000) modified the ETS assay in order to measure the activity *in vivo* (no addition of substrates), obtaining a linear relationship between respiration and ETS activity in *Praunus flexuosus* ($r^2 = 0.82$). By using the new assay, Båmstedt (2000) found that the variability in respiration/ETS ratios was reduced by 50–70%, as compared with the traditional ETS assay. In summary, a successful application of ETS assay and other enzymatic methods as indices of zooplankton respiration depends on the validity of the respiration/enzyme activity ratios of zooplankton under study.

5.2.4 Respiratory Quotient (RQ)

Despite the fact that measurements of carbon dioxide production, rather than oxygen consumption by zooplankton, are more relevant to carbon-cycle studies in the ocean, this has rarely been attempted, because of a strong buffering action of seawater. Oxygen-derived rates must be thus converted to carbon equivalents using the respiratory quotient (RQ, the molar ratios of carbon dioxide produced to O_2 consumed).

The RQ value, as its reciprocal the photosynthetic quotient, is a simple stoichiometric consequence of the elemental composition of the substrate being oxidized. The composition of the major biochemicals can be constrained well enough that theoretical RQ values can be calculated for each type of substrate (see Williams and del Giorgio, Chapter 1). Calculated values for carbohydrate, protein, nucleic acid, and lipid are, respectively, 1.0, 0.97, 1.24, and 0.67. These values span the range (0.67–1.24) of possible RQ values, with the greatest uncertainty associated with the value for lipid—the value given here is a minimum one. Although the majority of experimental determinations of RQ fall within this range, a few give a wider span, much of this is probably due to the difficulty of measuring the carbon dioxide changes in seawater.

The choice of RQ affects to the conversion of respiration rates to carbon units. Ikeda *et al.* (2000) suggested that a RQ value of 0.97 (protein metabolism resulting in ammonia as the end-product) would be appropriate for marine zooplankton. This is because marine zooplankton are primarily ammonotelic in nature, although a recent study shows that the proportion of excreted urea, ammonia, and dissolved amines by the copepod *Acartia tonsa* is highly variable (Miller and Glibert 1998). Ingvarsdottir *et al.* (1999) used an RQ of 0.75 assuming a predominance of lipid metabolism mixed with some protein for the calculation of carbon budgets of overwintering *Calanus finmarchicus*. The adoption of a fixed RQ results in an uncertainty of about 20% in the oxygen to carbon dioxide conversion. For more accurate conversion of respiration rates to carbon units information about likely metabolic substrates is a prerequisite, and concurrent measurements of ammonia production may help constrain the respiratory substrate (see Rodrigues and Williams 2001).

5.3 Metabolic scope

The respiration rate of a zooplankter in nature is not constant with time but changes in relation to its feeding and swimming activities, some of these often mediated by environmental cues. The respiration rate tends to increase with increasing swimming activity, and the rate extrapolated to zero swimming activity is usually defined as *standard* metabolism (R_s). The rate at maximum activity, beyond which the animal enters anaerobiosis and fatigue, is referred to as *active* metabolism (R_a). The metabolism of the animal in nature falls somewhere between R_s and R_a, and is loosely termed as *routine* metabolism (R_r). In most existing studies on zooplankton metabolism, swimming activity of specimens is uncontrolled, and the resulting respiration rates have been assumed to be close to R_r. As notable exceptions, R_s, R_a, and R_r have been determined (29, 89, and 50 nmol O_2 mg dry wt^{-1} h^{-1}, respectively), for the euphausiid *Euphausia pacifica* at 8°C and normal pressure (1 atm; Torres and Childress 1983), and 335, 2030, and 621 nmol

O_2 mg dry wt^{-1} h^{-1}, respectively, for the marine copepod *Dioithona oculata* (Buskey 1998).

The *metabolic scope* (MS) is defined as

$$MS = (R_a - R_s).$$

It is also of interest to assess the *factorial scope* (FS), the ratio between active and standard rates, which in invertebrates tends to have values of the range 2–10.

$$FS = R_a/R_s.$$

For *E. pacifica* and *D. oculator* mentioned above, normal values range between 60 and 1700 nmol O_2 mg dry wt^{-1} h^{-1} for MS and 3.1 and 6.1 for FS. These FS values are much lower than those reported for salmon (10–20; Brett 1964), humans (10–15), and flying insects (50–200; cf. Prosser 1961).

The determination of R_a and R_s for a given zooplankter in laboratory experiments, allows the calculation of the *metabolic power index* (MPI) from the measured *in situ* respiration rate (R) of the zooplankter, as a diagnostic of the metabolic condition of the zooplankter in the field:

$$MPI = (R - R_s)/(R_a - R_s).$$

Usually, R may be represented by R_r. MPI is dimensionless and varies between 0 and 1, being 0 if $R = R_s$ and 1 if $R = R_a$. Because zooplankton respiration rates in the field are never minimal or maximal, the MPI index provides an average picture of the metabolic status of the zooplankton relative to standard metabolism. For marine copepods, respiration rates of diel vertical migrators such as *Pleuromamma xiphias* and *Calanus euxinus* are known to be different depending on the time of the day, while this was not the case for other non-diel vertical migrators such as *Acartia clausi* and *Temora discaudata* (Pavlova 1994). In this regard, the MPI may be used as an useful index to compare metabolic characteristics among zooplankton species.

5.4 Factors affecting respiration rate

5.4.1 Physical factors

Temperature
From a global viewpoint, the most important factor affecting zooplankton respiration rates is probably

temperature, which ranges from freezing point sea-water ($\sim -2°C$) to warmth (ca. 30°C) in tropical regions.

The response of biological rate processes to temperature is defined according to the timescale involved, as "acclimation," "acclimatization," and "adaptation" (Clarke 1987). "Acclimation" is the adjustment of rate processes to a new temperature in the laboratory, and "acclimatization" is the adjustment of the rate processes to changes in environmental temperature (diurnal, seasonal). It should be noted that in laboratory acclimation, it is usual to modify only the temperature, keeping all other conditions constant. In contrast, acclimatization involves adjustment to a whole range of environmental variables characteristic of field situations. "Adaptation" is an evolutionary adjustment, or genetic change accomplished at the population level to a daily or seasonal variation in temperature requiring acclimatization. "Acclimated," "acclimatized," and "adapted" respiration rates thus defined may roughly correspond to temporal/spatial variation patterns at micro, meso, and large scales, respectively, of marine zooplankton. For the systematic analysis of the entire temperature range $-2°C$ to 30°C in marine environments, comparisons of adapted respiration rates of phylogenetically related zooplankters is a necessary approach, since there is no single zooplankton species that can be found along the entire temperature spectrum.

The relationship between biological rate processes and temperature has been described by the Arrehenius relationship and the van't Hoff rule. The Arrehenius relationship describes the relationship between reaction rate (k) and absolute temperature (T),

$$k = Ae^{-Ea/RT},$$

where A is constant, E_a is the Arrehenius activation energy, and R is the gas constant (8314 kJ mol^{-1}). A plot of $\ln(k)$ against $1/T$ is linear, with a slope of $-E_a/R$. Ivleva (1980) compiled a voluminous dataset on "adapted" respiration rates and body mass (expressed as wet, dry, or organic matter) relationships of pelagic crustaceans (plus some benthic ones) from high and low latitude seas, and concluded that E_a ranged from 54 to 63 kJ mol^{-1}.

A similar E_a has been reported for "acclimated" ETS activities of epipelagic zooplankton (67 kJ mol^{-1}), while Packard et al. (1975) obtained a value of 13 kJ mol^{-1} for bathypelagic marine zooplankton.

The van't Hoff rule, which is described by the Q_{10} approximation, is the most commonly used way to describe the relationship between respiration rates and temperature in zooplankton,

→ think of
$t_1 = 10°C$
$t_2 = 20°C$

$$Q_{10} = (k_1/k_2)^{10/(t_1 - t_2)},$$

where k_1 and k_2 are the respiration rates corresponding to temperatures t_1 and t_2. Q_{10} may be solved graphically, plotting the logarithm of k against t. By using temperatures T_1 and T_2 (absolute temperatures), E_a can be converted to Q_{10} (Ivleva 1980),

$$\log_{10}(Q_{10}) = 10 \times R \times (\log_{10} e) \times E_a/(T_1 \times T_2).$$

In the study of Ivleva (1980), mentioned above, the Q_{10} values for the "adapted" respiration rates of pelagic crustaceans ranged from 2.1 to 2.7. From global equations for the "adapted" respiration rates and body mass (expressed as dry, carbon, nitrogen, or phosphorus) as a function of habitat temperature for epipelagic marine zooplankton including diverse taxa (Ikeda 1985) and for copepods only (Ikeda et al. 2001), Q_{10} has been computed as 1.6–1.9 and 1.8–2.1, respectively. A simple calculation of Q_{10} from the relationships between specific respiration and temperature in Fig. 5.4 ($Q_{10} = E^{(\text{exponent} \times 10)}$) gives values for epipelagic (2.03) and mesopelagic (3.0) zooplankton in the range observed in the literature.

While there is no comprehensive compilation of respiration data for freshwater zooplankton comparable to those on marine zooplankton by Ivleva (1980), Ikeda (1985), and Ikeda et al. (2001), the agreement between the two under equivalent body mass and temperature (Fig. 5.1 and Section 5.4.2) suggests the possible application of the relationship developed for marine zooplankton to freshwater taxa. Q_{10} values derived from global compilation of respiration data by the above researchers range between 2 and 3, and the possible maximum increment in respiration rates of zooplankton

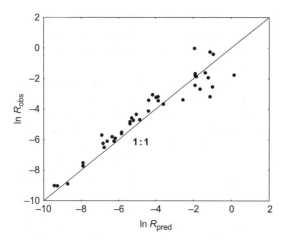

Figure 5.1 Comparison of observed and predicted respiration rates of freshwater zooplankton. Datasets of respiration rates, dry mass, and experimental temperature of freshwater zooplankton (R_{obs}: μmol individual^{-1} d^{-1}) are from Lampert (1984). Predicted rates (R_{pred}: μmol individual^{-1} d^{-1}) from empirical respiration rate–dry mass–temperature relationships of marine epipelagic zooplankton (Ikeda 1985). Note that most data are close to 1:1 line, indicating a good agreement. See text for details.

encountering to the temperature range of $-2°$C to 30°C is 8–27 fold.

Pressure

Hydrostatic pressure is, together with temperature, one of the most important physical factor in marine environments. Early studies focused on the effects of rapid changes in hydrostatic pressure on extensive diel vertical migrating zooplankton who experience large pressure changes (>20 atm), twice daily as a result of their vertical excursion. Laboratory experiments using pressurized respiration chambers revealed that the effect of the changes in pressure on the respiration rates of such the migrating zooplankters was small (cf. review of Ikeda *et al*. 2001). The lack of appreciable hydrostatic pressure effects on respiration rates has also been confirmed on bathypelagic chaetognaths, hydromedusae, polychaetes, and copepods by comparing rates at 1 atm and 100 atm in the laboratory (Childress and Thuesen 1993; Thuesen *et al*. 1998). King and Packard (1975) also noted that the ETS activity of marine microplankton was insensitive to increases in hydrostatic pressure. All these results suggest that determination of respiration rates of

zooplankton living at depth is experimentally sound at 1 atm, provided live specimens are recovered to the surface without damage. As an exception, reduced respiration rates have been observed in delicate mesopelagic gelatinous zooplankton recovered from deep waters, mainly due to detrimental decompression effects and subsequent losses in motor activity (Bailey *et al*. 1994).

Light

Zooplankton living in the euphotic zones are under the exposure to sunlight during daytime, and light is known to affect the respiration rates of zooplankton, but in the standard procedure of "sealed-chamber method," containers with animals are kept in the dark during incubation (Ikeda *et al*. 2000), so most experiments do not take into consideration the effect of light. Moshiri *et al*. (1969) found that respiration rates of a freshwater zooplankter in the light were twice the rates in the dark. According to Fernández (1977), marine zooplankton increased the respiration rates in proportion to the light level until a certain threshold, which varied species-specifically and in relation to the depths where the specimens were captured in the field. Contrary to these, no effect of light was observed in the experiments with mixed freshwater mesozooplankton (Bishop 1968), the cladoceran *Daphnia magna* (Schindler 1968) and the euphausiid *E. pacifica* (Pearcy *et al*. 1969).

Turbulence

Rothchild and Osborn (1988) postulated that zooplankton may be able to encounter more prey with increasing turbulence of the surrounding water. Turbulence exerts positive and negative effects on zooplankton feeding (Kiørboe and Saiz 1995), the latter effect being typical at high levels of turbulence because of the inability to detect the prey. In fact, Visser *et al*. (2001) observed a significant negative response of feeding activities of several copepods to increased turbulence in the surface layer. Alcaraz *et al*. (1994) studied the effect of small-scale turbulence on heartbeat rates of several zooplankton species, and observed an average increase of 5–93%. For zooplankton, a strong relationship between metabolic and heartbeat rates was reported in the study of Pavlova and Minkina

(1983) for marine copepods, while this was not the case in the study of Opalinski (1979) on crustacean plankton. Clearly, there is a need to develop new technologies to improve our understanding of metabolism–turbulence interactions in zooplankton. Turbulence decreases rapidly with increasing depth and may thus not be a factor affecting the respiration rates of meso and bathypelagic zooplankton.

5.4.2 Chemical factors

Oxygen

In marine and freshwater systems, oxygen is generally near saturation in the surface layers, but may be low or depleted at midlayers (see Codispoti *et al*., Chapter 12) or bottom layers (see Middelburg *et al*., Chapter 11). For zooplankton living in oxygen-rich environments, the respiration rates are nearly independent of the ambient oxygen saturation level (Davenport and Trueman 1985). In contrast, zooplankton inhabiting oxygen-poor environments maintain constant respiration rates until certain critical level of oxygen saturation (P_c), beyond which the rates decrease (Childress 1975). The former type is called "metabolic conformers" and the latter type "metabolic regulators" (Prosser 1961), but intermediate types also exist. As a practical rule for measuring respiration rates of zooplankton with "sealed-chamber method" mentioned above, Ikeda *et al*. (2000) set the oxygen concentration of 80% saturation or above at the end of incubation to avoid possible artifacts.

The midwater oxygen minimum zones (OMZ), characteristic of tropical and subtropical oceans, are produced as the results of high oxygen consumption by bacteria and protista and sluggish horizontal mixing of the water below the thermocline (Wyrtki 1962, and see Codispoti *et al*., Chapter 12). There are also anoxic layers at depth in closed basins such as the Baltic and Black Sea. In those areas, the upper aerobic zone is separated from the lower anaerobic layers (containing hydrogen sulfide) by the main pycnocline, which is found between 70 and 200 m depth. In the OMZ, oxygen concentration decreases to less than 10 μmol O_2 dm^{-3}, and zooplankton biomass is also low. Most zooplankton avoid the OMZ and live above or below (Vinogradov and

Voronina 1961; Sameoto 1986; Sameoto *et al*. 1987; Saltzman and Wishner 1997), except for only a few species of copepods and euphausiids (Vinogradov and Voronina 1961).

The capacity to take up oxygen at extremely low ambient oxygen concentrations has been demonstrated on variety of pelagic crustaceans from the OMZ off Southern California (Childress 1975), and in the euphausiid *Euphausia mucronata* from the OMZ, off the coast of Chile (Teal and Carey 1967). Some of the organisms survived from several to tens of hours anaerobically in the laboratory (Childress 1975). Such anaerobic ability seen in some residents of the OMZ appears to be poorly developed for Nordic krill *Meganyctiphanes norvegica*, which visit the OMZ as part of their diel vertical mixing (DVM) behavior (Spicer *et al*. 1999). Recently, Svetlichny *et al*. (2000) established daily energy budgets during vertical migration of a copepod C. *euximus* from warm, oxygen-rich surface layers to cold, oxygen-poor deep waters (ca. 100 m) in the Black Sea. According to their results, *C. euximus* is a metabolic conformer, and migration from the oxygenated warm surface layer (22.1°C, 237 μmol O_2 dm^{-3}) down to the hypoxic cold 100 m depth (7.5°C, 22 μmol O_2 dm^{-3}) caused a total reduction in the region of sevenfold in the respiration rates. From the point of view of daily energy budgets, energy needed for migration represents only 12% of the total energy expenditures, thus *Calanus euximus* achieves considerable energy savings by reducing respiration rates in deep hypoxic zones for 9 h every day.

pH

From the viewpoint of biology at low pH, the OMZ is of special interest. The pH is stabilized at 7.5–8.4 in oceanic waters, but the pH at the OMZ in the Pacific off the coasts of North and South America are lower (≤ 7.5, Park 1968). Mickel and Childress (1978) examined the effects of pH on respiration rates and pleopod movement, in a mysid *Gnathophausia ingens* collected from the OMZ off the coast of southern California and brought to a land laboratory. Comparing the results obtained at pH 7.1 and 7.9, these authors found no appreciable difference in respiration rates, pleopod movement, or any relationship between the two. *G. ingens* is different from most other zooplankton in that it has an extreme capacity to withdraw oxygen from a low oxygen environment (Teal and Carey 1967) and uses hemocyanin as an oxygen carrier (Freel 1978). Many marine planktonic crustaceans and other groups are known to lack oxygen carriers in their body fluids (Prosser 1961; Mangum 1983). Because of these differences, direct application of pH sensitivity of *G. ingens* to other marine zooplankton cannot be made. For recent accounts on the effects of low pH on marine zooplankton, see Yamada and Ikeda (1999).

Salinity

For zooplankton living in freshwater–marine interface regions, salinity changes in ambient water often lead to osmotic disequilibrium. Specimens exposed to extreme salinities are expected to be forced to use their osmoregulatory mechanisms to achieve the electrochemical gradient, which in turn affect their respiration rates. However, the results in the literature on the effects of salinity on zooplankton respiration rates are often conflicting. Both reduced and increased respiration rates have been observed on neritic copepods placed into salinity-lowered seawater (Anraku 1964; Lance 1965); the former was interpreted as a typical response of *stenohaline* species and the latter *euryhaline* species. The effect of salinity on respiration rates appear to vary depending on timescales involved (acclimation, acclimatization, and adaptation), as was the case in temperature–respiration responses mentioned above. In support of this, Pagano *et al.* (1993) found no differences between the respiration rates of mesozooplankton in the river plume (salinity changed from 11 to 33) and those in the plume vertical interface of the Rhône (normal Mediterranean water salinity of 38).

The question whether respiration rates of freshwater zooplankton differ systematically from those of comparable marine counterparts has not yet been answered. A difficulty lies in the fact that no single zooplankton species is distributed across both fresh and marine environments. The only way to answer the question is by a broad comparison of respiration rate–body mass relationships of zooplankton from these two habitats under comparable thermal conditions. For this purpose, 41 respiration rate–dry

body mass–temperature datasets of freshwater zooplankton in Lampert (1984; his table 10.6) were used. The 41 datasets were represented by 3 rotatorians, 12 copepods, 10 cladocerans, 2 amphipods, 1 isopod, 3 mysids, 1 molluscan, 1 plecopteran, 4 ephemenropterans, and 4 dipterans. From the data of dry body mass and temperature, the respiration rates of these 41 freshwater zooplankton species were predicted from Ikeda's (1985) respiration rate–dry body mass–temperature relationship for epipelagic marine zooplankton. The predicted rates were compared with the observed rates on a log–log graph (Fig. 5.1). The relationship established by Ikeda (1985) is a comprehensive compilation of the marine data, which cover various taxonomic animal groups from wide geographical locations of the ocean world. From Fig. 5.1, it is clear that while predicted rates differed from observed rates in some cases, the agreement of the two rates was good in general (paired *t*-test, $p > 0.52$). These results lead us to conclude that on the ground of comparable temperature and body mass zooplankton respiration rates do not differ significantly between freshwater and marine habitats. Winberg (1956) reached the same conclusion in the metabolic comparison of freshwater and marine fishes.

5.4.3 Biological factors

Body mass

Among the many biological factors affecting specific respiration rates (rate per individual) of zooplankton, the body mass (W) of the animals is of prime importance; this relationship is expressed as $R = aW^b$, where a and b are proportional constants. For aquatic organisms, the constant b or body mass exponent is <1, and usually falls into the range of 0.7–0.9. The weight-specific respiration rates ($R/W = aW^b/W = aW^{b-1}$) thus decrease with increasing body mass. This body mass effect on zooplankton respiration rates is commonly seen in both inter and intraspecific comparisons, but "recapitulation theory" appears not to be warranted anymore (Banse 1982). For the intra and interspecific comparison of respiration rates, the assumption of a constant mass exponent (b) is a prerequisite (if not, the conclusion will change depending on the

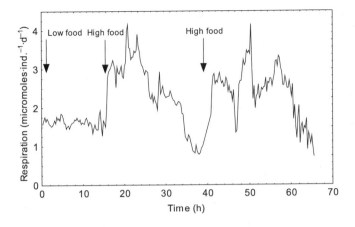

Figure 5.2 Observations of respiration rates of *D. magna* in a flow-through system. The arrows indicate the start of a short pulse of phytoplankton. Observe the three to fourfold increase in respiration rates after the introduction of food (from Schmoker and Hernández-León 2003).

choice of body mass). Ikeda (1988) proposed the so-called "adjusted metabolic rate" ($AMR = R/W^{-b}$) to facilitate metabolic comparison of zooplankters with different body masses. AMR assumes constant body mass exponent b, and converts the rate to constant body mass for valid comparison.

Feeding

Respiration rates are known to vary depending on feeding conditions (Lampert 1984; Ikeda *et al.* 2000). Starvation of zooplankton leads a reduction in their respiration rates, the magnitude of the reduction varying in relation to the duration of starvation as well as the level of metabolism (specimens with higher metabolism are less tolerant, cf. Ikeda 1974). Respiration rates increase during feeding and continue high for a certain period (Fig. 5.2); low respiration rates of *D. magna* maintained under low food availability increased rapidly following the addition of food (Schmoker and Hernández-León 2003). This increment in respiration rates by feeding is referred to as the "calorific effect" or "specific dynamic action" (SDA) (Prosser 1961), and represents the energy cost of protein synthesis (Jobling 1983; Kiørboe *et al.* 1985). Using a flow-through system, Kiørboe *et al.* (1985) found respiration rates of a marine copepod *A. tonsa* feeding at saturating food concentrations to be more than four times higher than those of starved specimens, and suggested that the cost of biosynthesis (growth) was the major component (50–116%) of observed SDA. Thor (2000, 2002) determined SDA in *C. finmarchicus*

and *A. tonsa*, as well as the incorporation of food carbon into protein, lipids, and polysaccharides in the body, concluding that the amount of carbon incorporated into body protein was correlated to the magnitude of SDA.

5.5 Temporal variability

5.5.1 Diel rhythms

Estimates of food demand of zooplankter in short-term experiments are susceptible to error due to the existence of diel rhythms in respiration. Diel metabolic rhythmicity in zooplankton is considered to be a complex phenomenon that couples feeding, swimming behavior, and spawning to diel cycles of light, temperature, oxygen concentration, and other environmental variables. It is noted, however, that biological rhythms are often difficult to distinguish from purely exogenously controlled behaviors without experiments in which environmental factors are strictly controlled (cf. DeCoursey 1983).

Duval and Geen (1976) reported bimodal respiration peaks (an average of 2.3 times higher than during mid-day) at dawn and dusk on mixed freshwater zooplankton and *Diaptomus kenai*. Bimodal peaks with similar timing but of greater amplitudes (six to eightfold) have also been observed in marine copepods (*P. xiphias, C. euxinus*) undertaking diel vertical migration (Pavlova 1994). Observed high respiration rates are thought to be related to the activation of swimming for vertical migration in nature, since in the same study such peaks were found to be lacking in the nonmigrant copepods *A. clausi, T. discaudata* (Pavlova 1994). In relation to diel vertical migratory behavior, nocturnal (Durbin *et al.* 1990) and postdawn/predusk feedings (Simard *et al.* 1985) are additional components of diel rhythm of zooplankton respiration. From this point of view, the close coupling of the diel cycle of respiration and that of feeding has been well documented in both freshwater and marine zooplankton (Ganf and Blazka 1974; Duval and Geen 1976; Gyllenberg 1981; Cervetto *et al.* 1993).

Not all reports, however, support the existence of endogenous diel cyclic rhythm in zooplankton respiration. Devol (1979) could not detect any diel cyclic patterns in the respiration rates of zooplankton from Lake Washington, and ascribed this to the lack of diel vertical migration behavior. Pearcy *et al.* (1969) placed individual *E. pacifica*, a species which is known as an oceanic diel vertical migrant, in Warburg flasks and monitored its respiration rate over 32 h in continuous darkness. The results showed no endogenous diel rhythm in the respiration rates of this euphausiid; this finding is further supported by the absence of light–dark effect on the respiration rate of this species. However, the failure to find the rhythm under constant laboratory conditions does not necessary mean the absence of diel respiration pattern of migrating specimens of this euphausiid in nature. This is because the euphausiid encounter temperature gradients in the course of migration and the respiration rate is known to change with temperature.

5.5.2 Seasonal changes

The amplitude of seasonal variations in the sea surface temperatures is 0–2°C in polar seas and tropical seas and as large as 7–8°C and 9–10°C in waters at midlatitudes of Northern (40–50°N) and Southern Hemispheres (30–50°S), respectively (cf. Figure 19 in Van Der Spoel and Heyman 1983). A large change in the sea surface temperatures in midlatitude waters implies accompanied seasonal changes in physical structures of the water column, then overall biological production cycle. It is therefore anticipated that the seasonal variations in respiration rates per unit zooplankton biomass would be the greatest

in midlatitude seas, particularly of the Northern Hemisphere.

Seasonal studies of respiration rates (standardized as weight-specific rates measured at 4–6°C) of largely herbivorous copepods (*C. finmarchicus, Calanus hyperboreus,* and *Metridia longa*) in the northern Atlantic Ocean revealed a rapid increase in the respiration rates during the spring bloom of phytoplankton, reaching a minimum during winter where phytoplankton reached the annual minimum, with overall ranges of the seasonal variations, in which the effect of temperature was removed, to be three to fivefold (Conover and Corner 1968; Marshall 1973). Some copepods enter in "diapause" at depth as an energy-saving overwintering mechanism (Hirche 1983; Ingvarsdottir *et al.* 1999). Christou and Moraitou-Apostolopoulou (1995) determined the relationship between respiration rates (specific rates at *in situ* temperatures) of mixed zooplankton, and environmental variables (temperature, chlorophyll) biweekly over 1 year in the eastern Mediterranean. The annual variation in the rates thus determined was approximately eightfold. Their regression analyses showed that temperature (annual range: ca. 12–26°C) and mean dry weight of the animals together explained 97% of the seasonal variability in the respiration rates of mixed zooplankton, whereas parameters such as chlorophyll concentration and other population composition parameters (copepod or cladoceran dominance index) were less important.

Polar and tropical waters are similar in terms of their relatively narrow annual temperature range, but these two regions are different in terms of seasonal stability in food supply to zooplankton due to very different environmental controls of primary production. While there may be massive phytoplankton blooming during the short summer in the light-limited polar waters, phytoplankton growth continues at low level in nutrient-limited tropical waters throughout the year, with small increases supported by pulses of nutrient supply from climate and hydrological events (cf. Lalli and Parsons 1993). In Antarctic waters, Torres *et al.* (1994) determined respiration rates of pelagic crustaceans (euphausiids, decapods, isopods, mysids, and ostracods) in summer and winter, and postulated that the summer rates of some of the species were approximately two times higher than the winter rates. They considered that the low winter rates reflected near starved conditions of the animals because of low phytoplankton abundance. In the same study, the only species that did not show any significant differences between summer and winter rates were all carnivores. Atkinson *et al.* (2002) found that respiration and ammonium excretion rates of the largely herbivorous krill *Euphausia superba* caught in autumn–winter were 60–80% of the rates of the same species in summer.

In the Gulf of Guinea (equatorial water), Le Borgne (1977) measured zooplankton respiration in March (warm season, water temperature: 28°C) and August (upwelling season, water temperature: 17°C), and found that the rates in the former was almost two times greater than the rates of the latter, which would suggest a Q_{10} of approximately 2 for zooplankton respiration (cf. Section 5.1). In subtropical waters around Canary Islands, where phytoplankton blooms occur in late winter due to weakened water column stability by surface cooling, Hernández-León (1986, 1987) observed a significant increase in zooplankton respiration rates during the phytoplankton bloom, a response seen in copepods in the waters at high latitudes mentioned above.

5.6 Spatial variability

Three spatial scales are in common use in biological oceanography: for example, micro meso, and macroscale (or large scale). The first level (microscale) comprises distribution of zooplankton within the epipelagic (0–150 m), mesopelagic (150–1000 m) and bathypelagic zones (1000–4000 m).[1] The information about zooplankton respiration at these zones is of paramount importance in order to understand the fate of the organic production in the ocean. Moreover, the carbon flux mediated by the mesopelagic vertical migrants feeding in the epipelagic zones and excreting at depth has not been fully understood in carbon flux studies in the ocean. Mesoscale phenomena in the sea also have a major effect on the role of the zooplankton in oceanic carbon cycle. The physical–biological couplings in frontal systems, cyclonic

and anticyclonic eddies, filaments of upwelling, etc. cause accumulation and dispersion of zooplankton and its prey organisms thereby creating spatial discontinuation in oxygen consumptions by zooplankton communities.

5.6.1 Vertical variability

Epipelagic zone (0–150 m)
The epipelagic zone is of central importance in carbon cycle in the ocean, because it is in this zone that inorganic carbon is converted to organic carbon by the photosynthetic activity of phytoplankton. Almost all the heterotrophic metabolism in the ocean depends on this organic carbon directly, and indirectly. Within the water column, zooplankton respiration is most intensive in this zone because of their greater biomasses, the high particulate organic carbon (POC) levels, and the immediacy of the organic supply and higher temperatures.

Bidigare *et al.* (1982) investigated vertical profiles of specific ETS activity of zooplankton community in the euphotic zone of the Gulf of Mexico, and found no consistent changes in the activities with increasing depth. A similar study conducted in the waters around Canary Islands by Hernández-León *et al.* (1998) showed a similar pattern in most seasons, but in some seasons the specific ETS activities were high in the upper 50 m layer where water temperature and primary production were elevated likewise.

The absence of any pattern of vertical distribution of biomass-specific zooplankton ETS activities in the euphotic zone may be the result of small vertical migrations of zooplankton, especially between the deep chlorophyll maximum and superficial layers. Vertical profiles of zooplankton ETS activities in temperate waters during the spring bloom and in tropical waters (Fig. 5.3) showed higher activities in not only the phytoplankton-rich layer but also the layer beneath it, suggesting active vertical migrations of zooplankton in the water column.

Since specific respiration rates are generally rather constant throughout the epipelagic zone, the vertical profile of zooplankton community respiration (specific respiration rates × biomass) reflects that of their biomass (see Bidigare *et al.* 1982). The vertical profile of zooplankton community respiration is characterized by a maximum above

[1] Terminology adopted from Parsons *et al.* (1984).

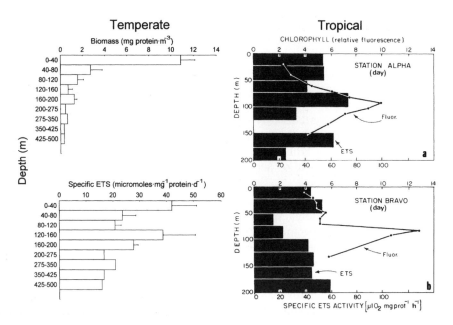

Figure 5.3 Vertical distribution of zooplankton biomass and respiration rates in temperate and tropical environments. The temperate area was located in an oceanic site south of Iceland during the spring bloom of 1989 (from Torres, S., unpublished). Values in the upper 200 m are the average (±SE) of three samplings while values below are single values at the same station. Vertical distribution in the tropical area (Gulf of Mexico) from Bidigare *et al.* (1982).

the deep chlorophyll maximum layer, or coinciding with the maximum of primary production or with the assimilation number (Roman *et al.* 1986; Herman 1989; Hernández-León *et al.* 1998). Most of the variability in zooplankton community respiration is associated with seasonal changes in temperatures in midlatitude waters and the presence of upwelling systems in subtropical waters (Fig. 5.4(a)). Published data on specific respiration rates reviewed by Hernández-León and Ikeda (unpublished) show a good agreement with the results from Ikeda's (1985) empirical equation, although seasonal variability is evident in all areas (see Fig. 5.9). The contribution of zooplankton community to global ocean respiration in the epipelagic zone of the world oceans is estimated to be in the order of 0.87 Pmol C a^{-1}.

Mesopelagic zone (150–1000 m)
Direct respiration measurements on mesopelagic zooplankton have rarely been made because of

difficulties to concentrate sufficient biomass of specimens from the mesopelagic zone. The use of submersibles and manipulation of zooplankton captured at depth appears to be a method of choice, but it requires considerable logistical and financial support. To date, respiration rates of mesopelagic zooplankton have been determined on the specimens brought into ship laboratory or extrapolated from epipelagic zooplankton on the premise that there are no changes in the rates with increasing depth.

Specific respiration rates of the deep fauna (Fig. 5.4(b)) are not significantly different to those found in the epipelagic zone in relation to ambient temperatures, although variability was much lower. Most of those measurements have been made on large-sized copepods and euphausiids only, which may explain less scattering of the data as compared with epipelagic zooplankton (Fig. 5.4(a)). The average value found for mesopelagic zooplankton is 1.36 μmol O$_2$ mg dry wt^{-1} d^{-1} (SD = ±1.22, $n = 136$, Hernández-León and Ikeda, unpublished).

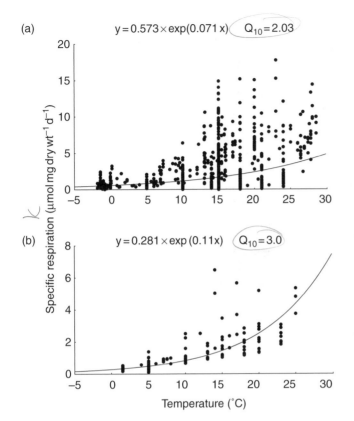

(a)

$y = 0.573 \times \exp(0.071\,x)$ $Q_{10} = 2.03$

(b)

$y = 0.281 \times \exp(0.11x)$ $Q_{10} = 3.0$

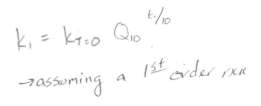

$k_1 = k_{T=0}\ Q_{10}^{t/_{10}}$

→assuming a 1st order rxn

Figure 5.4 Review of published values of specific respiration rates of zooplankton in relation to temperature. (a) Epipelagic zooplankton and (b) mesopelagic organisms. All the data are respiration rates obtained using the sealed-chamber method (or similar). Data derived from ETS activity in epipelagic zooplankton are those in which direct calibration from respiration measurements were obtained. Mesopelagic data are derived from direct estimations of respiration.

In the waters surrounding the Canary Islands, vertical distribution patterns of zooplankton biomass in the 0–1000 m stratum, as evaluated by means of net samplings (Figs. 5.5(a) and (b)), are characterized by a peak at about 500 m depth during daylight hours (i.e. deep scattering layer (DSL)), which disappears during the night partly due to the vertical migration of zooplankton. The DSL, however, is still visible at night in acoustic recordings. These shadow fauna are nonmigratory mesopelagic fishes and decapods, which avoid the nets. Zooplankton community respiration determined by ETS assays, decreased progressively with increasing depth, recovered slightly at about 500 m depth in the DSL, and then decreased again toward the bottom of the mesopelagic zone (Figs. 5.5(c) and (d)). At night, the vertical distribution pattern of zooplankton community respiration changed slightly, reflecting their diel vertical migration behavior. Daily zooplankton

community respiration in 200–900 m stratum in Canary Island waters represented 31.3% of the total community respiration in the upper 900 m layer, the remaining 68.7% the respiration occurring in the upper 0–200 m (Hernández-León et al. 2001a). In the eastern North Pacific, King et al. (1978) found that zooplankton respiration in the 200–1000 m layer was only 5.5% of that in the 0–1000 m layer, thus 94% of the respiration occurred in the the upper 200 m. The regional variability in mesopelagic zooplankton respiration appears to be appreciable in tropical and subtropical regions as well, as deduced from the profiles given by Longhurst et al. (1990). Hernández-León and Ikeda (unpublished) noted that zooplankton community respiration in the mesopelagic zone is highly variable (range 0.21–14 mmol C m^{-2} d^{-1}), largely because of a large variation in zooplankton biomass between regions. The average value found was 1.7 mmol C m^{-2} d^{-1} (± 2.1 SD, $n = 57$), which represents approximately

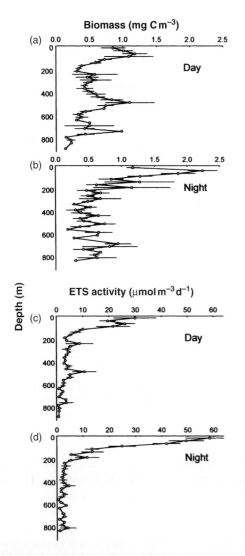

Figure 5.5 Vertical distribution of diel differences in zooplankton biomass and ETS activity in the Canary Island waters. The figure shows (a) the increase in biomass during day at 400–600 m depth, coinciding with the DSL and (b) biomass during night. The difference between (c) day and (d) night in ETS activity (μmol m^{-3} d^{-1}) at the mesopelagic zone is also shown. (Data from Hernández-León et al. 2001a.)

21% of the global epipelagic zooplankton respiration discussed above. These results indicate clearly that zooplankton living down to 1000 m cannot be ignored when assessing zooplankton respiration in the ocean. The global contribution of mesopelagic

zooplankton respiration is in the order of 0.18 Pmol C a^{-1} (Hernández-León and Ikeda, in preparation).

Bathypelagic zone (1000–4000 m)

Only a few respiration data on mixed bathypelagic zooplankton are found in the literature, and most of these are derived from ETS activity measurements (see Koppelmann and Weikert 1999). The reported specific respiration rates are very low (around 0.01 d^{-1}), reflecting cold temperatures (<5°C) and scarcity of food in bathypelagic environments. Using literature data, Childress and Thuesen (1992) estimated the respiratory carbon flux mediated by zooplankton and micronekton in the depth range from 1000 to 4000 m of a hypothetical central oceanic region to be 0.01 mmol C m^{-2} d^{-1}. In the temperate northeast Atlantic, Koppelmann and Weikert (1999) estimated a respiratory carbon flux in the depth range 1000–2250 m to be 0.11 (spring) and 0.32 (summer) mmol C m^{-2} d^{-1}. Below 2250 m depth the estimated flux was rather similar in spring (0.03 mmol C m^{-2} d^{-1}) and summer (0.03 mmol C m^{-2} d^{-1}). Other assessments of bathypelagic respiration have been made using published profiles of biomass and the average specific respiration rates at those depths (Koppelmann and Weikert 1999). The average value of the flux in the bathypelagic zone from 1000 to 4000 m depth is 0.32 mmol C m^{-2} d^{-1} (±0.40 SD, $n = 12$; Hernández-León and Ikeda, in preparation), which is less than 20% of the values observed in the mesopelagic zone, despite the much larger volume involved. The global contribution of bathypelagic zooplankton respiration is in the order of 0.03 Pmol C a^{-1}.

Active flux

The vertical excursions performed by zooplankton below the picnocline have the potential to inject dissolved organic and inorganic carbon into deep water, that would contribute to the so-called "biological pump" (Longhurst et al. 1989, 1990; Steinberg et al. 2000). In addition, zooplankton undertaking diel vertical migration feed in surface layers at night and defecate unassimilated POC at depth during day. In contrast to the passive or gravitational flux of particles sinking from

Table 5.1 Zooplankton active flux estimated by different authors in oceanic regions. POC stands for particulate organic carbon

Location	Migrating biomass (mmol C m^{-2})	Migratory flux (mmol C m^{-2} d^{-1})	% of POC flux	Reference
Subtropical September		0.46 (0.23–0.73)	6 (4–14)	Longhurst *et al.* (1990)
Bermuda March/April	16 (6.8–45)	1.21 (0.52–3.4)	34 (18–70)	Dam *et al.* (1995)
Eq. Pacific March/April	8.0	0.35	18	Zhang and Dam (1997)
Eq. Pacific October	13	0.61	25	Zhang and Dam (1997)
Eq. Pacific Oligotrophic	4	0.32	8	Le Borgne and Rodier (1997)
HNLC area September	4.4	0.66	4	Rodier and Le Borgne (1997)
North Atlantic (NABE)	0.4–40		19–40	Morales (1999)
Bermuda (year-round)	4.2 (0–10)	0.17 (0–0.83)	8 (0–39)	Steinberg *et al.* (2000)
Hawaii (year-round)	12	0.30 (0.08–0.77)	15 (6–25)	Al-Mutairi and Landry (2001)
Canary Islands (August)	16 (13–25)	0.22 (0.16–0.36)	25 (16–45)	Hernández-León *et al.* (2001a)
Eq. North Pacific (Winter)	25 (12–37)	1.40 (0.83–2)	31 (18–43)	Hidaka *et al.* (2001)
Canary Islands (July)	48–107	0.15–0.69	12–53	Yebra *et al.* (in preparation)

Note: Eq. stands for equatorial.
Source: Revised from Ducklow *et al.* 2001.

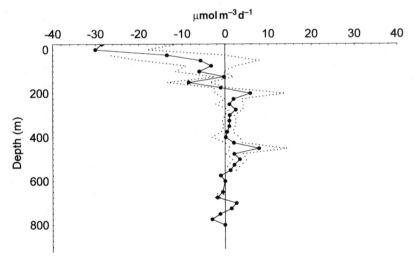

Figure 5.6 Day minus night values and standard deviation (dashed lines) of ETS activity in the Canary Islands waters. The Figure shows the diference between day and night in ETS activity (μmol m^{-3} d^{-1}). Positive values in the mesopelagic zone suggests a net flux of carbon by day to the mesopelagic zone (Hernández-León *et al.* 2001a).

the euphotic zone, this process mediated by the vertical migration of zooplankton has been called "active flux."

On a global scale, the vertical migration of zooplankton is responsible for a daily displacement of zooplankton biomass in the order of 10^{15} g wet wt (Longhurst 1976). While the evaluation of this active transport is of importance in the assessment of the biological pump in the ocean, information presently available is insufficient to integrate over the world ocean (see Table 5.1). The data on active transport of carbon by vertical movement of zooplankton (average 0.54 mmol C m^{-2} d^{-1}) fall into the range of 4–34% of the gravitational flux of organic particles. Associated with diel vertical movement of zooplankton biomass, the carbon produced in the form of carbon dioxide and excreted as dissolved organic carbon by zooplankton may contribute to active vertical transport of carbon in the ocean. As an example, Hernández-León *et al.* (2001*a*) determined vertical profile of zooplankton ETS activity both day and night in the waters of the Canary Islands, and examined the balance between the two periods of the day (cf. Fig. 5.6). The positive balance (day data minus night data) seen in the mesopelagic zone suggests net carbon transport to the mesopelagic zone by zooplankton respiration.

5.6.2 Mesoscale variability

Frontal systems
In frontal systems, shearing, convergence, or divergence promotes stability, mixing, upwelling, or sinking of water masses which may cause accumulation or dispersion of organisms. The effect of these physical phenomena on the metabolic activity of zooplankton has been assessed only in few studies. Packard (1985) developed an automated ETS assay system to map microplankton respiration in coastal fronts, and observed good relationships between the microplankton respiration and chlorophyll, and nutrient concentration and temperature. Arístegui and Montero (1995) also mapped microplankton ETS activity in the Bransfield Strait (Antarctica) showing increased values in the front between the Bellingshausen and Weddell water. The shearing between both water masses promotes intense meanders (García *et al.* 1994), which give rise to strong hydrodynamic activity, such as eddies and, therefore, suitable conditions for the development of planktonic organisms. Specific zooplankton ETS activity in this area showed a high variability and no significant differences were observed in the front and between both water masses (Hernández-León *et al.* 2000).

Eddies
Cyclonic and anticyclonic eddies are widespread physical structures in the oceans; they may enhance primary production or accumulate plankton due to the doming or downbowing effect, respectively.

Eddies induced by oceanic islands have been the subject of intense investigations during the 1990s, mainly in the Hawaiian and Canary Islands (see Olaizola *et al.* 1993; Arístegui *et al.* 1997). In the Canary archipelago, Arístegui *et al.* (1997) observed that chlorophyll increased as the result of island stirring or local upwelling near the flanks of the islands. At the same time, anticyclonic eddies showed a convergent effect, entraining adjacent waters and transporting chlorophyll downward. Zooplankton biomass was dramatically lower in the core of cyclonic eddies (Hernández-León *et al.* 2001*b*), probably due to the divergent effect produced by this physical structure. However, high biomass of large organisms was observed in relation to the anticyclonic eddy as the result of their inward motion. Specific zooplankton respiration rates estimated by the ETS method showed an increase in specific rates in the core of the cyclonic eddy (Fig. 5.7), where the highest values of primary production were observed, but not in the anticyclonic eddy, because the organisms were less active and simply accumulated there. The low biomass in the core of the cyclonic structure was composed of active organisms, coinciding with the area of intense phytoplankton growth. While specific respiration rates were lower, the community respiration was higher in the anticyclonic ones because of the higher zooplankton biomass there. The interplay of both cyclonic and anticyclonic structures shaped metabolic activities of zooplankton in these mesoscale structures, and mediated in the relationship between primary production and zooplankton metabolic rates.

Filaments from upwelling
Filaments shed from upwelling systems are colder and fresher than adjacent waters, and transport phytoplankton (Jones *et al.* 1991), zooplankton (Mackas *et al.* 1991; Hernández-León *et al.* 2002), and ichtyoplankton (Rodríguez *et al.* 1999) to more oligotrophic oceanic waters. In the northwest African upwelling, Navarro-Pérez and Barton (1998) found a considerable volume transport (in excess of 1Sv) flowing over the stratified subtropical waters with no evidence of mixing. Basterretxea and Arístegui

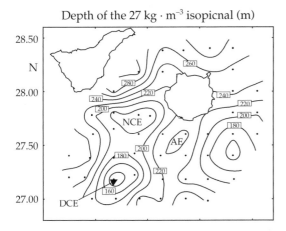

Depth of the 27 kg · m⁻³ isopicnal (m)

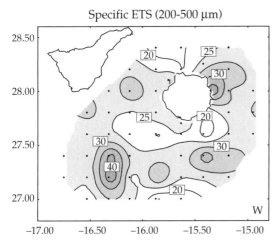

Specific ETS (200-500 µm)

Figure 5.7 Effect of cyclonic eddies. Topography of the 27 kg m⁻³ isopicnal showing the different hydrological structures found leeward of Gran Canaria Island (Canary Islands). Observe a newly formed cyclonic eddy (NCE), a drifting cyclonic eddy (DCE) and an anticyclonic eddy (AE). Observe the increase in specific ETS activity (μmol mg^{-1} protein d^{-1}) related to the difting cyclonic eddy (from Hernández-León *et al.* 2001*b*).

(2000) observed a sharp decrease in diatoms, chlorophyll, and primary production in the course of horizontal transport of newly upwelled water to offshore. However, Hernández-León *et al.* (2002) found zooplankton biomass to follow the signature of the filament, and as a consequence of their longer generation times, the biomass only decreased by 1/2–1/3 at the edge of the filament structure (Fig. 5.8). Zooplankton ETS activity and their gut pigment contents decreased gradually along the

filament, while the growth index used in this study (aspartate transcarbamylase (ATC), cf. Runge and Roff 2000) showed a slight increase. The calculated omnivory index from gut content and ETS-derived respiration rates showed that the sharp decrease in pigmented food, the slow decrease in ETS activity and the increase in the index of growth were due to a change in ingestion of phytoplankton to non-pigmented protista along the filament (Fig. 5.8). The general decrease in the indices of feeding and metabolism suggested that advection, rather than local enrichment processes inside the filament, was responsible for the high biomass values in this physical structure.

5.6.3 Large-scale and latitudinal variability

As mentioned earlier, specific respiration rates of zooplankton vary greatly with latitude, primarily following pronounced latitudinal gradients in water temperature and zooplankton community structure (or size structure) and biomass. As habitats of zooplankton, polar, temperate, subtropical, and tropical areas are characterized by different patterns of primary production, which in turn affect the life-cycle patterns and specific respiration rates of zooplankton (Section 5.4). At a given latitude, regional differences in temperature and zooplankton community composition are often the result of general circulation patterns (e.g. Labrador and Gulf Stream systems), or the presence of upwelling in the boundary areas of the warm water subtropical gyres. Large-scale divergences in tropical areas also cause variations in the thermal structure and productivity. Subtropical gyres are considered to be the most stable areas of the ocean due to the small seasonal changes in the physical structure of the water column. However, such situations are not always the case. Along a transect in the Atlantic Ocean at 21°N, Hernández-León *et al.* (1999) found that ETS and ATC activities of micro and mesozooplankton did not follow the patterns in chlorophyll and primary production, but rather matched the wave length of low frequency waves, suggesting that such waves were related to the observed two to threefold variability of zooplankton metabolism. Thus, studies of latitudinal changes in zooplankton respiration,

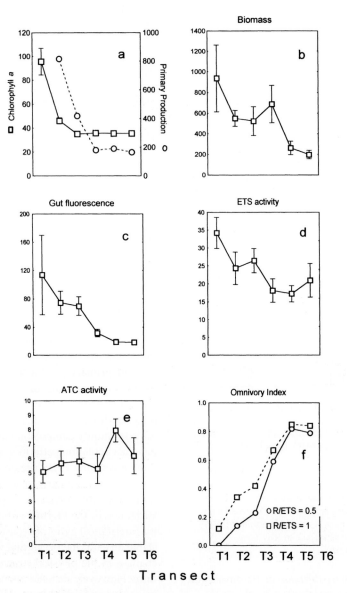

Figure 5.8 Effect of filaments of upwelling. Average values (±SE) for six sections (section T1 is close to the African coast; section T6 is the farthest offshore in the oceanic area) performed in a filament off the Northwest African upwelling area of (a) chlorophyll (μg l^{-1}) and primary production (mg C m^{-2} d^{-1}) (b) zooplankton biomass (mg protein m^{-2}), (c) specific gut fluorescence (ng pigments mg^{-1} protein), (d) specific ETS activity (μmol mg^{-1} protein d^{-1}), and (e) specific ATC activity (nmol carbamyl aspartate mg^{-1} protein min^{-1}), and (f) the index of omnivory (ingestion-grazing/ingestion). Observe the sharp decrease in chlorophyll followed by primary production and zooplankton biomass. While gut fluorescence and ETS activity decreased offshore, ATC activity remained constant and the calculated index of omnivory sharply increased towards the ocean. The index of omnivory was calculated using respiration/ETS ratios of 0.5 and 1.0, the range normally observed in the variability of this ratio (from Hernández-León et al. 2002).

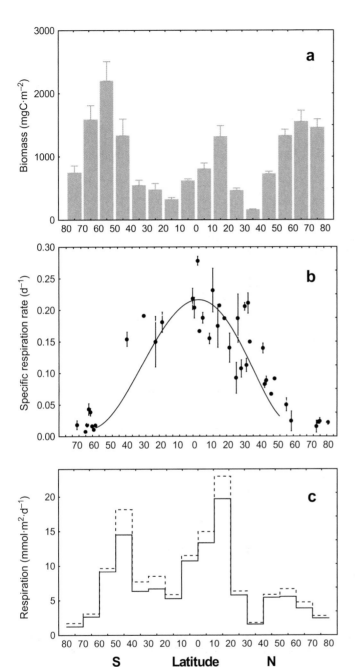

Figure 5.9 Global assesment of zooplankton respiration. (a) Latitudinal variations in zooplankton biomass, (b) specific respiration, and (c) community respiration (biomass × specific respiration) in the epipelagic zones from Arctic to Antarctic waters. (a) and (b) are compilation of published data. The line in (b) is the specific respiration derived from an empirical model of marine zooplankton respiration by Ikeda (1985) which is superimposed for comparison (from Hernández-León and Ikeda, in preparation). The error bar in (a) and the dashed line in (c) are the standard error.

and subsequent generalization of the results require caution, since the observed patterns result not only from seasonal events but often from mesoscale spatial variability such as low frequency waves, eddies, filaments, or frontal systems. Latitudinal variation in specific respiration rates in meso and bathypelagic zooplankton are assumed to be considered less, because of less marked regional differences in temperature in these layers and a reduced effect of physical features.

Compilations of specific respiration data of epipelagic zooplankton (Hernández-Leon and Ikeda, in preparation) show that the specific respiration rates increases from polar through equatorial areas (Fig. 5.9(b)). Such a pattern had been predicted by Ikeda (1985) based on latitudinal changes in the surface temperature and mean body mass of epipelagic zooplankton in the western Pacific Ocean. From the compiled specific respiration data, it is evident that the error bars associated with means are appreciable, partly because of fluctuations due to season and possible temporal events such as mesoscale phenomena or upwelling. Zooplankton community respiration in the epipelagic zones calculated across all latitudes as the product of specific respiration rate and biomass (Fig. 5.9(c)), showed a prominent peak at 10°S–20°N, a secondary peak at 40–50°S, and a third peak at 40–60°N. In both the Northern and Southern Hemispheres, midlatitudes (30–40°N, 10–40°S) and high polar regions (>60°S, >70°N) are characterized by lower zooplankton community respiration (Hernández-Leon and Ikeda, unpublished).

5.7 Concluding remarks

In designing respiration experiments on zooplankton, the requirements to make sensitive determinations and to obtain more realistic data (i.e. rates that are close to those *in situ*) are often not complementary, though not necessarily mutually exclusive. For whole animal respirometry using live specimens, sealed-chamber method appears to be inferior to flow-through methods from the viewpoint of difficulties to build-up metabolic waste products in environment and poor resolution of short-term changes in the rates. However, flow-through methods are not readily applicable to all zooplankton, and the calculation of respiration rates is not straightforward. The advent of enzymatic methods offered the possibility of overcoming many of the problems inherent in whole-animal respirometry, but nevertheless require whole animal respirometry to obtain conversion factors. Considering the high phylogenetic diversity of zooplankton

and the large range in body size, behavior, physiology, and nutrition, the appropriate method may vary from one zooplankton species to another (cf. Ikeda *et al*. 2000). While the choice of methods may seriously impede the comparison of respiration data among identical zooplankters under similar environmental conditions, these differences are less important in broad comparisons among zooplankters whose size differs several orders of magnitude (cf. Lampert 1984).

As mentioned above (Section 5.6.3), zooplankton community respiration in the epipelagic zone (0–200 m) of the world oceans has been estimated to be 0.87 Pmol C a^{-1} (Hernández-León and Ikeda, in preparation). The value of 0.87 Pmol C a^{-1} is an improvement on the earlier calculation (Ikeda 1985) because of better assessment of zooplankton biomass in the waters at >60°N and >70°S. The integration of zooplankton respiration over the entire water column and across latitudes requires better biomass estimates than currently available (Huntley and Lopez 1992). Regionally, the data on zooplankton biomass and respiration rates are particularly scarce in the Southern Hemisphere.

Zooplankton biomass in the mesopelagic zone represents about 50% of the observed in the 0–1000 m layer in subtropical waters. There are only a handful of studies of zooplankton respiration in this important layer of the ocean, and estimates vary from 1/20 to 1/4 of zooplankton community respiration in the upper 1000 m depth. The estimated value of respiration at the global scale in the mesopelagic is 0.18 Pmol C a^{-1}. The respiration rates in the bathypelagic zone at a global scale is even more uncertain, but the few published reports indicate very low values (average 0.03 Pmol C a^{-1}). Therefore, the total zooplankton respiration in the ocean can be estimated as the sum of their three rates that is, 1.1 Pmol C a^{-1}. These values are three to eightfold greater than the zooplankton respiration used in recently published assessments of total respiration in the ocean by del Giorgio and Duarte (2002).

Finally, micronekton biomass in the mesopelagic and bathypelagic zones has not been well evaluated, except for a few groups of fish micronekton. As such, the biomass of myctophid fishes has been estimated to be 0.7–18.5 g wet wt m^{-2} (mean: 7.2) in the world oceans (Gjøsaeter and Kawaguchi 1980; Balanov and Il'inskii 1992; Watanabe *et al.* 1999). Assuming that dry weight of myctophids is 20% of wet weight and carbon forms 40% of dry weight, myctophid biomass should be in the order of 48.3 mmol C m^{-2}. If one assumes a specific respiration rate of 0.01 d^{-1} for myctophids, their respiration integrated over the world oceans can be estimated as 0.48 mmol C m^{-2} d^{-1}, which is more than a quarter of the zooplankton respiration in the mesopelagic zone. Moreover, recent estimates of active flux due to micronekton in the western equatorial Pacific (Hidaka *et al.* 2001) shows values of the same magnitude, or even higher, than zooplankton. To advance our understanding of carbon cycle in the ocean further, better assessment of respiration of mesopelagic and bathypelagic fauna should be the focus of future research.

Acknowledgments

This work was funded, in part, by projects, Mesopelagic (MAR97-1036) and Pelagic (1FD97-1084) from the Spanish Ministry of Science and Technology and the European Union. We thank Peter Williams and Paul del Giorgio for comments which improved the text.

References

Alcaraz, M., Saiz, E., and Calbet, A. 1994. Small-scale turbulence and zooplankton metabolism: effects of turbulence on heartbeat rates of planktonic crustaceans. *Limnol. Oceanogr.*, **39**: 1465–1470.

Al-Mutairi, H. and Landry, M. R. 2001. Active export of carbon and nitrogen by diel migrant zooplankton at station ALOHA. *Deep-Sea Res. II*, **48**: 2083–2103.

Anraku, M. 1964. Influence of the Cape Cod Canal on the hydrography and on the copepods in Buzzards Bay and Cape Cod Bay, Massachusetts, II. Respiration and feeding. *Limnol. Oceanogr.*, **9**: 195–206.

Arístegui, J. and Montero, M. F. 1995. Plankton community respiration in Bransfield Strait (Antarctic Ocean) during austral spring. *J. Plankton Res.*, **17**: 1647–1659.

Arístegui, J., Tett, P., Hernández-Guerra, A., Basterretxea, G., Montero, M. F., Wild, K., Sangrá, P.,

Hernández-León, S., Cantón, M., García-Braun, J. A., Pacheco, M., and Barton, E. D. 1997. The influence of island-generated eddies on chlorophyll distribution: a study of mesoscale variation around Gran Canaria. *Deep-Sea Res., I*, **44**: 71–96.

Atkinson, A., Meyer, B., Stübing, D., Hagen, W., Schmidt, K., and Bathmann, U. 2002. Feeding and energy budgets of Antarctic krill *Euphaussia superba* at the onset of winter. II. Juveniles and adults. *Limnol. Oceanogr.*, **47**: 953–966.

Bailey, T. G., Torres, J. J., Youngbluth, M. J., and Owen G. P. 1994. Effect of decompression on mesopelagic gelatinous zooplankton: a comparison of *in situ* and shipboard measurements of metabolism. *Mar. Ecol. Prog. Ser.*, **113**: 13–27.

Balanov, A. A. and Il'inskii, E. N. 1992. Species composition and biomass of mesopelagic fishes in the Sea of Okhotsuk and the Bering Sea. *J. Ichtyol.*, **32**: 85–93.

Båmstedt, U. 2000. A new method to estimate respiration rate of biological material based on the reduction of tetrazolium violet. *J. Exp. Mar. Biol. Ecol.*, **251**: 239–263.

Banse, K. 1982. Mass-scaled rates of respiration and intrinsic growth in very small invertebrates. *Mar. Ecol. Prog. Ser.*, **9**: 267–283.

Basterretxea, G. and Arístegui, J. 2000. Mesoscale variability in phytoplankton biomass distribution and photosynthetic parameters in the Canary-NW African coastal transition zone. *Mar. Ecol. Prog. Ser.*, **197**: 27–40.

Berges, J. A., Roff, J. C., and Ballantyne, J. S. 1990. Relationship between body size, growth rate, and maximal enzyme activities in the brine shrimp, *Artemia franciscana*. *Biol. Bull.*, **179**: 287–296.

Bidigare, R. R., King, F. D., and Biggs, D. C. 1982. Glutamate dehydrogenase (GDH) and respiratory electron transport system (ETS) activities in Gulf of Mexico zooplankton. *J. Plankton Res.*, **4**: 895–911.

Bishop, J. W. 1968. Respiratory rates of migrating zooplankton in the natural habitat. *Limnol. Oceanogr.*, **13**: 58–62.

Brett, J. R. 1964. The respiratory metabolism and swimming performance of young sockeye salmon. *J. Fish Res. Board Can.*, **21**: 1183–1226.

Buskey, E. J. 1998. Energetic costs of swarming behavior for the copepod *Dioithona oculata*. *Mar. Biol.*, **130**: 425–431.

Cervetto, G., Gaudy, R., Pagano, M., Saint-Jean, L., Verriopoulos, G., Arfi, G., and Leveau, M. 1993. Diel variations in *Acartia tonsa* feeding, respiration and egg production in a Mediterranean coastal lagoon. *J. Plankton Res.*, **15**: 1207–1228.

Childress, J. J. 1975. The respiratory rates of midwater crustaceans as a function of depth of occurrence and relation to the oxygen minimum layer off southern California. *Comp. Biochem. Physiol.*, **50A**: 787–799.

Childress, J. J. and Thuesen, E. V. 1992. Metabolic potential of deep-sea animals: regional and global scales. In G. T. Rowe and V. Pariente (eds) *Deep-Sea Food Chains and Global Carbon Cycle*. NATO ASI Series, Kluwer Academic, Dordrecht, pp. 217–236.

Childress, J. J. and Thuesen, E. V. 1993. Effect of hydrostatic pressure on metabolic rates of six species of deep-sea gelatinous zooplankton. *Limnol. Oceanogr.*, **38**: 665–670.

Christou, E. D., and Moraitou-Apostolopoulou, M. 1995. Metabolism and feeding of mesozooplankton in the eastern Mediterranean (Hellenic coastal waters). *Mar. Ecol. Prog. Ser.*, **126**: 39–48.

Clarke, A. 1987. The adaptation of aquatic animals to low temperatures. In B. W. W. Grout and G. J. Morris (eds) *The Effects of Low Temperatures on Biological Systems*. Edward Arnold, London, pp. 315–348.

Conover, R. J. 1978. Feeding interactions in the pelagic zone. *Rapp. P-V. Reun. Cons. Int. Explor. Mer.*, **173**: 66–76.

Conover, R. J. and Corner, E. D. S. 1968. Respiration and nitrogen excretion by some marine zooplankton in relation to their life cycles. *J. Mar. Biol. Assoc. UK* **48**: 49–75.

Curl, H. J. and Sandberg, J. 1961. The measurement of dehydrogenase activity in marine organisms. *J. Mar. Res.*, **1209**: 123–138.

Dam, H. G., Roman, M. R., and Youngbluth, M. J. 1995. Downward export of respiratory carbon and dissolved inorganic nitrogen by diel-migrant mesozooplankton at the JGOFS Bermuda time-series station. *Deep-Sea Res. I*, **42**: 1187–1197.

Davenport, J., Trueman, E. R. 1985. Oxygen uptake and buoyancy in zooplanktonic organisms from the tropical eastern Atlantic. *Comp. Biochem. Physiol.*, **81A**: 857–863.

DeCoursey, P. J. 1983. Biological timing. In F. J. Vernberg and W. B. Vernberg (eds) *The Biology of Crstacea*. Academic Press, New York, pp. 107–162.

del Giorgio, P. A, and Duarte, C. M. 2002. Total respiration and the organic carbon balance of the open ocean. *Nature*, **420**: 379–384.

Devol, A. H. 1979. Zooplankton respiration and its relation to plankton dynamics in two lakes of contrasting trophic state. *Limnol. Oceanogr.*, **24**: 893–905.

Ducklow, H. W., Steinberg, D. K., and Buesseler, K. O. 2001. Upper ocean carbon export and the biological pump. *Oceanography*, **14**: 50–58.

Durbin, A. G., Durbin, E. G., and Wlodarczyk, E. 1990. Diel feeding behavior in the marine copepod *Acartia tonsa* in relation to food availability. *Mar. Ecol. Prog. Ser.*, **68**: 23–45.

Duval, W. S. and Geen, G. H. 1976. Diel feeding and respiration rhythms in zooplankton. *Limnol. Oceanogr.*, **21**: 823–829.

Fernández, F. 1977. Efecto de la intensidad de la luz natural en la actividad metabólica y en la alimentación de varias especies de copépodos planctónicos. *Inv. Pesq.*, **41**: 575–602.

Freel, R. W. 1978. Oxygen affinity of the hemolymph of the mesopelagic mysidacean *Gnathophausi ingens*. *J. Exp. Zool.*, **204**: 267–274.

Ganf, G. G. and Blazka, P. 1974. Oxygen uptake, ammonia and phosphate excretion by zooplankton of a shallow equatorial lake. *Limnol. Oceanogr.*, **19**: 313–325.

García, M. A., Piugdefábregas, J., Figa, J., Espino, M., and Arcilla, A. S. 1994. Dynamical oceanography of the Bransfield Strait (Antarctica) during austral summer 1992/93: an application of ERS-1 products. *Proceedings of the Second ERS-1 Symposium*, pp. 553–556.

Gifford, D. J. and Caron, D. A. 2000. Sampling, preservation, enumeration and biomass of marine protozooplankton. In R. Harris *et al.* (eds) *ICES Zooplankton Methodology Manual*. Academic Press, San Diego, pp. 193–221.

Gjøsaeter, J. and Kawaguchi, K. 1980. A review of the world resources of mesopelagic fish. *FAO Fish. Tech. Pap.*, **193**: 1–151.

Gyllenberg, G. 1981. Eudiaptomus gracilis (Copepoda, Calanoida): diel vertical migration in the field and diel oxygen consumption rhythm in the laboratory. *Ann. Zool.*, **18**: 229–232.

Herman, A. W. 1989. Vertical relationships between chlorophyll, production and copepods in the eastern tropical Pacific. *J. Plankton Res.*, **11**: 243–261.

Hernández-León, S. 1986. Nota sobre la regeneración de amonio por el mesozooplankton en aguas de Canarias. *Bol. Inst. Esp. Oceanogr.* **3**: 75–80.

Hernández-León, S. 1987. Actividad del sistema de transporte de electrones en el mesozooplancton durante un máximo primaveral en aguas del Archipielago Canario. *Inv. Pesq.*, **51**: 491–499.

Hernández-León, S. and Gómez, M. 1996. Factors affecting the Respiration/ETS ratio in marine zooplankton. *J. Plankton Res.*, **18**: 239–255.

Hernández-León, S. and Torres, S. 1997. The relationship between ammonia excretion and GDH activity in marine zooplankton. *J. Plankton Res.*, **19**: 587–601.

Hernández-León, S., Arístegui, J., Gómez, M., Torres, S., Almeida, C., Montero, M. F., and Ojeda, A. 1998. Mesozooplankton metabolism and its effect on chlorophyll and primary production in slope waters of the Canary Islands. *Ann. Instit. Océanogr. Paris*, **74**: 127–138.

Hernández-León, S., Postel, L., Arístegui, J., Gómez, M., Montero, M. F., Torres, S., Almeida, C., Kühner, E., Brenning, U., and Hagen, E. 1999. Large-scale and mesoscale distribution of plankton biomass and metabolic activity in the Northearstern Central Atlantic. *J. Oceanogr.*, **55**: 471–482.

Hernández-León, S., Almeida, C., Portillo-Hahnefeld, A., Gómez, M., and Montero I. 2000. Biomass and potential feeding, respiration and growth of zooplankton in the Bransfield Strait (Antarctic Peninsula) during austral summer. *Polar Biol.*, **23**: 697–690.

Hernández-León, S., Gómez, M., Pagazaurtundua, M., Portillo-Hahnefeld, A., Montero, I., and Almeida, C. 2001*a*. Vertical distribution of zooplankton in Canary Island waters: implications for export flux. *Deep-Sea Res. I*, **48**: 1071–1092.

Hernández-León, S., Almeida, C., Gómez, M., Torres, S., Montero, I. and Portillo-Hahnefeld, A. 2001*b*. Zooplankton biomass and indices of feeding and metabolism in island-generated eddies around Gran Canaria. *J. Mar. Syst.*, **30**: 51–66.

Hernández-León, S., Almeida, C., Portillo-Hahnefeld, A., Gómez, M., Rodriguez, J. M., and Arístegui, J. 2002. Zooplankton biomass and indices of feeding and metabolism in relation to an upwelling filament off northwest Africa. *J. Mar. Res.*, **60**: 327–346.

Hidaka, K., Kawaguchi, K., Murakami, M., and Takahashi, M. 2001. Downward transport of organic carbon by diel migratory micronekton in the wester equatorial Pacific: its quantitative and qualitative importance. *Deep-Sea Res. I*, **48**: 1923–1939.

Hirche, H. J. 1983. Overwintering of *Calanus finmarchicus* and *C. helgolandicus*. *Mar. Ecol. Prog. Ser.*, **11**: 281–290.

Huntley, M. E. and Lopez, M. D. G. 1992. Temperature-dependent production of marine copepods: a global synthesis. *Am. Nat.*, **140**: 201–242.

Ikeda, T. 1974. Nutritional ecology of marine zooplankton. *Mem. Fac. Fish Hokkaido Univ.*, **22**: 1–97.

Ikeda, T. 1985. Metabolic rates of epipelagic marine zooplankton as a function of body mass and temperature. *Mar. Biol.*, **85**: 1–11.

Ikeda, T. 1988. Metabolism and chemical composition of crustacean from the Antarctic mesopelagic zone. *Deep-Sea Res.*, **35**: 1991–2002.

Ikeda, T. and Motoda, S. 1978. Estimated zooplankton production and their ammonia excretion in the Kuroshio and adjacent seas. *Fish Bull. US*, **76**: 357–367.

Ikeda, T., Torres, J. J., Hernández-León, S., and Geiger, S. P. 2000. Metabolism. In R. P. Harris, P. H. Wiebe, J. Lenz, H. R. Skjoldal, and M. Huntley (eds) *ICES Zooplankton Methodology Manual*. Academic Press, San Diego, pp. 455–532.

Ikeda, T., Kanno, Y., Ozaki, K., and Shinada, A. 2001. Metabolic rates of epipelagic marine copepods as a function of body mass and temperature. *Mar. Biol.*, **139**: 587–596.

Ingvarsdottir, A., Houlihan, D. F., Hearth, M. R., and Hay, S. J. 1999. Seasonal changes in respiration rates of copepodite stage V *Calanus finmarchicus* (Gunnerrus). *Fish Oceanogr.*, **8** (Suppl. 1): 73–83.

Ivleva, I. V. 1980. The dependence of crustacean respiration rate on body mass and habitat temperature. *Int. Rev. Ges Hydrobiol.*, **65**: 1–47.

Jobling, M. 1983. Towards an explanation of specific dynamic action (SDA). *J. Fish Biol.*, **23**: 549–555.

Jones B. H., Mooers C. N. K., Rienecker M. M., Stanton T., and Washburn L. 1991. Chemical and biological structure and transport of a cool filament associated with a jet-eddy system off northern California in July 1986 (OPTOMA21). *J. Geophys. Res.*, **96**: 22 207–22 225.

King, F. D. and Packard, T. T. 1975. The effect of hydrostatic pressure on respiratory electron transport system activity in marine zooplankton. *Deep-Sea Res.*, **22**: 99–105.

King, F. D., Devol, A. H., and Packard, T. T. 1978. Plankton metabolic activity in the eastern tropical North Pacific. *Deep-Sea Res.*, **25**: 689–704.

Kiøerboe, T. and Saiz, E. 1995. Planktivorous feeding in calm and turbulent environments with emphasis on copepods. *Mar. Ecol. Prog. Ser.*, **122**: 35–45.

Kiøerboe, T., Mohlemberg, F., and Hamburger, K. 1985. Bioenergetics of the planktonic copepod *Acartia tonsa* : relation between feeding, egg production and respiration, and composition of specific dynamic action. *Mar. Ecol. Prog. Ser.*, **26**: 85–97.

Koppelman, R. and Weikert, H. 1999. Temporal changes of deep-sea mesozooplankton abundance in the temperature NE Atlantic and estimates of the carbon budget. *Mar. Ecol. Prog. Ser.*, **179**: 27–40.

Lalli, C. M. and Parsons, T. R. 1993. *Biological Oceanography*. Elsevier Science Ltd, Oxford.

Lampert, W. 1984. The measurement of respiration. In J. A. Downing and F. H. Rigler (eds) *A Manual on Methods for the Assessment of Secondary Productivity in*

Fresh Waters. Blackwell Scientific Publishers, Oxford, pp. 413–468.

Lance, J. 1965. Respiration and osmotic behavior of the copepod *Acartia tonsa* in diluted sea waters. *Comp. Biochem. Physiol.*, **14**: 155–165.

Le Borgne, R. 1977. Étude de la production pélagique de la zone équatoriale de L'Atlantique a 4° W. III. Respiration et excrétion d'azote et de phosphore du zooplancton. *Cah. ORSTOM.*, **15**: 349–362.

Le Borgne, R. and Rodier, M. 1997. Net zooplankton and the biological pump: a comparison between the oligotrophic and mesotrophic equatorial *Pacific. Deep-Sea Res. II*, **44**: 2003–2023.

Longhurst, A. R. 1976. Vertical migration. In D. H. Cushing and J. J. Walsh (eds) *The Ecology of the Seas*. Blackwell Science, London, pp. 116–137.

Longhurst, A. R., Bedo, A., Harrison, W. G., Head, E. J. H., Horne, E. P., Irwin, B., and Morales, C. 1989. NFLUX: a test of vertical nitrogen flux by diel migrant biota. *Deep-Sea Res. A*, **36**: 1705–1719.

Longhurst, A. R., Bedo, A. W., Harrison, W. G., Head, E. J. H., and Sameoto, D. D. 1990. Vertical flux of respiratory carbon by oceanic diel migrant biota. *Deep-Sea Res., A*, **37**: 685–694.

Mackas, D., Washburn, L., and Smith, S. L. 1991. Zooplankton community pattern associated with a California Current cold filament. *J. Geophys. Res.*, **96**: 14781–14797.

Mangum, C. P. 1983. Oxygen transport in the blood, In L. H. Mantel (ed.) *The Biology of Crustacea*, Vol. 5. Academic Press, New York, pp. 373–429.

Marshall, S. 1973. Respiration and feeding in copepods. *Adv. Mar. Biol.*, **11**: 57–120.

Mickel, T. J. and Childress, J. J. 1978. The effect of pH on oxygen consumption and activity in the bathypelagic mysid *Gnathophausia ingens. Biol. Bull.*, **154**: 138–147.

Miller, C. and Glibert, P. 1998. Nitrogen excretion by the calanoid copepod *Acastia tonsa*: results of mesocosm experiments. *J. Plankton Res.*, **20**: 1767–1780.

Morales, C. E. 1999. Carbon and nitrogen fluxes in the oceans: the contribution by zooplankton migrants to active transport in the North Atlantic during the Joint Global Ocean Flux Study. *J. Plankton Res.*, **21**: 1799–1808.

Moshiri, G. A., Cummins, K. W., and Costa, R. R. 1969. Respiratory energy expenditure by the predaceous zooplankter *Leptodora kindtii* (Focke). *Limnol. Oceanogr.*, **14**: 475–484.

Návarro-Pérez, E. and Barton, E. D. 1998. The physical structure of an upwelling filament off the north-west African Coast during August 1993. *S. Afr. J. Mar. Sci.* **19**: 61–73.

Olaizola, M., Ziemann, D. A., Bienfang, P. K., Walsh, W. A., and Conquest, L. D. 1993. Eddy-induced oscillations of the pycnocline affect the floristic composition and depth distribution of phytoplankton in the subtropical Pacific. *Mar. Biol.*, **116**: 533–542.

Omori, M. and Ikeda, T. 1984. *Methods in Marine Zooplankton Ecology*. John Wiley and Sons., New York, 332 pp.

Opalisnki, K. W. 1979. Heartbeat rate in two antarctic crustaceans: *Euphausia superba* and *Parathemisto gaudichaudi. Pol. Arch. Hydrobiol.*, **26**: 91–100.

Packard, T. T. 1971. The measurement of respiratory electron transport activity in marine phytoplankton. *J. Mar. Res.*, **29**: 235–244.

Packard, T. T. 1985. Oxygen consumption in the ocean: measuring and mapping with enzyme analysis. In A. Zirino (ed) *Mapping Strategies in Chemical Oceanography*. American Chemical Society, Washington DC, pp. 177–209.

Packard, T. T., Devol, A. H., and King, F. D. 1975. The effect of temperature on the respiratory electron transport system in marine plankton. *Deep-Sea Res.*, **22**: 237–249.

Packard, T. T., Berdalet, E., Blasco, D., Roy, S. O., St-Amand, L., Lagacé, B., Lee, K., and Gagné, J. P. 1996. Oxygen consumption in the marine bacterium *Pseudomonas nautica* predicted from ETS activity and bisubstrate enzyme kinetics. *J. Plankton Res.*, **18**: 1819–1835.

Pagano, M., Gaudy, R., Thibault, D., and Lochet, F. 1993. Vertical migrations and feeding rhythms of mesozooplanktonic organisms in the Rhône River Plume Area (North-west Mediterranean Sea). *Estuar. Coast. Shelf Sci.*, **37**: 251–269.

Park, K. 1968. Alkalinity and pH off the coast of Oregon. *Deep-Sea Res.*, **15**: 171–183.

Parsons, T. R., Takahashi, M., and Hargrave, B. 1984. *Biological Oceanographic Processes*, 3rd edition Pergamon Press, Oxford.

Pavlova, E. V. 1994. Diel changes in copepod respirationrates. *Hydrobiologica*, **292/293**: 333–339.

Pavlova, E. V. and Minkina, N. I. 1983. Evaluation of basal energy metabolism in marine copepods. *Dokl. Biol. Sci.*, **265**: 406–408.

Pearcy, W. G., Theilacker, T. H., and Lasker, R. 1969. Oxygen consumption of *Euphausia pacifica*. The lack of a diel rhythm or light–dark effect, with a comparison of experimental techniques. *Limnol. Oceanogr.*, **14**: 219–223.

Prosser, C. L. 1961. Oxygen: respiration and metabolism. In C. L. Prosser and F. A. Brown (eds) *Comparative Animal Physiology*. WB Saunders, Philadelphia, PA, pp. 165–211.

Rodier, M. and Le Borgne, R. 1997. Export flux of particles at the equator in the western and central Pacific ocean. *Deep-Sea Res. II*, **44**: 2085–2113.

Rodríguez, J. M., Hernández-León, S. and Barton, E. D. 1999. Mesoscale distribution of fish larvae in relation to an upwelling filament off Northwest Africa. *Deep-Sea Res. I*, **46**: 1969–1984.

Rodrigues, R. M. N. V. and Williams, P. J. le B. 2001. Heterotrophoc bacterial utilization of nitrogenous and non-nitrogenous substrates, determined from ammonia and oxygen fluxes. *Limnol. Oceanogr.*, **46**: 1673–1683.

Roman, M. R., Yentsch, C. S., Gauzens, A. L., and Phinney, D. A. 1986. Grazer control of the fine-scale distribution of phytoplankton in warm-core Gulf Stream rings. *J. Mar. Res.*, **44**: 795–813.

Rothchild, B. J. and Osborn, T. R. 1988. Small-scale turbulence and plankton contact rates. *J. Plankton Res.*, **10**: 465–474.

Roy, S. O. and Packard, T. T. 1998. NADP-isocitrate deshydrogenase from *Pseudomonas nautica*: kinetic constant determination and carbon limitation effects on the pool of intracellular substrates. *Appl. Environ. Microbiol.*, **64**: 4958–4964.

Roy, S. O., Packard, T. T., Berdalet, E., and St-Amand, L. 1999. Impact of acetate, pyruvate, and physiological state on respiration and respiratory quotients in *Pseudomonas nautica. Aquat. Microb. Ecol.*, **17**: 105–110.

Runge, J. A. and Roff, J. C. 2000. The measurement of growth and reproductive rates. In R. P. Harris, P. H. Wiebe, J. Lenz, H. R. Skjoldal, and M. E. Huntley (eds) *ICES Zooplankton Methodology Manual*. Academic Press, San Diego, pp 401–454.

Saltzman, J. and Wishner, K. F. 1997. Zooplankton ecology in the eastern tropical Pacific oxygen minimum zone above a seamount: 1. General trends. *Deep-Sea Res. I*, **44**: 907–930.

Sameoto, D. D. 1986. Influence of the biological and physical environment on the vertical distribution of mesozooplankton and micronekton in the eastern tropical Pacific. *Mar. Biol.*, **93**: 263–279.

Sameoto, D., Guglielmo. L., and Lewis, M. K. 1987. Day-night vertical distribution of euphausiids in the eastern tropical Pacific. *Mar. Biol.*, **96**: 235–245.

Schindler, D. W. 1968. Feeding, assimilation and respiration rates of *Daphnia magna* under various environmental conditions and their relation to production estimates. *J. Anim. Ecol.* **37**: 369–385.

Schmoker, C. and Hernández-León, S. 2003. The effect of food on the respiration rates of Daphnia magna using a flow-through system. *Sci. Mar.* **67**: 361–365.

Simard, Y., Lacroix, G., and Legendre, L. 1985. *In situ* twilight grazing rhythm during diel vertical migrations of a scattering layer of *Calanus finmarchicus. Limnol. Oceanogr.* **30**: 598–606.

Skjoldal, H. R., Båmstedt, U., Klinken, J., and Laing A. 1984. Changes with time after capture in the metabolic activity of the carnivorous copepod *Euchaeta norvegica. J. Exp. Mar. Biol. Ecol.*, **83**: 195–210.

Spicer, J. I., Thomasson, M. A., and Stromberg, J. O. 1999. Pocssessing a poor anaerobic capacity does not prevent the diel vertical migration of Nordic krill Meganyctiphanes norvegica into hypoxic waters. *Mar. Ecol. Prog. Ser.*, **185**: 181–187.

Steinberg, D. K., Carlson, C. A., Bates, N. P., Goldthwait, S. A., Madin, L. P., and Michaels, A. F. 2000. Zooplankton vertical migration and the active transport of dissolved organic and inorganic carbon in the Sargasso Sea. *Deep-Sea Res.*, **47**: 137–158.

Svetlichny, L. S., Hubareva, E. S., Erkan, F., and Gucu, A. C. 2000. Physiological and behavioral aspects of *Calanus euxinus* females (Copepodda: Calanoida) during vertical migration across temperature and oxygen gradients. *Mar. Biol.*, **137**: 963–971.

Teal, J. M. and Carey, F. G. 1967. Respiration of a euphausiid from the oxygen minimum layer. *Limnol. Oceanogr.*, **12**: 548–550.

Thor, P. 2000. Relationship between specific dynamic action and protein deposition in calanoid copepods. *J. Exp. Mar. Biol. Ecol.*, **245**: 171–182.

Thor, P. 2002. Specific dynamic action and carbon incorporation in *Calanus finmarchicus* copepodites and females. *J. Exp. Mar. Biol. Ecol.*, **272**: 159–169.

Thuesen, E. V. and Childress, J. J. 1993. Enzymatic activities and metabolic rates of pelagic chaetognaths: lack of depth-related declines. *Limnol. Oceanogr.*, **38**: 935–948.

Thuesen, E. V., Miller, C. B., and Childress, J. J. 1998. Ecophysiological interpretation of oxygen consumption rates and enzymatic activities of deep-sea copepods. *Mar. Ecol. Prog. Ser.*, **168**: 95–107.

Torres, J. J. 2000. Oxygen consumption-electrode. In R. H. Harris P. H. Wiebe, J. Lenz, H. R. Skjoedal, and M. E. Huntley, (eds) *ICES Zooplankton Methodology Manual*. Academic Press, San Diego, pp. 499–506.

Torres, J. J. and Childress, J. J. 1983. Relationship of oxygen consumption to swimming speed in *Euphausia pacifica*. I. Effects of temperature and pressure. *Mar. Biol.*, **74**: 79–86.

Torres, J. J., Aarset, A. V., Donnelly, J., Hopkins, T. L., Lancraft, T. M., and Ainley, D. G. 1994. Metabolism of Antarctic micronektonic Crustacea as a function of depth of occurrence and season. *Mar. Ecol. Prog. Ser.*, **113**: 207–219.

Van Der Spoel, S. and Heyman, R. P. 1983. *A Comparative Atlas of Zooplankton. Biological Patterns in the Oceans*, Berlin, Springer-Verlag, 186pp.

Vinogradov, N. E. and Voronina, N. M. 1961. Influence of the oxygen deficit on the distribution of plankton in the Arabian Sea. *Okeanologia*, **1**: 670–678.

Visser, A. W., Saito, H., Saiz, E., and KiØrboe, T. 2001. Observations of copepod feeding and vertical distribution under natural turbulent in the North Sea. *Mar. Biol.*, **138**: 1011–1019.

Watanabe, Y., Moku, M., Kawaguchi K., Ishimaru, K., and Akinori, O. 1999. Diel vertical migration of myctophyd fishes (Family Myctophidae) in the transitional waters of the western North Pacific. *Fish. Oceanogr.*, **8**: 115–127.

Winberg, G. G. 1956. Rate of metabolism and food requirements of fishes. Belorrussian State University Minsk, USSR *Fisheries Research Board of Canada, Translation Series*, 194.

Wyrtki, K. 1962 The oxygen minima in relation to ocean circulation. *Deep-Sea Res.*, **9**: 11–23.

Yamada, Y. and Ikeda, T. 1999. Acute toxicity of lowered pH tosome oceanic zooplankton. *Plankton Biol. Ecol.*, **46**: 62–67.

Zhang, X. and Dam, H. G. 1997. Downward export of carbon by diel migrant mesozooplankton in the central equatorial Pacific. *Deep-Sea Res. II*, **44**, 2191–2202.

CHAPTER 6

Respiration in wetland ecosystems

Charlotte L. Roehm

Département des sciences biologiques, Université du Québec à Montréal, Canada

Outline

This chapter reviews the current state of knowledge of respiration and emissions of carbon from freshwater wetlands to the atmosphere. Data are drawn from an array of studies addressing carbon dioxide and methane production and fluxes in wetland soils both *in situ* and in laboratory experiments. Regional estimates of wetland coverage are used to estimate the annual contribution of carbon dioxide-C and methane-C from each latitudinal zone and wetland type. The estimated mean rates of carbon dioxide and methane respiration from freshwater wetlands range between 0.1 and 0.6 $molC\,m^{-2}\,d^{-1}$ and between 0.6 and 12 $mmolC\,m^{-2}\,d^{-1}$, respectively. The mean rate for methane from rice paddies is 28 $mmolC\,m^{-2}\,d^{-1}$. The global estimates for total carbon respired from freshwater wetlands are 966 $Tmol\,a^{-1}$ from carbon dioxide and 23 $Tmol\,a^{-1}$ from methane. Rice paddies contribute an additional 16 $Tmol\,a^{-1}$. Although these respiratory fluxes are large, they are mostly offset by primary production, and many wetlands are still considered sinks for atmospheric carbon. The data do suggest, nonetheless, that wetland ecosystems are very sensitive to change, implying that these ecosystems may easily shift to being significant sources of carbon to the atmosphere. Despite the advances that have been made with annual carbon flux studies in wetlands with the use of eddy-covariance, the chapter concludes that large gaps still exist in our knowledge: uncertainty of the processes controlling the carbon cycle; uncertainty in the spatial distribution and aerial coverage of wetlands; lack of process studies in tropical regions.

6.1 Introduction

Wetlands inhabit a transitional zone between terrestrial and aquatic habitats, and are influenced to varying degrees by both. They are often ignored when addressing both terrestrial and aquatic ecosystem processes due to the confusing definition and poor classification of these ecosystems. The most current and commonly used definition is that provided by Ramsar (2000): *"where water is at or near the soil surface for a significant part of the growing season."* These ecosystems, therefore, range from northern peatlands to tropical swamps and marshes, differing in function and character primarily due to the differing pH, base cation concentrations, vegetation, and hydrologic regimes (Vitt *et al.* 1995; Mullen *et al.* 2000).

Wetlands cover only between 4% and 9% of the global surface area (Matthews and Fung 1987; Spiers 1999; see also comment in Section 6.5), but store nearly 37% of the global terrestrial carbon (Bolin and Sukamar 2000). Covering 3% of the northern latitude surface area (Maltby and Immirzi 1993) but storing up to 30% of the soil carbon pool (~455 Pg; Gorham 1995) Northern peatlands are particularly sensitive to climate change. There is a large discrepancy in the literature as to the extent and aerial contribution of tropical wetlands (Matthews and Fung 1987; Aselmann and Crutzen 1989; Gorham 1991). The role of wetlands in terrestrial carbon cycling is particularly complex as these environments are intimately associated with all aspects of the production and consumption of carbon dioxide and methane. Wetland biogeochemistry is largely controlled by the availability of oxygen as these ecosystems are characterized by the dominance of anaerobic processes. Anaerobic conditions and poor nutrient status promote carbon accumulation. Wetlands are currently perceived as small sinks for carbon dioxide (Gorham 1995) and large sources of methane, contributing nearly 40% of global atmospheric methane annually (Fung *et al.* 1991; Khalil and Sheariv 1993), of which up to 60% is from tropical wetlands (Wang *et al.* 1996). Although a smaller amount of carbon as methane is released to the atmosphere compared with carbon dioxide, methane remains an important greenhouse gas as it has a 21-times greater global warming potential relative to carbon dioxide.

Due to the high spatial and temporal variability observed in trace gas flux studies, it is often difficult to make regional estimates of exchanges based on local measurements (Whalen and Reeburgh 2000). To date many studies have not only found that different systems act either as sources or sinks of carbon dioxide, but that individual ecosystem units can act as both sources and sinks in consecutive years (Shurpali *et al.* 1995; Oechel *et al.* 1998; Griffis *et al.* 2000; Lloyd *et al.* 2001). While the latter phenomenon has been attributed to the differences in environmental conditions among years, it may not always hold true (Lafleur *et al.* 2001).

The production of carbon dioxide stems from both organic matter mineralization and plant respiration processes, which are, to date, still inherently hard to separate. Methane is produced under strictly anaerobic conditions as the terminal step in organic matter mineralization. The rate of mineralization varies considerably and reflects the assorted array of wetland types. The rate is controlled by water table fluctuations, temperature of the soil and the water column, botanical composition and chemical characteristics (Johnson and Damman 1991; Updegraff *et al.* 1995). The subsequent release of carbon dioxide and methane to the atmosphere is controlled mostly by redox conditions and vegetation patterns, which influence the diffusivity of the gases within the soil and water

columns and the oxidation potential of methane (Whiting and Chanton 1993; Bubier *et al.* 1995). As a result, the biogeochemical process rates are often decoupled from one another, making it inherently difficult to predict efflux rates from environmental conditions (Moore *et al.* 1998; Bellisario *et al.* 1999). This is due to the interaction between production, consumption, storage, and transport of gases.

Emissions from wetlands show high variability and can only be partly explained by correlations with environmental variables (Shannon and White 1994; Moore *et al.* 1998; Bellisario *et al.* 1999). At the field level, poor relationships between temperature/water tables and effluxes have been, in part, attributed to microtopographical differences (Waddington and Roulet 1996; Saarnio *et al.* 1997) and botanical composition (Johnson and Damman 1991; Bubier *et al.* 1995; Carroll and Crill 1997). Botanical composition is particularly important in tropical regions and in wetlands with longer hydroperiods, because emergent vegetation can provide a conduit for methane escape to the atmosphere or methane oxidation from oxygen transport to the roots (Schimel 1995; Jespersen *et al.* 1998) and because methane is closely related to net primary production (Whiting and Chanton 1993).

A large hindrance to date in quantifying global respiration estimates for wetland ecosystems lies in a multitude of factors, including both the lack of agreement as to what constitutes a wetland (Finlayson *et al.* 1999), a bias in ecosystem process studies largely omitting tropical regions (particularly of South America and Africa), and the rate of conversion of wetlands for agricultural and industrial purposes. The objective of this chapter is to compile a global respiration figure for the respiration of wetland ecosystems based on data available to date. The chapter is divided into a methodological section; a process section; a section addressing issues of conflict in the global recognition and distribution of wetlands; a section that combines this information to derive global estimates of wetland respiration.

6.2 Measurement techniques

The gas flux measurements provide unique fundamental mechanistic process and environmental data for evaluating ecosystem models and for assessing the role of terrestrial and aquatic ecosystems in the global carbon balance. Defining rates of respiration in wetland ecosystems falls into two main categories: (i) measurement of production rates and/or concentration profiles within the sediments and calculating effluxes to the atmosphere through mathematical diffusion equations; (ii) direct measurement of efflux to the atmosphere through chamber (plot scale and short term) and/or eddy-covariance towers (ecosystem scale and long-term continuous).

A common approach to measuring potential rates of respiration within the soils of drier wetlands (i.e. peatlands) is to incubate soil from different depths over a period of time (ranging between 24 h to 3 weeks). The rates of production are calculated as the change in carbon dioxide and methane concentrations in the headspace or dissolved in the water over time. The rates are then integrated over depth to calculate a flux to the atmosphere (Scanlon and Moore 2000). This approach can be used to determine both aerobic and anaerobic rates. Incubation experiments are, however, somewhat limited in that the procedure often involves disturbance, particularly when slurry techniques are used. Disturbance effects include: break up of the soil macro-structure and microbial communities, redox-induced chemical breakdown, and enhanced recycling of biomass (Aller 1994; Kieft et al. 1997) and activation of exoenzymes due to short-term exposure to oxygen (Freeman et al. 1996). One approach to overcome these disturbances is to take intact soil cores and subdivide them under anaerobic conditions and incubate them in intact sections.

Calculation of carbon dioxide and methane efflux to the atmosphere has also been made using the carbon dioxide concentration profiles combined with a Fickian diffusion model (Dueñas et al. 1995, 1999; Risk et al. 2002) and through the use of mass-balance equations (Dabberdt et al. 1993; Denmead and Raupach 1993). The former technique has not been used extensively in wetland ecosystems (Davidson and Trumbore 1995). Gaseous concentration profiles would be particularly useful in wetlands as surface carbon dioxide fluxes are often limited by diffusivity within the soil matrix, which varies as

a function of water filled pore space (Washington et al. 1994). A mass-balance approach could also be used whereby changes in the concentration of dissolved gases are measured in intact mesocosms and production/consumption rates are calculated from the rate of turnover of a variety of constituents (Blodau and Moore 2003).

Chamber techniques are a widely-used means for measuring the exchange of gases between the biosphere and the atmosphere at the small plot-scale. The different types of chamber systems used, however, can produce varying results and are not necessarily comparable with one another quantitatively, although the patterns obtained with the different approaches do correlate to one-another (Janssens et al. 2000). Although the large systematic differences highlight uncertainties in comparing fluxes from various sites obtained with different techniques, inter-site comparisons are possible if techniques are properly cross calibrated (Janssens et al. 2000).

Chamber methods include static and dynamic systems that can be either closed or open. Static techniques are the least used form at present as they tend to overestimate small fluxes and underestimate large fluxes (Nay et al. 1994). Dynamic chambers can be used with a closed path where air is circulated between the chamber and the infra-red gas analyzer (IRGA) and fluxes are calculated from a change in concentration over time, or with an open path which has a constant air flow that is vented to the atmosphere and the IRGA measures the difference between the concentration of the ambient air entering the chamber and that exiting the chamber. Due to the longer sampling time periods required for a closed chamber system, problems arise with changes in air temperature, soil moisture, and build-up of carbon dioxide in the chamber headspace, making the open chamber system more favored (Janssens et al. 2000). A problem that results from these chamber techniques, particularly with the open-chamber, is the sensitivity to pressure differences between the chamber and the atmosphere and the partial elimination of ambient turbulence (Rayment and Jarvis 1997; Lund et al. 1999). In all cases, methane cannot be measured with the IRGA. The standard method for measuring methane, therefore, is by taking a set of air samples over time with either syringes attached to a tube entering the chamber, or with evacuated glass vials with a needle inserted through a rubber stopcock into the chamber. The methane concentrations in the syringes or in the vials are then measured by gas chromatography.

The eddy-covariance technique has been used for several decades, but only in the past decade has it become a more reliable tool for determining continuous long-term carbon dioxide fluxes between the soil surface and the atmosphere. Eddy-covariance has enabled a great improvement in defining the magnitude of carbon dioxide fluxes and net ecosystem production on timescales ranging from hourly to seasonal, annual, and inter-annual (Wofsy and Hollinger 1998; Valentini *et al*. 2000). This technique measures the rate of exchange of carbon dioxide across an interface by measuring the covariance between fluctuations in the vertical wind velocity (eddies) and the carbon dioxide mixing ratio (Baldocchi 2003). Eddy-covariance, therefore, is most accurate when the flow below the detector is nearly horizontally homogenous and other terms of mass conservation can be ignored (Massman and Lee 2002).

Eddy-covariance is restricted in wetland studies that seek to determine respiration rates directly, as it measures the net flux. Direct carbon dioxide efflux rates can only be measured at nighttime when the photosynthesis factor is removed. This technique is also limited during stable conditions (particularly at night-time) when exchange is governed by slow and very intermittent air movement. As a result, other terms need to be included in the equations used to estimate the carbon dioxide flux: a storage term; an integrated form of the quasi-advective term; a mean-horizontal advective term; a vertically integrated horizontal flux divergence term; and a measured mean plus turbulent flux term (Massman and Lee 2002). In order to further increase the accuracy of the data, certain location conditions need to be met: a homogenous vegetation surface, a relatively flat terrain (see Massman and Lee 2002 for a detailed review), and a large and homogenous footprint (Baldocchi 2003; Werner *et al*. 2003).

The concentration profile and mass-balance approaches are useful for understanding the processes at a small-scale but represent difficulties when attempting to scale up to whole ecosystems. It would, therefore, be best to combine the above techniques so as to understand the true nature of the small-scale processes and temporal variations that affect carbon dioxide and methane production and control diffusion to the surface, and relate these to temporal and spatial patterns observed with chamber and/or eddy-covariance techniques.

6.3 Freshwater wetlands: biogeochemistry and ecology

6.3.1 Peatlands

Bogs and fens comprise the majority of peatland types predominating in northern latitudes. Bogs are particularly poor in nutrients as they receive inputs solely from precipitation. Peatlands are characteristically more acidic, have low primary production ranging between 12 and 30 (mean 20) $mol\,C\,m^{-2}\,a^{-1}$ and 30 and 58 (mean 44) $mol\,C\,m^{-2}\,a^{-1}$ in boreal and tropical peatlands, respectively (Campbell *et al*. 2000; Mitsch and Gosselink 2000) (Table 6.1). Decomposition rates are even lower, resulting in high rates of peat accumulation of between 0.6 and 15 $mol\,C\,m^{-2}\,a^{-1}$ (Tolonen *et al*. 1992; Gorham 1995; Yu *et al*. 2001). The interdependence between vegetation and nutrient and mineral availability is particularly strong in peatlands, with pH values ranging between 2.6 and 6.5 (Pjavchenko 1982; Nicholson *et al*. 1996). Low pH results primarily from the decomposition of *Sphagnum* species, which produce uronic acids (Verhoeven and Toth 1995), whose own decomposition is slowed due to the phenolic nature of the compound. Organic matter decomposition studies have shown typical exponential decay constants of between −0.05 and −1.2 a^{-1} (Aerts and Ludwig 1997; Öquist and Sundh 1998), which are lower than those observed for labile organic matter of terrestrial soils (−0.1 to −4.2 a^{-1}, Aiwa *et al*. 1998) and coastal marine sediments (−0.5 to −8.8 a^{-1}, Berner 1980).

Potential rates of carbon dioxide and methane production are low in peatlands ranging between 0.04 and 0.9 $mol\,CO_2\,m^{-2}\,d^{-1}$ (average ∼0.37 $mol\,CO_2\,m^{-2}\,d^{-1}$) and between 0.42 and 118 $mmol\,CH_4\,m^{-2}\,d^{-1}$ (average ∼22.3 $mmol\,CH_4\,m^{-2}\,d^{-1}$) (Blodau 2002; Roehm 2003; van den Bos and van de Plassche 2003)

Table 6.1 Mean and ranges of primary productivity and nutrient loading in freshwater wetlands (ranges are in brackets)

Wetland type	NPP[a] (mol C m^{-2} a^{-1})	Aerial coverage (10^{12} m^2)	Total (Tmol a^{-1})
Northern bogs and fens[b]	20 (12–30)	3.3[c] (2.6–3.8)	66
Northern marshes and swamps[b]	44 (30–58)	1.1[c] (0.65–1.50)	48.4
Tropical peatlands[d]	60 (45–155)	3.4 (1.7–5.1)	204
Tropical marshes and swamps[d]	92 (26–480)	2.4 (2.1–2.8)	221
Total all			540

[a] NPP: net primary production.
[b] Adapted from Campbell *et al.* (2000).
[c] Includes Boreal and Temperate estimates.
[d] Adapted from Mitsch and Gosselin (2000).

depending on the temperature of incubation and water table position. Carbon dioxide and methane surface fluxes from peatlands range between 0.02 and 2.6 mol C m^{-2} d^{-1} (average $\sim$0.4 mol C m^{-2} d^{-1}) and 0.06–120 mmol m^{-2} d^{-1} (average $\sim$4.1 mmol C m^{-2} d^{-1}), respectively (Table 6.2). The higher methane values are from a few measurements that are restricted to ponds (Hamilton *et al.* 1994; Roulet *et al.* 1997; Scott *et al.* 1999). Values of between 0.4 and 4 mmol m^{-2} d^{-1} are, however, more typical for Northern peatlands. It has been shown that respiration will initially dramatically increase as a result of both redox-cycles of wetting and drying (Clein and Schimel 1994; Aerts and Ludwig 1997), which may be a result of enzyme activation after short-term exposure to oxygen (Freeman *et al.* 1997) and freeze-thaw cycles (three to five fold) (Schimel and Clein 1996; Prieme and Christensen 1997). For a comprehensive review of carbon cycling in northern peatlands refer to Blodau (2002).

6.3.2 Freshwater marshes and swamps

Inland freshwater marshes are characterized by high pH ($\sim$5.1–8.6), available nutrients, primary productivity ($>$83 mol C m^{-2} a^{-1}), and high microbial

activity resulting in rapid decomposition and recycling and low peat accumulation due to large part of the carbon being bound in the vegetation (Weller 1994; Brady 1997; Mitsch and Gosselin 2000). Unlike peatlands, freshwater marshes do not accumulate large quantities of organic matter and are characterized by mineral soils overlain by autochthonous organic matter from the vegetation. While the vegetation of freshwater marshes is typically characterized by gramminoids, such as *Typha* and *Phragmites*, and grasses, sedges, monocots, and floating plants, freshwater swamps also include trees, particularly cypress.

Typical rates for productivity can range according to the type of marsh and the botanical composition. Net primary production (NPP) can range between 30 and 58 (mean 44) mol C m^{-2} a^{-1} and between 26 and 480 (mean 92) mol C m^{-2} a^{-1} in northern marshes and swamps and tropical marshes and swamps, respectively (Campbell *et al.* 2000; Mitsch and Gosselin 2000). Higher NPP values were found in wetlands dominated by reeds and grasses and lower in wetlands dominated by broad-leaved monocots. These ecosystems are closely tied to the hydrologic regime where strong nonlinear correlations with NPP are observed (Mitsch and Gosselin 2000). Perennial species accumulate more below-ground biomass but have smaller root:shoot production ratios indicating that the root system is long-lived (Mitsch and Gosselin, 2000). Below-ground productivity was also seen to decrease from 82 to 5 mol C m^{-2} a^{-1} with increasing flooding in freshwater swamps (Powell and Day 1991). Decreases are also connected with increasing nutreint availability (Müllen *et al.* 2000). Typical exponential decay constants range between -0.73 and -8.76 a^{-1} (Davis and Van der Valk 1978; Nelson *et al.* 1990) and between -0.23 and -1.40 a^{-1} (Deghi *et al.* 1980; Brinson *et al.* 1981) for freshwater marshes and swamps, respectively. Estimates and measures of carbon dioxide production rates from freshwater swamp and marsh soils are scarce and most studies have focused on surface effluxes. Published carbon dioxide surface fluxes range between 0.02 and 0.87 mol C m^{-2} d^{-1} (Murayama and Bakar 1996; Jauhiainen *et al.* 2001; Trettin and Jurgensen 2003) and methane fluxes

Table 6.2 Estimated methane and CO_2 respiration from freshwater wetlands (Mean value with ranges in brackets)

Type	Boreal area (10^{12} m^2)	Boreal flux (mmol C m^{-2} d^{-1})	Temperate area (10^{12} m^2)	Temperate flux (mmol C m^{-2} d^{-1})	Tropical area (10^{12} m^2)	Tropical flux (mmol C m^{-2} d^{-1})
(a) Methane						
Peat	3.1	0.7	0.17	6	3.4	9
	(2.6–3.6)	(0.06–120)	(0.17)	(0.24–25)	(1.7–5.1)	(0.19–15)
Total (Tmol a^{-1})		0.79		0.37		11.14
		(0.57–157)		(0.02–1.6)		(0.12–28)
Marsh and	1.1	0.56	0.004	6	2.4	11.80
swamp	(0.65–1.5)	(0.40–6.4)	(0.004)	(0.01–40)	(2.1–2.8)	(2.3–35)
Total (Tmol a^{-1})		0.24[a]		0.01[a]		10.50[a]
		(0.10–3.5)		(0.00–0.06)		(1.8–36)
	Total global flux	Total boreal flux		Total temperate flux		Total tropical flux
Total (Tmol a^{-1})	23	1.03		0.38		22
		(0.67–161)		(0.02–1.6)		(1.8–65)

Type	Boreal area (10^{12} m^2)	Boreal flux (mol C m^{-2} d^{-1})	Temperate area (10^{12} m^2)	Temperate flux (mol C m^{-2} d^{-1})	Tropical area (10^{12} m^2)	Tropical flux (mol C m^{-2} d^{-1})
(b) CO_2						
Peat	3.1	0.4	0.17	0.6	3.4	0.24
	(2.6–3.6)	(0.02–2.6)	(0.17)	(0.01–1.2)	(1.7–5.1)	(0.13–1.54)
Total (Tmol a^{-1})		451[a]		37[a]		297[a]
		(19–3400)		(0.62–75)		(82–2839)
Marsh and	1.1	0.21	0.004	0.21	2.4	0.11*
swamp	(0.65–1.50)	(0.04–0.54)	(0.004)	(0.04–0.54)	(2.1–2.8)	(0.02–0.87)
Total (Tmol a^{-1})		83[a]		0.31[a]		98[a]
		(9.5–296)		(0.06–0.79)		(15–889)
	Total global flux	Total boreal flux		Total temperate flux		Total tropical flux
Total (Tmol a^{-1})	966	534		37		395
		(29–3700)		(0.68–76)		(97–3728)

* Taken from Raich and Schlesinger (1992)–but likely an underestimate.

[a] These numbers represent the calculated total aerial fluxes (mean and ranges) for each wetland type and region. The lower and upper ranges in all the tables are calculated by multiplying the lowest flux by the lowest aerial coverage and the highest flux by the highest aerial coverage.

between 0.01 and 40 mmol C m^{-2} d^{-1} (Mitsch and Wu 1995).

6.3.3   Rice paddies

Rice is the world's most important wetland crop. Conversion of natural temperate and tropical wetlands to rice paddies has prompted attention on these ecosystems due to the potential for large emissions of methane that are caused from permanent flooding. The warm, waterlogged soil of rice paddies provides ideal conditions for methanogenesis, and though some of the methane produced is usually oxidized by methanotrophs in the shallow overlying water, the vast majority is released into the atmosphere. Rice paddies contribute approximately 10–13% to the global methane emission (Neue 1997; Crutzen and Lelieveld 2001) and 65% of natural wetland emissions. Large regional variations in methane fluxes from

Table 6.3 Estimated methane respiration from rice paddies (Mean values and ranges in brackets)

	Global area (10^{12} m^2)	CH$_4$ flux (mmol C m^{-2} d^{-1})
Rice paddies	1.5	28 (0.29–144)
Total (Tmol a^{-1})	16	16[a] (0.16–79)

[a] These numbers represent the calculated total aerial fluxes (mean and ranges) for each wetland type and region.

rice paddies are controlled by the interplay between average temperature, water depth, and the length of time that the rice paddy soil is waterlogged. Methane emissions range between 0.3 and 144 mmol C m^{-2} d^{-1} (Miura *et al*. 1992; Lindau 1994; Table 6.3).

Methane emissions in rice paddies can be stimulated by nitrogen fertilization (Singh *et al*. 1996), although other studies have found that methane can be inhibited due to competition from alternative electron acceptors (Cai *et al*. 1997). The methane emission rates from paddy fields generally decrease during the ripening stage (Kruger and Frenzel 2003). However, it has not been well documented that the decrease is due to the decrease in methane production rate or the decrease in the flux rate of methane in the soil (Watanabe and Kimura 1995). The type of cultivar plays a critical role in methane emission and oxidation in rice paddies, as the ability to transfer oxygen to the rhizosphere varies, thus altering the redox potential of the soil system or modifying the bacterial response of the rhizosphere (Parashar *et al*. 1991). The type of cultivar also determines the quality of the organic substrate, which is critical to methane production (Conrad 1989).

6.4 Carbon dioxide and methane dynamics

Net ecosystem exchange (NEE) is the difference between carbon dioxide photosynthetic uptake (PSN) and respiration (R) (root and heterotrophic). This term is equivalent to net ecosystem production

used elsewhere in this book, except that as it deals with CO$_2$ rather than organic production, it carries the opposite sign. The contribution of root respiration is still poorly defined but has been shown to range from 10% to 63% with a mean of 50% of the total respiration (Backeus 1990; Shurpali *et al*. 1995; Silvola *et al*. 1996; Stewart and Moore, personal communication). Effluxes measured at the soil/water-atmosphere, therefore, represent both plant and soil respiration. Methane is produced in anaerobic zones of the wetlands but can easily be re-oxidized at the water–air interface and lost to the atmosphere as carbon dioxide (Fig. 6.1). In rice fields and the surface waters of the Everglades it was found that 80–90% of produced methane was oxidized to carbon dioxide (Conrad and Rothfuss 1991; Frenzel *et al*. 1992).

The magnitude of the carbon sink or source issue of wetlands is driven, to a degree, by latitudinal gradient. Colder systems of the northern latitudes, namely peatlands, store greater amounts of carbon in the peat due to the slow decomposition of organic material caused by the colder temperatures and low evaporation rates, higher acidity and lower nutrient content. Hence, PSN > R and results in a negative NEE (mean long-term annual emission to the atmosphere of 1.92 mol C m^{-2} a^{-1} (Gorham 1995)). In tropical systems, although the PSN is generally greater than in northern systems, the rates of respiration are also greater due to the higher temperatures and less nutrient limitation. Hence, the storage of carbon in the soil is not as important as in more northern latitudes. While below ground biomass in both boreal and tropical wetlands is approximately 50% of above ground biomass (Campbell *et al*. 2000; Moore *et al*. 2002), the range in tropical wetlands is much greater (15–223%; Gill and Jackson 2000). The fine root turnover is much greater when C:N ratios are lower which is important for NPP calculations and C allocation (Jackson *et al*. 1997). Methane comprises a larger portion of the total carbon efflux to the atmosphere in tropical regions due to increased anaerobic conditions, higher annual temperatures promoting greater microbial activity, the presence of plants able to transport methane past the oxidizing layer (Jespersen *et al*. 1998; Boon 1999), and greater ebullition due to longer periods of standing water (Sorrel and Boon 1994).

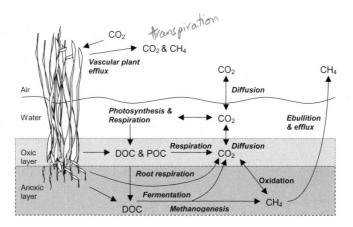

Figure 6.1 Carbon cycling and transformations in wetlands (adapted from Mitsch and Gosselink 2000).

At a larger scale, soil temperature, redox conditions, plant community structure, and chemistry of the soil can explain much of the variation observed in fluxes (Bubier *et al.* 1993; Yavitt *et al.* 1997; Moore *et al.* 1998). At a smaller scale these relationships often breakdown and less of the observed variability can be explained (Aerts and Ludwig 1997; Bellisario *et al.* 1999).

6.4.1 Variables controlling production

Carbon dioxide production results from plant respiration and the mineralization of soil organic carbon. Decomposition includes both intracellular and extracellular enzyme-mediated breakdown of complex molecules, and particular chemical bonds, with subsequent intracellular oxidation. In anaerobic environments, extracellular enzymatic hydrolysis predominates, producing by-products of alcohols, organic acids, and acetate (Fenchel *et al.* 1998). In anaerobic environments, enzymatic hydrolysis is often the rate-limiting step for the provision of fermentative products (Valentine *et al.* 1994). Organic matter can be oxidized with a number of terminal electron acceptors including O_2, NO_3^-, Mn^{2+}, Fe^{3+}, and SO_4^{2-}. Methane production is the terminal process of organic matter mineralization. Methanogenic bacteria can use a limited number of substrates of which acetate and CO_2 reduction by

H_2 are the most important in freshwater ecosystems (Yavitt and Lang 1990; Peters and Conrad 1996, and see King, Chapter 2). The acetate pathway becomes more favorable at higher temperatures as methanogenic populations become more competitive. Therefore, in tropical wetlands with higher primary production and root exudation, and higher annual temperatures, acetoclastic methanogens may predominate resulting in a greater rate of methane production (Brooks-Avery *et al.* 1999).

Aerobic respiration is more efficient in the transfer of energy, however, anaerobic conditions often predominate in wetland ecosystems. Aerobic–anaerobic ratios for carbon mineralization and subsequent carbon dioxide production range between 1.2 and 6 (Moore and Dalva 1993; Updegraff *et al.* 1995; Aerts and Ludwig 1997), but ratios of up to 200 have been observed in incubations of older previously anaerobic material exposed to oxygen (Hogg 1993). Fluctuations in redox conditions may have a net effect on carbon dioxide production rates. For example, increased cumulative carbon dioxide production rates (by a factor of 1.5–3) were observed in laboratory experiments where water tables and aeration were fluctuated (Aerts and Ludwig 1997).

Carbon dioxide and methane production rates are coupled to temperature, with an observed mean increase by a factor of 2 to 3 (CO_2) and 4.1 (CH_4) for every 10°C temperature increase (Q_{10}). Values reported in the literature for Q_{10}, however, vary between 1 and 16 for carbon dioxide (Moore and Dalva 1993; Bubier *et al*. 1998; McKenzie *et al*. 1998) and between 1.8 and 28 for methane (Dunfield *et al*. 1993; Von Hulzen *et al*. 1999; Frenzel and Karofeld, 2000). The variability found in these values may be a result of several factors: the temperature dependent lags imparted by alternative electron acceptors prior to methane and carbon dioxide production (Von Hulzen *et al*. 1999), by the sensitivity of substrate type (Abraham *et al*. 1998), by the differing temperature optima of different microbial populations (Westermann 1996; Segers 1998), and by declining Q_{10} values with increasing measurement–temperature intervals (Lloyd and Taylor 1994; Tjoelker *et al*. 2001).

Methanogenesis, the least energetically favorable step, depends in part on the presence of alternative electron acceptors (Achtnich *et al*. 1995; Roden and Wetzel 1996; Dise and Verry 2001). Methanogens are inferior competitors for substrate when other electron acceptors are abundant (Achtnich *et al*. 1995, and see King, Chapter 2). The presence and quantity of alternative electron acceptors can explain some of the variation in carbon dioxide and methane production rates observed between wetland types and can have important consequences for carbon cycling in wetlands (Roden and Wetzel 1996). In tropical and other southern wetlands where there is a higher presence of ferric iron oxides, iron reduction will predominate over sulfate reduction and methanogenesis (Achtnich *et al*. 1995; Roden and Wetzel 1996). In northern systems, the concentration of iron is low enough that, due to thermodynamic and kinetic reasons, sulfate reduction can predominate over methanogenesis when high sulfate atmospheric deposition occurs (Watson and Nedwell 1998). It has, however, been recently observed that even at low sulfate concentrations, sulfate can be recycled within days due to reoxidation of organic and inorganic sulfur following water table fluctuations (Mandernack *et al*. 2000). Enhanced carbon dioxide production has been subsequently observed from syntrophic acetate oxidation coupled to sulfate reduction (Nüsslein *et al*. 2001).

Methane production has been correlated to carbon dioxide production in anaerobic environments, and although stoichiometrically it can

contribute up to 50% of carbon mineralization, methane usually contributes between 0.5% and 12.1% (Bridgham *et al*. 1998). The lower rate of methane production in northern peatlands as compared with tropical wetlands may be due to both the presence of organic electron acceptors and/or the higher rate of methane consumption. Organic electron acceptors (namely quinones contained in humic substances) can oxidize both H_2S and acetate (Curtis and Reinhard 1994; Loveley *et al*. 1996; Scott *et al*. 1998) providing an alternative pathway for carbon mineralization and sulfate recycling (see also King, Chapter 2). Methane consumption is undertaken by a single class of microorganisms. A large part of methane produced can be consumed in the oxic upper layer (i.e. as in many northern peatlands) or in the oxic rhizosphere (Epp and Chanton 1993; Happell *et al*. 1993; Denier van der Gon and Neue 1996; King 1996). While anaerobic methane oxidation has not been well studied, it has been observed in paddy soils and it appears that ferrous iron and sulphate may be involved (Miura *et al*. 1992; Murase and Kimura 1994). However, it is unlikely this is an important process in wetlands, as sulfate and iron concentrations tend to be lower than the upper limit of affinity and because it likely only contributes a very minor amount of total oxidation (Segers 1998). Rates of methane oxidation are highest not in the oxic layer, but within 25 cm of the oxic/anoxic interface due to the affinity of methanotrophs to anaerobic conditions (Roslev and King 1994). It has been observed that oxidation is negatively correlated to distance to the water table (Moore and Dalva 1993; Sundh *et al*. 1995). Oxidation is also found in the oxic root rhizosphere and is dependent on plant type and root oxygen release (Calhoun and King 1997; Dannenberg and Conrad 1999). The importance of methane oxidation in wetlands will vary with wetland type and as a result of an interplay of the above factors.

6.4.2 Variables controlling effluxes

Numerous studies have addressed carbon dioxide and methane surface–atmosphere exchanges over the past couple of decades with a strong focus on methane dynamics. Both carbon dioxide and methane exchanges show marked spatial and seasonal variations. Seasonal variations are correlated to soil temperature and soil moisture fluctuations. Experimental evidence and field observations indicate that ecosystem respiration increases with lower water table level and drier soil conditions (Oechel et al. 1993, 1995). The lowering of water tables resulted in an increase in carbon dioxide effluxes from 25 to 42 mmol CO_2 m^{-2} d^{-1} (with water table at -10 cm) to 500–783 mmol CO_2 m^{-2} d^{-1} (water table at -70 cm) (Moore and Knowles 1989). This may be attributed to the stimulation of phenol oxidase activity and hydrolase mobilization resulting from exposure to oxygen (Freeman et al. 1996). Dry conditions, however, cannot sustain increased carbon dioxide production and emissions as decomposition is highest at intermediate moisture levels (Skopp et al. 1990). At low moisture levels the rate of diffusion of substrates to microbes is reduced due to the removal of water films between the organic particles (Stark and Firestone 1995). Temporary increases in methane emissions have also been observed following water table draw-down (Moore et al. 1990; Shannon and White 1994) and have been ascribed to short-term degassing of stored methane.

In tropical peatlands, water level has a more important role than temperature in the abiotic control of gas fluxes between the peat and the atmosphere (Jauhiainen et al. 2001). In more northern latitude wetlands, temperature and soil moisture influence on soil effluxes can be overridden by seasonal characteristics. Longer or shorter growing seasons may explain the differences observed in fluxes between two environmentally similar years (Griffis et al. 2000; Lafleur et al. 2001). This is not the case in tropical regions, where seasonal differences are not so pronounced and effluxes are likely more controlled by diffusional constraints due to soil moisture and botanical composition. A large portion of methane emissions also occur as a result of ebullition (Fechner-Levy and Hemond 1997).

Photosynthetically active radiation can also be an explanatory parameter particularly in northern latitude wetlands during the leafing-out period (Bubier et al. 1998; Griffis et al. 2000). It has been shown that a portion of carbon dioxide respired to the atmosphere can be recycled and re-assimilated mid-morning when leaf area index, respiration rates and incident radiation are at their highest. The recycling ranges between 2% and 8% of carbon dioxide of net ecosystem exchange (Lloyd et al. 1996; Brooks et al. 1997; Clark et al. 1999a). There is, however, a lag between primary production and the rate of respiration, as soil respiration, comprises carbon dioxide that has been produced at different times and from different sources. Hence, it represents the diffusion of gas originating from carbon pools of different turnover times.

Methane emissions may be linked to the dynamics of short-term carbon dioxide fluxes and annual NEE (Whiting and Chanton 1993; Bellisario et al. 1999). At the field scale, and in wetter communities, methane emissions were closely related to both the NEE and plant communities (Whiting and Chanton 1993; Bubier et al. 1995). Net average carbon dioxide exchange was found to be strongly correlated to methane flux, with methane emissions approximating 4% of carbon dioxide uptake (Bellisario et al. 1999). In this same study it was observed that the sites with large methane emissions had the largest carbon dioxide exchanges and were $^{13}CH_4$ enriched, indicating that acetate fermentation was the dominant methane production pathway. In bogs this may not pertain as a large portion of the acetate produced can either be used for carbon dioxide production or it may accumulate in its dissociated toxic form (acetic acid, Blodau et al. 2002). If most of methane is produced from recent organic material as indicated by ^{14}C signatures (Popp et al. 1999), then methane production should be linked to NPP and decomposition of relatively new material (Valentine et al. 1994) and carbon dioxide production. Panikov and Dedysh (2000), however, found that carbon dioxide and methane fluxes only correlated during the cold season. During the summer season carbon dioxide fluxes were dependent on plant community structure and function (diurnal photosynthetic cycles) and methane fluxes, which had no diurnal cycles, were more correlated to soil temperature and moisture conditions as confirmed by higher respiratory quotients for methanogenic populations ($Q_{10} = 4$ for methane; ~ 2 for carbon dioxide, Panikov et al. 1995).

It has been recognized that measurements of net ecosystem exchange confined to the growing season often overestimate the strength of the carbon sink of wetlands. Peatlands, in particular, emit a considerable amount of carbon dioxide during the non-growing season (2.5–42 mmol C m^{-2} d^{-1}) (Alm *et al*. 1999; Lafleur *et al*. 2001; Roehm and Roulet 2003). Although these rates are substantially smaller than both summer carbon dioxide effluxes and night-time respiration, the length of the non-growing season in northern and temperate latitudes signifies that winter effluxes can offset 3–50% of the growing season NEE.

6.5   Extent and distribution of global wetlands

The most widely used databases for the aerial distribution of freshwater wetlands are those proposed by Matthews and Fung (1987) (5.3 × 10^{12} m^2) and Aselmann and Crutzen (1989) (5.7 × 10^{12} m^2) for their methane emission studies. These estimates have remained the basis of current mapping efforts but they only pertain to freshwater wetlands. As a result new proposals have been put forward which include coastal and salt water wetlands with estimates varying between 5.6 and 13 × 10^{12} m^2 (Darras *et al*. 1999; Spiers 1999; Ramsar Wetland Convention 2000). There is, obviously, a great inconsistency in the information for review. Due to these discrepancies, I have calculated the global respiration contribution from wetlands, providing a mean estimate and a range for varying aerial coverage (Table 6.2). For the purpose of this chapter, the calculations are based on freshwater wetlands only, of which lakes, rivers, and estuaries have been omitted as these ecosystems have already been accounted for in other chapters of this book (Pace and Prairie, Chapter 7 and Hopkinson and Smith, Chapter 8). The freshwater wetlands dealt with include northern peatlands, freshwater marshes, and freshwater swamps. This implies that the figures of aerial coverage found in Tables 6.2 and 6.3 (3% of the earth's surface) are lower than the figure cited in the introduction (4–9% of the earth's surface).

The data used to compile the estimated global coverage of freshwater wetlands (Table 6.2) was taken from databases provided by Spiers (1999). One of the databases encompassed various sources from the literature and is found in the national wetland area estimate by wetland types from the information presented in "Extent and distribution of wetlands" and GRoWI databases; the second was based on the work by Taylor (1983) and other inventory sources. The databases were divided by country, which I then classed into boreal, temperate, or tropical and further subdivided wetland types into peatlands or marshes and swamps. The resulting table, therefore, provides a range of coverage for each region and wetland type due to the discrepancies encountered in the literature. Boreal regions indicate a global coverage of 3 to 5 × 10^{12} m^2; temperate regions 0.2 × 10^{12} m^2; and tropical regions 4 to 8 × 10^{12} m^2. The lower range of these estimates are only slightly higher than those previously suggested by Matthews and Fung (1987) and Aselmann and Crutzen (1989), but the total amount is not out of the range of the figures provided by the Ramsar Wetland Convention (2000). The estimates may be higher due to better coverage that has been provided since those studies were made, and partially because it was broken down into national inventories. This undoubtedly leaves space for error in that overlapping can occur and the definition of the classifications used by each study and each country vary.

6.6   Global respiration estimates

6.6.1   Data sources

The large and already compiled dataset of the literature from Bartlett and Harriss (1993) was used as the main source for methane fluxes. Data from approximately 12 more recent studies were also incorporated in order to update the dataset. The dataset for carbon dioxide fluxes was compiled from the literature. A large number of data were available for carbon dioxide fluxes in peatlands of boreal and temperate regions (>40 studies). Fewer studies were available for marshes and swamps (~15) of these regions and for tropical regions (~5) overall. The literature on rice paddies is vast, therefore, the data for rice paddies were also taken mostly from one main study (IPCC 1996) and complimented with several other sources.

6.6.2 Discussion

Global respiration rates by region and wetland type, show variability both within biome types and between regions (Table 6.2). This impedes the ability to accurately estimate the global contribution of carbon dioxide and methane fluxes to the atmosphere, but it allows one to gain a better understating of some of the regional diversity. Although methane fluxes only represent a minor portion of the total carbon respired from wetlands (~2.5%), methane flux is globally significant, and for example, is equivalent to over 20% of global lake respiration (Pace and Prairie, Chapter 7). As methane has a 21-fold greater global warming potential than carbon dioxide, even minor fluxes are of some significance.

The figures strongly indicate that both tropical and boreal regions can respire potentially large and equivalent amounts of carbon dioxide (395 and 538 $TmolCa^{-1}$, respectively). These figures alone are almost equivalent to current emissions from fossil fuels and terrestrial land-use changes (460–500 $TmolCa^{-1}$, Roulet 2000). Rice paddies, although not natural wetlands, provide a substantial input of atmospheric methane and alone account for the equivalent of 65% of the methane fluxes emitted from natural wetlands.

While these figures seem large, it must be taken into account that gross primary production strongly offsets these potential fluxes. On an annual basis many northern peatlands are considered sinks for carbon with long-term carbon sequestration rates of between 0.67 and 4.2 $molCm^{-2}a^{-1}$ (Ovenden 1990; Gorham 1995; Malmer and Wallén 1996). However, seasonal estimates of net ecosystem exchange indicate that large annual fluctuations do occur with growing season ranges of between sinks of −7.9 to sources of +5.9 $molCm^{-2}$ (Shurpali et al. 1995; Griffis et al. 2000; Lafleur et al. 2001) implying that indeed respiration plays a primary role in the shifting of ecosystems between sources and sinks. The net degree of carbon release is substantial as

compared with other ecosystems (del Giorgio and Williams, Chapter 14). The magnitude of the respiration factor stems from the timing of the seasons, where longer growing seasons tend to increase the springtime respiration value. Warmer winters also may promote more winter and spring respiration when the soils are not frozen to a great extent. Little is also still known about the role *Sphagnum* mosses play in the net ecosystem production and respiration balance.

Contrasting hydrology and botanical characteristics often explain the fundamental differences between wetland types in the way that carbon is fixed, stored, and cycled, an understanding of which is imperative for defining processes and carbon balances at the landscape level (Clark et al. 1999b). Although gross productivity and decomposition in the tropics are higher than in northern wetlands, these former systems appear to be similar in terms of carbon sinks. Recent evidence, however, has indicated that tropical wetlands may actually be in a state of carbon balance, signifying that they are also more sensitive to changes in climate than previously thought (Richey et al. 2002). Continued conversion of tropical wetlands for agricultural purposes (i.e. rice paddies) is likely to greatly increase the global methane contribution from wetlands. The difference between northern and southern wetlands is seen in the sensitivity of these systems to changes in environmental parameters. Tropical wetland carbon balances are mostly affected by variations in the moisture regime, whereas northern wetlands are affected by changes in moisture, temperature, and length of the growing season.

One of the primary hindrances encountered to date in studies of wetland carbon cycles is the difficulty of extrapolating process rates at the plot scale to ecosystem scale fluxes. The use of eddy-covariance and long-term continuous studies is currently allowing us to better define the inter- and intra-annual variability of fluxes, and to more accurately quantify the net carbon fluxes in these ecosystems. One of the recent advances made in carbon flux studies in wetlands is the recognition that measurements confined to the growing season introduce a bias in the accounting of carbon dioxide exchange, and losses equivalent to 3–50%

of the net growing season uptake have now been observed in the non-growing season (Alm *et al.* 1999; Lafleur *et al.* 2001, Roehm and Roulet 2003). This is of particular importance in view of climate change in northern wetlands, where the changes in the timing and extent of the growing season will have a profound effect on the rate of carbon uptake.

The response of wetlands to climate change is to date largely unknown and confounded by our limited understanding of the complexity of the processes involved in the cycling of carbon. It seems likely that the greatest impact will occur in northern wetlands where the greatest increase in temperature is expected (1–5.7°C, IPCC 2001) particularly during the winter. Precipitation is also expected to increase, again, mostly during the winter. This may mean warmer and wetter winters and drier and hotter summers. The impacts of this are to date uncertain and could be more complicated than a simple carbon release (Camill 1999). While increasing decomposition due to an increase in the thickness of the oxic layer is likely (Silvola *et al.* 1996; Bubier *et al.* 1998; Moore *et al.* 1998, Christensen *et al.* 2002) the amount of carbon dioxide released may be balanced or offset by an equivalent increase in primary productivity resulting from lags in nitrogen stimulation (Oechel *et al.* 2000).

Water-level drawdown is likely to cause a decrease in methane emissions as consumption of methane in the thicker aerobic layer is enhanced (Glenn *et al.* 1993; Roulet *et al.* 1993; Martikainen *et al.* 1995). However, warmer and moister winters may also enhance methane production and in some regions increased photosynthetic activity of deep-rooted wetland plants, such as sedges, may enhance substrate availability for methanogenesis leading to higher methane emissions (Valentine *et al.* 1994; Segers, 1998).

One area that remains of great concern is the thawing of peatlands underlain by permafrost. These ecosystems may become net carbon sources rather than sinks (Lal *et al.* 1998). Oechel *et al.* (1993) estimated that Artic systems are now net sources of 20 $TmolC\,a^{-1}$, caused mainly by melting of permafrost and lowering of the water table due to increased temperatures. Similar observations have been made in peatlands of the former Soviet Union,

now thought to be net sources of 6 $TmolC\,a^{-1}$ (Botch *et al.* 1995). Melting of permafrost may, however, increase water tables and lead to changes in community composition and enhance carbon sequestration (Camill 1999) overriding temperature responses and increased dryness (Hobbie 1996).

Should northern ecosystems become stronger sources of carbon dioxide as a result of climate change, a positive feedback will occur due to the significant soil carbon pool relative to the atmospheric carbon pool (29 000 $TmolC\,a^{-1}$ and 62 000 $TmolC\,a^{-1}$, respectively, Gorham 1991).

6.7 Conclusions and future directions

While this chapter has estimated the possible role of freshwater wetlands in the global carbon cycle in terms of gross respiratory contribution, large gaps in our knowledge and datasets do exist. Here we have provided a basis upon which to stimulate future research on freshwater wetlands. Key areas to be addressed are as follows:

1. Better and more verifiable methods for accounting for wetland sources and sinks are required. This has become a main focus in current studies undertaking continuous measurements of carbon dioxide net ecosystem exchange with eddy-covariance. This will create long-term datasets, which will enhance our understanding of ecosystem flux variability on decadal rather than annual timescales. By taking this approach, problems encountered with overestimation of carbon sinks due to the omission of non-growing season fluxes, can be overcome.

2. To enhance the long-term continuous measurements it is imperative we accompany these large-scale studies with process based research at smaller scales. Of particular importance is the requirement for more basic process studies in tropical regions. Currently we have a limited understanding of the complexity of the carbon cycle and hence a limited capability of predicting changes and feedbacks in view of climate change. In this Chapter, we have seen that this is particularly pertinent to Northern regions where the change and the impact of

change are likely greatest. Simple relationships are not accurate enough to allow us to predict processes and hence limit our capability to predict carbon cycling patterns. This may be addressed through mesocosm studies which allow for manipulation of specific components without fully disturbing the original structure of the soils. Further small-scale process studies responsible for methane production and consumption are required, as well as long-term continuous measurements of methane effluxes and dynamics in view of changing environmental parameters. We have a good understanding of the range of variability of methane fluxes, but limited understanding of the processes governing the interplay between production, storage, consumption, and emission.

3. Better identification of wetland regions and functions need to be made. This can be achieved by using remote sensing tools to identify the extent of wetland regions. It is also critical to identify and monitor land-use changes and account for these in global carbon balances. These changes may include both destruction of natural wetlands for agricultural and industrial purposes and restoration of wetlands habitats.

References

Abraham, W. R., Hesse, C., and Pelz, O. 1998. Ratios of carbon isotopes in microbial lipids as an indicator of substrate usage. *Appl. Environ. Microbiol.*, **64**: 4202–4209.

Achtnich, C., Bak, F., and Conrad, R. 1995. Competition for electron donors among nitrate reducers, ferric iron reducers and methanogens in an anoxic paddy soil. *Biol. Fert. Soil.*, **19**: 65–72.

Aerts, R. and Ludwig, F. 1997. Water-table changes and nutritional status affect trace gas emissions from laboratory columns of peatland soils. *Soil Biol. Biochem.*, **29**(11/12): 1691–1698.

Aiwa, H. A., Rice, C. W., and Sotomayor, D. 1998. Carbon and nitrogen mineralization in tallgrass prairie and agricultural soil profiles. *Soil Sci. Soc. Am. J.*, **62**: 942–951.

Aller, R. C. 1994. Bioturbation and mineralization of sedimentary organic matter: effects of redox oscillation. *Chem. Geol.*, **114**: 221–345.

Alm, J., Saarnio, S., Nykänen, H., Silvola, J., and Martikainen, P. J. 1999. Winter CO_2, CH_4 and N_2O fluxes on some natural and drained boreal peatlands. *Biogeochemistry*, **44**: 163–183.

Aselmann, I. and Crutzen, P. J. 1989. Global distribution of natural freshwater wetlands and rice paddies, their net primary productivity, seasonality and possible methane emissions. *J. Atmos. Chem.*, **8**: 307–358.

Backeus, I. 1990. Production and depth distribution of fine roots in a boreal bog. *Ann. Bot. Fennici*, **27**: 261–365.

Baldocchi, D. D. 2003. Assessing the eddy covariance technique for evaluating carbon dioxide exchange rates: Past, present and future. *Glob. Change Biol.*, **9**(4): 479.

Bellisario, L. M., Bubier, J. L., Moore, T. R., and Chanton, J. P. 1999. Controls on CH_4 emissions from a northern peatland. *Glob. Biogeochem. Cyc.*, **13**(1): 81–91.

Bartlett, K. B. and Harriss, R. C. 1993. Review and assessment of methane emissions from wetlands. *Chemosphere* **26**: 261–320.

Berner, R. A. 1980. A rate model for organic matter decomposition during bacterial sulfate reduction in marine sediments. In *Biogeochemistry of Organic Matter at the Sediment–Water Interface*. CNRS Int. Colloq, pp. 35–45.

Blodau, C. 2002. Carbon cycling in peatlands—a review of processes and controls. *Environ. Revi.* **10**: 111–134.

Blodau, C. and Moore, T. R. 2003. Experimental response of peatland carbon dynamics to a water table fluctuation. *Aquat. Sci.*, **65**: 47–62.

Blodau, C., Roehm, C. L., and Moore, T. R. 2002. Iron, sulphur and dissolved organic carbon dynamics in northern peatland. *Arch. Hydrobiol.*, **154**: 561–583.

Bolin, B. and Sukumar, R. 2000. Global Perspective. In *Land Use, Land-Use Change and Forestry*. R. T. Watson, I. R. Noble, B. Bolin, N. H. Ravindranath, D. J. Verardo, and D. J. Dokken (eds) A Special Report of the IPCC. Cambridge University Press, Cambridge, UK, pp. 23–51.

Boon, P. I. 1999. Carbon cycling in Australian wetlands: the importance of methane. *Verh. Int. Verein. Limnol.*, **27**: 1–14.

Botch, M. S., Kobak, K. I., Vinson, T. S., and Kolchunia, T. P. 1995. Carbon pools and accumulation in peatlands of the former Soviet Union. *Glob. Bioceochem. Cyc.*, **9**: 37–46.

Brady, M. A. 1997. Effects of vegetation changes on organic matter dynamics in three coastal peat deposits in Sumatra, Indonesia. In J. O. Rieley, and S. E. Page (eds) *Tropical Peatlands*. Samara Publishing Limited, Cardigan, UK, pp. 113–114.

Bridgham, S. D., Updegraff, K., and Pastor, J. 1998. Carbon, nitrogen, and phosphorus mineralization in northern wetlands. *Ecology*, **79**: 1545–1561.

Brinson, M. M., Lugo, A. E., and Brown, S. 1981. Primary productivity, decomposition and consumer activity in freshwater wetlands. *Annu. Revi. Ecol. Syst.*, **12**: 123–161.

Brooks, J. R., Flanagan, L. B., Varney, G. T., and Ehleringer, J. R. 1997. Vertical gradients in photosynthetic gas exchange characteristics and refixation of respired CO_2 within Boreal Forest canopies. *Tree Physiol.*, **17**: 1–12.

Brooks-Avery, G. B., Shannon, R. D., White, J. R., Martens, C. S., and Alperin, M. J. 1999. Effect of seasonal changes in the pathways of methanogenesis on the ^{d13}C values of pore water methane in a Michigan peatland. *Glob. Biogeochem. Cyc.*, **13**: 475–484.

Bubier, J. L., Moore, T. R., and Roulet, N. T. 1993. Methane emissions from wetlands in the midboreal region of northern Ontario, Canada. *Ecology*, **74**: 2240–2254.

Bubier, J. L., Moore, T. R., Bellisario, L., and Comer, N. T. 1995. Ecological controls on CH_4 emissions from a northern peatland complex in the zone of discontinuous permafrost, Manitoba, Canada. *Glob. Biogeochem. Cyc.*, **9**(4): 455–470.

Bubier, J. L., Crill, P. M., Moore, T. R., Savage, K., and Varner, R. 1998. Seasonal patterns and controls on net ecosystem CO_2 exchange in a boreal peatland complex. *Glob. Biogeochem. Cyc.*, **12**(4): 703–714.

Cai, Z. C., Xing, G. X., and Yan, X. Y. 1997. Methane and nitrous oxide emissions from rice paddy fields as affected by nitrogen fertilizers and water management. *Plant Soil*, **196**: 7–14.

Calhoun, A. and King, G. M. 1997. Regulation of root-associated methanotrophy by oxygen availability in the rhizosphere of two aquatic macrophytes. *Appl. Environ. Microbiol.*, **63**(8): 3051–3058.

Camill, P. 1999. Peat accumulation and succession following permafrost thaw in the boreal peatlands of Manitoba, Canada. *Ecoscience*, **6**: 592–602.

Campbell, C., Vitt, D. H., Halsey, L. A., Campbell, I. D., Thormann, M. N., and Bayley, S. E. 2000. Net primary production and standing biomass in Northern continental wetlands. *Northern Forestry Centre Information Report*, NOR-X-369.

Carroll, P. and Crill, P. 1997. Carbon balance of a temperate poor fen. *Glob. Biogeochem. Cyc.*, **11**: 349–356.

Christensen, T. R., Lloyd, D., Svensson, B., Martikainen, P. J., Harding, R., Oskarsson, H., Friborg, T., Soegaard, H., and Panikov, N. 2002. Biogenic controls on trace gas fluxes in northern wetlands. *Glob. Change News Lett.*, **51**: 9–11.

Clark, H., Newton, P. C. D., and Barker, D. J. 1999a. Physiological and morphological responses to elevated CO_2 and a soil moisture deficit of temperate pasture species growing in an established plant community. *J. Exp. Bot.*, **50**: 233–242.

Clark, K. L., Gholz, H. L., Moncrieff, J. B., Cropley, F., and Loescher, H. W. 1999b. Environmental controls over net exchanges of carbon dioxide from contrasting ecosystems in north Florida. *Ecol. Appl.*, **9**: 936–948.

Clein, J. S. and Schimel, J. P. 1994. Reduction in microbial activity in birch litter due to drying and rewetting events. *Soil Biol. Biochem.*, **26**: 403–406.

Conrad, R. 1989. Control of methane production in terrestrial ecosystems. In M. O. Andreae and D. S. Schimel (eds) *Exchange of Trace Gases Between Terrestrial Ecosystems and the Atmosphere*. John Wiley, New York, pp. 39–58.

Conrad, R. and Rothfuss, F. 1991. Methane oxidation in the soil surface layer of a flooded rice field and the effect of ammonium. *Biol. Fert. Soil.*, **12**: 28–32.

Crutzen, P. J. and Lelieveld, J. 2001. Human impacts on atmospheric chemistry. *Annu. Rev. Earth Planet. Sci.*, **29**: 17–45.

Curtis, G. P. and Reinhard, M. 1994. Reductive dehalogenation of hexachloroethane, carbon tetrachloride, and bromoform by anthrahydroquinone disulfonate and humic acid. *Environ. Sci. Technol.*, **28**: 193–202.

Dabberdt, W. F., Lenschow, D. H., Horst, T. W., Zimmermann, S. P., Oncley, S. P., and Delany, A. C. 1993. Atmosphere-surface exchange measurements. *Science*, **260**: 1472–1481.

Dannenberg, S. and Conrad, R. 1999. Effect of rice plants on methane production and rhizospheric metabolism in paddy soil. *Biogeochemistry*, **45**: 53–71.

Darras, S., Michou, M., and Sarrat, C. 1999. IGBP-DIS Wetland Data Inititiative. A first step towards identifying a global delineation of wetlands. *IGBP-DIS Working Paper*, No. 19.

Davidson, E. A. and Trumbore, S.E. 1995. Gas diffusivity and production of CO_2 in deep soils of the eastern Amazon. *Tellus*, **47B**: 550–565.

Davis, C. B. and Van der Valk, A. G. 1978. The decomposition of standing and fallen litter of *Typha glauca* and *Scirpus fluviatilis*. *Can. J. Bot.*, **56**: 662–675.

Deghi, G. S., Ewel, K. C., and Mitsch, W. J. 1980. Effects of sewage effluent application on litterfall and litter decomposition in cypress swamps. *J. Appl. Ecol.*, **17**: 397–408.

Denier van der Gon, H. A. C., and Neue, H. U. 1996. Oxidation of methane in the rhizosphere of rice plants. *Biol. Fert. Soil.*, **22**: 359–366.

Denmead, O. T. and Raupach, M. R. 1993. Methods for measuring atmospheric trace gas transport in agricultural and forest systems. In *Agricultural Ecosystem Effects on Trace Gases and Global Climate Change*. ASA Special Publication, Switzerland, pp. 19–43.

Dise, N. B and Verry, E. S. 2001. Suppression of peatland methane emissions by cumulative sulfate deposition in simulated acid rain. *Biogeochemistry*, **53**: 143–160.

Dueñas, C., Fernandez, M. C., Carretero, J., Liger, E. and Perez, M. 1995. Emissions of CO_2 from some soils. *Chemosphere*, **30**(10): 1875–1889.

Dueñas, C., Fernandez, M. C., Carretero, J., and Liger, E. 1999. Methane and carbon-dioxide fluxes in soils evaluated by RN-222 flux and soil air concentration profiles. *Atmos. Environ.*, **333**: 4495–4502.

Dunfield, P., Knowles, R., Dumont, R. and Moore, T. R. 1993. Methane production and consumption in temperate and subarctic peat soils—Response to temperature and pH. *Soil Biol. Biochem.*, **25**(3): 321–326.

Epp, M. A. and Chanton, J. P. 1993. Application of the methyl fluoride technique to the determination of rhizospheric methane oxidation. *J. Geophys. Res.*, **98**: 18, 413–18, 422.

Fechner-Levy, E. and Hemond, H. 1997. Trapped methane volume and potential effects on methane ebullition in a Northern Peatland. *MIT Center for Global Change Science Bulletin*, 5.

Fenchel, T., King, G. M., and Blackburn, T. H. 1998. *Bacterial Biogeochemistry*. Academic Press, New York.

Freeman, C., Liska, G., Ostle, N. J., Lock, M. A., Hughes, S., Reynolds, B., and Hudson, J. 1996. Microbial activity and enzymatic decomposition processes following peatland water table drawdown. *Plant Soil*, **180**: 121–127.

Freeman, C., Liska, G., Ostle, N. J., Lock, M. A., Hughes, S., Reynolds, B., and Hudson, J. 1997. Enzymes and biogeochemical cycling in wetlands during a simulated drought. *Biogeochemistry*, **39**: 177–187.

Frenzel, P., Rothfuss, F., and Conrad, R. 1992. Oxygen profiles and methane turnover in a flooded rice microcosm. *Biol. Fert. Soil.*, **14**: 84–89.

Frenzel, P. and Karofeld, E. 2000. CH_4 emission from a hollow-ridge complex in a raised bog: the role of CH_4 production and oxidation. *Biogeochemistry*, **51**: 91–112.

Finlayson, C. M., Davidson, N. C., Spiers, A. G., and Stevenson, N. J. 1999. Global wetland inventory—current status and future priorities. *Mar. Freshwater Res.*, **50**: 717–727.

Fung, I., John, J., Lerner, J., Matthews, M., Prather, M., Steele, L., and Frazer, P. 1991. Global budgets of atmospheric methane: results from three dimensional global model synthesis. *J. Geophys. Res.*, **6**: 13 033–13 065.

Gill, R. A. and Jackson, R. B. 2000. Global patterns of root turnover for terrestrial ecosystems. *New Phytol.*, **147**: 13–31.

Glenn, S., Heyes, A., and Moore, T. R. 1993. Carbon dioxide and methane fluxes from drained peat soils, Southern Quebec. *Glob. Biogeochem. Cyc.*, **7**(2): 247–257.

Gorham, E. 1991. Northern peatlands: role in the carbon cycle and probable responces to climatic warming. *Ecol. Appl.*, **1**: 183–195.

Gorham, E. 1995 The biogeochemistry of northern peatlands and its possible responce to global warming. In G. M. Woodwell and F. T. McKenzie (eds) *Biotic Processes and Potential Feedbacks*. Oxford University Press, UK, pp. 169–187.

Griffis, T. J., Rouse, W. R., and Waddington, J. M. 2000. Interannual variability of net ecosystem exchange at a subarctic fen. *Glob. Biogeochem. Cyc.*, **14**: 1109–1121.

Hamilton, J. D., Kelly, C. A., Rudd, J. W. M., Hesslein, R. H., and Roulet, N. T. 1994. Flux to the atmosphere of CH_4 and CO_2 from wetland ponds on the Hudson Bay lowlands (HBLs). *J. Geophys. Res.*, **99**: 1495–1510.

Happell, J. D., Whiting, G. J., Showers, W. S., and Chanton, J. P. 1993. A study contrasting methane emission from marshlands with and without active methanotrophic bacteria. *J. Geophys. Res.*, **98**: 14 771–14 782.

Hobbie, S. E. 1996. Temperature and plant species control over carbon and nitrogen cycling through litter and soil in Alaskan tundra. *Ecol. Monogr.*, **66**: 503–522.

Hogg, E. H. 1993. Decay potential of hummock and hollow *Sphagnum* peats at different depths in a Swedish raised bog. *Oikos*, **66**: 269–278.

IPCC. 1996. Methane emissions from rice cultivation: fooded rice fields. In *The Revised 1996 IPCC Guidelines for National Greenhouse Gas Inventories: Reference Manual* www.ipcc-nggip.iges.or.jp/public/gl/guidelin/ch4ref5.pdf.

IPCC. 2001. *Climate change 2001: The Scientific basis*. www.ipcc.ch/pub/taroldest/wg1/012.htm.

Jackson, R. B., Mooney, H. A., and Schulze, E-D. 1997. A global budget for fine root biomass, surface area and nutrient contents. *Proc. Natl. Acad. Sci. USA*, **94**: 7362–7366.

Janssens, I. A., Kowalski, A. S., Longdoz, B., and Ceulemans, R. 2000. Assessing forest soil CO_2 efflux: an *in situ* comparison of four techniques. *Tree Physiol.*, **20**: 23–32.

Jauhiainen, J., Heikkinen, J., Martikainen, P. J., and Vasander, H. 2001. CO_2 and CH_4 fluxes in pristine peat swamp forest and peatland converted to agriculture in central Kalimantan, Indonesia. *Int. Peat J.*, **11**: 43–49.

Jespersen, D. N., Sorrell, B. K., and Brix, H. 1998. Growth and oxygen release by *Typha latifolia* and its effects on sediment methanogenesis. *Aqua. Bot.*, **61**: 165–180.

Johnson, L. C. and Damman, A. W. H. 1991. Species-controlled *Sphagnum* decay on a south Swedish raised bog. *Oikos*, **61**: 234–242.

Khalil, M. A. K. and Shearer, M. J. 1993. Sources of methane: an Overview. In M. A. K. Khalil (ed) *Atmospheric Methane, Sources, Sinks and Role in Global Change*. Springer-Verlag, Germany.

Kieft, T. L., Soroker, E. and Firestone, M. K. 1997. Microbial biomass response to a rapid increase in water potential when dry soil is wetted. *Soil Biol. Biogeochem.*, **19**: 119–126.

King, G. M. 1996. Regulation of methane oxidation: contrasts between anoxic sediments and oxic soils. In M. E. Lidstrom, and F. R. Tabita (eds) *Microbial Growth on C_1 Compounds*. Kluwer Academic Publishers, Dordrecht, The Netherlands, pp. 318–328.

Kruger, M. and Frenzel, P. 2003. Effects of N-fertilisation on CH_4 oxidation and production, and consequences for CH_4 emissions from microcosms and rice fields. *Glob. Change Biol.*, **9**: 773–784.

Lafleur P. M., Roulet, N. T., and Admiral, S. 2001. The annual cycle of CO_2 exchange at a boreal bog peatland. *J. Geophys. Res.*, **106**: 3071–3081.

Lal, R., Kimble, J. M., Cole, C. V. and Follet, R. F. 1998. *The Potential of U.S. Cropland to Sequester Carbon and Mitigate the Greenhouse Effect*. Ann Arbor Press, Chelsea, MI.

Lindau, C. W. 1994. Methane emissions from Louisiana rice fields amended with nitrogen fertilizers. *Soil Biol. Biochem.*, **26**(3): 353–359.

Lloyd, J. and Taylor, J. A. 1994. On the temperature dependence of soil respiration. *Funct. Ecol.*, **8**: 315–323.

Lloyd, J., Kruijt, B., Hollinger, D. Y., Grace, J., Francey, R. J., Wong, S. C., Kellihe, F. M., Miranda, A. C., Farquhar, G. D., Gash, J. H., Vygodskaya, N. N., Wright, I. R., Miranda, H. S., and Schulze, E. D. 1996. Vegetation effects on the isotopic composition of atmospheric CO_2 at local and regional scales: theoretical aspects and a comparison between rain forest in Amazonia and boreal forest in Siberia. *Aust. J. Plant Physiol.*, **23**: 371–399.

Lloyd, J., Francey, R. J., Sogachev, A., Byers, J. N., Mollicone, D., Kelliher, F. M., Wong, S. C., Arneth, A., Bauer, G., McSeveny, T. M., Rebmann, C., Valentini, R., and Schulze, E. D. 2001. Boundary layer budgets and regional flux estimates for CO_2, its carbon and oxygen isotopes and for water vapour above a forest/bog mosaic in central Siberia. *J. Geophys. Res.*, **104**: 6647–6660.

Lovley, D. R., Coates, J. D., Blunt-Harris, E. L., Phillips, E. J. P., and Woodward, J. C. 1996. Humic substances as electron acceptors for microbial respiration. *Nature*, **382**: 445–448.

Lund, C. P., Riley, W. J., Pierce, L. L., and Field, C. B. 1999. The effects of chamber pressurization on soil-surface CO_2 flux and the implication for NEE measurements under elevated CO_2. *Glob. Change Biol.*, **5**: 269–281.

Malmer, N. and Wallen, B. 1996. Peat formation and mass balance in subarctic ombrotrophic peatlands around Abisko, Northern Scandinavia. *Ecol. Bull.*, **45**: 79–92.

Maltby, E. and Immirzi, C. P. 1993. Carbon dynamics in peatlands and other wetland soils—regional and global perspectives. *Chemosphere*, **27**(6): 999–1023.

Mandernack, K. W., Lynch, L., Krouse, H. R., and Morgan, M. D. 2000. Sulfur cycling in wetland peat of the New Jersey Pinelands and its effect on stream water chemistry. *Geochim. Cosmochim. Acta*, **64**: 3949–3964.

Martikainen, P. J., Nykänen, H., Alm, J., and Silvola, J. 1995. Changes in fluxes of carbon dioxide, methane and nitrous oxide due to forest drainage of mire sites of different trophy. *Plant Soil*, **168–169**: 571–577.

Massman, W. J and Lee, X. 2002. Eddy covariance flux correction and uncertainties in long term studies of carbon and energy exchanges. *Agric. Forest Met.*, **113**: 121–144.

Matthews, E. and Fung, I. 1987. Methane emissions from natural wetlands: global distribution, area and environmental characteristics of sources. *Glob. Biogeochem. Cycl.*, **1**: 61–86.

McKenzie, C., Schiff, S., Aravena, R., Kelly, C., and St. Louis, V. 1998. Effect of temperature on production of CH_4 and CO_2 from peat in a natural and flooded boreal forest wetland. *Clim. Change*, **40**(2): 247–267.

Mitsch, W. J. and Wu, X. 1995. Wetlands and global change. In R. Lal, J. Kimble, E. Levine, and B. A. Stewart (eds) *Advances in Soil Science, Soil Management and Greenhouse Effect*. Lewis Publishers, Boca Raton, FL, pp. 205–230.

Mitsch, W. J. and Gosselink, J. G. 2000. *Wetlands*, 3rd edition, John Wiley & Sons, Inc., New York, p. 920.

Miura, Y., Watanabe, A., Murase, J., and Kimura, M. 1992. Methane production and its fate in paddy fields. *Soil Sci. Plant Nutr.*, **38**: 673–678.

Moore, T. R. and Knowles, R. 1989. The influence of water table levels on methane and carbon dioxide emissions from peatland soils. *Can. J. Soil. Sci.*, **69**: 33–38.

Moore, T. R. and Dalva, M. 1993. The influence of temperature and water table position on carbon dioxide and methane emissions from laboratory columns of peatland soils. *J. Soil Sci.*, **44**: 651–664.

Moore, T. R., Roulet, N. T., and Knowles, R. 1990. Spatial and temporal variations in methane flux from subarctic/northern boreal fens. *Glob. Biogeochem. Cyc.*, **4**: 29–46.

Moore, T. R., Roulet, N. T., and Waddington, J. M. 1998. Uncertainty in predicting the effect of climatic change on the carbon cycling of Canadian peatlands. *Clim. Change*, **40**(2): 229–246.

Moore, T. R., Bubier, J., Lafleur, P., Frolking, S., and Roulet, N. T. 2002. Plant biomass, production and CO_2 exchange in an ombrotrophic bog. *J. Ecol.*, **90**: 25–36.

Müllen, S. F., Janssen, J. A., and Gorham, E. 2000. Acidity of and the concentration of major and minor metals in the surface waters of bryophyte assemblages from 20 North American bogs and fens. *Can. J. Bot.*, **78**: 718–727.

Murase, J. and Kimura, M. 1994. Methane production and its fate in paddy fields. IV. Sources of microorganisms and substrates responsible for anaerobic CH_4 oxidation in subsoil. *Soil Sci. Plant Nutr.*, **40**(1): 57–61.

Murayama, S. and Bakar, Z. A. 1996. Decomposition of tropical peat soils, 2. Estimation of *in situ* decomposition by measurement of CO_2 flux. *Jpn. Agric. Res. Quart.*, **30**: 153–158.

Nay, S. M., Mattson, K. G., and Bormann, B. T. 1994. Biases of chamber methods for measuring soil CO_2 efflux demonstrated with a laboratory apparatus. *Ecology*, **75**: 2460–2463.

Nelson, J. W., Kadlec, J. A., and Murkin, H. R. 1990. Seasonal comparisons of weight loss of two types of *Typha glauca* Godr. Leaf litter. *Aquat. Bot.*, **37**: 299–314.

Neue, H. U. 1997. Fluxes of methane from rice fields and potential for mitigation. *Soil Use Manage.*, **13**: 258–267.

Nicholson, B. J., Gignac, L. D., and Bayley, S. E. 1996. Peatland distribution along a north-south transect in the McKenzie River Basin in relation to climatic and environmental gradients. *Vegetatio*, **126**: 119–133.

Nüsslein, B., Chin, K. J., Eckert, W., and Conrad, R. 2001. Evidence for anaerobic syntrophic acetate oxidation during methane production in the profundal sediment of subtropical Lake Kinneret (Isreal). *Environ. Microbiol.*, **3**(7): 460–470.

Oechel, W. C., Hastings, S. J., Vourlitis, G., Jenkins, M., Riechers, G., and Grulke, N. 1993. Recent change of Arctic tundra ecosystems from a net carbon dioxide sink to a source. *Nature*, **361**: 520–523.

Oechel, W. C., Vourlitis, G. L., Hastings, S. J., and Bochkarev, S. A. 1995. Change in Arctic CO_2 flux over two decades: Effects of climate change at Barrow, Alaska. *Ecol. Appl.*, **5**: 846–855.

Oechel, W. C., Vourtilis, G. L., Hastings, S. J., Ault, R. P., and Bryant, P. 1998. The effects of water table manipulation and elevated temperature on the net CO_2 flux of wet sedge tundra ecosystems. *Glob. Change Biol.*, **4**(1): 77–90.

Oechel, W. C., Vourlitis, G. L., Hastings, S. J., Zulueta, R. C., Hinzman, L., and Kane, D. 2000. Acclimation of ecosystem CO_2 exchange in the Alaskan Arctic in response to decadal climate warming. *Nature*, **406**: 978–981.

Öquist, M. and Sundh, I. 1998. Effects of transient oxic period on mineralisation of organic matter to CH_4 and CO_2 in anoxic peat incubations. *Geomicrobiology*, **15**: 325–333.

Ovenden, L. 1990. Peat accumulation in northern wetlands. *Quart. Res.*, **33**: 377–386.

Panikov, N. S. and Dedysh, S. N. 2000. Cold season CH_4 and CO_2 emission from boreal peat bogs (West Siberia): winter fluxes and thaw activation dynamics. *Glob. Biogeochem. Cyc.*, **14**(4): 1071–1080.

Panikov, N. S., Sizova, M. V., Zelenev, V. V., Mahov, G. A., Naumov, A. B., and Gadzhiev, I. M. 1995. Emission of CH_4 and CO_2 from wetland in Southern part of West Siberia: spatial and temporal flux variation. *J. Ecol. Chem.*, **4**: 9–26.

Parashar, D. C., Rai, J., Gupta, P. K., and Singh, N. 1991. Parameters affecting methane emission from paddy fields. *Indian J. Radio Space Phys.*, **20**: 12–17.

Peters, V. and Conrad, R. 1996. Sequential reduction processes and initiation of CH_4 production upon flooding of oxic upland soils. *Soil Biol. Biochem.*, **28**: 371–382.

Pjavchenko, N. J. 1982. Bog ecosystems and their importance in nature. In D. O. Logofet and N. K. Luckyanov (eds) *Ecosystem Dynamics in Freshwater Wetlands and Shallow Water Bodies, Vol. 1 SCOPE and UNEP Workshop*. Center for International Projects, Moscow, pp. 7–21.

Powell, S. W. and Day, F. P. 1991. Root production in four communities in the Great Dismal Swamp. *Am. J. Bot.*, **78**: 288–297.

Prieme, A. and Christensen, S. 1997. Seasonal and spatial variation of methane oxidation in a Danish spruce forest. *Soil Biol. Biochem.*, **29**(8): 1165–1172.

Popp, T. J., Chanton, J. P., Whiting, G. J. and Grant, N. 1999. The methane stable isotope distribution at a *Carex* dominated fen in North Central Alberta. *Glob. Biogeochem. Cyc.*, **13**(4): 1063–1078.

Raich, J. W. and Schlesinger, W. H. 1992. The global carbon dioxide flux in soil respiration and its relationship to vegetation and climate. *Tellus* **44B**: 81–99.

Ramsar Convention on Wetlands. 2000. www.ramsar.org/index_key_docs.htm.

Rayment, M. B. and Jarvis, P. G. 1997. An improved open chamber system for measuring soil CO_2 effluxes of a boreal black spruce forest. *J. Geophys. Res.*, **102**: 28779–28784.

Richey, J. E., Melack, J. M., Aufdenkampe, A. K., Ballester, V. M., and Hess, L. L. 2002. Outgassing from

Amazonian rivers and wetlands as a large tropical source of atmospheric CO_2. *Nature*, **416**: 617–620.

Risk, D., Kellman, L., and Beltrami, H. 2002. Soil CO_2 production and surface flux at four climate observatories in eastern Canada. *Glob. Biogeochem. Cyc.*, **16**(4): 1122–1133.

Roden, E. R. and Wetzel, R. G. 1996. Organic carbon oxidation and suppression of methane production by microbial Fe(III) oxide during reduction in vegetated and unvegetated freshwater wetland sediments. *Limnol. Oceanogr.*, **41**: 1733–1748.

Roehm, C. L. 2003. Carbon Dynamics in Northern Peatlands, Canada. PhD Thesis, McGill University, Montreal, Canada.

Roehm, C. L. and Roulet, N. T. 2003. Seasonal contribution of CO_2 fluxes in the annual carbon budget of a northern bog. *Glob. Biogeochem. Cyc.*, **17**(1): 1–9.

Roslev, P. and King, G. M. 1994. Survival and recovery of methanotrophic bacteria starved under oxic and anoxic conditions, *Appl. Environ. Microbiol.*, **60**(7): 2602–2608.

Roulet, N. T. 2000. Peatlands, carbon storage, greenhouse gases, and the Kyoto Protocol: prospects and significance for Canada. *Wetlands*, **20**(4): 605–615.

Roulet, N. T., Ash, R., Quinton, W. and Moore, T. R. 1993. Methane flux from drained northern peatlands: effect of a persistent water table lowering on flux. *Glob. Biogeochem. Cyc.* **7**: 749–769.

Roulet, N. T., Crill, P. M., Comer, N. T., Dove, A., and Boubonniere, R. A. 1997. CO_2 and CH_4 flux between a boreal beaver pond and the atmosphere. *J. Geophys. Res.*, **102**(24): 29 313–29 321.

Saarnio, S., Alm, J., Silvola, J., Lohila, A., Nykanen, H., and Martikainen, P. J. 1997. Seasonal variation in CH_4 emissions and production and oxidation potentials at microsites on an oligotrophic pine fen. *Oecologia*, **110**(3): 414–422.

Scanlon, D. and Moore, T. R. 2000. Carbon dioxide production from peatland soil profiles: the influence of temperature, oxic/anoxic conditions and substrate. *Soil Sci.*, **165**: 153–160.

Schimel, J. P. 1995. Plant transport and methane production as controls in methane flux from Arctic wet meadow tundra. *Biogeochemistry*, **28**: 183–200.

Schimel, J. P. and Clein, J. S. 1996. Microbial response to freeze-thaw cycles in tundra and taiga soils. *Soil Biol. Biochem.*, **28**(8): 1061–1066.

Scott, D. T., McKnight, D. M., Blunt-Harriss, E. L., Kolesar, S. E., and Lovley, D. R. 1998. Quinone moieties act as electron acceptors in the reduction of humic substances by humics-reducing microorganisms. *Environ. Sci. Technol.*, **32**: 2984–2989.

Scott, K. T., Kelly, C. A., and Rudd, J. W. M. 1999. The importance of relating peat to methane fluxes from flooded peatlands. *Biogeochemistry*, **47**: 187–202.

Segers, R. 1998. Methane production and methane consumption: a review of processes underlying wetland methane fluxes. *Biogeochemistry*, **41**: 23–51.

Shannon, R. D. and White, J. R. 1994. A three year study of controls on methane emissions from two Michigan peatlands. *Biogeochemistry*, **27**: 35–60.

Shurpali, N. J., Verma, S. B., Kim, J., and Arkebauer, T. J. 1995. Carbon dioxide exchange in a peatland ecosystem. *J. Geophys. Res.*, **100**(D7): 14 319–14 326.

Silvola, J., Alm, J., Ahlholm, U., Nykänen, and Martikainen P. J. 1996. The contribution of plant roots to CO_2 fluxes from organic soils. *Biol. Fert. Soil.*, **23**: 126–131.

Singh, J. S., Singh, S., and Raghubanshi, A. S. 1996. Methane flux from rice/wheat agroecosystem as affected by crop phenology, fertilization and water level. *Plant Soil*, **183**: 323–327.

Skopp, J., Jawson, M. D., and Doran, J. W. 1990. Steady-state aerobic microbial activity as a function of soil water content. *Soil Sci. Soc. Am. J.*, **54**: 1619–1625.

Sorrell, B. K. and Boon, P. I. 1994. Convective gas flow in *Eleocharis sphacelata* R. Br.: methane transport and release from wetlands. *Aquat. Bot.*, **47**: 197–202.

Spiers, A. G. 1999. Review of international/continental wetland resources. In C. M. Finlayson and A. G. Spiers (eds) *Global Review of Wetland Resources and Priorities Inventory*. Canberra, No. 144, pp. 63–104.

Stark, J. M. and Firestone, M. K. 1995. Mechanisms for Soil moisture effects on activity of nitrifying bacteria. *Appl. Environ. Microbiol.*, **61**(1): 218–221.

Sundh, I. Mikkelä, C., Nilsson, M., and Svensson, B. H. 1995. Potential aerobic methane oxidation in a *Sphagnum* dominated peatland—controlling factors and relation to methane emission. *Soil Biol. Biochem.*, **27**: 829–837.

Taylor, J. A. 1983. Peatlands of Great Britain and Ireland. In A. J. P. Gore (ed) *Ecosystems of the World 4B: mires: swamp, Bog, Fen and Moor: regional studies*. Elsevier Scientific Publishing Company, Amsterdam, The Netherlands, pp. 1–16 .

Tjoelker, M. G., Oleksyn, J., and Reich, P. B. 2001. Modelling respiration of vegetation: evidence for a general temperature-dependent Q10. *Glob. Change Biol.*, **7**: 223–230.

Tolonen, K., Vasander, H., Damman, A. W. H., and Clymo, R. S. 1992 Preliminary estimate of long-term carbon accumulation and loss in 25 Boreal peatlands. *Suo (Mires and Peat)*, **43**: 277–280.

Trettin, C. C. and Jurgensen, M. F. 2003. Carbon cycling in wetland forest soils. In J. M. Kimble, L. S. Heath, R. A.

Birdsey, and R. Lal (eds) *The Potential of U.S. Forest Soils to Sequester Carbon and Mitigate the greenhouse Effect*, CRC Press, New York, pp. 311–331.

Updegraff, K., Pastor, J., Bridgham, S. D., and Johnston, C. A. 1995. Environmental and substrate controls over carbon and nitrogen mineralization in northern wetlands. *Ecol. Appli.*, **5**(1): 151–163.

Valentine, D. W., Holland, E. A., and Schimel, D. S. 1994. Ecosystem physiological control over CH_4 production in northern wetlands. *J. Geophys. Res.*, **99**: 1563–1544.

Valentini, R., Matteucci, G., Dolman, A. J., Schulze, E. D., Rebmann, C., Moors, E. J., Granier, A., Gross, P., Jensen, N. O., Pilegaard, K., Lindroth, A., Grelle, A., Bernhofer, C., Grunwald, T., Aubinet, M., Ceulemans, R., Kowalski, A. S., Vesala, T., Rannik, U., Berbigier, P., Loustau, D., Gudmundsson, J., Thorgeirsson, H., Ibrom, A., Morgenstern, K., Clement, R., Moncrieff, J., Montagnani, L., Minerbi, S., and Jarvis, P. G. 2000. Respiration as the main determinant of carbon balance in European forests. *Nature*, **404**: 861–865.

Van den Bos, R. and Van de Plassche, O. 2003. Incubation experiments with undisturbed cores from coastal peatlands (Western Netherlands): carbon dioxide fluxes in response to temperature and water table changes. In Human Influence in Carbon Fluxes in Coastal Peatlands; Process Analysis, Quantification and Prediction. PhD thesis, Free University Amsterdam, The Netherlands, pp. 11–34.

Verhoeven, J. T. A. and Toth, E. 1995. Decomposition of *Carex* and *Sphagnum* litter in fens: effect of litter quality and inhibition by living tissue homogenates. *Soil Biol. Biochem.*, **27**(3): 271–275.

Vitt, D. H., Bayley, S. E., and Jin, T. L. 1995. Seasonal variation in water chemistry over a bog-rich fen gradient in Continental Western Canada. *Can. J. Fish. Aquat. Sci.*, **52**: 587–606.

Von Hulzen, J. B., Segers, R., van Bodegom, P. M., and Leffelaar, P. A. 1999. Temperature effects on soil methane production: an explanation for observed variability. *Soil Biol. Biochem.*, **31**: 1919–1929.

Waddington, J. M. and Roulet, N. T. 1996. Atmosphere-wetland carbon exchanges: scale dependency of CO_2 and CH_4 exchange on the development topography of a peatland. *Glob. Biogeochem. Cycl.*, **10**: 233–245.

Wang, Z. P., Zeng, D. and Patrick, W. H. 1996. Methane emissions from natural wetlands. *Environ. Monit. Assess.*, **42**: 143–161.

Washington, J. W., Rose, A. W., Ciolkosz, E. J., and Dobos, R. R. 1994. Gaseous diffusion and permeability in four soils in central Pennsylvania. *Soil Sci.*, **157**: 65–76.

Watanabe, A. and Kimura, M. 1995. Methane production and its fate in paddy fields. VIII. Seasonal variations in the amount of methane retained in soil. *Soil Sci. Plant Nutr.*, **41**: 225–233.

Watson, A. and Nedwell, D. B. 1998. Methane production and emission from peat: the influence of anions (sulphate, nitrate) from acid rain. *Atmos. Environ.*, **32**(19): 3239–3245.

Weller, M. W. 1994. *Freshwater Marshes*, 3rd edition, University of Minnesota Press, Minneapolis, MN, p. 192.

Werner, C., Davis, K., Bawkin, P., Chuixiang, Y., Hurst, D., and Lock, L. 2003. Regional-scale measurements of methane exchange from a tall tower over a mixed temperate-boreal lowland forest. *Glob. Change Biol.*, (in press).

Westermann, P. 1996. Temperature regulation of anaerobic degradation of organic matter. *World J. Microbi. Biotechnol.*, **12**: 497–503.

Whalen, S. C. and Reeburgh, W. S. 2000. Methane oxidation, production, and emission at contrasting sites in a boreal bog. *Geomicrobiol. J.*, **17**: 237–251.

Whiting, G. J. and Chanton, J. P. 1993. Primary production control over methane emissions from wetlands. *Nature*, **364**: 794–795.

Wofsy, S. C. and Hollinger, D. Y. 1998. Measurements and Modeling Net Ecosystem CO_2 Flux from Major Terrestrial Biomes Using Continuous, Long-Term Flux Measurements at the Ecosystem Scale. Executive Summary "Science Plan for AmeriFlux: Long-term flux Measurement Network of the Americas". www.esd.ornl.gov/programs/NIGEC/scif.htm.

Yavitt, J. B. and Lang, G. E. 1990. Methane production in contrasting wetland sites: response to organic-chemical components of peat and to sulfate reduction. *Geomicrobiol. J.*, **8**: 27–46.

Yavitt, J. B., Williams, C. J., and Wieder R. K. 1997. Production of methane and carbon dioxide in peatland ecosystems across North America: effects of temperature, aeration, and organic chemistry of peat. *Geomicrobiol. J.*, **14**: 299–316.

Yu, Z., Turetsky, M. R., Campbell, I. D., and Vitt, D. H. 2001. Modelling long-term peatland dynamics. II. Processes and rates as inferred from litter and peat-core data. *Ecol. Model.*, **145**(2–3): 159–173.

CHAPTER 7

Respiration in lakes

Michael L. Pace[1] and Yves T. Prairie[2]

[1] Institute of Ecosystem Studies, USA
[2] Département des sciences biologiques, Université du Québec à Montréal, Canada

Outline

This chapter reviews, from a quantitative perspective, estimates and models of planktonic and benthic respiration in lakes, as derived from bottle and core incubations. Using data gleaned from the literature, empirical models of planktonic and benthic respiration are presented in which respiration is shown to depend primarily on lake trophic conditions and, secondarily, on temperature and other factors such as dissolved organic carbon loading and community structure. The possibilities and limits of alternative methods of measuring lake respiration and its components are discussed. Extrapolations of the findings to the globe suggest that about 62–76 Tmol of carbon are respired annually in the world's lakes, exceeding the estimated gross primary production of these ecosystems. Lakes thus contribute significantly to the oxidation of land-derived organic carbon.

7.1 Introduction

Respiration in lakes recycles organic carbon arising from photosynthesis back to inorganic carbon. Prior to this transformation, the organic carbon is potentially available to support secondary production. The efficiency of primary and secondary production relative to respiration is an important feature of lakes and other aquatic ecosystems. The relative rates and locations of primary production and respiration also influence oxygen concentrations. Intensive respiration in sediments and isolated waters, such as the hypolimnion or in ice-capped lakes, often leads to oxygen depletion and even anoxia. Thus, the rates and controls of respiration are of central importance in lake ecosystems.

Respiration affects net balances of carbon in lakes. Many lakes are net heterotrophic, meaning respiration exceeds gross primary production (del Giorgio and Peters 1994). In this case, the partial pressure of carbon dioxide exceeds that in the atmosphere and so carbon dioxide diffuses out of the lake while oxygen is undersaturated and diffuses into the lake (Prairie *et al.* 2002). The loss of carbon dioxide to the atmosphere, however, does not imply that storage of carbon in net heterotrophic lakes is zero. Quite the contrary, lakes accumulate organic matter in sediments (Dean and Gorham 1998). This seeming contradiction that lakes are sources of carbon to the atmosphere and sedimentary sinks for organic carbon is reconciled by respiration of terrestrial organic inputs in lakes (del Giorgio *et al.* 1999). Thus, the carbon that supports respiration in lakes is derived from within-lake primary production as well as allochthonous sources in the watershed. Some lakes are net autotrophic, where gross primary production exceeds respiration. These draw inorganic carbon from the atmosphere and store sediment organic carbon. Thus, both net autotrophic and net heterotrophic lakes are sinks for atmospheric carbon when considered on a watershed scale.

The purpose of this chapter is to summarize measured rates, variation in rates among lakes, and factors regulating respiration in lakes. Methods of measuring respiration are not reviewed in detail but common approaches are mentioned briefly to give the reader context for the discussion of rates. Rates of planktonic and sediment respiration are presented from literature data on oxygen consumption obtained from studies using bottle and core incubations. The data are used to develop general empirical models of planktonic and sediment respiration as a function of lake trophic conditions. These models are applied to generate a global estimate of respiration in lakes. Anaerobic respiration is also important in many lakes but only considered briefly here. Fenchel (1998) and King (Chapter 2) provide an excellent introduction to anaerobic processes. Wetzel (2001) reviews the general significance of anaerobic respiration in lakes. A limitation of traditional methods for estimating respiration is that changes in oxygen or carbon dioxide of enclosed communities are measured and extrapolated over time and space for an entire lake. Diel cycles of oxygen can be measured using *in situ* instruments. We provide an example of respiration derived from this "free water" approach and compare estimates from *in situ* and enclosure approaches.

7.2 Planktonic respiration

7.2.1 Methods

Consumption of oxygen in dark bottles has been the main approach to measuring planktonic respiration. Alternative methods have been used, including measuring activity of the electron transport system (del Giorgio 1992), changes in the partial pressure of carbon dioxide (Davies *et al.* 2003), and changes in oxygen isotope ratios (Bender *et al.* 1987). We primarily discuss dark bottle oxygen consumption because the data synthesis provided below is based on measurements made with this method. We also briefly consider the application of some of the other methods to lake systems.

In the dark bottle method, plankton communities are enclosed in bottles and replicate measurements of the initial and final concentrations of oxygen are made. Because the method depends on measuring small differences in oxygen concentrations, precision and accuracy are important.

Much effort has focused on improving standard oxygen measuring techniques to enhance accuracy (e.g. Carignan *et al.* 1998) and facilitate the ease of replicate measurements in order to improve precision (e.g. Roland *et al.* 1999). In addition, the availability of very sensitive instruments for oxygen measurement, such as dual-inlet mass spectrometry, allows shorter incubation intervals (del Giorgio and Bouvier 2002). With improvements in methods, incubations of a few hours rather than a full day are sufficient to estimate rates even in oligotrophic lakes (e.g. Carignan *et al.* 2000). Short incubation reduces artifacts associated with enclosing microbial communities in bottles and provides results over timescales commensurate with other measures of microbial activity such as primary and bacterial production assays.

Differences in oxygen concentration at the beginning and end of a dark bottle assay represent net oxygen consumption from all processes occurring in the bottle over time. Since bottles are incubated in the dark, it is assumed no oxygen production occurs. Oxygen consumption may arise from abiotic chemical oxidation as well as biotic reactions. Chemical oxidation reactions are primarily light driven. The aggregate of these reactions is referred to as photooxidation, and this process is presumed to cease during dark incubations. Graneli *et al.* (1996) directly compared measurements of photooxidation and dark respiration in Swedish lakes that varied in dissolved organic matter and found that photooxidation was <10% of dark respiration. Thus, measurements of dark respiration are probably not compromised by oxygen consumption due to photooxidation reactions initiated in the light that might continue during some portion of the dark incubation.

Extrapolations of bottle respiration measurements typically make the assumption that oxygen consumption measured in the dark is equivalent to oxygen consumption that occurs in the light. However, oxygen consumption in the light can be higher than in the dark because of photorespiration, the Mehler reaction (a photoreduction of O_2), and light-stimulation of normal cytochrome oxidase respiration (also referred to as dark or mitochondrial respiration). In addition, a process known as the alternative oxidase respiration may consume oxygen in the light and the dark (Luz *et al.* 2002, see also Raven and Beardall, Chapter 3). Photorespiration occurs because of oxygen accumulation and a relative decline of carbon dioxide as photosynthesis proceeds such that the primary carbon fixation enzyme (ribulose-1,5-*bis*-phosphate carboxylase/oxygenase) no longer fixes carbon dioxide but instead consumes oxygen (see also Raven and Beardall, Chapter 3). The Mehler reaction involves electron tranport among photosystems 1 and 2 with concurrent oxygen reduction (Falkowski and Raven 1997). Finally, enhanced cytochrome oxidase respiration occurs in the light for actively growing phytoplankton in cultures (Grande *et al.* 1989; Weger *et al.* 1989). The overall effect of these mechanisms that result in light-enhanced oxygen consumption on measurements of respiration is poorly understood and is an area of active research (Luz *et al.* 2002). Underestimates of respiration in dark incubations will likely be related to the relative importance of autotrophs and their growth rate in a given sample as well as to contemporaneous environmental conditions such as light exposure and temperature (see also Williams and del Giorgio, Chapter 1).

Following increases in carbon dioxide is a feasible means for measuring respiration but is complicated by potential changes in the components of dissolved inorganic carbon (DIC)—aqueous carbon dioxide, bicarbonate, and carbonate. Davies *et al.* (2003) describe a method applicable to lakes for measuring changes in the partial pressure of carbon dioxide (pCO$_2$) in a headspace created in a sealed sample. The technique involves measuring pCO$_2$ over time along with at least one measurement of the total DIC. If carbonate alkalinity is constant during sample incubation, DIC can be calculated from pCO$_2$ for every time point. Respiration is then estimated from the change in DIC over time. This method has not yet been extensively applied, but offers the opportunity to estimate gross primary production, respiration, and net community production (Davies *et al.* 2003) as currently done with light and dark bottle measurements of oxygen. Combining this method with oxygen techniques would, in theory, allow measurement of both carbon and oxygen changes as a consequence of photosynthesis and respiration and provide direct estimates of photosynthetic and respiratory quotients.

Another approach to production and respiration measurements involves spiking samples with H$_2$^{18}O. This method has been used in a few lakes (Luz *et al.* 2002). The advantage of this technique is that gross primary production and respiration can be estimated in the light allowing evaluation of the assumption necessary with other methods that respiration in the light equals respiration in the dark. The method is technically demanding and requires measuring oxygen isotopes and estimating fractionation for a number of processes related to biological oxygen consumption and production as well as abiotic exchanges of oxygen (e.g. Luz *et al.* 2002).

7.2.2 Data compilation

We compiled data from the literature for lake studies that reported rates of planktonic respiration based on dark bottle oxygen consumption along with the independent variables chlorophyll *a*, total phosphorus, and dissolved organic carbon. Some studies used other methods to measure respiration such as DIC accumulation in dark bottles (e.g. Graneli *et al.* 1996), and we excluded these to obtain a methodologically consistent dataset. We only used data from studies more recent than those of del Giorgio and Peters (1993) who summarized data from the literature on mean respiration in 36 lakes. We found seven studies that reported mean rates for lakes along with at least some of the independent variables—del Giorgio and Peters (1994), Pace and Cole (2000), Duarte *et al.* (2000), Carignan *et al.* (2000), Biddanda *et al.* (2001), Biddanda and Cotner (2002), and Berman *et al.* (2004). Our search was not exhaustive but was extensive. Keywords in titles and abstracts were searched electronically and numerous individual articles were reviewed. In addition, we visually reviewed all article titles and abstracts over several years in a few journals. A number of the studies noted above measured respiration in multiple lakes and so an aggregate 70 values of mean planktonic respiration were derived from the literature.

The studies were all of north temperate lakes sampled during the May–September period with the exception of Duarte *et al.* (2000) where lakes in northeast Spain were sampled from December through May. Rates in the literature studies are all derived from surface, mixed-layer samples, and means are based on as few as one sampling time (e.g. Duarte *et al.* 2000), 3–4 samples from approximately monthly observations (e.g. del Giorgio and Peters 1994; Carignan *et al.* 2000), 15–17 samples based on weekly observations (Pace and Cole 2000). Reservoirs and saline lakes sampled by Duarte *et al.* (2000) were excluded from the final data, because respiration in these systems tended to be substantially lower and higher, respectively than in the freshwater lakes. Means from Pace and Cole (2000) were derived from one reference and three lakes manipulated with additions of nitrogen and phosphorus over a number of years. We averaged all data for the reference lake to generate a single mean (years 1991–7). We averaged data for 2 years prior to nutrient loading manipulations for the other three lakes (1991–2) to generate three means and then treated each year of nutrient manipulation (loading rates varied among years) as a separate mean. This generated 19 mean values overall representing 28 lake-years of measurements.

Typically, comparisons of production, nutrient cycling, or biomass among lakes reveal ranges in variation of over a 1000-fold. Mean values of planktonic respiration only varied from 0.029 to 6.73 mmol O_2 m^{-3} h^{-1}, a factor of 265. This modest range must underrepresent actual variation among lakes, because of the limited number of studies and the focus on temperate lakes in summer. A more global survey of lakes would probably reveal systems with lower respiration because of low productivity and low temperatures, and some with higher respiration because hypereutrophic lakes (chlorophyll less than 75 mg m^{-3}) are not well represented in the dataset. However, the limited range indicates an important feature of planktonic respiration. Respiration is less variable than other variables measured in aquatic systems.

7.2.3 Models of planktonic respiration

There are few models of planktonic respiration that are comparable to well-developed models

of primary production based on photosynthesis—irradiance relationships (Platt and Jassby 1976) as well as approaches that combine photosynthetic parameters with depth distributions of primary producers to estimate rates over large scales such as the global ocean. Models for planktonic respiration are largely empirical and are based on features such as primary productivity as, for example, in the global analysis of aquatic ecosystems inclusive of lakes provided by Duarte and Agusti (1998).

Temperature
Because lake planktonic respiration studies have focused on surface water in summer of temperate lakes, there has been relatively little consideration of the effects of temperature on rates. Nevertheless, Carignan *et al.* (2000) determined a relationship between volumetric respiration and several variables, including temperature. This multiple regression relationship (their equation (10) of table 3) is useful in that it allows one to isolate the influence of temperature on respiration while statistically keeping all other variables constant. Over their observed temperature range of 11–22.5°C, the log–log slope coefficient of temperature is 1.28, implying a Q_{10} of about 2.3 in that temperature range. If this slope coefficient of 1.28 remains valid even for lower temperatures, it would suggest an even higher Q_{10} at low temperatures. This prediction needs to be confirmed by additional observations across a broader temperature range, but the same pattern has been observed for sediment respiration, as discussed below.

Lake trophic conditions and organic matter loading
Plankton respiration (PR) increases in concert with increases in chlorophyll, phosphorus, and dissolved organic carbon (Fig. 7.1). Mean lake chlorophyll *a* accounts for 71% of the variance in respiration (Fig. 7.1(a)). A similar relationship is observed with total phosphorus although r^2 was higher (0.81) for this relationship based on fewer observations ($n = 62$, Fig. 7.1(b)). The equations relating PR (mmol O_2 m^{-3} h^{-1}) to chlorophyll *a* (Chl, mg m^{-3}) and total phosphorus (TP mg m^{-3}) are:

$$\log_{10} \text{PR} = -0.81 + 0.56 \log_{10} \text{Chl} \qquad (1)$$

$$\log_{10} \text{PR} = -1.27 + 0.81 \log_{10} \text{TP} \qquad (2)$$

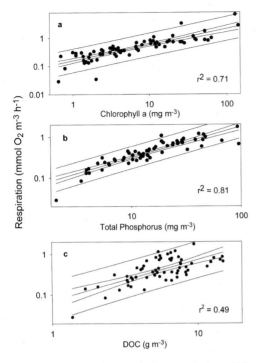

Figure 7.1 Relationships between planktonic respiration and (a) chlorophyll *a* (b) total phosphorus, and (c) dissolved organic carbon (DOC) for lakes. Regressions are model 1 least square plots with 95% confidence intervals for the regression and for individual predictions.

These two relationships reflect a typical pattern of lakes where primary production is strongly nutrient limited and closely related to total phosphorus inputs and concentrations. The large-scale comparison in Fig. 7.1 indicates that planktonic respiration is strongly coupled with variation in phytoplankton biomass and primary production.

Equation 1 compares well with the relationship established by del Giorgio and Peters (1993) from earlier literature data. They reported a regression for PR in units of $mg\,C\,m^{-3}\,d^{-1}$ based on chlorophyll *a* with a slope of 0.65 and an intercept of 1.65. Transforming the data underlying equation (1) into commensurable units yields the same intercept of 1.65 observed by del Giorgio and Peters (1993). Thus the two independent relationships overlap closely, differing only slightly in slope (0.65 versus 0.56) and give similar predictions, especially in the range of 0.5 to 5 mg chlorophyll m^{-3}.

Dissolved organic carbon (DOC) in lakes is derived both from watershed sources and within-lake primary production. If allochthonous inputs of DOC are significant to lake carbon cycling and supplement ecosystem metabolism, DOC should be positively related to PR. This is the case for 63 lakes (Fig. 7.1(c)) where there is a positive relationship between DOC ($g\,m^{-3}$) and PR ($r^2 = 0.49$).

$$\log_{10} PR = -1.15 + 1.01 \log_{10} DOC \qquad (3)$$

The positive relationship of PR with DOC reflects, in part, increased microbial metabolism associated with increased inputs of DOC to lakes. However, DOC is not simply the result of watershed inputs but also arises from phytoplankton and littoral primary producers, the relative contributions of which are still not well known. Hence, the relationship is complex and a more interesting test of the positive effect of watershed-derived DOC on PR would be to compare lakes with differing DOC concentrations (and loadings) but similar levels of total phosphorus or chlorophyll *a*.

Both chlorophyll and DOC are significant variables in a multiple regression ($r^2 = 0.76$, $n = 63$).

$$\log_{10} PR = -0.92 + 0.41 \log_{10} Chl$$
$$+ 0.30 \log_{10} DOC \qquad (4)$$

The *t*-values associated with each variable indicate chlorophyll is highly significant while DOC explains a modest but also significant amount of the residual variation (chlorophyll $t = 7.96$, $p < 0.0001$; DOC $t = 2.35$, $P = 0.022$). Chlorophyll and DOC, however, are correlated ($r = 0.71$, $p < 0.0001$) and so this relationship may be less suitable for predicting PR and associated uncertainty than the single variable relationship with chlorophyll (equation 1).

Other factors influencing planktonic respiration
There have been relatively few detailed studies on respiration and depth in lakes. The primary reason is that respiration becomes difficult to measure at low rates. With the declines in temperature and primary productivity associated with increases depth in lakes, respiration rates probably must also decline. However, bacterial metabolism contributes significantly to respiration and partially compensates for the expected declines with depth. Numerous efforts have been made to partition oxygen consumption in the hypolimnion of lakes among

water and sediments because of the importance of hypolimnetic oxygen deficits during seasonal stratification. These studies either involve modeling (e.g. Livingstone and Imboden 1996) or, when direct measurements of hypolimnetic PR have been attempted, long incubations of organism concentrates (e.g. Cornett and Rigler 1987). Refinements in methods for measuring respiration discussed above provide an opportunity to now make direct measurement over reasonable incubation times.

While planktonic respiration is strongly related to nutrient and organic matter loading as well the biomass of primary producers, food web structure can also influence respiration. Pace and Cole (2000) reported strong food web effects on respiration from a series of whole lake manipulation studies. Fish communities in three lakes were shifted to promote the dominance of either piscivorous or planktivorous fish and then the lakes were fertilized with nutrients. Predator–prey interactions arising from the top of the food web either strongly limited large crustacean grazers or alternatively tended to promote dominance of large zooplankton through trophic cascades (Carpenter *et al.* 2001). Nutrient additions were carried out over 5 years in each lake and dark bottle respiration was measured. Planktonic respiration differed among the fertilized lakes at the same nutrient loading and, over the 15 lake-years represented in this study, was inversely related to the average size of crustacean zooplankton (Fig. 7.2). Gross primary production, system respiration, and net ecosystem production as measured

by free water methods (discussed below) were also strongly affected by differences in food web structure in these experimental lakes (Cole *et al.* 2000).

7.3 Sediment respiration

The sediment–water interface is the site of some of the most intense biological activity of freshwater systems (Krumbein *et al.* 1994). The large exchange surface area provided by loosely compacted sediment particles, the organic nature of the sediments, and its physical stability all act synergistically to produce an extremely active milieu. In comparison to the water column, sediments are typically 1000-fold richer in both nutrients and carbon, and thereby provide a highly concentrated growth medium for heterotrophic bacteria. Correspondingly, bacterial abundances and production rates are typically about 2–3 orders greater than in the water column, a phenomenon recognized nearly half a century ago (Kuznetsov 1958). Depending on the transparency of the water column above the sediments, the high benthic metabolic activity is differently divided between anabolic and catabolic processes. However, because most of the sediments are below the euphotic zone, respiration is by far the dominant metabolic process in sediments. In his treatise, Hutchinson (1957) argued that oxygen consumption within lake hypolimnia can be accounted for by water column respiration, and therefore, concluded that sediment respiration represented only a minor portion of the hypolimnetic metabolism, a view at odds with recent literature. For example, Lund *et al.* (1963) estimated that about half of the hypolimnetic metabolism of Lake Windermere was benthic, a conclusion upheld by Cornett and Rigler (1987) who found between 30 and 80% of the total hypolimnetic oxygen consumption was attributable to sediment respiration in a series of Québec lakes.

7.3.1 Methods

Historically, sediment respiration has been determined mostly using measurements of oxygen consumption (Hayes and Macaulay 1959; Pamatmat 1965) although some early work also used carbon

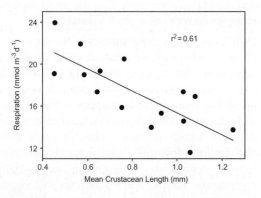

Figure 7.2 Relationship between mean crustacean length and mean planktonic respiration in nutrient fertilized experimental lakes.

dioxide accumulation (Ohle 1952). Oxygen consumption and carbon dioxide release are not always well coupled (Rich 1975, 1979), particularly in dystrophic lakes and peat bogs (see Roehm, Chapter 6), and the degree of decoupling is usually attributed to anaerobic respiration pathways (methanogenesis or use of alternate terminal electron acceptors such as sulfates, nitrates, and metal oxides, see also King, Chapter 2). In systems where anaerobic metabolism is negligible, oxygen consumption and inorganic carbon release are both adequate measures of benthic respiration, provided carbon dioxide release and consumption induced by $CaCO_3$ precipitation and dissolution, respectively, are taken into account (Graneli 1979; Andersen *et al.* 1986, and see Hopkinson and Smith, Chapter 6, and Middelburg *et al.*, Chapter 11). Most reports of benthic respiration are based on oxygen consumption.

Benthic respiration rates has been estimated by several direct and indirect techniques including: sediment core incubations (Hargrave 1969*a*; Rich 1975; den Heyer and Kalff 1998; Ramlal *et al.* 1993; Schallenberg 1993), benthic chambers (Pamatmat 1965, 1968; James 1974; Cornett 1982), vertical profiles of decomposition products in interstitial sediment water (Carignan and Lean 1991), or as the difference between total and water column metabolism in lake hypolimnia (Cornett and Rigler 1987). Because these methods encompass very different degrees of spatial integration and because of the notorious spatial heterogeneity of lake sediments, it is difficult to compare rates of benthic respiration obtained from these different methods.

James (1974) compared sediment respiration rates obtained from cores and benthic chambers and concluded that cores tended to underestimate respiration rates by an average of 35%, although Campbell (1984) showed that core incubations can estimate whole-lake sediment respiration within 10%. Given that most literature values of benthic respiration were derived from core incubations, models of benthic respiration, therefore, carry the potential bias of being slight underestimates.

A more spatially integrative measure of sediment respiration has recently been proposed by Livingstone and Imboden (1996). The effects of

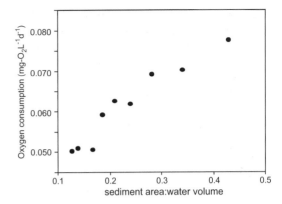

Figure 7.3 Relationship between oxygen consumption as a function of the sediment area: water volume ratio for each 1-m hypolimnetic strata in Lac Fraser, Québec (Canada).

sediment respiration can be observed through the nearly ubiquitous nonlinearities of hypolimnetic oxygen profiles, with lower concentrations near the sediments. The extent of these sharper reductions close to the sediments is largely determined by the increasing area of sediments in the deeper strata of the hypolimnion. Livingstone and Imboden (1996) proposed a simple modeling approach to estimate the relative contributions of benthic and water column metabolism to the total oxygen consumption of the hypolimnion by relating the rate of decline of oxygen at a given depth to the sediment surface area: water volume at that particular depth. Application of this method to a vertical oxygen profile from Lac Fraser (Québec) shows a very strong relationship, the intercept and slope of which correspond to the water column and benthic respiration rates, respectively (Fig. 7.3). If the general applicability of this approach can be firmly established, it may provide a more useful measure of benthic respiration, by implicitly integrating differences in respiration at different depths within the hypolimnion and by avoiding potential experimental artifacts inherent to incubations in closed containers.

7.3.2   Data compilation

The literature on sediment respiration in freshwater systems is much sparser than for marine or estuarine waters, particularly in recent years. As for pelagic respiration, we synthesized the published literature and obtained values on sediment respiration rates from a variety of freshwater systems, located mostly in North America or Europe. The lakes spanned the entire trophic range, from the arctic oligotrophic Char Lake (Canada, Welch and Kalff 1974) to eutrophic Lake Alserio (northern Italy, Provini 1975) and encompassed very different types of sediments (organic matter content varied from 10 to 40%). Sediment respiration varied only 20-fold, from 1.6 to 33 mmoles $O_2\,m^{-2}\,d^{-1}$ (average is 10.9), similar to the range observed by den Heyer and Kalff (1998). For consistency, models developed from the literature data were restricted to measurements made in core incubations.

7.3.3 Models of benthic respiration

Existing models of benthic respiration can be broadly categorized as those based on sediment properties and those based on water column or whole system properties. In comparison with models of planktonic respiration, the predictive ability of existing sediment respiration models has been rather poor and their generality tested more regionally.

Effect of temperature
The earliest general model of benthic respiration was originally proposed by Hargrave (1969*a*), who argued that temperature was the main driving variable. Graneli (1978) suggested that temperature and oxygen concentration were more important factors in regulating sediment respiration than differences in lake trophic status and its associated lake primary productivity. Our compilation of sediment respiration rates from the literature gives, at first sight, credibility to this claim. Temperature alone has been shown to induce changes in respiration rates of nearly one order of magnitude (Hargrave 1969*b*; Graneli 1978; Ramlal *et al.* 1993) within the 4–25°C temperature range, in sediments incubated at different temperatures or followed through an annual temperature cycle. Thus, temperature-induced variations in sediment respiration of a given sediment can be nearly as large as among-lake differences (1.5 and 1 orders of magnitude range, respectively). Several authors have shown that the temperature effect

is more pronounced at low temperatures than in warm environments (Hargrave 1969*b*; Graneli 1978; den Heyer and Kalff 1998). These relationships are often best described on a double logarithmic scale, and therefore, the relative increase in respiration rate with temperature (i.e. a Q_1) decreases with increasing temperature and can be expressed by the simple function:

$$Q_1 = 1 + b/t \tag{5}$$

where b is the slope of the log–log relationship between respiration rate and temperature (t in °C). An analysis of b values derived from the literature yields an average value of 0.65 ± 1 and Fig. 7.4 illustrates how the relative rate of increase in respiration rate per degree centigrade decreases with temperature. Most of the nonlinearity in this relationship occurs below 10°C, suggesting that changes in temperature are much more important for hypolimnetic sediments than in warm littoral zones. For example, increasing the temperature of hypolimnetic sediments from 4 to 5°C has the same relative effect on benthic respiration as increasing the temperature of epilimnetic sediments from 20 to 25°C. For comparison, Fig. 7.4 also illustrates the inferred influence of temperature on planktonic respiration derived in the previous section. It appears that, at least for warm waters, the influence of temperature on respiration is greater for the plankton than it is for sediments.

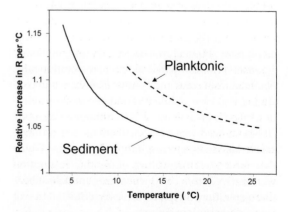

Figure 7.4 Temperature dependence of respiration expressed as relative increase in respiration rate per degree Celsius. Solid line is for sediment respiration, dashed line corresponds to planktonic respiration.

Assuming that the temperature relationship holds generally, sediment respiration (SR) rates can be normalized to a standard temperature (e.g. 10°C) by using

$$SR_{std\ 10°C} = SR/(Temp/10)^{0.65} \qquad (6)$$

to allow more direct comparisons among sediments of different systems. Application of this standardization to our data compilation showed that while individual values changed as much as threefold, it affected the overall range of observed benthic respiration rates very little, suggesting that the strong relationship between temperature and benthic respiration rates of Hargrave (1969a) may be more the result of cross correlations between temperature and other system properties (such as lake productivity) than only the metabolic enhancement of respiration at high temperatures.

Organic matter loading, primary production, and lake trophic status

Ultimately, benthic respiration below the photic zone is fueled by organic material sinking from the pelagic zone or transported laterally from shallow sediments. Despite this functional coupling, the link between short-term organic carbon flux and sediment respiration has not been well established largely because of the apparent delay between when the organic particles settle and decompose. Indeed, the relative invariance of benthic respiration over the seasons when large changes occur in the organic matter flux has led several authors to argue that benthic respiration is the reflection of the organic matter loading integrated over several years rather than the rapid decomposition of freshly sedimented material (Lasenby 1975; Graneli 1978; Linsey and Lasenby 1985). The implications of this dampening effect of the accumulated surface sediments are important: reductions in lake productivity following eutrophication control measures may only translate into a response in benthic metabolism after several years (Graneli 1978; Carignan and Lean 1991). In lakes with shallow hypolimnia, which constitute the majority of eutrophic lakes (Kalff 2002), sediment respiration is responsible for a large fraction of the total hypolimnetic oxygen consumption (Cornett

and Rigler 1987). Hypolimnetic anoxia may, therefore, continue long after nutrient loading reduction targets have been achieved, a phenomenon unfortunately observed in many lake restoration efforts (Sas 1989). Baines and Pace (1994) have shown that organic particle flux is well correlated to lake trophic status as measured by epilimetic chlorophyll concentration. In steady-state conditions, benthic respiration should therefore be well correlated with primary production and its determinants, particularly given that the fraction of the primary production carbon exported to the deeper layers appears only modestly reduced in highly productive systems (Baines and Pace 1994). A closely related hypothesis was first proposed by Hargrave (1973), who suggested that benthic respiration should be expressed as a power function of the ratio of areal primary production (PP) to mixing depth (Z_m)

$$SR = \alpha(PP/Z_m)^{\beta} \qquad (7)$$

Graneli (1978) further tested this pattern in a series of five Swedish lakes and showed that while the relationship was strong, Hargrave's model systematically overpredicted the measured sediment respiration. This compound variable (PP/Z_m) is equivalent to the volumetric primary production averaged over the epilimnion, a variable known to be well predicted from phosphorus concentration (e.g. del Giorgio and Peters 1994). Because phosphorus concentration data are more generally available compared to primary production estimates, we further examined the generality of the relationship between sediment respiration and variables indicative of lake trophic status.

The analysis of our data compilation revealed that total phosphorus concentration was a very good predictor of sediment respiration (standardized to 10°C according to equation (6)) explaining 69% of the variance in sediment respiration and yielded the following empirical equation:

$$\log_{10} SR_{std\ 10°C} = 0.17 + 0.58 \log_{10} TP, \qquad (8)$$

where SR is in mmoles $O_2\ m^{-2}\ d^{-1}$ and total phosphorus (TP) is in milligram per meter cube (Fig. 7.5). Because total phosphorus is also a well-known predictor of chlorophyll concentrations and primary production in lakes (Dillon and Rigler 1974 among

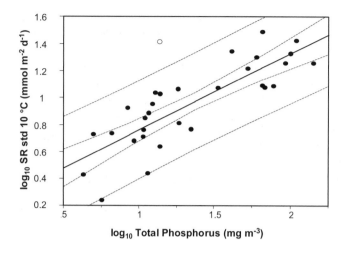

Figure 7.5 Log–log relationship between standardized sediment respiration rates (SR, mmoles m^{-2} d^{-1}) and total phosphorus concentrations (mg m^{-3}). Individual data points were obtained from Provini (1975), Ramlal *et al.* (1993), Schallenberg (1993), Hayes and MacAulay (1959), Graneli (1978), Welch and Kalff (1974), Lasenby (1975), and Hargrave (1972). Open circle (Lake Southport, Hayes and MacCauley 1959) excluded from regression line.

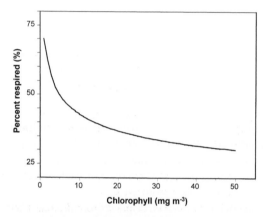

Figure 7.6 Inferred relationship between the fraction of carbon flux (as %) to sediments that is eventually respired, as a function of chlorophyll concentration (mg m^{-3})

many others, Smith 1979, del Giorgio and Peters 1994), sediment respiration can also be expressed as a function of chlorophyll (mg m^{-3}) or primary production:

$$\log_{10} SR_{std\ 10°C} = 0.63 + 0.40 \log_{10} Chl, \qquad (9)$$

$$\log_{10} SR_{std\ 10°C} = 0.093 + 0.47 \log_{10} PP, \qquad (10)$$

where primary production (PP, mg C m^{-3} d^{-1}) is volumetric primary production averaged over the euphotic zone (del Giorgio and Peters 1994). The shallow slopes of all three relationships imply that SR does not increase proportionately with any measure of lake trophic status. The consequences

of this disproportionality can be further explored by combining these trends with the work of Baines and Pace (1994) who showed that the vertical flux of carbon can be well estimated ($r^2 = 0.83$) from chlorophyll concentration. Assuming a respiratory quotient of 0.9 (Graneli 1979, although the trend is not affected by the particular choice of RQ), the fraction of the vertical flux of carbon that, at steady-state, will be respired in the sediments (%C_{resp}) will be a declining function of lake trophic status. Expressed as a function of chlorophyll, the resulting equation is:

$$\%C_{resp} = 71.4\ Chl^{-0.22} \qquad (11)$$

This inferred relationship, illustrated in Fig. 7.6, suggests that, in oligotrophic lakes, about 70% of the carbon raining down to the sediments will be respired while this fraction is considerably reduced in more eutrophic systems (to about 40% in lakes with 20 mg m^{-3} chlorophyll). Long-term carbon burial is, therefore, increasingly important in more productive systems, a conclusion also reached by Hargrave (1973).

Other factors influencing sediment respiration
While lake productivity and temperature are clearly the main determinants of sediment oxygenic respiration, other factors are known to influence sediment respiration in lakes, albeit to lesser extents. Sample depth, sediment organic matter content, ambient oxygen concentrations are among the

most cited factors influencing sediment respiration although general tests of their importance are rare. Lake acidification has also been shown to reduce respiration in some instances (Andersson *et al.* 1978).

Unlike water column respiration processes, sediment respiration is suppressed when ambient oxygen concentrations decrease (e.g. Edberg 1976; Graneli 1978; Cornett and Rigler 1984), particularly when they drop below 30–60 mmol O_2 m^{-3} (Anderson *et al.* 1986). This is probably the result of the supply of oxygen to the sediment/water interface becoming limited by the reduced diffusion. Whether this reduced oxygen consumption is fully compensated by increased anaerobic decomposition is unclear (Bédard and Knowles 1991). However, we suggest that the failure of empirical models such as equations (4)–(6) to incorporate this first-order concentration effect may not overly compromise their utility.

The influence of the depth at which sediment cores are taken has also been noted by some authors (e.g. Campbell 1984; den Heyer and Kalff 1998) but the effect appears significant only when comparing sites differing in depth by orders of magnitude. Given that very deep lakes are usually oligotrophic, it may be that the depth effect is a proxy for a trophic status effect. Indeed, variations among depths within individual lakes are nonexistent or modest (Lasenby 1975; Newrkla 1982). In our data compilation, adding site depth as an additional independent variable was not significant ($p > 0.05$) once lake trophic status (as total phosphorus) was entered in the model, providing further indication that the depth acted as a surrogate in earlier models.

7.4 Free water respiration

By frequent *in situ* measurements of oxygen dynamics over diel cycles, it is possible to calculate respiration. This so-called "free water" approach (Odum 1956) has the advantage of not requiring enclosure and incubation. The method also measures the contribution of all components, from bacteria to fish, to total system respiration and oxygen consumption. An open question discussed further below is the vertical and horizontal scales over which the estimate of respiration is made. This technique potentially

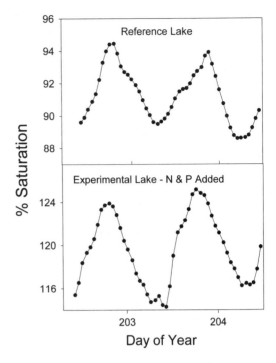

Figure 7.7 Diel oxygen dynamics as percent saturation in a reference and experimental lake (Paul and Peter Lakes, respectively, Michigan, USA). Details in text.

provides an integrated measurement of pelagic and sediment respiration in lakes.

Respiration can be estimated from the night-time decline in oxygen once external exchanges (e.g. with the atmosphere) are accounted for. Figure 7.7 presents an example from a reference lake and an experimental lake fertilized with nitrogen and phosphorus of diel oxygen dynamics measured *in situ* with autonomous instruments such as rapid-pulsed oxygen electrodes (details of study in Cole *et al.* 2000). Oxygen is undersaturated in the reference lake and diel dynamics are modest while in the experimental lake oxygen is ovesaturated and there is a large diel excursion (Fig. 7.7). Under these conditions, the decline of oxygen at night is a function of respiration and net diffusion to or from the atmosphere. Oxygen diffusion at the base of the mixed layer can be ignored (Cole and Pace 1998). Atmospheric gains or losses of oxygen were calculated as the difference between measured and saturated oxygen times a coefficient of gas exchange. The gas exchange coefficient is either modeled or measured directly (e.g. Clark *et al.* 1995; Cole and Caraco 1998; Wanninkhof and McGillis 1999). Diel cycles of oxygen can also be used to estimate gross primary production based on increases in oxygen during the day and the assumption that night and day respiration are equal.

Respiration calculated from diel oxygen dynamics in the mixed layer of lakes should represent an integrated response of the surface mixed layer and sediments on the edge of the lake that lie within the mixed layer. Hence, the pelagic and sediment respiration are included in the estimate. In the small lakes (1–3 ha) illustrated in Fig. 7.8, free water respiration is greater than dark bottle respiration. For example in the unfertilized, reference lake, weekly measures of dark bottle respiration were made during the same time as diel oxygen dynamics were measured (Fig. 7.8). Dark bottle respiration averaged 8.8 mmol m^{-3} d^{-1} while free water respiration averaged 30.6 mmol m^{-3} d^{-1}. While dark bottle respiration might underestimate planktonic respiration, the free water method indicates that benthic respiration must be about twofold higher than planktonic respiration in the reference lake. Thus, in small water bodies much of the overall system metabolism can occur in sediments, particularly shallow littoral sediments where benthic respiration is stimulated by primary production of macrophytes and periphyton.

Such discrepancies between littoral and pelagic respiration can create spatial heterogeneities in gas concentrations that must be contended with. Thus, while the free-water method offers a greater integration of all the respiring components than bottle measurements, the approach is not without difficulties. For instance, continuous mapping of surface P_{CO_2} shows that the spatial heterogeneity in the distribution of respiration can be substantial, even in small lakes. Figure 7.9 illustrates the spatial distribution of the night-time increase in surface P_{CO_2} in a small embayment (0.53 km^2) of Lake Memphremagog (Québec). Over the bay, the increase averaged a modest 28 μatm P_{CO_2} but some littoral areas increased by as much as 140 μatm. As expected, the amplitude in the daily variation is much greater in the shallow littoral zones. Clearly, horizontal diffusivity is not sufficiently large to homogenize gas distribution at the scale necessary to use the free-water method without consideration of spatial heterogeneities. More interestingly, the night-time increase was significantly smaller in the deepest zone of the bay where samples are normally taken or autonomous sondes deployed. Thus, significant biases may be incurred if the autonomous sondes are placed only in the deepest area of the lake, because the metabolism originating from the more active littoral areas does not completely homogenize even over small horizontal areas.

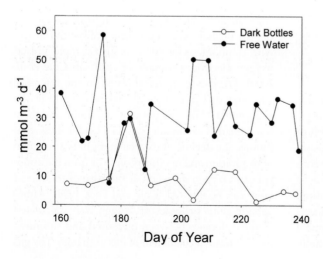

Figure 7.8 Comparison of free water and bottle estimates of respiration over the summer season in Paul Lake (Michigan, USA). Details in text.

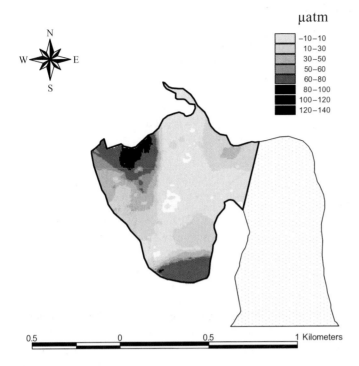

μatm

	−10–10
	10–30
	30–50
	50–60
	60–80
	80–100
	100–120
	120–140

N
W · E
S

0.5   0   0.5   1 Kilometers

Figure 7.9 Spatial distribution of the night-time increase in $p\mathrm{CO}_2$ in a small bay of L. Memphremagog (Québec) showing large heterogeneities.

7.5   Global estimate of lake respiration

Although freshwater systems occupy a small area relative to the surface area of the world's oceans, freshwaters are comparatively more productive, leading to higher pelagic and sediment respiration rates per unit volume and area, respectively. In addition, lakes have a much larger relative contact area with surrounding terrestrial ecosystems than oceans, and substantial terrestrially derived organic matter is oxidized during its freshwater transit (e.g. Richey *et al.* 2002). The planktonic and sediment respiration models presented above can be used to roughly estimate annual carbon processing in lakes at a global scale. While such calculations are necessarily fraught with large uncertainties and untested assumptions, they can nevertheless provide order-of-magnitude estimates useful for casting the role of freshwaters in the larger context of global carbon cycling.

Our calculations are based on several simple allometric relationships with lake surface area. Lake size is negatively related to lake abundance

(Meybeck 1995) and positively related to average depth (Patalas 1984). Large lakes also tend to have longer water residence times (Kalff 2002), a variable associated with reduced productivity on average (del Giorgio and Peters 1994). Figure 7.10 depicts changes in lake abundance, depth, and production (as chlorophyll *a*) across the logarithmic lake size classes of Meybeck (1995). While about two-thirds of the total lake surface area is occupied by lakes less than 320 km², more than 70% of total volume is in very large lakes (>320 km²). This asymmetry has large implications for the distribution of respiration across the lake size classes.

For shallow lakes (less than or equal to 15-m mean depth), we calculated total water column respiration as the product of mean depth and the volumetric respiration rate estimated from the chlorophyll values (equation 2), to which we added sediment respiration, also a function of chlorophyll (equation 9). For large and deep lakes (greater than 15-m mean depth, we integrated planktonic respiration only for the upper 15 m, because respiration deeper in the water column is considerably lower than respiration

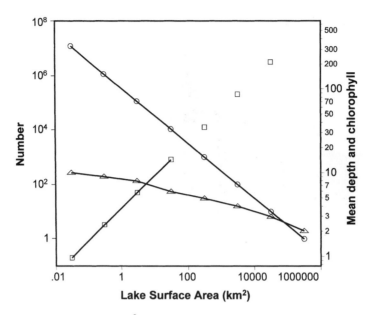

Figure 7.10 Global distribution of lake surface areas (km^2) used in the global estimate of lake respiration. Lake surface areas are arranged in logarithmic size classes following Meybeck (1995) and its relationships are shown, on the left axis, with lake total abundance (open circles, in numbers), and on the right axis, with lake mean depth (open squares, in m), and chlorophyll (open triangles, in mg m^{-3}), as used in our model estimates of global respiration. Note the line break for mean depth indicates that for large lakes unconnected mean depths points in the figure were not used in the calculations, see text for details.

in the uppermost layers. For example, in the deep (100 m) central basin of Lake Memphremagog, epilimnetic plankton respiration has been measured at 10.5 mmol m^{-3} d^{-1} (del Giorgio and Peters 1994) and average hypolimnetic plankton respiration at about 0.4 mmol m^{-3} d^{-1} (Cornett and Rigler 1987). Thus, our method of estimation, while empirical, has the advantage of providing a conservative estimate of respiration by not inflating respiration in the vast volume of water contained in the deep, cold layers of the relatively few very large lakes that dominate global lake volume (e.g. Lake Baikal). This is, however, a poor assumption in a number of ways as we discuss further below and results in what is almost certainly an underestimate.

Integration of the estimate of respiration over lake size classes yields a value of 62 TmolC a^{-1}, about one quarter of which occurs in the two largest lake size classes (Fig. 7.11 and Table 7.1). Globally, sediments are responsible for only about 6% of total lake R. Although our global R estimate is rather

small when compared to the total respiration of the oceans (del Giorgio and Duarte 2003), the importance of this figure is more revealing when compared to the gross primary production (GPP) that occurs in lakes. There are several published models that relate lake primary production to total phosphorus and chlorophyll. We used the relationship of Carignan *et al.* (2000), because it yields higher values for oligotrophic lakes than all other models, and thus provides an upper estimate of pelagic primary production. Assuming then that water column GPP follows the relationship of Carignan *et al.* (2000) with respect to chlorophyll, and that the depth of the euphotic zone is either equal to the mean depth of small lakes or 15 m for larger lake size classes, global pelagic GPP is estimated to be 51 TmolC a^{-1}. In all lakes there is some additional primary production from benthic and littoral communities, and this littoral production is especially important in small and oligotrophic lakes (Vadeboncouer *et al.* 2003). If we assume benthic primary production is

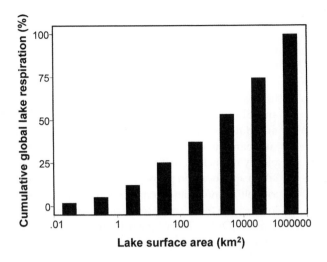

Figure 7.11 Cumulative global lake respiration (%) with increasing lake size classes (km^2).

Table 7.1 Range and averages for planktonic and sediment respiration and of global estimates derived from lake size classes. For the conversion of oxygen to carbon the calculation assumes $PQ = RQ = 1$

	Range	Average (median)
Field observations		
Planktonic respiration		
Volumetric (mmol O_2 m^{-3} d^{-1})	0.7–162	15.7 (10.3)
Areal (mmol O_2 m^{-2} d^{-1})	10–184	71.3 (61.5)
Sediment respiration (mmol O_2 m^{-2} d^{-1})		
Primary data	1.6–32	10.3 (9.7)
Standardized at the rate of 10°C	1.8–31	10.7 (9.2)

Global estimates

Lake surface area (km^2)	Global surface area (10^{12} m^2)	Respiration (Tmoles C a^{-1})		GPP (Tmoles C a^{-1})
		Planktonic	Sediment	
0.01–0.1	0.38	1.3	0.7	1.5
0.1–1	0.37	2.8	0.6	2.9
1–10	0.35	6.2	0.6	5.8
10–10^2	0.34	12.5	0.5	11.4
10^2–10^3	0.33	11.4	0.4	10.3
10^3–10^4	0.32	9.7	0.4	8.9
10^4–10^5	0.30	7.9	0.3	7.5
10^5–10^6	0.29	6.1	0.3	6.0
Sum	2.69	58	3.8	54

half of pelagic primary production in the smallest lakes and diminishes by a factor of 0.5 for successive lake size classes, then we obtain a global estimate of gross primary production of 54 Tmol C a^{-1}. The estimated difference between global lake respiration and global lake production using these assumptions is 8 Tmol C a^{-1}. This value of excess respiration is reasonably close to the 12 Tmol a^{-1} of global lake CO_2 evasion estimated by Cole et al. (1994) using the average partial pressure of carbon dioxide in lakes. This correspondence, however, is not fortuitous as we constrained our calculations to fall in the realm of the Cole et al. estimate. There are two aspects of comparing our respective estimates that need to be made explicit. The CO_2 evasion from lakes estimated by Cole et al. may have other sources beyond excess respiration. This would suggest the excess respiration should be lower than CO_2 evasion. The global lake area used by Cole et al. (1994) excluded the very large inland seas (e.g. Caspian, Aral) and is, therefore, not directly comparable with our estimate. Excluding our two largest lake size classes (yielding a total lake area of 2.1 × 10^{12} m^2 more commensurate with the Cole et al. value of 2 × 10^{12} m^2) further reduces the gap between global lake primary production and respiration in our estimate to 6 Tmol C a^{-1}.

The greatest problem with the assumptions we use is that they exclude respiration in the deeper waters of large lakes. This respiration is probably quite significant. For example, McManus et al. (2003) recently estimated the annual respiration for the hypolimnion of Lake Superior as about 0.9 Tmol of C, or 30 mmol m^{-2} d^{-1}. This estimate points to the significance of large lakes in the overall calculation of global lake respiration. If we apply the areal estimate for Lake Superior to all large lakes (the four largest lake size classes), this would increase our estimate of global lake respiration by about 14 Tmol C, to a total of 76 Tmol C a^{-1}. This may be a better estimate, but there are problems with concluding lake respiration is much higher. Indeed, our assumptions tend to provide a realistic but upper limit to global lake GPP because the relationship we used yields very high estimates of pelagic production relative to other models. Thus, greatly increasing respiration would result in an unreasonable excess of respiration over production. One possible explanation is that GPP is too low. We cannot, however, justify a larger value. A second and related problem with increasing the estimate is that the respiration increases mainly in the larger lake size classes. Respiration of >75 Tmol C a^{-1} leads to P:R ratios for large lakes of about 0.65 on an annual scale, which may not be consistent with estimated fluxes and concentrations of O_2 and CO_2. In summary, our calculations cannot at present be completely reconciled. Better global data and models of lake respiration and primary production are needed as well as information on fluxes that can be used to constrain these processes such as carbon dioxide evasion. Our estimates point to the critical lack of understanding of respiration in large lakes and how respiration varies with depth in lakes.

The global fluxes of carbon related to lake respiration are relatively small when compared to rates for the ocean and land. However, inland waters may still be significant as sites of active carbon cycling and carbon losses. Thus, the fluxes of carbon associated with lake respiration need to be accounted for when estimating net ecosystem production and carbon storage as well as evasion of the combined terrestrial and freshwater landscape especially in regions rich in lakes.

References

Andersson, G., Fleischer, S., and Granéli, W. 1978. Influence of acidification processes in lake sediments. *Verh. Int. Verein. Limnol.*, **20**: 802–807.

Anderson, G. L., Hall, P. O. J., Iverfeldt, Å., Rutgers van der Loeff, M. M., Sundby, B., and Westerland, S. F. G. 1986. Benthic respiration measured by total carbonate production. *Limnol. Oceanogr.*, **31**: 319–329.

Baines, S. B. and Pace, M. L. 1994. Sinking fluxes along a trophic gradient: patterns and their implications for the fate of primary production. *Can. J. Fish. Aquat. Sci.*, **51**: 25–36.

Bédard, C. and Knowles, R. 1991. Hypolimnetic O_2 consumption, denitrification, and methanogenesis in a thermally stratified lake. *Can. J. Fish. Aquat. Sci.*, **48**: 1048–1054.

Bender, M., Grande, K., Johnson, K., Marra, J., Hunt, C., Donaghay, P., and Heinemman, C. 1987. A comparison of four methods for the determination of planktonic community metabolism. *Limnol. Oceanogr.*, **32**: 1087–1100.

Berman, T., Parparov, A., and Yacobi, Y. Z. 2004. Lake Kinneret: Planktonic community production and respiration and the impact of bacteria of photic zone carbon cycling. *Aquat. Microb. Ecol.* **34**: 43–55.

Biddanda, B. A. and Cotner, J. B. 2002. Love handles in aquatic ecosystems: the role of dissolved organic carbon drawdown, resuspended sediments, and terrigenous inputs in the carbon balance of Lake Michigan. *Ecosystems*, **5**: 431–445.

Biddanda, B., Ogdahl, M., and Cotner, J. 2001. Dominance of bacterial metabolism in oligotrophic relative to eutrophic waters. *Limnol. Oceanogr.*, **46**: 730–739.

Campbell, P. J. 1984. Laboratory Measurement and Prediction of Sediment Oxygen Consumption. MSc Thesis, McGill University, Montreal, 229 pp.

Carignan, R. and Lean, D. R. S. 1991. Regeneration of dissolved substances in a seasonally anoxic lake: the relative importance of processes occuring in the water column and in the sediments. *Limnol. Oceanogr.*, **36**: 683–707.

Carignan, R., Blais, A.-M., and Vis, C. 1998. Measurement of primary production and community respiration in oligotrophic lakes using the Winkler method. *Can. J. Fish. Aquat. Sci.*, **55**: 1078–1084.

Carignan, R., Planas, D., and Vis, C. 2000. Planktonic production and respiration in oligotrophic Sheild lakes. *Limnol. Oceanogr.*, **45**: 189–199.

Carpenter, S. R., Cole, J. J., Hodgson, J. R., Kitchell, J. F., Pace, M. L., Bade, D., Cottingham, K. L., Essington, T. E., Houser, J. N., and Schindler, D. E. 2001. Trophic cascades, nutrients, and lake productivity: experimental enrichment of lakes with contrasting food webs. *Ecol. Monogr.*, **71**: 163–186.

Clark, J. F., Schlosser, P., Wanninkhof, R., Simpson, H. J., Schuster, W. S. F., and Ho, D. T. 1995. Gas transfer velocities for SF_6 and ^{3}He in a small pond at low wind speeds. *Geophys. Res. Lett.*, **22**: 93–96.

Cole, J. J. and Caraco, N. F. 1998. Atmospheric exchange of carbon dioxide in a low-wind oligotrophic lake measured by the addition of SF_6. *Limnol. Oceanogr.*, **43**: 647–656.

Cole, J. J. and Pace, M. L. 1998. Hydrologic variability of small, northern lakes measured by the addition of tracers. *Ecosystems*, **1**: 310–320.

Cole, J. J., Caraco, N. F., Kling, G. W., and Kratz, T. K. 1994. Carbon dioxide supersaturation in the surface waters of lakes. *Science*, **265**: 1568–1570.

Cole, J. J., Pace, M. L., Carpenter, S. R., and Kitchell, J. F. 2000. Persistence of net heterotrophy in lakes during nutrient addition and food web manipulation. *Limnol. Oceanogr.*, **45**: 1718–1730.

Cornett, R. J. 1982. Prediction and interpretation of rates of hypolimnetic oxygen depletion. PhD Thesis, McGill University, Montreal, 273 pp.

Cornett, R. J. and Rigler, F. H. 1984. Dependence of hypolimnetic oxygen consumption on ambient oxygen concentration: Fact or artefact? *Water Resour. Res.*, **20**: 823–830.

Cornett, R. J. and Rigler, F. H. 1987. Decomposition of seston in the hypolimnion. *Can. J. Fish. Aquat. Sci.*, **44**: 146–151.

Davies, J. M., Hesslein, R. H., Kelly, C. A., and Hecky, R. E. 2003. pCO2 method for measuring photosynthesis and respiration in freshwater lakes. *J. Plankton Res.*, **25**: 385–395.

Dean, W. and Gorham, E. 1998. Magnitude and significance of carbon burial in lakes, reservoirs, and peatlands. *Geology*, **26**(6): 535–538.

del Giorgio, P. A. 1992. The relationship between ETS (electron transport system) activity and oxygen consumption in lake plankton: a cross-system calibration. *J. Plankton. Res.*, **14**: 1723–1741.

del Giorgio, P. A. and Bouvier, T. C. 2002. Linking the physiologic and phylogenetic successions in free-living bacterial communities along an estuarine salinity gradient. *Limnol. Oceanogr.*, **47**: 471–486.

del Giorgio, P. A. and Duarte, C. M. 2003. Total respiration in the global ocean. *Nature*, **420**: 379–384.

del Giorgio, P. A. and Peters, R. H. 1993. Balance between phytoplankton production and plankton respiration in lakes. *Can. J. Fish. Aquat. Sci.*, **50**: 282–289.

del Giorgio, P. A. and Peters, R. H. 1994. Patterns in planktonic P:R ratios in lakes: influence of lake trophy and dissolved organic carbon. *Limnol. Oceanogr.*, **39**: 772–787.

del Giorgio, P. A., Cole, J. J., Caraco, N. F., and Peters, R. H. 1999. Linking planktonic biomass and metabolism to net gas fluxes in northern temperate lakes. *Ecology*, **80**: 1422–1431.

den Heyer, C. and Kalff, J. 1998. Organic matter mineralization in sediments: a within- and among-lake study. *Limnol. Oceanogr.*, **43**: 695–705.

Dillon, P. J. and Rigler, F. H. 1974. The phosphorus–chlorophyll relationship in lakes. *Limnol. Oceanogr.*, **19**: 767–773.

Duarte, C. M. and Agusti, S. 1998. The CO_2 balance of unproductive aquatic ecosystems. *Science*, **281**: 234–236.

Duarte, C. M., Agusti, S., and Kalff, J. 2000. Particulate light absorption and the prediction of phytoplankton biomass and planktonic metabolism in northeastern Spanish aquatic ecosystems. *Can. J. Fish. Aquat. Sci.*, **57**: 25–33.

Edberg, N. 1976. Oxygen consumption of sediment and water in certain selected lakes. *Vatten*, **31**: 330–340.

Falkowski, P. G. and Raven, J. A. 1997. *Aquatic Photosynthesis*. Blackwell Science, Malden, MA.

Fenchel, T. 1998. *Bacterial Biogeochemistry: The Ecophysiology of Mineral Cycling*, 2nd edition. Academic Press, 307 pp.

Grande, K., Marra, D. J., Langdon, C., Heinemann, K., and Bender, M. L. 1989. Rates of respiration in the light measured in marine phytoplankton using an [18]O isotope-labelling technique. *J. Exp. Mar. Biol. Ecol.*, **129**: 95–120.

Granéli, W. 1978. Sediment oxygen uptake in south Swedish lakes. *Oikos*, **30**: 7–16.

Granéli, W. 1979. A comparison of carbon dioxide production and oxygen uptake in sediment cores from four south Swedish lakes. *Holarct. Ecol.*, **2**: 51–57.

Graneli, W., Lindell, M., and Tranvik, L. 1996. Photo-oxidative production of dissolved inorganic carbon in lakes of different humic content. *Limnol. Oceanogr.*, **41**: 698–706.

Hargrave, B. T. 1969*a*. Similarity of oxygen uptake by benthic communities. *Limnol. Oceanogr.*, **14**: 801–805.

Hargrave, B. T. 1969*b*. Epibenthic algal production and community respiration in the sediments of Marion Lake. *J. Fish. Res. Board Can.*, **26**: 2003–2026.

Hargrave, B. T. 1972. A comparison of sediment oxygen uptake, hypolimnetic oxygen deficit and primary production in Lake Esrom, Denmark. *Verh. Int. Verein. Limnol.*, **18**: 134–139.

Hargrave, B. T. 1973. Coupling carbon flow through some pelagic and benthic communities. *J. Fish. Res. Board Can.*, **30**: 1317–1326.

Hayes, F. R. and MacAulay, M. A. 1959. Lake water and sediment. V. Oxygen consumed in water over sediment cores. *Limnol. Oceanogr.*, **4**: 291–298.

Hutchinson, G. E. 1957. *Treatise of Limnology. Vol I: Geography, Physics and Chemistry*. John Wiley and Sons, New York.

James, A. 1974. The measurement of benthal respiration. *Water Res.*, **8**: 955–959.

Kalff, J. 2002. *Limnology*. Prentice-Hall, USA.

Krumbein, W. E., Paterson, D. M., and Stal, L. J. (eds). 1994. *Biostabilization of Sediments*. BISVerlag, Oldenburg, 526 pp.

Kuznetsov, S. I. 1958. A study of the size of bacterial populations and of organic matter formation due to photo- and chemosynthesis in water bodies of different types. *Verh. Int. Verein. Limnol.*, **13**: 156–169.

Lasenby, D. C. 1975. Development of oxygen deficits in 14 southern Ontario lakes. *Limnol. Oceanogr.*, **20**: 993–999.

Linsey, G. A. and Lasenby, D. C. 1985. Comparison of summer and winter oxygen consumption rates in a temperate dimictic lake. *Can. J. Fish. Aquat. Sci.*, **42**: 1634–1639.

Livingstone, D. M. and Imboden, D. M. 1996. The prediction of hypolimnetic oxygen profiles: a plea for a deductive approach. *Can. J. Fish. Aquat. Sci.*, **53**: 924–932.

Lund, J. W. G., MacKereth, F. J. H., and Mortimer, C. H. 1963. Changes in depth and time of certain chemical and physical conditions and of the standing crop of *Asterionella formosa* Hass. in the north basin of Windermere in 1947. *Philos. Trans. R. Soc. Lond. Ser. B*, **246**: 255–290.

Luz B., Barkan, E., Sagi, Y., and Yacobi, Y. Z. 2002. Evaluation of community respiratory mechanisms with oxygen isotopes: a case study in Lake Kinneret. *Limnol. Oceanogr.*, **47**: 33–42.

McManus, J., Heinen, E. A., and Baehr, M. H. 2003. Hypolimnetic oxidation rates in Lake Superior: role of dissolved organic material on the lake's carbon budget. *Limnol. Oceanogr.*, **48**: 1624–1632.

Meybeck, M. 1995. Global distribution of lakes. In A. Lerman, D. M. Imboden, and J. R. Gat (eds) *Physics and Chemistry of Lakes*, 2nd edition, Springer-Verlag, Berlin, pp. 1–35.

Newrkla, P. 1982. Annual cycles of benthic community oxygen uptake in a deep oligotrophic lake (Attersee, Austria). *Hydrobiologia*, **94**: 139–147.

Odum, H. T. 1956. Primary production in flowing waters. *Limnol. Oceanogr.*, **1**: 103–117.

Ohle, W. 1952. Die hypolimnische Kohlen-dioxid-Akkumulation als produktions-biologischer Indikator. *Arch. Hydrobiol.*, **46**: 153–285.

Pace, M. L. and Cole, J. J. 2000. Effects of whole-lake manipulations of nutrient loading and food web structure on planktonic respiration. *Can. J. Fish. Aquat. Sci.*, **57**: 487–496.

Pamatmat, M. M. 1965. A continuous-flow apparatus for measuring metabolism of benthic communities. *Limnol. Oceanogr.*, **10**: 486–489.

Pamatmat, M. M. 1968. An instrument for measuring subtidal benthic metabolism *in situ*. *Limnol. Oceanogr.*, **13**: 537–540.

Patalas, K. 1984. Mid-summer mixing depths of lakes of different latitudes. *Verh. Int. Verein. Limnol.*, **22**: 97–102.

Platt, T. and Jassby, A. D. 1976. The relationship between photosynthesis and light for natural assemblages of coastal marine phytoplankton. *J. Phycol.*, **12**: 421–430.

Prairie, Y. T., Bird, D. F., and Cole, J. J. 2002. The summer metabolic balance in the epilimnion of southeastern Quebec lakes. *Limnol. Oceanogr.*, **47**: 316–321.

Provini, A. 1975. Sediment respiration in six Italian lakes in different trophic conditions. *Verh. Int. Verein. Limnol.*, **19**: 1313–1318.

Ramlal, P. S., Kelly, C. A., Rudd, J. W. M., and Furitani, A. 1993. Sites of methyl mercury production in remote Canadian shield lakes. *Can. J. Fish. Aquat. Sci.*, **50**: 972–979.

Rich, P. H. 1975. Benthic metabolism in a soft-water lake. *Verh. Int. Verein. Limnol.*, **19**: 1023–1028.

Rich, P. H. 1979. Differential CO_2 and O_2 benthic community metabolism in a soft-water lake. *J. Fish. Res. Board Can.*, **36**: 1377–1389.

Richey, J. E., Melack, J. M., Aufdenkampe, A. K., Ballester, V. M., and Hess, L. L. 2002. Outgassing from Amazonian rivers and wetlands as a large tropical source of atmospheric CO_2. *Nature*, **416**: 617–620.

Roland, F., Caraco, N. F., Cole, J. J., and del Giorgio, P. A. 1999. Rapid and precise determination of dissolved oxygen by spectrophotometry: evaluation of interference from color and turbidity. *Limnol. Oceanogr.*, **44**: 1148–1154.

Sas, H. 1989. *Lake Restoration by Reduction of Nutrient Loading: Expectations, Experiences, Extrapolations.* Academia Verlag Richarz, St Augustin, 407 pp.

Schallenberg, M. 1993. The ecology of sediment bacteria and hypolimnetic catabolism in lakes: The relative importance of autochthonous and allochthonous organic matter. PhD Thesis, McGill University, Montreal, 259 pp.

Smith, V. H. 1979. Nutrient dependence of primary productivity in lakes. *Limnol. Oceanogr.*, **24**: 1051–1064.

Vadeboncoeur, Y., Jeppesen, E., Vander Zanden, M. J., Schierup, H.-H., Christoffersen, K.C., and Lodge, D. M. 2003. From Greenland to green lakes: cultural eutrophication and the loss of benthic pathways in lakes. *Limnol. Oceanogr.*, **48**: 1408–1418.

Wanninkhof, R. and McGillis, W. R. 1999. A cubic relationship between air-sea CO_2 exchange and wind speed. *Geophys. Res. Lett.*, **26**: 1889–1892.

Weger, H. G., Herzig, R., Falkowski, P. G., and Turpin, D. H. 1989. Respiratory losses in the light in a marine diatom: measurements by short-term mass spectrophometry. *Limnol. Oceanogr.*, **34**: 1153–1161.

Welch, H. E. and Kalff, J. 1974. Benthic photosynthesis and respiration in Char Lake. *J. Fish. Res. Board Can.*, **31**: 609–620.

Wetzel, R. G. 2001. *Limnology: Lake and River Ecosystems*, 3rd edition. Academic Press, San Diego.

CHAPTER 8

Estuarine respiration: an overview of benthic, pelagic, and whole system respiration

Charles S. Hopkinson, Jr[1] and Erik M. Smith[2]

[1] *Ecosystems Center, Marine Biological Laboratory, USA*
[2] *Department of Biological Sciences, University of South Carolina, USA*

Outline

This chapter reviews rates of benthic, pelagic, and whole system respiration in estuaries. We define estuaries as semi-enclosed coastal bodies of water with some degree of mixing between fresh and salt water. Rates of respiration in these locations are high, reflecting high rates of organic loading from both autochthonous and allochthonous sources. Areal rates of pelagic respiration (58–114 mmol C m^{-2} d^{-1}) are 2–4 times higher than benthic respiration rates (34 mmol C m^{-2} d^{-1}), consistent with estimates that only about 24% of total organic inputs to estuaries are respired by the benthos. Estimates of whole system respiration derived from open-water techniques (294 mmol C m^{-2} d^{-1}) are substantially higher than those obtained by summing component rates (92–148 mmol C m^{-2}d^{-1}), most likely due to the different spatial scales sampled by the two different approaches. The fundamental limit on benthic, pelagic, and whole system respiration appears to be the supply of organic matter, and in many locations allochthonous inputs fuel a major portion of estuarine respiration. Nonetheless, information on the factors that affect benthic respiration is far greater than it is for pelagic respiration, and knowledge of whole system respiration is particularly lacking. This prevents a full understanding of the fate of the vast amount of organic carbon that is imported and produced in estuarine ecosystems.

8.1 Introduction

Estuaries are those regions at the interface of the terrestrial and oceanic realms where seawater is measurably diluted by freshwater runoff from land. Whether they be drowned river valleys, sandbar-built lagoons, or fjords, estuaries have one thing in common, they are the loci through which all products eroded or washed from land pass on their way to the ocean. On average, 0.08–0.17 mol organic C m^{-2} of land per year are exported from the terrestrial biosphere (Schlesinger and Melack 1981; Meybeck 1982; Mulholland and Watts 1982) and transported by the world's rivers to estuaries. Some

estuaries receive further sources of organic carbon, including wastewater effluents and adjacent oceanic upwelling (Smith and Hollibaugh 1997). For such a small region to be the recipient of such a mass of allochthonous organic matter is equivalent to local production on the order of 8 mol C m^{-2} year^{-1}. This is half the average areal rate of gross primary production for the entire biosphere (Odum 1971). Estuaries are also sites of tremendous inorganic nutrient loading, however, rivaling that of intensively fertilized agroecosystems (Howarth *et al*. 2000). High rates of nutrient loading in combination with a tidal-energy subsidy (Odum 1971) and a diversity of functional groups of primary producers, including macrophytes, benthic macrophytes, and phytoplankton, contribute to rates of organic production in estuaries that are among the highest in the entire biosphere, including tropical rain forests and agroecosystems (Kelly and Levin 1986). Indeed, nutrient-enhanced eutrophication is arguably the most severe, present-day threat to the integrity of estuaries with extensive consequences including anoxic and hypoxic waters, reduced fishery harvests, toxic algal blooms, and loss of biotic diversity (Howarth *et al.* 2000).

The fate of allochthonous and autochthonous inputs of organic matter to estuaries reflects the balance between the inputs and consumption by heterotrophs (measured as both heterotroph respiration and heterotroph biomass growth), harvest of fish and shellfish, burial in sediments and export to the ocean. Net ecosystem production (NEP = $P - R$) is the balance between all forms of production (P) and respiration (R) by all organisms. NEP is a measure of ecosystem trophic state and represents the extent to which an ecosystem is a net source or sink of atmospheric carbon dioxide.

Estuaries are complex, open systems that experience large inputs of both organic matter and inorganic nutrients from land. Thus they have the potential to be either autotrophic or heterotrophic systems (i.e. $P > R$ or $P < R$). During primary production inorganic nutrients are taken up from the environment and during respiration they are released. In autotrophic systems there is net assimilation of inorganic nutrients and net production of organic matter: such systems must receive external sources of nutrients. Heterotrophic systems are net remineralizers of organic matter and thus are net exporters of inorganic nutrients but also net importers of organic matter. Estuaries are sites of high secondary production and much research has focused on mechanisms that control the production of commercially important fisheries (Houde and Rutherford 1993) and the efficiency of trophic transfer (Nixon 1988). For estuaries to sustain high levels of commercial fisheries harvest, we expect NEP to be positive, that is, $P - R > 0$ or for there to be allochthonous organic matter inputs to subsidize the harvest. Thus this autotrophic characteristic could be supported by net assimilation of inorganic

nutrients from watershed drainage, or it could also be supported by the excess organic matter inputs from land. A key to quantifying the overall fate of allochthonous organic matter inputs to estuaries and estuarine primary production is to directly measure the rate of ecosystem respiration.

In this chapter we review rates of estuarine respiration for aquatic portions of estuaries. We consider benthic and pelagic components of overall system respiration. Due to the lack of standardized approaches, we do not consider portions of estuaries dominated by macrophytes, such as sea grass beds and intertidal marshes as there are few measures of respiration in these habitats, relative to open-water habitats. We have also opted not to focus on respiration "hotspots," such as oyster and mussel reefs, as information necessary to integrate these regions into the larger estuary (e.g. areal extent) is often lacking. This is not to say these areas are unimportant. In many estuaries, benthic filter feeders such as clams, oysters, or mussels control overall levels of water column productivity (e.g. Peterson 1975; Dame and Patten 1981; Kautsky 1981). Focusing on estuarine open-water habitats where a great number of studies have been conducted in a wide variety of locations facilitates comparisons across systems.

8.2 Measuring estuarine respiration

There are a variety of ways to estimate estuarine respiration. The most common approach is to isolate various *components* of the system, such as the benthos or the water column, in containers and to measure concentration changes in metabolic reactants (e.g. oxygen) or products (e.g. carbon dioxide) over time. A second approach is to measure diel changes in concentrations of metabolic reactants or products in the entire water column. This *open-water whole system* approach typically involves following a specific water mass, identified either by its unique salinity signature or by an added tracer, such as rhodamine dye. This approach is not frequently employed in estuaries, as advective and dispersive transport can be large in estuaries and mask signals resulting from primary production and respiration (Kemp and Boynton 1980). A rarely used third approach relies on calculating respiration indirectly as the difference between independent measures of P and $P - R$. Water, salt, and biogeochemical models have been used successfully to calculate $P - R$ directly (e.g. Nixon and Pilson 1984; Smith and Hollibaugh 1997). The modeling approach for measuring $P - R$ can be superior to methods involving the summation of each component of gross primary production (GPP) and R, because measures of GPP and R often have variances as large as or greater than the difference between P and R (Smith and Hollibaugh 1997).

8.2.1 Benthic respiration

The most direct approach to measure benthic respiration is to place an opaque chamber over the sediment so as to isolate a small volume of water over a known bottom area. Benthic respiration is calculated as the rate of change of dissolved gas (CO_2, oxygen) or inorganic nutrient concentrations measured over time, accounting for sediment surface area and enclosed water volume considerations. The rate of respiration in the water above the sediment (measured in separate bottles) is subtracted from the chamber rate to estimate the sediment contribution alone. Water is stirred in the chambers by a variety of methods, including battery-powered propellers. Benthic chambers usually have a wide flange that rests on the sediment surface and a skirt that penetrates the sediment. The flange and skirt insure constancy of internal volume, prevent erosion from water currents and help isolate the chamber water mass. Benthic chambers result in minimal sediment disturbance, but the logistics of deployment and sampling prevent their common usage.

More commonly benthic respiration is measured in cores collected from the field and incubated in the laboratory under controlled conditions. As with benthic chambers, respiration is calculated from the rate of change of metabolic products over time, accounting for surface area to volume relations and correcting for water column respiration. The duration of incubation depends on sediment activity rates, the surface area to volume ratio, and analysis detection limits. Solute changes are typically linear for well in excess of 24 h, indicating the presence of large, relatively labile organic matter

stores. Water overlying cores is usually exchanged with fresh estuarine water prior to incubation, as the coring and transporting process usually result in some disturbance of the sediment surface and the overlying water. The advantage of core incubations is that conditions can be experimentally manipulated, easily enabling controls of benthic respiration to be evaluated (e.g. temperature or organic matter amendments).

Benthic respiration is most easily measured as the consumption of dissolved oxygen in water overlying the sediment. To convert oxygen-based measures of respiration to C-based units however, requires knowledge of the respiratory coefficient ($RQ = \Delta CO_2 / -\Delta O_2$),[1] which is generally not known but typically assumed to be 1.0. In situations where there is considerable anaerobic respiration, oxygen will underestimate organic carbon mineralization to the extent that reduced end-products of anaerobic metabolism are not reoxidized. Even when end-products are reoxidized, there is often a considerable time lag before oxygen is consumed. The direct method for measuring organic carbon respiration is to measure inorganic carbon concentration (dissolved inorganic carbon—DIC) over time. DIC, which we will hereafter simply label carbon dioxide, can be measured directly as the sum of all inorganic carbon species dissolved in estuarine water. It can also be calculated from measures of pH and alkalinity, after correcting for alkalinity changes associated with nutrients and organic acids. Unfortunately, even carbon dioxide flux is not without problems, as carbon dioxide change can also be associated with carbonate dissolution and precipitation and chemoautotrophic growth.

A final consideration in benthic respiration measurements is the effect of water movement on solute fluxes (including oxygen and carbon dioxide). In an attempt to mimic *in situ* water current fields in chambers or cores, investigators typically generate circular flow fields (Boynton *et al.* 1981). It has been shown however that solute fluxes can be artificially enhanced by circular flow fields, especially in permeable sediments (Glud *et al.* 1996).

---

[1] See Chapter 1 for a discussion of the basis of the value of the RQ.

8.2.2 Pelagic respiration

Respiration in the water column is far simpler to measure than benthic respiration, but not without its challenges. Typically, a water sample is enclosed in a bottle or a nonpermeable bag (to translate natural turbulence) and incubated under controlled conditions until there is sufficient change in solute concentrations. Chamber size generally ranges from 60 cm^3 BOD bottles to 20 dm^3 carboys. The advantage of large chambers is the ability to capture a larger percentage of the plankton community. The statistical probability of capturing macrozoo-plankton increases with sample size (Sheldon *et al.* 1972). Often, 200-μm screening is used to screen out the large zooplankton from small incubation vessels, so as to decrease variability. For most of the twentieth century, oxygen was the solute measured to estimate organic carbon degradation. As there is seldom anaerobic metabolism occurring in the water column, the carbon equivalent of oxygen consumption can be determined if nitrification is measured simultaneously (or the change in NH_4^+ and NO_3^- concentrations).

Plankton respiration is generally much lower than benthic respiration and incubation intervals are typically much longer (benthic respiration can be amplified by maximizing the sediment surface area to overlying water volume ratio). The primary reason why the oxygen technique for measuring primary production was dropped during the 1950s and 1960s in favor of the ^{14}C technique was greatly increased sensitivity. As a result, measures of plankton respiration, which are a component of the oxygen technique, were seldom made thereafter. Only in the past decade or two has the precision and sensitivity of instrumentation for measuring carbon dioxide and oxygen concentrations increased sufficiently to enable relatively short incubations. The precision (as measured by the standard error of the mean of replicates) of modern, computer-controlled instrumentation for measuring carbon dioxide and oxygen concentration is currently on the order of 0.5 μM DIC and 0.02 μM O_2.

Minimum planktonic incubation times must reflect not only the precision of instrumental analysis, but also variation between bottles. With increased precision of oxygen and carbon dioxide measurements, planktonic incubations can be quite short. Based on a standard error for initial and final dissolved oxygen replicates of about 0.05 μM and an average rate of pelagic respiration of 10 mmol O_2 m^{-3} d^{-1}, the minimum incubation time required for the standard error of the rate calculation (slope of $\Delta O_2 / \Delta$ time) to be 10% or better is less than 2 h. This is a best case scenario, however, and based on oligotrophic systems that seem to have little bottle to bottle variation. In eutrophic estuarine systems, we find that bottle to bottle variation for the second time point is substantially higher than it is for oligotrophic systems. Incubation times are likely to increase as a result. For carbon dioxide, incubation time is considerably longer. Recently there has been an interest in using membrane inlet mass spectrometry to measure dissolved oxygen. In addition to greatly shortened analysis times (<1 min), simultaneous measures of oxygen and O_2/Ar ratios show the potential to increase the measurement precision, thereby affording reduced incubation times.

What is an acceptable duration of incubation? There is strong evidence that heterotrophic plankton respiration is closely coupled to the production of fresh photosynthate and incubations in the dark that require more time than it takes for the depletion of fresh photosynthate are likely to lead to underestimated rates of respiration. High temporal resolution analyses of plankton respiration often show a non-linear trajectory developing within hours or even minutes of incubation initiation (Sampou and Kemp 1994; Hopkinson *et al.* 2002). On the other hand, the average night has 12 h of darkness when photosynthate production is halted. So while planktonic respiration might be underestimated during daylight periods, when there is a close coupling between photosynthate production and heterotrophic respiration, there should be no such bias during night. Thus there will be on some scale a rundown in respiration in the dark, but if we try to avoid this and measure instantaneous rates then we might end up overestimating respiration in natural systems.

8.2.3 Open-water whole system respiration

Open-water approaches are another option for measuring respiration (Odum and Hoskin 1958;

Odum and Wilson 1962; Balsis *et al.* 1995). In this approach the total mass of oxygen or carbon dioxide in the water column is measured over time. Without being contained however, changes in these constituents are also due to exchange with the atmosphere and mixing with adjacent water masses. Problems are exacerbated in stratified systems because advection and mixing with additional water masses must be quantified. To address these problems in stratified systems, Swaney *et al.* (1999) used statistical regressions of oxygen with salinity, temperature, and time to calculate diel changes in oxygen attributable to biological activity. Atmospheric exchange is difficult to quantify accurately. Exchange is controlled by the saturation deficit or excess and the exchange coefficient (piston velocity). Factors controlling piston velocity include wind speed and current velocity. Time course changes in the mass of sulfur hexafloride (SF_6) gas added to estuarine waters has proven to be an effective means of integrating wind, rainfall, and current velocity effects on gas exchange (Carini *et al.* 1996). Recently a new technique was developed to measure exchange coefficients that is based on the rate of dissipation of heat applied to micropatches on the water surface (Zappa *et al.* 2003).

While the open-water, whole-system diel change approach is not without its own significant methodological problems, it does avoid problems associated with container approaches that may underestimate respiration for a variety of reasons, including reduced turbulence and altered heterotrophic communities. Unfortunately, it is difficult to apply in large estuarine systems and is not often employed. Further it is not clear how to extrapolate rates of respiration measured during night to rates on a daily basis. While it is often assumed that respiration rates at night are time invariant, there is evidence to suggest they are not (e.g. rates that vary even at night). Isotopic approaches that involve the analysis of isotopes of carbon dioxide, water, and oxygen could help resolve some of these scaling issues.

8.3 Benthic respiration

There have been a great number of studies on estuarine benthic respiration over the past 50 years. We have compiled data from approximately 50 locations, primarily in temperate regions, where there was information for an annual cycle on benthic respiration as well as ancillary information on other system attributes such as system productivity, water temperature, and depth. We did not include studies with less than annual coverage.

8.3.1 The data

The mean annual respiration of all the studies we compiled is 34 mmol C m^{-2} d^{-1} and ranges from a low of 3 mmol C m^{-2} d^{-1} to a high of 115 mmol C m^{-2} d^{-1} (Fig. 8.1). The highest rate is from Corpus Christi estuary, Texas, which

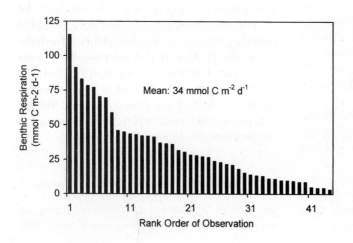

Figure 8.1 Average annual rate of benthic respiration from 48 sites around the world. The data are presented in rank order. While rates are presented in terms of mmol C m^{-2} d^{-1}, data may have been measured as oxygen consumption. These rates were converted to carbon equivalents either by the original author(s) or by us assuming an RQ of 1. See text for problems associated with this conversion approach. Many of these values represent averages of multiple stations within single regions.

has allochthonous organic matter inputs from sea grasses as well as fringing salt marshes. We have not included data from estuarine reefs habitats, which can have substantially higher rates of respiration. For instance, average annual respiration of oyster reefs in the Duplin River estuary in the southeastern United States is 780 mmol C m^{-2} d^{-1}, over an order of magnitude higher than the mean estuarine rate (Bahr 1976). There are less than two orders of magnitude that range in the rates we have compiled. This is less than the actual range because we averaged rates within study areas, thus removing the highs and lows that occur over an annual cycle. Ranges of rates even within single study areas can be as large as that reported across all our study areas. For instance, along the salinity gradient in the Plum Island Sound estuary, mean respiration over a 12 month period ranges from 21 mmol C m^{-2} d^{-1} at a sandy, euryhaline site, to 171 mmol C m^{-2} d^{-1} at a muddy, mesohaline site with abundant clams (Hopkinson et al. 1999). In Boston Harbor benthic respiration ranges from 24 to 111 from site to site (Giblin et al. 1997). On first glance, it appears that sediment respiration is lowest in sandy sediments and highest where there is an abundant benthic filter feeding community. However, not all sandy sites have low rates, witness the high rates of respiration (65–83 mmol C m^{-2} d^{-1}) in sandy Georgia coastal sediments (Smith 1973; Hopkinson 1985).

8.3.2 Carbon dioxide versus oxygen measures of respiration

Since respiration is typically measured as either oxygen consumption or carbon dioxide production, we might ask whether the two approaches measure the same processes and give similar results. Lacking simultaneous measures of oxygen and carbon dioxide, most investigators assume an RQ of 1 and Redfield C : N : P stoichiometry. The validity of this assumption is often incorrect however. It had been assumed that most benthic respiration was aerobic and that oxygen utilization by nitrification could be accounted for easily. It was also assumed that under anaerobic conditions when sulfate reduction was the dominant respiratory pathway, any reduced sulfur produced during respiration would

eventually be reoxidized and that carbon dioxide production associated with sulfate reduction would be balanced by oxygen consumption when reduced sulfur was reoxidized, that is, no net reduced sulfur storage.

The RQ of benthic systems ranges widely across studies and also across sites within study areas. Intensive, multiyear studies in Plum Island Sound, Boston Harbor, and Tomales Bay illustrate the extent to which oxygen and carbon dioxide fluxes differ (Fig. 8.2(a)). In Tomales Bay carbon dioxide flux is 2.2 times greater than oxygen flux (Dollar et al. 1991) on average (RQ = 2.15). In Plum Island estuary, average annual carbon dioxide flux for three sites is 1.35 × O$_2$ flux, ranging from 0.88 to 1.71 across sites. Carbon dioxide production was less than oxygen consumption only in the permanently freshwater site where porewater sulfate concentrations are low and methanogenesis is a major metabolic pathway (Hopkinson et al. 1999). In Boston Harbor, mean annual carbon dioxide flux is 1.21 × O$_2$ consumption, but ranges from 1.05 to 1.49 across sites (Giblin et al. 1997). For all these sites, with the exception of the freshwater site, oxygen flux substantially underestimates total organic carbon metabolism because of non-O$_2$ based anaerobic metabolism.

Dissolved inorganic carbon flux exceeds oxygen consumption when organic matter is degraded anaerobically, that is, without oxygen as the terminal electron acceptor. During anaerobic sulfate reduction 1 mole of SO$_4^{2-}$ oxidizes 2 moles of organic carbon, leading to the production of 2 equivalents of alkalinity, 1 mole of reduced sulfur (HS$^-$) and no consumption of oxygen. The reduced sulfur can be stored in the sediment, typically as pyrite (FeS$_2$) or it can diffuse into oxidized regions and be reoxidized. When reoxidized, 2 moles of oxygen are reduced and 2 equivalents of alkalinity are consumed. The net rate of sulfate reduction can be estimated from the net alkalinity flux. In Tomales Bay, the difference between carbon dioxide flux and oxygen consumption was shown to be due to the net reduction of sulfur. There was a net production of alkalinity that was proportional to the imbalance between carbon dioxide and oxygen. With oxygen flux "corrected" for alkalinity flux, the carbon dioxide/"corrected" oxygen flux was reduced to

(a)

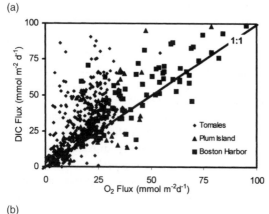

(b)

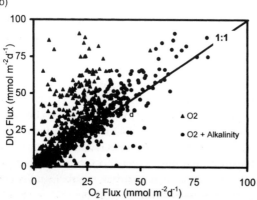

Figure 8.2 The relation between rates of benthic respiration measured as either oxygen consumption or carbon dioxide production. (a) The relations for several stations within three estuaries: Tomales Bay (Dollar *et al.* 1991), Boston Harbor (Giblin *et al.* 1997), and Plum Island Sound (Hopkinson *et al.* 1999). (b) The effect of "correcting" oxygen consumption for anaerobic metabolism that does not result in the consumption of oxygen but which produces alkalinity. Diamonds represent oxygen flux alone, while triangles represent the sum of oxygen and alkalinity flux (from Tomales Bay website and Dollar *et al.* 1991).

1.09 (from 2.15—Fig. 8.2(b)). Independent measures of net pyrite formation in Tomales Bay sediments were in balance with net alkalinity flux. Thus oxygen-based estimates of benthic respiration can underestimate total organic carbon mineralization substantially.

In estuarine systems benthic respiration is generally high (averaging 30 mmol C m^{-2} d^{-1} for the studies examined) and sulfate is in abundance. As a result, sulfate reduction is one of the dominant

metabolic pathways in sediments. In experimental mesocosms, Sampou and Oviatt (1991) showed the importance of sulfate reduction to increase with increased organic matter loading to sediments: the greater the loading the smaller the aerobic zone in sediments. In mesocosms with added sewage sludge, sediment carbon metabolism was dominated by sulfate reduction (75% of total). Total carbon dioxide flux from the sediments was in excess of sediment oxygen consumption but could be balanced by sedimentary sulfide storage.

One further consideration in measuring benthic respiration is that carbon dioxide flux is not exclusively due to respiration. Carbonate incorporation into shells of benthic macrofauna, and carbonate precipitation and dissolution can also contribute to measurable carbon dioxide fluxes. In Norsminde Fjord, Denmark, calcium carbonate deposition by mollusks was an important carbon dioxide flux during late summer, as revealed by positive relationships between carbon dioxide flux and mollusk density (Therkildsen and Lomstein 1993). In order to obtain a better understanding on the potential of carbonate precipitation or dissolution in sediments, pH and concentrations of carbonates and calcium in porewater should be measured.

8.3.3 Controls of benthic respiration

There is no single explanation for what controls the magnitude and seasonal pattern of benthic respiration that holds for all the systems we examined. However, there are some general patterns that emerged from our analysis. The magnitude of organic matter supply explains the largest percentage of the variance across systems when dealing with annual rates. Temperature explains a large percentage of the seasonal variance within systems. While these patterns hold in general, there are exceptions to each. In the section that follows, we will discuss the major controls and how they operate and interact.

Spatial patterns
There have been numerous attempts and approaches to determine the effect of the rate of supply organic matter on benthic metabolism

(e.g. Hargrave 1973; Nixon 1981). Unfortunately, it is extremely difficult and seldom possible to quantify input. Inputs can be based on primary production, primary production plus allochthonous organic matter inputs, or organic matter deposition rates onto benthic sediments. Estimates of organic matter deposition have been attempted in shallow coastal areas using sediment trap approaches, but resuspended bottom sediments can contribute more to the traps than deposition of fresh particles from the water column. Proxy measures of organic matter inputs have also been used, including sediment organic content and chlorophyll content. Bulk descriptors of sediments, such as organic content and grain size, generally explain little of the variation in benthic respiration rates, however, at neither local nor regional scales. Others have attempted to model inputs on the basis of overlying water primary production and loss of organic matter during settling. Ideally we would like to know the total rate of organic matter input to a system (primary production and allochthonous inputs), the percentage of inputs that settle to the bottom (and the controls on the percentage that settles), and the relation between inputs to the bottom and the magnitude and timing of benthic respiration.

We compiled data from 20 sites with information on both benthic respiration and water column net primary production and allochthonous organic matter inputs to the estuarine ecosystem (Fig. 8.3(a)). There is little relation between only autochthonous production and benthic respiration (data not shown). This is to be expected in estuarine systems where allochthonous inputs represent a substantial portion of total organic matter inputs. van Es (1982) considered the large intercept (30 mmol C m^{-2} d^{-1}) for a regression between net primary production and benthic respiration in the Ems-Dollard estuary to be a measure of the importance of allochthonous organic matter inputs to the benthic system. Our data show a strong relation between total organic matter inputs (allochthonous and autochthonous) and benthic respiration: the greater the organic input to an estuarine system, the greater the benthic respiration. The R^2 of the regression between total system production and benthic respiration indicates

that this parameter explains 44% of the variation in respiration. From the slope of the regression, we see that on average 24% of the total system production is respired in benthic sediments in estuarine systems. Experimental approaches have also been used to show the importance of organic matter inputs in controlling the magnitude of benthic respiration. In large mesocosms (supposedly mimicking Narragansett Bay), Kelly and Nixon (1984) showed that benthic respiration increased with increased organic matter deposition. Their studies also showed that the effects of fresh organic deposition could be short lived, however, as small inputs of very labile organic matter caused dramatic increases in benthic respiration that lasted only a few days.

We find little relation between the magnitude of total organic matter inputs to an estuary and the percentage respired by the benthos (Fig. 8.3(b)). While on average the benthos respires 24% of inputs, at low input rates the benthos respires anywhere from about 10% to nearly 100% of the production. With increasing inputs, there is less variability so that at the highest levels of production about 20–30% of inputs are respired. It is not clear whether this pattern is the result of inadequate sampling of higher production systems or whether benthic system efficiency (or vice versa pelagic efficiency) varies with input rates (see Hargrave 1973, 1985).

Water column depth and turbulence have also been used, independently, to explain spatial patterns in rates of benthic respiration. Depth can be expected to play a role, as the residence time of particles in the water column prior to settling should be proportional to depth. The greater the depth of the water, the longer the particle is in the water column, the greater the particle degradation while in the water column and hence less organic matter reaches the benthos. Turbulence can be expected to play a role if it delays the time for particle sedimentation. We find that depth is poorly correlated with benthic respiration, however (Fig. 8.4(a) and (b)). At depths shallower than 10 m there is absolutely no relationship between either benthic respiration and depth or the percentage of total system production that is respired by the benthos. It is only in the presumably rare estuarine systems, greater than 10 m deep,

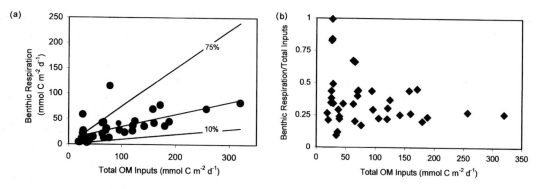

Figure 8.3 The relation between organic matter inputs to a system and benthic respiration. Organic matter inputs include both primary production within the system (e.g. planktonic primary production) and allochthonous inputs from outside the system (e.g. riverine input of organic matter to an estuary). (a) Relation between inputs and benthic respiration. Included are lines that describe the best fit relation as well as the amount of benthic respiration to be expected if either 10% or 75% of total organic inputs were respired by the benthos. (b) The relation between total system organic matter inputs and the portion that is respired by the benthos.

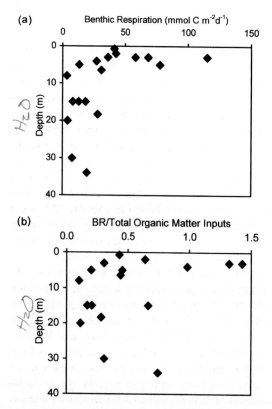

Figure 8.4 The relation between water column depth and benthic respiration. (a) The figure shows benthic respiration versus depth. (b) The figure shows how the relative portion of organic matter inputs respired by the benthos varies with depth.

where there is the tendency for absolute or relative rates of benthic respiration to decrease with increasing depth ($R^2 < 0.11$). The estuarine systems contrast with their deep-water oceanic cousins where there is a strong correlation between system depth and benthic respiration (see Middelburg, Chapter 11).

Hargrave (1973) showed that mixed layer depth was related to the relative importance of benthic respiration. In shallow estuarine systems and those that are stratified, turbulence from tidal currents and winds retards particle sedimentation thereby increasing residence time in the pelagic zone and the extent of organic matter degradation prior to settling. However we have insufficient data to examine rigorously the effect of turbulence. The general lack of a relationship between benthic respiration and depth might be expected, as there is tremendous variability in the amount of organic matter potentially available to the benthos. Given sufficient data, it would be interesting to evaluate the effect of depth and/or turbulence when corrected for total organic matter inputs.

Temporal patterns

The supply rate of organic matter to the benthos as estimated from measures of autochthonous and allochthonous inputs to the estuary are our best predictors of spatial patterns in benthic respiration.

What factors control variability over time? Factors often shown to be important include temperature (e.g. Nixon *et al.* 1980) and animal activity (e.g. Aller 1982).

At regional scales, average annual temperature explains none of the variability in annual average

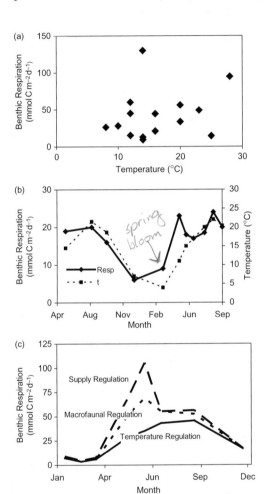

Figure 8.5 Factors controlling spatial and temporal variation in benthic respiration. (a) The relation between mean annual temperature and mean annual benthic respiration rate for a variety of temperate estuaries. There is no significant relationship. (b) The seasonal pattern of benthic respiration and temperature in Buzzards Bay. Note the strong rise in benthic respiration in late winter when temperature is still declining. Banta *et al.* (1995) showed this asynchrony was due to deposition of the spring bloom. (c) The seasonal pattern of benthic respiration in Chesapeake Bay as it relates to variations in temperature, macrofaunal biomass, and organic matter supply.

benthic respiration (Fig. 8.5(a)). Even within certain estuaries, temperature generally explains little of the difference in benthic respiration between sites (e.g. Fisher *et al.* 1982; Giblin *et al.* 1997; Hopkinson *et al.* 1999, 2001). At the site level, however, patterns of benthic respiration over an annual cycle are often strongly related to temperature, with respiration increasing with temperature. While Giblin *et al.* (1997) observed no correlation between temperature and benthic respiration across sites in Boston Harbor, up to 90% of the annual pattern in benthic respiration at individual sites could be explained by temperature.

Different responses to temperature are often seen across seasons and an asynchrony between an increase in benthic respiration and temperature in spring. Earlier we mentioned that benthic respiration in mesocosms can respond very quickly to fresh inputs of labile organic matter, such as would be expected following phytoplankton blooms. Banta *et al.* (1995) showed an overall strong relation between seasonal temperature and benthic respiration patterns, but a pronounced departure from this pattern in early spring. In fact, while temperatures were still dropping during late winter, benthic respiration reached its second highest level over the annual period (Fig. 8.5(b)). Banta *et al.* related the increased respiration to the deposition of the spring phytoplankton bloom. While temperature alone explained 42% of the annual pattern, 72% could be explained when sediment chlorophyll *a* content (indicator of fresh plankton inputs) and temperature were regressed against benthic respiration.

Other investigators have demonstrated the importance of fresh organic matter inputs in interpreting temporal dynamics as well (e.g. Fisher *et al.* 1982; Graf *et al.* 1982; van Es 1982; Grant 1986; Dollar *et al.* 1991). It is interesting that there are often differences in the timing of response to fresh organic matter inputs. Whereas Banta *et al.* showed a nearly simultaneous response, Hargrave (1978) showed a 1–2 month lag between deposition and respiration. To some extent the timing of the response can be attributed to benthic community composition. When benthic respiration is entirely microbial, temperature plays a larger role and lags are greater, whereas when benthic macrofauna are a major component of the community, lags can be shorter.

Benthic animals, which change in abundance and activity over an annual cycle, also have been shown to exert an influence on seasonal patterns of benthic respiration. Kemp and Boynton (1981) showed that seasonal patterns of benthic respiration in Chesapeake Bay sediments could be attributed to temperature and macrofaunal biomass (see Fig. 8.5(c)) as well as substrate supply. On the basis of macrofaunal size–class distributions and temperatures, they were able to predict benthic respiration with close agreement to measured rates. Residuals as large as 40% could not be explained during spring, however. While at the time the unexplained residual was attributed to the non-linear effect of macrofaunal "microbial gardening," it is more likely the result of deposition of the spring bloom (Boynton and Kemp 1985).

Other controls on benthic respiration
While major spatial and temporal patterns in benthic respiration are largely explained by variations in organic loading rates, temperature, and macrofaunal activity, they do not explain other patterns such as organic matter preservation in sediments or respiration enhancement in shallow water systems with permeable sediments. Organic matter preservation in the sea is an active area of research (Hedges and Keil 1995). There is a critical interaction between organic and inorganic materials with a direct relationship between organic matter content in sediments and mineral surface area. More than 90% of total sediment organic matter can not be physically separated from its mineral matrix and is thus unavailable to benthic microbes and macrofauna (Mayer 1994a, b). This may partially explain the lack of correlation generally found between sediment organic content and benthic respiration. In fine grained sediments, which have the greatest surface area and hence high organic content, organic matter may be unavailable.

In contrast, course sediments in shallow water systems while typically having low organic content can have extremely high respiration rates. Investigations of this paradox are an active area of benthic research, which focuses on the role of advective water movement and particle trapping in course, permeable sandy sediments. Here sedimentary organic content is low due to low mineral surface area, but because of strong advective flushing of bottom water through the course sediments, particulate organic matter from the water column can be efficiently filtered by the sediments and then decomposed by benthic organisms (Huettel and Rusch 2000).

8.4 Pelagic respiration

In contrast to that of the benthos, there have been far fewer studies of respiration in estuarine pelagic communities. We identified 22 estuarine locations for which direct measures of respiration in the water column are available. All but one of these locations is in the Northern Hemisphere. Most are in the temperate climatic region. All of the studies reported here focused on the open water portion of estuarine environments. Due to the paucity of data, we include all available data regardless of the extent to which they represent full annual coverage within a location.

8.4.1 The data

Mean pelagic respiration rates among locations range from 1.7 to 84 mmol C m^{-3} d^{-1} over their study periods, although variability in rates within any one location can often far exceed this range. (Table 8.1). The lowest mean rate is observed in the Gulf of Finland, the northern-most estuary, and the highest mean rate is observed in the one tropical, Southern Hemisphere location. Other than these extremes, however, there is no apparent trend in respiration rates with location or latitude. Minimum respiration rates tend to be rather similar among most locations, whereas maximum rates are substantially more variable. Although it is typical to report arithmetic mean values, a convention we have followed here, frequency distributions of respiration rates tend to be highly skewed, rather than normally distributed, which can greatly bias calculated mean values. This pattern is readily seen in the combined dataset of surface respiration rates (some 700 observations) from all locations where

Table 8.1 Arithmetic mean and range of pelagic respiration rates (mmol C m^{-3} d^{-1}) reported for the open–water portion of estuaries

Location	Latitude	Longtitude	Coverage	Mean	Min.	Max.	N	Reference
Gulf of Finland	59.83	−23.25	Jan.–Nov.	1.7	0.1	3.8	14	Kuparinen 1987
Loch Ewe	57.78	5.60	Not available	5.9	0.5	20.2	43	Williams 1984
Gulf of Riga	57.30	−23.85	May–Jul.	14.6	2.9	35.6	18	Olesen et al. 1999
Tweed estuary	55.77	2.00	Apr.–Aug.	11.5	0.7	31.9	8	Shaw et al. 1999
Roskilde Fjord	55.75	−12.08	May–Sep.	44.8	10.4	123.8	57	Jensen et al. 1990
Gulf of Gdansk	54.50	−19.25	Feb.–Nov.	7.4	0.4	32.2	149	Witek et al. 1999
Southampton estuary	50.90	1.40	Feb.–Sep.	9.3	0.1	24.6	49	de Souza Lima and Williams 1978
Urdaibai estuary	43.37	2.67	Aug.	59.0	4.8	214.3	38	Iriarte et al. 1996
Urdaibai estuary	43.37	2.67	Feb.–Nov.	41.3	5.9	227.3	36	Revilla et al. 2002
Ria de Vigo	42.24	8.76	Apr.–Nov.	15.1	1.9	47.2	31	Moncoiffe et al. 2000
Bay of Blanes	41.67	−2.80	Jan.–Dec.	5.1	0.1	45.0	25	Satta et al., 1996
Ria de Aveiro	40.63	8.65	Oct.–Jan.	15.8	2.3	82.7	20	Cuhna et al. 1999
Chesapeake Bay	39.33	76.18	Feb.–Nov.	14.3	1.6	57.8	64	Smith and Kemp 1995
Tomales Bay	38.13	122.87	Jan.–Nov.	16.9	3.8	45.8	22	Fourqurean et al., 1997
San Francisco Bay	37.84	122.67	Feb.–Dec.	5.8	0.4	25.5	32	Rudek and Cloern 1996
San Francisco Bay	37.58	122.20	Feb.–Apr.	10.7	2.5	25.2	23	Caffrey et al. 1998
Savanah River	32.03	80.85	Feb., Jul., Oct.	9.1	0.5	34.8	24	Pomeroy et al. 2000
Ogeechee River	31.92	81.18	Feb., Jul., Oct.	11.0	0.2	28.6	15	Pomeroy et al. 2000
Georgia estuary	31.90	80.98	Dec.–Jan.	13.8	0.1	17.0	149	Turner 1978
Altamaha River	31.38	81.33	Feb., Jul., Aug., Oct.	17.5	0.5	53.5	28	Pomeroy et al. 2000
Satilla River	30.96	81.68	Feb., Jul., Aug, Oct.	14.3	0.7	66.7	47	Pomeroy et al. 2000
St. Mary's River	30.71	81.40	Jul., Oct.	14.9	5.8	33.1	12	Pomeroy et al. 2000
Celestun Lagoon	20.75	90.25	Mar.–Mar.	26.5	8.9	74.5	13	Herrer-Silveira 1998
Fly River Delta, PNG	−8.75	−143.50	Feb.	84.0	23.4	150.2	14	Robertson et al. 1993

Note: Summarized are only those studies where plankton community respiration was measured directly and could be extracted from the original source. All rates were measured as *in vitro* changes in oxygen concentrations and converted to C equivalents either by the original author(s), or by us assuming an RQ of 1. *N* is the number of reported rate measurements of each study. Negative longitudes are degrees eastward from Greenwich.

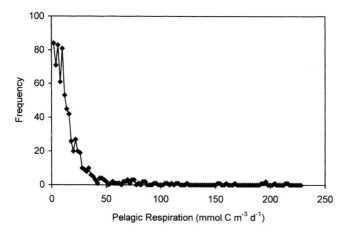

Figure 8.6 Frequency distribution of individual pelagic respiration rates measured in 21 estuarine sites from around the world. While rates are presented as mmol C m^{-3} d^{-1}, original data were measured as *in vitro* oxygen consumption and converted to carbon equivalents either by the original author(s) or by us assuming an RQ of 1. The data ($n = 707$) follow a highly lognormal distribution, such that the arithmetic mean is 17.8 while the geometric mean is only 9.1.

individual measurements could be extracted from the original literature (Fig. 8.6). Respiration rate, which ranges from 0.05 to 227 mmol C m^{-3} d^{-1}, is best described by the lognormal distribution (chi-square goodness of fit test, $p < 0.01$). As a result, while the arithmetic mean is 17.8, the geometric mean is only 9.1 and the mode is just 4.0 mmol C m^{-3} d^{-1}.

8.4.2 Variability in pelagic respiration— effects of temperature and substrate supply

Respiration rates of the pelagic community tend to be much more variable than those measured in the benthos (Kemp *et al.* 1992; Rudek and Cloern 1996; Pomeroy *et al.* 2000). In comparing studies of pelagic respiration, we conclude that there are few, if any, consistent patterns in the variability of respiration among estuaries. Large seasonal variations appear to be one prominent feature of water column respiration in most estuaries, although diel variability can be as much as 50% of seasonal variability (e.g. Sampou and Kemp 1994) and in many systems spatial variability is often larger than seasonal variability (e.g. Jensen *et al.* 1990; Smith and Kemp 1995; Iriarte *et al.* 1996). Attempts at explaining variability in pelagic respiration usually focus on the effects of temperature and substrate supply as regulatory mechanisms.

Temperature sensitivity of respiration rate should be expected due to the profound physiological effects of temperature on cellular metabolism (e.g. Li and Dickie 1987). Sampou and Kemp (1994) investigated the effect of temperature on plankton respiration in the Chesapeake Bay by conducting a series of temperature-manipulation experiments (from 5–30°C) in both spring and summer. In these experiments, relationships between respiration and manipulated temperature were not significantly different between seasons, nor were they different from the relationship obtained from *in situ* measurements of respiration and temperature over the annual cycle. This supports the notion of a temperature sensitivity of pelagic respiration, but suggests an absence of significant physiological adaptation and/or selection for temperature optima in the plankton community over the annual cycle.

Can variations in ambient water temperatures explain differences in respiration rates among or within estuaries? When we combine the data from all locations, there is a significant, logarithmic relationship with temperature (Fig. 8.7). The R^2 of the regression, however, indicates that temperature can explain only 28% of the variability in pelagic respiration rates among estuaries. Within individual locations, attempts at correlating pelagic respiration and ambient water temperatures have produced mixed results. Strongly positive relationships between respiration and temperature have been observed in a number of locations, such as Tomales Bay (Forqurean *et al.* 1997), the Gulf of Gdansk (Witek *et al.* 1999), and all of the Georgia river

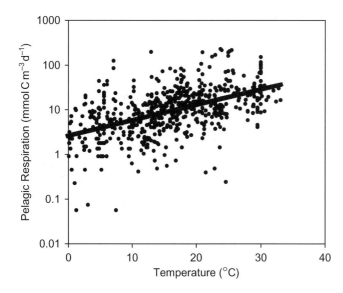

Figure 8.7 The relationship between pelagic respiration rate (mmol C m^{-3} d^{-1}) and water temperature (°C) for the surface waters of estuaries. The fitted line is the ordinary least squares regression log $R = 0.89 + 0.08 \times$ Temp; $n = 648$, $R^2 = 0.28$, $p < 0.001$.

estuaries (Turner 1978; Pomeroy *et al.* 2000). In the Urdaibai estuary respiration was observed to be highly sensitive to the 5°C temperature changes that occurred on timescales of just a few days (Iriarte *et al.* 1996). One difficulty in inferring causality from these statistical correlations, however, is the fact that in temperate estuaries other factors (e.g. increased organic production by algae) often tend to co-vary with water temperature. Several estuaries, in fact, show no apparent relationship between temperature and respiration. Prominent among these locations are Roskilde Fjord (Jensen *et al.* 1990), San Francisco Bay (Rudek and Cloern 1996), and the Bay of Blanes (Satta *et al.* 1996). Iriarte *et al.* (1996) found that observed temporal relationships between respiration and temperature in the Urdaibai broke down in a region of the estuary that receives substantial inputs of organic matter from a sewage treatment plant. The utility of temperature as an explanation for variability in respiration within and among estuaries thus remains equivocal.

Several lines of evidence suggest that availability of organic substrates explains a large part of the variability in pelagic respiration rates. This is to be expected, as pelagic respiration must ultimately be dependent on the supply of organic matter, just as in the case of the benthos. Strong positive relationships between respiration in the euphotic zone and phytoplankton biomass (as measured by chlorophyll *a* concentrations) or productivity have been observed at seasonal timescales in many estuaries. These include all those locations mentioned above where respiration and temperature showed no significant relationship. For example, in the eutrophic Roskilde Fjord, Jensen *et al.* (1990) found phytoplankton biomass to be the single best predictor of variations in respiration across both seasonal and spatial scales within the estuary. In both the Chesapeake Bay (Smith and Kemp 1995) and the Urdaibai estuary (Iriarte *et al.* 1996), seasonal relationships between pelagic respiration and phytoplankton production appear to vary along spatial trophic gradients, with the strongest relationships occurring in areas experiencing the lowest levels of primary production. Interestingly, the spatial patterns in the strength of production–respiration relationships are opposite each other in these estuaries. In Chesapeake Bay, lowest productivities and the tightest relationship occur in the upper reaches of the estuary. In the Urdaibai, this situation occurs at the seaward end of the estuary. Nonetheless, these results are consistent with the idea of a greater degree of autotrophic–heterotrophic coupling, and a higher degree of dependence on autotrophic production by heterotrophs, in less productive areas, relative to higher productive areas.

On shorter timescales, Sampou and Kemp (1994) observed a diel periodicity in surface water respiration rates, for periods of both low (spring) and high (summer) respiratory activity, that exhibited a characteristic pattern of peak rates just after midday and decreasing to a night-time minimum. They attributed this pattern to a tight coupling between respiration and the daily pattern of primary production, where enhanced respiration during the day corresponded to peak rates of phytoplankton exudation of dissolved organic matter. In contrast to the tightly coupled diel cycles in the surface waters, these authors found no such cycles in respiration occurring in the deeper, aphotic, layers of the water column. This layer was separated from the euphotic layer by a strong pycnocline, which effectively broke this link between production and respiration over diel timescales.

While a strong relationship between plankton production and respiration appears to be a common feature of most estuarine ecosystems, it should be noted that such a close coupling does not immediately imply causality by either variable. The question of control in autotrophic–heterotrophic coupling may be largely circular. Indeed, it has been suggested that the high rates of primary production in estuaries may be attributable, in part, by the high rates of nutrient regeneration associated with heterotrophic respiration (Smith and Hollibaugh 1993). Further, it may be that a strong relationship between production and respiration is indicative of regulation of both rates by a common variable. There is growing evidence that pelagic respiration in some estuaries can be strongly stimulated by nutrient enrichment. For example, in the Georgia rivers estuaries (Pomeroy *et al.* 2000), the most frequent positive response in respiration rate to enrichment was to glucose, but in a number of cases inorganic nitrogen, or occasionally inorganic phosphorus, stimulated a significant response. In the Chesapeake Bay (Smith and Kemp 2003), enrichment experiments showed organic carbon (as glucose) to be primarily limiting to respiration in the upper, oligohaline region of the Bay and inorganic nutrients (primarily phosphorus) to become limiting in the lower, polyhaline region. The interacting effects of carbon and nutrient substrates in controlling pelagic respiration, and its coupling to primary production, is an area of research that would greatly benefit from further study.

Most of the studies of planktonic respiration we compiled (Table 8.1) also include measures of phytoplankton biomass (as estimated by chlorophyll *a* concentration), allowing us to make comparisons across locations (Fig. 8.8). In the combined dataset, chlorophyll *a* does no better than temperature in

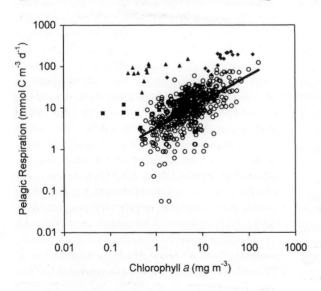

Figure 8.8 The relationship between pelagic respiration rate (mmol C m^{-3} d^{-1}) and chlorophyll *a* (mg m^{-3}) in the surface waters of estuaries. Closed symbols represent sites receiving substantial allochthonous organic input and excluded from the regression analysis. The closed squares are data from the Georgia rivers, the closed triangles are from the Fly River Delta, and the closed diamonds are from the Urdaibai Estuary in the vicinity of a sewage treatment plant. The open symbols are for the remainder of the data, for which allochthonous inputs are not so dominant. The fitted line is the ordinary least squares regression, using the open symbol data only, log R (mmol C m^{-3} d^{-1}) = 1.19 + 0.63 × log chl (mg m^{-3}); $n = 450$, $R^2 = 0.38$, $p < 0.001$. The regression for the entire dataset, line not shown, is log R (mmol C m^{-3} d^{-1}) = 1.45 + 0.54 × log chl (mg m^{-3}); $n = 531$, $R^2 = 0.25$, $p < 0.001$.

explaining variations in respiration rates among estuaries (R^2 of 0.25 versus 0.28, respectively). It is evident, however, that estuaries receiving substantial allochthonous inputs of organic matter, such as the Georgia rivers, the Fly River, and that portion of the Urdaibai estuary in the vicinity of a sewage treatment plant, all tend to separate out from the pattern displayed by the remainder of the data. In these locations, relationships between respiration and chlorophyll a tend to be rather flat. In contrast, the combined data for the remaining locations show a fairly reasonable relationship, of the form $\log R$ (mmol C m^{-3} d^{-1}) = 1.19 + 0.63 × log chl (mg m^{-3}); $n = 450$, $R^2 = 0.38$, $p < 0.001$, and thus now explains close to 40% of the variability in respiration. It is interesting to note that the slope of this relationship is significantly less than one, with pelagic respiration rates increasing proportionately less than chlorophyll a concentrations for this subset of estuaries. Thus, at high algal biomass proportionately more of the primary production associated with this biomass will remain unrespired within the pelagic community.

The data points that fall below the predicted values at the low end of the relationship in Fig. 8.8 are primarily those from the Gulf of Finland (Kuparinen 1987). This is the northern-most location in the dataset and exhibits the lowest seasonal water temperatures. Based on this, we combined temperature and chlorophyll a in a multiple linear regression for all estuarine locations. The resulting regression equation is: $\log R$ (mmol C m^{-3} d^{-1}) = 0.39 + 0.08 × temp (°C) + 0.45 × log chl (mg m^{-3}); $n = 502$, $R^2 = 0.49$, $p < 0.001$. Temperature and chlorophyll a are themselves poorly, though significantly, related (log chl (mg m^{-3}) = 1.08 + 0.03 × temp(°C); $n = 502$, $R^2 = 0.05$, $p < 0.001$), but together they explain 49% of the variability in respiration across all locations, with 80% of all observed values falling within 50% of predicted values. The even distribution of residual indicates this equation is an unbiased predictor of pelagic respiration rates. The 51% of the variability unexplained by temperature and chlorophyll is presumably due, in large part, to the influence of allochthonous inputs of organic matter in many of these estuaries. We have insufficient data, however, to include such a parameter in the predictive equation.

8.4.3 Contribution of various communities to pelagic respiration

The relative contribution of various metabolic groups to total respiration in estuarine waters remains an active research area. Iriarte et al. (1991) postulated that relationships between respiration and chlorophyll a should be strongest at high levels of phytoplankton biomass because the algae themselves would tend to dominate total respiration. At lower phytoplankton biomass levels the situation should be reversed due to a predominance of microheterotrophic respiration. Support for this was seen in San Francisco Bay (Rudek and Cloern 1996), but this pattern is not readily apparent from the data compiled here (Fig. 8.8). The importance of microheterotrophic communities to pelagic respiration has been inferred from significant positive relationships between respiration and bacterial abundance and/or substrate uptake rates in many estuaries (Jensen et al. 1990; Satta et al. 1996; Smith 1998, Smith and Kemp 2003; Witek et al. 1999, Revilla et al. 2002). Fourqurean et al. (1997), on the other hand, found no significant relationship between respiration and bacterial abundance or uptake rates in Tomales Bay, which also has very high phytoplankton to bacterioplankton biomass ratios. These authors thus concluded that phytoplankton were responsible for the bulk of total respiration rates in this estuary.

Several investigators have addressed the relative contributions of the various functional groups present in estuarine waters by quantifying the size distribution of respiration rates. This work has largely been confined to the contributions of the various microplankton groups. It is generally assumed that incubations conducted in 300 ml BOD bottles, largely considered the standard incubation vessel for pelagic respiration rate measurements, do not capture the contribution of macrozooplankton communities. Of course, a commonly assumed corollary to this is that respiration by macrozooplankton is relatively insignificant component of total pelagic community respiration rates, although this is not well tested in estuarine environments. Caution must be taken in the interpretation of microplankton respiration rates subject to filtration (Hopkinson et al. 1989), but results with this approach may be useful in a comparative sense. In the eutrophic Roskilde Fjord, Sand-Jensen *et al.* (1990) found that microbial respiration (operationally defined as cells passing through a 1 μm pore-size filter) accounted for 45%, on average, of total pelagic respiration rates. Similarly, in the productive waters of the Chesapeake Bay, Smith and Kemp (2001) observed microbial respiration (<3 μm cells) averaged 54% of total respiration rates. In contrast, contributions of microbial respiration in a turbid, moderate productivity estuary of the Georgia coast, Griffith *et al.* (1990) found microbial respiration (<1 μm cells) to account for a higher proportion of total respiration, with a mean of 73%. Robinson and Williams (Chapter 9), on the other hand, find about 40% of planktonic respiration in oceanic waters appears to be associated with the bacterial (<1μm) fraction. Although a trend of decreasing importance of microbial respiration along gradients of increasing productivity in estuaries is intuitively appealing, there are exceptions to this pattern. In the Gulf of Finland, a much less productive system than either the Roskilde or the Chesapeake, contributions of the <3 μm size fraction amounted to only 36%, on average, of total pelagic respiration rates. In the Urdaibai estuary (Revilla *et al.* 2002), contributions of the <5 μm size-fraction represented, on average, 99% of the total community respiration, although these numbers are biased by the fact that in most of the <5 μm size-fractioned samples respiration actually exceeded that measured in the whole-water fraction, suggesting an enhancement of microbial activity upon filtration (Hopkinson *et al.* 1989).

8.5   Open-water whole system respiration

It has only been in the past 10 years or so that total system metabolism has been measured with the open-water technique in enough estuaries to warrant an analysis of metabolic patterns. New measurements are primarily from estuaries within the US NOAA National Estuarine Research Reserve program (Caffrey 2003). Data are available for North American estuaries in both tropical and temperate regions. Unfortunately, few ancillary data on estuarine conditions (e.g. temperature,

chlorophyll *a*, benthic respiration, pelagic respiration, depth) have been reported for most of these sites, which prevents a rigorous analysis of controls on whole system respiration.

8.5.1   The data

Whole system measures of respiration range from 69 mmol C m^{-2} d^{-1} in the Newport River, NC, USA to 631 mmol C m^{-2} d^{-1} in Bojorquez lagoon, Mexico and average 294 mmol C m^{-2} d^{-1} (Fig. 8.9(a)). While the highest rate of respiration is from the southernmost, warmest site, the lowest rate is not from the northernmost, coldest site. Thus temperature and geographic latitude do not fully explain the variability in rates across sites.

8.5.2   Relation between whole system respiration and gross production

The methodology for measuring whole system respiration provides concurrent measures of system gross production (Pg). Estimates of Pg range from 60 to 870 mmol C m^{-2} d^{-1} with the lowest and highest rates being from the same sites where the extremes in respiration rate were observed. Mean Pg is 262 mmol C m^{-2} d^{-1}.

There is a nonlinear, logarithmic relation between Pg and system respiration, with decreasing increases in respiration per unit increase in Pg (Fig. 8.9(b)). The R^2 for the logarithmic relationship indicates that Pg explains 77% of the variability in respiration across sites. The high R^2 also indicates very close coupling between primary production and system respiration across all systems in this compilation. This nonlinear relation further suggests that allochthonous organic matter inputs are most important at low rates of Pg and that as Pg increases the relative importance of allochthonous organic matter inputs to estuaries decreases. As such, there is a tendency for estuarine systems to be heterotrophic at lower rates of Pg (<400 mmol C m^{-2} d^{-1}) and autotrophic at higher rates of Pg (>500 mmol C m^{-2} d^{-1}—Fig. 8.9(c)).

The ratio between Pg and system respiration demonstrates the trophic status of an ecosystem. Systems with Pg in excess of respiration (positive net ecosystem production) are autotrophic systems: more organic carbon is produced than consumed

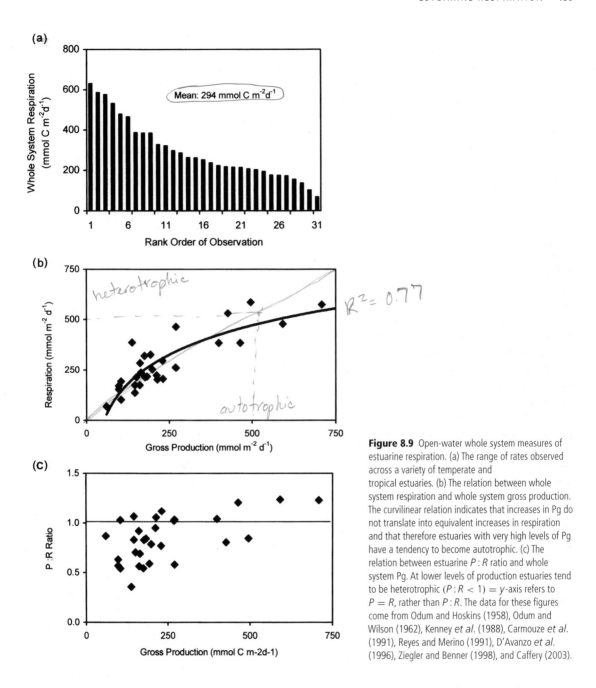

Figure 8.9 Open-water whole system measures of estuarine respiration. (a) The range of rates observed across a variety of temperate and tropical estuaries. (b) The relation between whole system respiration and whole system gross production. The curvilinear relation indicates that increases in Pg do not translate into equivalent increases in respiration and that therefore estuaries with very high levels of Pg have a tendency to become autotrophic. (c) The relation between estuarine $P:R$ ratio and whole system Pg. At lower levels of production estuaries tend to be heterotrophic ($P:R < 1$) = y-axis refers to $P = R$, rather than $P:R$. The data for these figures come from Odum and Hoskins (1958), Odum and Wilson (1962), Kenney et al. (1988), Carmouze et al. (1991), Reyes and Merino (1991), D'Avanzo et al. (1996), Ziegler and Benner (1998), and Caffery (2003).

in respiration. In heterotrophic systems (negative NEP), R exceeds Pg, indicating the importance of allochthonous organic carbon inputs to a system (assuming the system does not consume capital produced and stored in previous times). The $P:R$ ratio ranges from as low as 0.36:1 to as high as 1.38:1. The average $P:R$ ratio is 0.86:1, implying that these estuaries are generally net heterotrophic ecosystems. This suggests that, on average, the sum of total respiration plus organic matter export

stoichiometry - terr vs. marine

from these estuaries must depend on an input of allochthonous organic material equivalent to at least 14% of Pg.

Many factors can influence the autotrophic–heterotrophic nature of estuaries, including water residence time, lability of allochthonous organic matter and the ratio of inorganic to organic nitrogen load (Hopkinson and Vallino 1995). Kemp *et al.* (1997) showed that estuarine NEP was largely controlled by the relative balance between inputs of inorganic nutrients and allochthonous organic carbon loading. Terrestrial organic matter sources are C-rich, relative to planktonic material, and therefore release proportionally lower quantities of inorganic nutrients than that from decomposing planktonic material. When these terrestrial inputs are respired in the estuary, the Pg resulting from this nutrient source will be substantially less than the *R* associated with the release of these nutrients. Hence, when organic matter loading is dominated by terrestrial sources, estuaries tend towards negative NEP as a direct result of the low C:N ratio of estuarine organic matter relative to that of terrestrial. It would appear that in the future we can expect estuaries to become increasingly autotrophic (Kemp *et al.* 1997) as the long-term trend in inputs is for organic loading to decrease (due to decreased wetlands and sewage treatment that removes BOD) and inorganic loading to increase (due to intensification of agricultural N fertilization and increasingly N-rich diet for an increasing global population).

8.5.3 Comparison of component-derived and open-water whole system-derived measures of respiration

Rates of whole system respiration are very high relative to measures of planktonic and benthic respiration. The average rate of pelagic respiration we observed in our synthesis is $17.8 \, \text{mmol} \, C \, m^{-3} \, d^{-1}$ (geometric mean is 9.1). Assuming an average depth for all pelagic sites of 6.4 m (the average depth reported for benthic studies in Fig. 8.4), average depth-integrated pelagic respiration is $114 \, \text{mmol} \, C \, m^{-2} \, d^{-1}$. This is 4 times higher than benthic respiration ($34 \, \text{mmol} \, C \, m^{-2} \, d^{-1}$).

#'5

This calculated relative dominance by the pelagic system seems reasonable and is consistent with our previous estimate that on average 24% of total system production (autochthonous and allochthonous) is respired by the benthos (Fig. 8.3). The sum of benthic and pelagic components of system respiration are not equivalent to directly measured rates of system-determined respiration, however ($148 \, \text{mmol} \, C \, m^{-2} \, d^{-1}$ benthic and pelagic versus $294 \, \text{mmol} \, C \, m^{-2} \, d^{-1}$ for system-determined respiration).

Few studies have attempted to explain the disparity between component-derived and whole system-derived measures of respiration (e.g. Odum and Hoskin 1958; Kemp and Boynton 1980; Ziegler and Benner 1998). The comparisons we make in this synthesis are not based on simultaneous measures of benthic, pelagic and whole system respiration. Thus our conclusions may be spurious because of sampling bias. Significantly higher estimates of respiration based on whole system measures, relative to those based on component measures, have, however, been reported in a few specific locations where concomitant measures of each component were made. These locations include Chesapeake Bay, USA (Kemp and Boynton 1980), Laguna Madra, USA (Ziegler and Benner 1998) and the Plum Island Sound estuary, USA (C. Hopkinson, unpublished data), and suggest this difference is real. Clearly, container and whole system approaches operate on different spatial, and sometimes temporal, scales. This may be a large part of the explanation for the disparity between the two estimates of respiration. For example, it is likely that whole system-derived measures are "seeing" the effect of respiration in adjacent ecosystems, such as intertidal marshes (e.g. Cai *et al.* 1999), or in components within the estuary that are not adequately sampled by containers, such as floating rafts of senescent vascular plant material (e.g. Ziegler and Benner 1998). Alternatively, the lack of agreement may also, in part, be explained by methodological uncertainties inherent in each of the two approaches. For example, respiration may be decreased in containers when the contained community is removed from fresh organic matter inputs (organic matter deposition in cores or fresh phytoplankton photosynthate in dark bottles). In addition, respiration may be reduced in containers because of reduced turbulence. In fact, there have been many reports about the effect of stirring when measuring benthic respiration in cores (e.g. Boynton *et al.* 1981; Huettel and Gust 1992). As discussed in Section 8.2.3, however, the open-water approach to measuring total system respiration is not without its share of significant methodological problems as well. The application of the method suffers in estuaries in particular, where physical processes can often dominate open-water oxygen dynamics. This is particularly true for estuaries that are strongly stratified and experience pronounced short-term variability in tidal and wind-induced currents. Failure to account for the influence of physical processes on the open-water approach can result in unrealistically high estimates of biological rates (Kemp and Boynton 1980). There is currently no clear consensus on which approach is generally more appropriate for the estimation of total system respiration in estuaries, although we suspect that this will become an area of increased scientific interest as the disparity becomes more widely recognized.

8.6 Conclusion and synthesis

This review has shown rates of estuarine respiration are high, reflecting the high rates of organic matter loading to estuaries from both autochthonous and allochthonous sources. Direct field measurements of respiration suggest that the average rate of benthic respiration is 34 mmol C m^{-2} d^{-1} while the average rate of pelagic respiration is between 9.1 and 17.8 mmol C m^{-3} d^{-1}, depending on whether a geometric or arithmetic mean for the data is used. Assuming an average depth for all pelagic sites of 6.4 m, the depth-integrated pelagic respiration is between 58–114 mmol C m^{-2} d^{-1}. The areal rates of pelagic respiration are thus, on average, 2–4 times higher than benthic respiration rates in estuaries. This is consistent with estimates that only 24% of total organic inputs (allochthonous plus autochthonous) are respired by the benthos in these systems.

Combining the two direct measurements of respiration rate gives a range from 92 to 148 mmol C m^{-2} d^{-1}, depending on whether one adopts the algebraic or geometric mean for pelagic respiration rate. In contrast, estimates of whole system respiration obtained by the open-water approach averaged 294 mmol C m^{-2} d^{-1} for the locations compiled in this review. While this disparity may be attributable to methodological uncertainties inherent in each of the two techniques, it is more likely due to the different spatial scales sampled by the two different approaches. For instance there is the inclusion of respiration by other estuarine communities in the open-water techniques that are excluded in container approaches, such as respiration in adjacent intertidal marshes.

From this review, it is clear that the fundamental limit on benthic, pelagic and whole system respiration is the supply of organic matter. While major differences in benthic respiration among locations are best explained by variation in organic loading rates (44% explained), temporal patterns at specific sites are controlled by a combination of factors, primarily temperature, organic matter supply, and macrofaunal biomass and activity. We have a better understanding of what controls temporal variation in benthic respiration at single sites than we do of what controls spatial variation in respiration. This probably reflects the difficulty of quantifying organic matter inputs to benthic systems. Variations in pelagic respiration, within and among sites, are largely controlled by differences in the supply of organic matter and temperature. Allochthonous supply, as estimated from phytoplankton biomass (chlorophyll *a*) and temperature each explain about 25% of the variation in rates among estuaries. The 50% or so of the variability unexplained by temperature and chlorophyll *a* is presumably due, in large part, to the influence of allochthonous organic matter inputs. We lack sufficient information to quantify this relation however. For estuaries, in general, both allochthonous and autochthonous sources of organic matter fuel estuarine pelagic and benthic metabolism, and in some locations allochthonous inputs appear to be a major source of organic matter fueling estuarine respiration. There is a tendency for estuarine systems to be net heterotrophic at rates of primary production < 400 mmol $C\,m^{-2}\,d^{-1}$.

Given the estimates of respiration derived in this review and an estimated global area of estuaries of $1.4 \times 10^{12}\,m^{-2}$ (Gattuso *et al.* 1998) we can make an initial estimation of total annual respiration in estuaries as 76–150×10^{12} mol $C\,a^{-1}$ (i.e. 76–150 Tmol $C\,a^{-1}$). The wide range is due to the difference in component and whole system-derived estimates of respiration. Estuarine area does not include salt marshes or other wetlands. Woodwell *et al.* (1973) estimated that estimates of estuarine area are accurate to $\pm 50\%$. Global estuarine respiration is distinctly higher than the magnitude of estimated total carbon delivery from land to the ocean, 34 Tmol $C\,a^{-1}$, and total estuarine planktonic and benthic primary production, 35 T mol $C\,a^{-1}$ (Smith and Hollibaugh 1993), suggesting that most estuaries are generally net heterotrophic zones. We have not accounted for primary production or respiration of salt marshes or mangroves within estuaries, but acknowledge that whole water estimates of estuarine respiration probably reflect some salt marsh contribution (wetlands occupy approximately $0.4 \times 10^{12}\,m^2$ Woodwell *et al.* 1973).

This global estimate of estuarine respiration is subject to a great deal of uncertainty. Our knowledge of benthic, pelagic, and whole system respiration in estuaries is confined largely to the temperate environments of North America and Europe, and generally biased towards river-dominated coastal plain estuaries. The magnitude and factors that regulate respiration in the many estuaries of Asia, Africa, Australia, and South America are essentially unknown. In addition, our knowledge about benthic respiration and the factors that affect its variability is far greater than it is for pelagic respiration. We think the reasons for this are twofold. First, benthic rate measurements are technologically easier, given the larger rates of oxygen decline that occur in benthic incubations, relative to those occurring in pelagic incubations. Second, in the field of estuarine ecology, there has been an emphasis on controlling eutrophication in estuaries and the importance of nutrient loading in causing eutrophication. Benthic nutrient regeneration is a major source of internally regenerated nutrients and benthic denitrification is the primary nitrogen sink in estuaries. Measures of benthic respiration are thus typically made concomitant with measures of benthic nutrient fluxes. This is in contrast to measures of pelagic respiration, which are typically the objective of a study in and of themselves, and have only relatively recently been conducted on a routine basis as a result of methodological improvements in precision oxygen measurements. Finally, our current state of knowledge on whole system respiration in estuaries is particularly lacking. This paucity of information on whole system respiration prevents us from knowing the fate of the vast amount of organic carbon that is imported and produced in estuarine systems. Thus we lack information on the overall role of estuarine ecosystems in the overall global carbon budget. Increased research should be focused on quantifying estuarine respiration and in understanding differences in estimates of respiration derived from "container" and open-water techniques.

Acknowledgments

The authors greatly appreciate help from several scientists who provided data that contributed to this manuscript, W. Michael Kemp, Zbigniew Witek, and Marta Revilla. This work was supported by the National Science Foundation Long-Term Ecological Research program (OCE-9726921), the National Oceanic and Atmospheric Administration Woods Hole Oceanographic Sea Grant program, and the Environmental Protection Agency Science to Achieve Results program (R82-867701).

References

Aller, R. 1982. The effects of macrobenthos on chemical properties of marine sediment and overlying water. In P. McCall and J. Tevexz (eds) *Animal-Sediment Relations*. Plenum Press, New York, pp. 53–102.

Bahr, L. 1976. Energetic aspects of the intertidal oyster reef community at Sapelo Island, Georgia, USA. *Ecology*, **57**: 121–131.

Balsis, B., Alderman, D., Buffam, I., Garritt, R., Hopkinson, C., and Vallino, J. 1995. Total system metabolism in the Plum Island Sound estuary. *Biol. Bull.*, **189**: 252–254.

Banta, G., Giblin, A., Hobbie, J., and Tucker, J. 1995. Benthic respiration and nitrogen release in Buzzards Bay, Massachusetts. *J. Mar. Res.*, **53**: 107–135.

Boynton, W. and Kemp, W. 1985. Nutrient regeneration and oxygen consumption by sediments along an estuarine salinity gradient. *Mar. Ecol. Prog. Ser.*, **23**: 45–55.

Boynton, W., Kemp, W., Oxborne, K., Kaumeyer, K., and Jenkins, M. 1981. The influence of water circulation rate on *in situ* measurements of benthic community respiration. *Mar. Biol.*, **65**: 185–190.

✓Cai, W. J., Pomeroy, L. R., Moran, M. A., and Wang, Y. C. 1999. Oxygen and carbon dioxide mass balance for the estuarine–intertidal marsh complex of five rivers in the southeastern US. *Limnol. Oceanogr.*, **44**: 639–649.

✓✐Caffrey, J. M. 2003. Production, respiration and net ecosystem metabolism in US estuaries. *Environ. Monit. Asses.*, **81**: 207–219.   *no SAB sites*

Caffrey, J. M., Cloern, J. E., and Grenz. C. 1998. Changes in production and respiration during a spring phytoplankton bloom in San Francisco Bay, California, USA: implications for net ecosystem metabolism. *Mar. Ecol. Prog. Ser.*, **172**: 1–12.

Carini, S., Weston, N., Hopkinson, C., Tucker, J., Giblin, A., and Vallino, J. 1996. Gas exchange rates in the Parker River Estuary, Massachusetts. *Biol. Bull.*, **191**: 333–334.

Carmouze, J. P., Knoppers, B., and Vasconcelos, P. 1991. Metabolism of a subtropical Brazilian lagoon. *Biogeochemistry*, **14**: 129–148.

Cunha, M. A., Almeida, M. A., and Alcantara, F. 1999. Compartments of oxygen consumption in a tidal mesotrophic estuary (Ria de Aveiro, Portugal). *Acta Oecol. Int. J. Ecol.*, **20**: 227–235.

Dame, R. F. and Patten, B. C. 1981. Analysis of energy flows in an intertidal oyster reef. *Mar. Ecol. Prog. Ser.*, **5**: 115–124.

de Souza Lima, H. and Williams, P. J. le B. 1978. Oxygen consumption by the planktonic population of an estuary—Southampton water. *Estuar. Coast. Shelf Sci.* **6**: 515–521.

Dollar, S., Smith, S., Vink, S., Obrebski, S., and Hollibaugh, J. 1991. Annual cycle of benthic nutrient fluxes in Tomales Bay, California, and contribution of the benthos to total ecosystem metabolism. *Mar. Ecol. Prog. Ser.*, **79**: 115–125.

Fisher, T., Carlson, P., and Barber, R. 1982. Sediment nutrient regeneration in three North Carolina estuaries. *Estuar. Coast. Shelf Sci.*, **14**: 101–116.

Fourqurean, J. W., Webb, K. L., Hollibaugh, J. T., and Smith, S. V. 1997. Contributions of the plankton community to ecosystem respiration, Tomales Bay, California. *Estuar. Coast. Shelf Sci.*, **44**: 493–505.

✓Gattuso, J-P, Frankignoulle, M., and Wollast, R. 1998. Carbon and carbonate metabolism in coastal aquatic ecosystems. *Annu. Rev. Ecol. Syst.*, **29**: 405–434.

Giblin, A., Hopkinson, C., and Tucker, J. 1997. Benthic metabolism and nutrient cycling in Boston Harbor, Massachusetts. *Estuaries*, **20**: 346–364.

Glud, R, N., Forster, S., and Huettel, M. 1996. Influence of radial pressure gradients on solute exchange in stirred benthic chambers. *Mar. Ecol. Prog. Ser.*, **141**: 303–311.

Graf, G., Bengtsson, W., Diesner, U., Schulz, R., and Theede, H. 1982. Benthic response to sedimentation of a spring bloom: process and budget. *Mar. Biol.* **67**: 201–208.

Grant, J. 1986. Sensitivity of benthic community respiration and primary production to changes in temperature and light. *Mar. Biol.*, **90**: 299–306.

Griffith, P. C., Douglas, D. J., and Wainright, S. C. 1990. Metabolic activity of size-fractionated microbial plankton in estuarine, nearshore, and continental shelf waters of Georgia. *Mar. Ecol. Prog. Ser.*, **59**: 263–270.

Hargrave, B. T. 1973. Coupling carbon flow through some pelagic and benthic communities. *J. Fish. Res. Board Can.*, **30**: 1317–1326.

Hargrave, B. T. 1978. Seasonal changes in oxygen uptake by settled particulate matter and sediments in a marine bay. *J. Fish. Res. Board Can.*, **35**: 1621–1628.

Hargrave, B. T. 1985. Particle sedimentation in the ocean. *Ecol. Model.*, **30**: 229–246.

Hedges, J. I., and Keil, R. 1995. Sedimentary organic matter preservation: an assessment and speculative synthesis. *Mar. Chem.*, **49**: 81–115.

Herrer-Silveira, J. A. 1998. Nutrient-phytoplankton production relationships in a groundwater-influenced tropical coastal lagoon. *Aquat. Ecosyst. Health Manage.*, **1**: 373–385

✓Hopkinson, C. S. 1985. Shallow water benthic and pelagic metabolism. *Mar. Biol.*, **87**: 19–32.

Hopkinson, C. S., and Vallino, J., 1995. The nature of watershed perturbations and their influence on estuarine metabolism, *Estuaries*, **18**: 598–621.

Hopkinson, C. S., Sherr, B., and Wiebe, W. J. 1989. Size-fractionated metabolism of coastal microbial plankton. *Mar. Ecol. Prog. Ser.*, **51**: 155–166.

Hopkinson, C., Giblin, A., Tucker, J., and Garritt, R. 1999. Benthic metabolism and nutrient cycling along an estuarine salinity gradient. *Estuaries*, **22**: 825–843.

Hopkinson, C. S., Giblin, A. E., and Tucker, J. 2001. Benthic metabolism and nutrient regeneration on the continental shelf off eastern Massachusetts, USA. *Mar. Ecol. Prog. Ser.*, **224**: 1–19.

Hopkinson, C. S., Vallino, J., and Nolin, A. 2002. Decomposition of dissolved organic matter from the continental margin. *Deep-Sea Res. II*, **49**: 4461–4478.

Houde, E., and Rutherford, E. 1993. Recent trends in estuarine fisheries: predictions of fish production and yield. *Estuaries*, **16**: 161–176.

Howarth, R. W., Anderson, D., Church, T., Greening, H., Hopkinson, C., Huber, W., Marcus, N., Naiman, R., Segerson, K., Sharpley, A., and Wiseman, W. Committee on the Causes and Management of Coastal Eutrophication. 2000. *Clean Coastal Waters–Understanding and Reducing the Effects of Nutrient Pollution*. Ocean Studies Board and Water Science and Technology Board, Commission on Geosciences, Environment, and Resources, National Research Council. National Academy of Sciences, Washington DC, 405 pp.

Huettel, M. and Gust, G. 1992. Solute release mechanisms from confined sediment cores in stirred benthic chambers and flume flows. *Mari. Ecol. Prog. Ser.*, **82**: 187–197.

Huettel, M., and Rusch, A. 2000. Transport and degradation of phytoplankton in permeable sediment. *Limnol. Oceanogr.*, **45**: 534–549.

Iriarte, A., Daneri, G., Garcia, V. M. T., Purdie, D. A., and Crawford, D. W. 1991. Plankton community respiration and its relationship to chlorophyll *a* concentration in marine coastal waters. *Oceanol. Acta*, **14**: 379–388.

Iriarte, A., deMadariaga, I., DiezGaragarza, F., Revilla, M., and Orive, E. 1996. Primary plankton production, respiration and nitrification in a shallow temperate estuary during summer. *J. Exp. Mar. Biol. Ecol.*, **208**: 127–151.

Jensen, L. M., Sand-Jensen, K., Marcher, S., and Hansen, M. 1990. Plankton community respiration along a nutrient gradient in a shallow Danish estuary. *Mar. Ecol. Prog. Ser.*, **61**: 75–85.

Kautsky, N. 1981. On the trophic role of the blue mussel (*Mytilus edulis* L.) in a Baltic coastal ecosystem and the fate of the organic matter produced by the mussels. *Kiel. Meeresforsch*, **5**: 454–461.

Kelly, J. R. and Levin, S. A. 1986. A comparison of aquatic and terrestrial nutrient cycling and production processes in natural ecosystems, with reference to ecological concepts of relevance to some waste disposal issues. In G. Kullenberg (ed.) *The Role of the Oceans as a Waste Disposal Option*. D. Reidel Publishing Company, New York, p. 300.

Kelly, J. R. and Nixon, S. 1984. Experimental studies of the effect of organic deposition on the metabolism of a coastal marine bottom community. *Mar. Ecol. Prog. Ser.*, **17**: 157–169.

Kemp, W. M. and Boynton, W. 1980. Influence of biological and physical factors on dissolved oxygen dynamics in an estuarine system: implications for measurement of community metabolism. *Estuar. Coas. Mar. Sci.*, **11**: 407–431.

Kemp, W. M. and Boynton, W. 1981. External and internal factors regulating metabolic rates of an estuarine benthic community. *Oecologia*, **51**: 19–27.

Kemp, W. M., Sampou, P. A., Garber, J., Tuttle, J. and Boynton, W. R. 1992. Seasonal depletion of oxygen from bottom waters of Chesapeake Bay—roles of benthic and planktonic respiration and physical exchange processes. *Mar. Ecol. Prog. Ser.*, **85**: 137–152.

Kemp, W. M., Smith, E. M., Marvin-DiPasquale, M., and Boynton, W. R. 1997. Organic carbon balance and net ecosystem metabolism in Chesapeake Bay. *Mar. Ecol. Prog. Ser.*, **150**: 229–248.

Kenney, B. E., Litaker, W., Duke, C. S., and Ramus, J. 1988. Community oxygen metabolism in a shallow tidal estuary. *Estuar. Coast. Shelf Sci.*, **27**: 33–43.

Kuparinen, J. 1987. Production and respiration of overall plankton and ultraplankton communities at the entrance to the Gulf of Finland in the Baltic Sea. *Mar. Biol.*, **93**: 591–607.

Li, W. and Dickie, P. 1987. Temperature characteristics of photosynthetic and heterotrophic activities: seasonal variations in temperate microbial plankton. *Appl. Environ. Microbiol.*, **53**: 2282–2295.

Mayer, L. M. 1994*a*. Relationships between mineral surfaces and organic carbon concentrations in soils and sediments. *Chem. Geol.*, **114**: 347–363.

Mayer, L. M. 1994*b*. Surface area control of organic carbon accumulation in continental shelf sediments. *Geochim. Cosmochim. Acta*, **58**: 1271–1284.

Meybeck, M. 1982. Carbon, nitrogen, and phosphorus transport by world rivers. *Am. J. Sci.*, **282**: 401–450.

Moncoiffe, G., Alvarez-Salgado, X. A., Figueiras, F. G., and Savidge, C. 2000. Seasonal and short-time-scale dynamics of microplankton community production and respiration in an inshore upwelling system. *Mar. Ecol. Prog. Ser.*, **196**: 111–126.

Mulholland, P. and Watts, J. 1982. Transport of organic carbon to the oceans by rivers of North America. A synthesis of existing data. *Tellus*, **34**: 176–186.

Nixon, S. W. 1981. Remineralization and nutrient cycling in coastal marine ecosystems. In B. Neilson and L. Cronin (eds) *Nutrient Enrichment in Estuaries*. Humana Press, Clifton, NJ, pp. 111–139.

Nixon, S. W. 1988. Physical energy inputs and the comparative ecology of lake and marine ecosystems. *Limnol. Oceanogr.*, **33**: 1005–1025.

Nixon, S. W. and Pilson, M. E. Q. 1984. Estuarine total system metabolism and organic exchange calculated from nutrient ratios—an example from Narragansett Bay. In V. Kennedy (ed.) *The Estuary as a Filter*. Academic Press, New York, pp. 261–290.

Nixon, S. W., Kelly, J. R., Furnas, B. N. Oviatt, C. A., and Hale, S. S. 1980. Phosphorus regeneration and the metabolism of coastal marine bottom communities. In K. Tenore and B. Coull (eds) *Marine Benthic Dynamics*. University of South Carolina Press, Columbia, SC.

Odum, E. P. 1971. *Fundamentals of Ecology*. W.B. Saunders, New York.

Odum, H. T. and Hoskin, C. M. 1958. Comparative studies on the metabolism of marine waters. *Publ. Inst. Mar. Sci., Univ. Tex.*, **5**: 16–46.

Odum, H. T. and Wilson, R. F. 1962. Further studies on reaeration and metabolism of Texas Bays, 1958–1960. *Publ. Inst. Mar. Sci., Univ. Tex.*, **8**: 23–55.

Olesen, M., Lundsgaard, C., and Andrushaitis, A. 1999. Influence of nutrients and mixing on the primary production and community respiration in the Gulf of Riga. *J. Mar. Syst.*, **23**: 127–143.

Peterson, D. H. 1975. Sources and sinks of biologically reactive oxygen, carbon, nitrogen and silican in northern San Francisco Bay. In T. Conomos (ed.) *San Francisco Bay: The Urbanized Estuary*. California Academy of Sciences, San Francisco, pp. 175–194.

Pomeroy, L. R., Sheldon, J. E., Sheldon, W. M., Blanton, J. O., Amft, J., and Peters, F. 2000. Seasonal changes in microbial processes in estuarine and continental shelf waters of the south-eastern USA. *Estuar. Coast. Shelf Sci.*, **51**: 415–428.

Revilla, M., Ansotegui, A., Iriarte, A., Madariaga, I., Orive, E., Sarobe, A., and Trigueros, J. M. 2002. Microplankton metabolism along a trophic gradient in a shallow-temperate estuary. *Estuaries*, **25**: 6–18.

Robertson, A. I., Daniel, P. A., Dixon, P., and Alongi, D. M. 1993. Pelagic biological processes along a salinity gradient in the Fly delta and adjacent river plume (Papua New Guinea). *Conti. Shelf Res.*, **13**: 205–224.

Rudek, J. and Cloern, J. 1996. Plankton respiration rates in San Francisco Bay. In J. Hollibaugh (ed.) *San Francisco Bay: The Ecosystem*. AAAS Pacific Division, San Francisco, pp. 289–304.

Sampou, P. and Kemp, W. M. 1994. Factors regulating plankton community respiration in Chesapeake Bay. *Mar. Ecol. Prog. Ser.*, **110**: 249–258.

Sampou, P. and Oviatt, C. A. 1991. A carbon budget for a eutrophic marine ecosystem and the role of sulfur metabolism in sedimentary carbon, oxygen and energy dynamics. *J. Mar. Res.*, **49**: 825–844.

Sand-Jensen, K., Jensen, L. M, Marcher, S., and Hansen, M. 1990. Pelagic metabolism in eutrophic coastal waters during a late summer period. *Mar. Ecol. Prog. Ser.*, **65**: 63–72.

Satta, M. P., Agusti, S., Mura, M. P., Vaque, D., and Duarte, C. M. 1996. Microplankton respiration and net community metabolism in a bay on the NW Mediterranean coast. *Aquat. Microb. Ecol.*, **10**: 165–172.

Schlesinger, W. H., and Melack, J. 1981. Transport of organic carbon in the world's rivers. *Tellus*, **33**: 172–187.

Shaw, P. J., Chapron, C., Purdie, D. A., and Rees, A. P. 1999. Impacts of phytoplankton activity on dissolved nitrogen fluxes in the tidal reaches and estuary of the Tweed, UK. *Mar. Pollut. Bull.*, **37**: 280–294.

Sheldon, R. W., Prakash, A., and Sutcliffe, W. 1972. The size distribution of particles in the ocean. *Limnol. Oceanogr.*, **17**: 327–340.

Smith, E. M. 1998. Coherence of microbial respiration rate and cell-specific bacterial activity in a coastal planktonic community. *Aquat. Microb. Ecol.*, **16**: 27–35.

Smith, E. M. and Kemp, W. M. 1995. Seasonal and regional variations in plankton community production and respiration for Chesapeake Bay. *Mar. Ecol. Prog. Ser.*, **116**: 217–231.

Smith, E. M. and Kemp, W. M. 2001. Size structure and the production/respiration balance in a coastal plankton community. *Limnol. Oceanogr.*, **46**: 473–485.

Smith, E. M. and Kemp, W. M. 2003. Planktonic and bacterial respiration along an estuarine gradient: responses to carbon and nutrient enrichment. *Aquat. Microb. Ecol.*, **30**: 251–261.

Smith, K. L. Jr. 1973. Respiration of a sublittoral community. *Ecology*, **54**: 1065–1075.

Smith, S. V. and Hollibaugh, J. 1993. Coastal metabolism and the oceanic organic carbon balance. *Rev. Geophys.*, **31**: 75–89.

Smith, S. V. and Hollibaugh, J. 1997. Annual cycle and Inter-annual variability of ecosystem metabolism in a temperate climate embayment. *Ecol. Monogr.*, **67**: 509–533.

Swaney, D., Howarth, R., and Butler, T. 1999. A novel approach for estimating ecosystem production and respiration in estuaries: application to the oligohaline and mesohaline Hudson River. *Limnol. Oceanogr.*, **44**: 1509–1521.

Therkildsen, M. W. and Lomstein, B. 1993. Seasonal variation in net benthic C-mineralization in a shallow estuary. *FEMS Microbiol. Ecol.*, **12**: 131–142.

Turner, R. 1978. Community plankton respiration in a salt marsh estuary and the importance of macrophytic leachates. *Limnol. Oceanogr.*, **23**: 442–451.

van Es, F. 1982. Community metabolism of intertial flats in the Ems-Dollard Estuary. *Mar. Biol.*, **66**: 95–108.

Williams, P. J. le. B., 1984. A review of measurements of respiration rates of marine plankton populations. In J. Hobbie and P. J. le. B. Williams (eds) *Heterotrophic Activity in the Sea*. Plenum Press, New York, pp. 357–389.

Witek, Z., Ochocki, S., Nakonieczny, J., Podgorska, B., and Drgas, A. 1999. Primary production and decomposition of organic matter in the epipelagic zone of the Gulf of Gdansk, an estuary of the Vistula. *ICES J. Mar. Sci.*, **56**: 3–14.

Woodwell, G. M., Rich, P., and Hall, C. 1973. Carbon in estuaries. In G. Woodwell and E. Pecan (eds) *Carbon and the Biosphere*. US Atomic Energy Commission Report CONF-720510, Technical Information Center, Office of Information Services, Washington DC, pp. 221–240.

Zappa C., Raymond, P., Terray, E., and McGillis, W. 2003. Variation in surface turbulence and the gas transfer velocity in estuaries over a tidal cycle in a macro-tidal estuary. *Estuaries*, **26**: 1401–1415.

Ziegler, S. and Benner, R. 1998. Ecosystem metabolism in a subtropical, seagrass-dominated lagoon. *Mar. Ecol. Prog. Ser.*, **173**: 1–12.

CHAPTER 9

Respiration and its measurement in surface marine waters

Carol Robinson[1] and Peter J. le B. Williams[2]

[1] *Plymouth Marine Laboratory, UK*
[2] *School of Ocean Sciences, University of Wales, Bangor, UK*

Outline

This chapter reviews the current state of knowledge of the process and measurement of microplankton respiration in marine surface waters. The principal approaches are outlined and their potentials and limitations discussed. A global database, containing 1662 observations has been compiled and analyzed for the spatial and temporal distribution of surface water respiration. The database is tiny compared to that of photosynthesis and biased with respect to season, latitude, community structure, and depth. Measurements and models show that the major portions of respiration lies in that attributable to bacteria (12–59%) and to algae (8–70%). The mean of the volumetric rates of respiration in the upper 10 m of the open ocean is $3.3 \pm 0.15 \text{ mmol O}_2 \text{ m}^{-3} \text{ d}^{-1}$ and that of depth-integrated open-ocean respiration $116 \pm 8.5 \text{ mmol O}_2 \text{ m}^{-2} \text{ d}^{-1}$. A global estimate of $13.5 \text{ Pmol O}_2 \text{ a}^{-1}$ is derived from the mean depth-integrated rate, which significantly exceeds contemporary estimates of ocean plankton production ($2.3–4.3 \text{ Pmol O}_2 \text{ a}^{-1}$). This difference is at variance with the results of mass-balance calculations, which suggest a small difference (ca. 0.18) between oceanic production and respiration. The reasons for this are discussed.

9.1 Introduction

This chapter reviews our present understanding of the process of marine planktonic respiration, and the methodologies used for its determination. We collate the global database of measured respiration and use it to assess how far we can determine seasonal, regional, and latitudinal patterns of distribution. Measurements of respiration are not yet routine within national and international marine biogeochemical programs. A significant relationship between respiration and more routinely measured parameters would substantially progress our ability to determine its spatial and temporal variability. We investigate whether such a relationship exists. Linked to this is the importance of an appreciation of the distribution of respiration within the microbial food web, and so improved food web model parameterization and verification. When considered on comparable time and space scales, the balance between plankton respiration and photosynthesis indicates the potential amount of photosynthetically fixed carbon available for export from the upper mixed layer. We assume that the epipelagic zone of the open oceans approaches this ideal, and analyze the distributions of respiration and photosynthesis in order to assess the global balance between photosynthesis and respiration.

9.2 Available approaches and their constraints

Rates of planktonic respiration can be derived in a number of ways: (i) from the measurement of the rate of production/consumption of a product or reactant, (ii) the assay of an appropriate respiratory enzyme or enzyme system, (iii) predictions from biomass, and (iv) from inverse models of the community composition and activity.

9.2.1 Flux of products and reactants

Principle of the approach

This approach is based on the measurement of the change of a reactant (e.g. oxygen) or product (e.g. carbon dioxide) of respiration over a period of time (characteristically 12 or 24 h) in the dark. The basic requirement is the ability to analyze the sample, a replicate or a representative body of water at point(s) during the incubation period. The most common method is the *in vitro* dark bottle incubation, which was originally developed as a means of accounting for respiration during the determination of photosynthesis within the "light/dark" bottle protocol. For reasons of sensitivity, oxygen is the usual determinant and comprises most (>90%) of the global database. Carbon dioxide (either pCO_2 or total dissolved inorganic carbon, DIC) has also been used (Johnson *et al.* 1983, 1987; Robinson and Williams 1999; Robinson *et al.* 1999). Dark incubations are necessary for the measurement of respiration via oxygen or carbon dioxide flux to eliminate concurrent counter fluxes of the same compounds by photosynthesis. The dark bottles are incubated at *in situ* temperature, either deployed on an *in situ* rig (usually alongside samples incubated for determination of primary production), or held in temperature controlled incubators. In order to measure respiration occurring in both the light and the dark, an extension of the oxygen flux method has been developed. This entails subtracting measurements of net community production (derived from the oxygen flux during a "light" bottle incubation) from concurrent measurements of gross photosynthesis (derived from $^{18}O_2$ production from a "light" bottle incubation of a sample spiked with ^{18}O-labeled water, Bender *et al.* 1987; Grande *et al.*, 1989*a*,*b*).

The *in vitro* nature of the incubation potentially can give rise to errors, which has prompted attempts to determine respiration by measuring oxygen and carbon dioxide changes *in situ*. This relies on the ability to sample consistently within the same water mass and to account for non-respiratory processes which would alter the concentration of the chosen reactant. Depending on the timescale, these latter processes could include air–sea exchange, evaporation and precipitation, and production/dissolution of calcium carbonate. The use of sulfur hexafluoride (Upstill-Goddard *et al.* 1991) significantly increases the ability to track a particular water mass over several days, thereby allowing consecutive estimates of community respiration to be made. However, in order to obtain a satisfactory signal to noise ratio the spatial variability of the reactant within the labeled area needs to be small in comparison with the time-dependent change in the dark period due to respiration. Low latitude, low energy gyre systems with long well-defined dark periods offer the most favorable physical circumstances, however respiration rates in such systems are characteristically low (see Williams and Purdie 1991). High latitude, high energy systems with short and poorly delineated dark periods and high spatial variability offer the least promising circumstances.

Limit of detection

The limit of detection of the *in vitro* approach is determined by a combination of (i) the precision of the analytical technique, (ii) the number of replicates, (iii) the ambient concentration, and (iv) the incubation time. The analytical methods for oxygen and carbon dioxide work near the practical limits of volumetry (estimated to be ≈0.02%, Robinson and Williams 1991), so the ambient concentration is the main determinant of the sensitivity of the analytical method. In Table 9.1 the theoretical limits of oxygen and carbon dioxide derived rate measurements have been estimated. As oxygen solubility is strongly temperature dependant in the oceans, the calculation of this gas has been made for a range of temperatures. These are estimates of best performance; the precision of the analyses falls away in areas of high particulate load (see Hopkinson and Smith, Chapter 8) and under fieldwork conditions. The final determinant is the time of incubation. In the euphotic zone, where there is a diel cycle of oxygen and carbon dioxide flux, then 12 or 24 h incubations

Table 9.1 Calculation of the precision limits for the measurement of changes in oxygen and total inorganic carbon concentrations

Determinant	Water temperature (°C)	Ambient concentration[a] (mmol m^{-3})	Coefficient of variation of the method[b] (%)	Limit of analytical detection[c] (mmol m^{-3})
CO_2		2100	0.02	0.6
O_2	0	350	0.02	0.1
O_2	15	250	0.02	0.07
O_2	25	200	0.02	0.06

[a] Rounded off value, calculated from the specified temperature and a salinity of 35.

[b] From Robinson and Williams (1991).

[c] Calculated, assuming four replicates, as $2 \times (\sqrt{2} \times \text{ambient concentration} \times \text{coefficient of variation}/(100 \times \sqrt{4}))$.

have a logic. In deeper water, the diel cycle has little or no relevance, and the important consideration here is the continued linearity of the rates.

Random and systematic errors

The measurement of respiration will be subject to a number of random and potential systematic errors. Random errors derive from two sources: (i) the analytical method for the measurement of the reactant and (ii) a form of time-dependent randomness between the incubated replicates. The former can be determined from the precision of the zero time measurements, the latter by the difference between the precision of the zero time replicates and that of the incubated replicates. These errors need to be reported, as they set the precision and therefore the significance of the observations. Typical median coefficient of variation of the zero time dissolved oxygen replicates during fieldwork conditions fall in the region of 0.015–0.07% while that of dark incubated dissolved oxygen replicates are 0.08–0.15% (Robinson *et al.* 2002*a,b*). Typical means of the standard errors of respiration measurements derived from dissolved oxygen flux range from 0.06 to 0.5 mmol O_2 m^{-3} d^{-1} (Williams and Purdie 1991; Robinson and Williams 1999; Robinson *et al.* 2002*a,b*; Williams *et al.* 2004). The median standard error of the oxygen flux derived rates of respiration collated for this review is 0.2 mmol O_2 m^{-3} d^{-1} ($n = 1012$). The determination of respiration from the coulometric analysis of dissolved inorganic carbon flux is less precise and the analytical method requires frequent calibration to maintain its accuracy. Reported means of the standard errors of the combined

respiration and net community production determinations derived from DIC flux during fieldwork are 1–1.5 mmol C m^{-3} d^{-1} ($n = 19$), and of respiration measurements alone, 0.8 mmol C m^{-3} d^{-1} ($n = 11$) (Robinson and Williams 1999; Robinson *et al.* 1999, 2002*a*).

Errors of accuracy (systematic errors) are always difficult to assess, particularly when no reference is available. We have categorized these nonrandom errors under three headings: *procedural errors, errors of containment,* and *errors of interpretation.* The separation is not perfect but suffices for the purpose of discussion.

Procedural errors

The necessity to derive respiration measurements from oxygen and similar parameter changes in the dark gives rise to two forms of error: (i) omission in the determined rate of light-associated respiration and (ii) a time-dependent run down of respiration due to a decrease in the concentration of the substrates that fuel respiration.

There are two forms of respiration that occur exclusively or predominantly in the light—notably the Mehler and the RUBISCO oxidase/photorespiration reactions. These, and their implications for the measurement of respiration, have been discussed by Raven and Beardall (Chapter 3) and Williams and del Giorgio (Chapter 1). They cannot be measured by the conventional dark bottle approach. Whether this is a problem, or a blessing, depends upon the purpose of the measurements. The argument is made in Chapter 1 that these reactions have little to do with organic metabolism—particularly the Mehler reaction—and that for much ecological work its omission from the respiration measurement is more likely a gain rather than a loss. This would not be the case for fundamental physiological work directed to photon capture and quantum yield when these two processes would need to be taken into account. The circumstances surrounding the RUBISCO oxidase/photorespiration reactions are less straightforward in this respect.

The run down of respiration due to the exhaustion of substrates has received the attention of a number of workers. Of 34 published time courses (Williams and Grey 1970; Williams 1981; Harrison 1986; Hopkinson et al. 1989; Kruse 1993; Biddanda et al. 1994; Pomeroy et al. 1994; Sampou and Kemp 1994; Blight et al. 1995; Arístegui et al. 1996; Carlson et al. 1999; Robinson et al. 1999; Robinson 2000; Robinson et al. 2002b), more than 80% show a linear rate of decrease in oxygen concentration (i.e. no observable change in respiration rate) for at least the first 6 days of incubation. In only two inshore studies (Williams 1981; Pomeroy et al. 1994), where respiration rates were high (from 5 to 11 mmol O_2 m^{-3} d^{-1}) were monotonic decreases or increases in oxygen concentration seen during incubations of less than 1 day. Where nonlinearities have been encountered it is possible to calculate the error that would have been incurred by using the simple two time point 24 h incubation and this varies from 2 to 20%; the very high proportion of linear time courses suggests that the lower estimate is the more representative one.

Long dark incubations give rise to more subtle problems in that they may disrupt any inherent diel cycle. Recent studies have highlighted the diel synchrony of growth of photosynthetic prokaryotes in both culture and open-ocean situations (Zubkov et al. 2000; Jacquet et al. 2001). A 24 h dark incubation of a picophytoplankton dominated community would therefore disrupt this light : dark cell cycle. The consequences and importance of such an effect are unknown.

Errors of containment

All *in vitro* approaches (be they in 50 cm^3 bottles or 10 m^3 mesocosms) involve containment. Containment may cause two categories of error; (i) the sample size will involve omission of part of the heterotrophic community, (ii) the container itself may impact the enclosed population giving rise to so-called "bottle" effects.

The omission of organisms will have two consequences: (i) there will be loss of a component of respiration, (ii) the absence of a predator will remove grazing pressure and so control on the abundance of the prey will be lost. Larger organisms occur at sufficiently low abundances that they are poorly sampled by *in vitro* procedures, and so community respiration is in this respect underestimated. It is possible to quantify the effect in order to establish its significance. Distributions of individual abundances and cumulative respiration rates can be derived from observed or theoretical profiles of biomass. Figure 9.1 (based on Platt and Denman 1977) shows the theoretical distribution of both abundance and cumulative respiration with size. The intersection of the two straight lines shows the individual size (ca. 50 μm) above which the organisms are poorly represented (less than 10 dm^{-3}) in samples taken for *in vitro* studies. However, by this size, in excess of 99% of calculated respiration is accounted for by the smaller size fractions (see circles in the figure). An extensive analysis is summarized in Fig. 9.15, and suggests a figure of 3% of respiration associated with organisms greater than 100 μm. Hernandez-Leon and Ikeda (Chapter 5), on the other hand, suggest a larger contribution of zooplankton to total respiration, in the order of 6 to 10% of the total water column respiration. Thus, although the omission of large organisms does not introduce a substantial error to estimates of community respiration, the larger size classes cannot be ignored.

Omission of a predator will have a potential secondary effect on respiration deriving from the increase of biomass of the prey and its associated respiration. Pomeroy et al. (1994) showed the increase in bacterial numbers and substrate assimilation rates which can occur during bottle incubations, which they concluded was due to the disruption of the grazer/prey link within the food web. These effects may be expected to be greater in size fractionation studies in which the size fraction greater than for example, 1 μm is removed. Sherr et al. (1999) found a 2 to 8 fold increase in bacterial activity of a 1 μm filtrate incubated for 2–3 days.

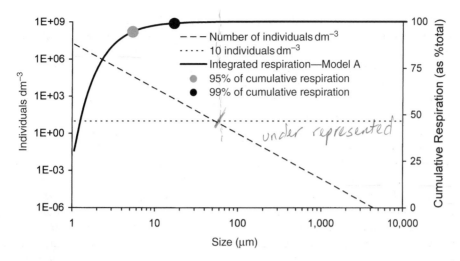

Figure 9.1 Calculated size distribution of numbers of individuals and cumulative respiration. The circles are the points of 99% and 95% cumulative respiration. Details are given in Fig. 9.14.

Where concomitant measurements have been made of changes in bacterial abundance and respiration rate, it has been found that whereas the former may increase, this surprisingly is not accompanied with an increase in community respiration rate (Pomeroy *et al.* 1994; Blight *et al.* 1995).

Aside from purported bottle-induced increases in bacterial numbers, there are well-defined instances of contamination effects, notably from heavy metals. This was one of the foci of the Planktonic Rate Processes in the Oligotrophic Oceans (PRPOOS) study and it was found (Marra and Heinemann 1984) that with careful attention to detail and cleanliness these effects could be overcome.

Errors of interpretation

The oxygen technique determines the decrease of oxygen in the dark and makes the assumption that this can be attributed to the physiological process of respiration. This assumption has been challenged by Pamatmat (1997), who argued that the decomposition of hydrogen peroxide, which gives rise to the production of oxygen, could aliaise the respiration measurements, depending upon the technique used to measure oxygen. Polarographic oxygen sensors and mass spectrometers will accurately record the changes in oxygen concentration; as a consequence the decomposition of hydrogen peroxide

would be recorded, quite properly, as the production of oxygen in the dark and therefore give rise to an underestimation of respiration. This would not occur if the Winkler reaction were used, as the reagents will react with hydrogen peroxide and so would be falsely but fortuitously recorded as the equivalent amount of oxygen—hence recording no net change due to the decomposition of hydrogen peroxide.

Pamatmat further discussed a cycle of hydrogen peroxide, which he claimed would give rise to the net production of 1/2 mole of oxygen per cycle. Pamatmat's proposed hydrogen peroxide/oxygen cycle can only give rise to net oxygen production at the expense of reduced structures. Whereas such a supply of organic proton and electron donors may exist in the types of environment he studied (tide pools, saline ponds), they would not accumulate in significant quantities in either shelf or offshore waters.

Significantly, where hydrogen peroxide concentrations are reported for coastal and offshore areas they are characteristically in the range of 2–50 nmol dm^{-3} (Herut *et al.* 1998; Hanson *et al.* 2001; Yuan and Shiller, 2001). The flux rates are in the range, 8–230 nmol dm^{-3} d^{-1} (Moffett and Zafiriou 1993; Yocis *et al.* 2000; Yuan and Shiller 2001). Both concentrations and fluxes of hydrogen peroxide are low in relation to their oxygen counterparts, and close to, or more often below, the sensitivity of the methods used to measure *in vitro* oxygen fluxes. In polluted waters, hydrogen peroxide concentrations are higher (8–100 nmol dm^{-3}; Herut *et al.* 1998), but even here they are near to or beyond the limit of the Winkler oxygen technique and so are a minor source of error in relation to the high respiratory oxygen fluxes observed in such waters.

Thus, present understanding of hydrogen peroxide cycling implies that the rates would be beyond the sensitivity of the Winkler incubation technique in open-ocean environments. Thus, Pamatmat's concerns over the use of the oxygen technique for both respiration and production would appear to be unfounded.

Anomalous observations and reconciliation with other measures of plankton activity and in situ *derived respiration rates*
One virtue of the *in vitro* oxygen flux technique is the ability to detect anomalous results; for example, apparent "negative" respiration rates or oxygen production in the dark. Williams (2000) analyzed a dataset of 293 observations after filtering the data for occasions where the gross production or respiration rate was less than twice its standard error or where "impossible" values of negative gross production or respiration occurred. This filtering process reduced the dataset by 5–10%, but the frequency of the negative respiration rates was not differentiated within this. We have access to the original titration data of 859 of the *in vitro* dark oxygen flux data collected for this review. Within this dataset, on only nine occasions did the final concentration of oxygen exceed the initial concentration, and so produce an anomalous "negative" respiration rate.

A different anomaly reported in the literature (Robinson and Williams 1999) is the determination of higher than expected respiratory quotients ($\Delta CO_2 / -\Delta O_2$) derived from the concurrent measurement of oxygen consumption and dissolved inorganic carbon production during a dark incubation. The authors were unable to account for these measurements without invoking the occurrence of metabolic processes such as methanogenesis in the aerobic euphotic zone of the Arabian Sea.

Several studies have addressed the question of whether *in vitro* estimates of plankton activity are representative of *in situ* rates. With regard to primary production, differences of ~25% between *in vitro* and *in situ* derived rates have been found (Williams and Purdie 1991; Daneri 1992; Chipman *et al.* 1993; Rees *et al.* 2001). Bearing in mind the difficulty of recreating an appropriate light field for *in vitro* estimates of primary production and the practical limitations to *in situ* work, it seems reasonable to suggest that the difference between *in vitro* and *in situ* derived rates of plankton respiration would be of a comparable or lesser magnitude.

9.2.2 Electron transport system (ETS) assay

The use of enzymatic indicators to estimate "potential" respiration is gaining acceptance due to their sensitivity (<0.1 mmol O_2 m^{-3} d^{-1}), especially for mesopelagic waters (del Giorgio 1992; Harrison *et al.* 2001; Arístegui and Harrison 2002; Arístegui *et al.* 2002; and see Hernández-León and Ikeda, Chapter 5). These methods are not yet routinely used in biogeochemical studies. ETS activity measurements have to be converted to *in situ* respiration rates by empirically determined relationships between ETS activity and respiration derived from dissolved oxygen flux. Such data interpretation faces problems, but no more so than those of other commonly used rate process techniques (^{14}C and ^{3}H-thymidine; del Giorgio 1992) or extrapolations of remotely sensed ocean color data to sea surface chlorophyll and primary production estimates (Sullivan *et al.* 1993).

The ETS method estimates the maximum activity of the enzymes associated with the respiratory ETSs in both eukaryotic and prokaryotic organisms. The rate of reduction of a tetrazolium salt (2-(*p*-iodophenyl)-3-(*p*-nitrophenyl)-5-phenyl tetrazolium chloride (INT)) is used as an indicator of electron transport activity, and so oxygen consumption or carbon dioxide production (Packard 1971; Kenner and Ahmed 1975; Arístegui and Montero 1995). The enzymatic rates are corrected to *in situ* temperature using the Arrhenius equation. Activation energies of between 46 and 67 kJ mol^{-1} have been measured (Arístegui and

Montero 1995; Arístegui *et al.* 2002). The ETS technique measures the maximum capacity of the terminal oxidation system *in vitro* (in the physiological sense). *In vivo* rates are controlled by the rates of oxidative phosphorylation and so the maximum rate is not thought to be achieved in the whole cell. The suppression is not constant, varying with taxonomic group and physiological state (Packard 1985), and so needs to be established empirically. Typical algorithms used are $\log R(\mathrm{mg}\,O_2\,\mathrm{m}^{-3}\,\mathrm{d}^{-1}) = 0.357 + 0.750 \log$ ETS ($r^2 = 0.75$; $n = 197$; Arístegui and Montero 1995) for oceanic surface waters or $R = $ ETS $\times 0.086$ (Christensen *et al.* 1980) for 200–1000 m depths. The error associated with such conversions is estimated to be $\sim \pm 30\%$ (Arístegui and Montero 1995, and see Hernández-León and Ikeda, Chapter 5).

9.2.3 Derivation from biomass

Given measurements or estimates of biomass within the planktonic community and appropriate biomass-specific rates of metabolism, a rate of respiration may be calculated. Methods are available to determine or estimate the biomass of most components of the planktonic community, although they are often very demanding of time and skill. An alternative to determining the biomass of individual organisms or groups is to use generalized derivations of biomass, either from observations (Sheldon *et al.* 1972) or theory (Platt and Denman 1978), within the size spectrum found in marine plankton. These can be coupled with allometric equations for respiration versus weight or size of the organism (Blanco *et al.* 1998). There is some consensus over the rate versus size relationship for respiration for larger plankton (see Fenchel, Chapter 4, and Hernández-León and Ikeda, Chapter 5), but as size decreases the basic assumption that one can specify metabolic rates becomes less secure: bacteria for instance probably exhibit the greatest metabolic range of all free-living organisms. Despite its obvious limitations the approach offers a valuable constraint to field observations of the distribution of respiration within the planktonic size spectrum (Robinson *et al.* 1999, 2002*b*).

9.2.4 Derivation from activity and growth efficiencies

Rates of primary production, bacterial production, and to a lesser extent microzooplankton herbivory and bacterivory are much more frequently determined than rates of plankton community respiration. Therefore, if the growth efficiency of each of these plankton groups can be sufficiently constrained, the respiration rate of each group can be calculated and the community respiration rate derived by summation. Unfortunately growth efficiencies, particularly those of the bacterial fraction, are not well constrained. While this approach has been used with some success—the summed estimated respiration of each trophic group falling within 50% of the measured community respiration (Robinson and Williams 1999; Robinson *et al.* 1999, 2002*b*)—the reliance upon an estimate of the growth efficiencies is the greatest limitation to this technique.

9.2.5 Inverse analysis

A different approach to the determination of community respiration or the respiration of a particular size class or group of the plankton, is to use inverse analysis (e.g. Vezina and Platt 1988). The procedure incorporates observations of rates or states and searches for the best fit of the unknown metabolic rate to the field data. It is potentially a very powerful approach as it is substantially, but not wholly, objective. However, the analysis is only as good as the data incorporated, and with many datasets, several parameters will not be known, requiring subjective interpretation. Fasham *et al.* (1999) used a size structured ecosystem model forced by measured water column integrated primary production and heterotrophic bacterial production to derive the size distribution of community respiration during a diatom bloom in the north east Atlantic. Ducklow *et al.* (2000) used equations of carbon flow through phytoplankton, zooplankton, and bacteria, along with the constraint of measured community respiration to estimate bacterial respiration, and Anderson and Ducklow (2001) used a simple steady-state model of the microbial loop to examine the reliability of measured bacterial production, growth efficiency, and hence respiration. As more datasets are collected which encompass concurrent rate and state measurements of most of the plankton groups, this procedure will become more commonplace.

9.2.6 Future analytical methods

The importance of plankton respiration measurements to an understanding of carbon flow through the microbial food web, and the relatively few observations currently available from traditional methods, lead us to continually seek new approaches which could improve our understanding of the spatial and temporal variability of community respiration as well as that of components of the plankton community.

The activity of the enzyme isocitrate dehydrogenase (IDH) has been investigated in batch cultures of marine bacteria in relation to carbon dioxide production and oxygen consumption (Packard *et al.* 1996). Rates of carbon dioxide production were predicted from models of second-order enzyme kinetics throughout the different phases of the cultures with an r^2 greater than 0.84 (Packard *et al.* 1996; Roy and Packard 2001). The potential to use this enzyme as an index of bacterial respiratory production of carbon dioxide in the ocean awaits further investigation and improvement.

Luz and Barkan (2000) have recently introduced a method to estimate time-integrated rates of respiration and photosynthesis from the difference between the triple isotope (^{16}O, ^{17}O, and ^{18}O) composition of atmospheric and dissolved oxygen and the rate of air–sea oxygen exchange. Their estimates of gross oxygen production agreed well with rates determined by bottle incubations in the Sea of Galilee and with seasonal cycles of production determined in upper ocean waters near Bermuda (Bender 2000). The advantage of the method is that it integrates over week to month periods, something that is impossible to achieve with bottle incubations. The disadvantages are the reliance on knowledge of gas exchange rates, (an associated error of about 30%) and of the fractionation by photosynthesis and respiration, and the cost and skill required for mass spectrometric

analysis of oxygen isotopes. However these latter problems are likely to diminish with time.

9.3 Community respiration rates

9.3.1 Overview of the data

Previous collations of community respiration data have gathered a few hundred measurements worldwide. Williams (2000) analyzed a dataset of 293 observations and assessed the interconnections between plankton community respiration, net community production, and gross production. Rivkin and Legendre (2001) compiled community respiration data from 12 studies ($n = 100$) in order to test the predictive capability of their equation relating bacterial growth efficiency to temperature and their assertion that the major part of community respiration is attributable to bacteria. Robinson *et al.* (2002*a*) collated data from six oligotrophic studies and four upwelling studies in order to calculate a mean respiration rate for oligotrophic regions of 2.4 mmol O_2 m^{-3} d^{-1} (<0.7–12.7; $n = 51$) and a mean respiration rate in upwelling areas of 6.5 mmol O_2 m^{-3} d^{-1} (0–33.4; $n = 132$).

In the present study, we have combined these datasets and augmented them with recently published data, unpublished data of our own, and unpublished and published data kindly supplied by colleagues (Arístegui, J., Gonzalez, N., Lefevre, D., and Morris, P., personal communication). The data can be accessed at www.pml.ac.uk/amt/data/Respiration.xls. This database will continue to be developed and maintained. We anticipate that we have acquired ca. 80% of the existing surface water respiration data. Data have not been used in the present analysis when the rate is less than twice the standard error (95% confidence limits). The combined dataset consists of 1662 respiration data (693 separate stations), collected between 1980 and 2002, from open ocean, coastal upwelling and shelf sea environments (Robinson and Williams 1993, 1999; Kiddon *et al.* 1995; Pomeroy *et al.* 1995; Cota *et al.* 1996; Lefevre *et al.* 1997; Eissler and Quinones 1999; Robinson *et al.* 1999, 2002*a,b*; Sherry *et al.*, 1999; Daneri *et al.* 2000; Hitchcock *et al.* 2000; Moncoiffe

Figure 9.2 Global database of respiration measurements positioned on map of SeaWiFs ocean color.

et al. 2000; Dickson and Orchardo, 2001; Dickson *et al.* 2001; Gonzalez *et al.* 2001, 2002). Where possible the respiration data are supported by concurrent measurements of *in situ* temperature, chlorophyll *a* concentration, bacterial abundance, particulate organic carbon, light attenuation, and gross production. Of the observations, 91% of the data are derived from Winkler oxygen titrations, 3% from oxygen consumption measured by oxygen electrode, 4% are "daily" respiration rates derived from $^{18}O_2$ (i.e. incorporating both "light" and "dark" respiration; Bender *et al.* 1999), and <2% of the data are derived from ETS activity. Open-ocean observations account for 71% of the measurements; the remainder have been collected on the shelf (<200 m). Figure 9.2 highlights just how spatially restricted and depauperate the global database is.

The mean respiration rate of this complete volumetric dataset is 3.5 ± 0.13 mmol O_2 m^{-3} d^{-1} (mean $\pm$ standard error, the standard deviation is 5.3; range of the mean is 0.02–75 mmol O_2 m^{-3} d^{-1}), and of surface water (<10 m) alone, is 4.9 ± 0.23 mmol O_2 m^{-3} d^{-1}. The mean surface water (<10 m) respiration rate measured in coastal areas is 7.4 ± 0.54 mmol O_2 m^{-3} d^{-1} ($n = 323$; range 0.1–75 mmol O_2 m^{-3} d^{-1}), and the mean of surface

water measurements made in open-ocean areas is 3.3 ± 0.15 mmol O_2 m^{-3} d^{-1} ($n = 502$; range 0.02–33.8 mmol O_2 m^{-2} d^{-1}). Where respiration was measured at three or more depths we have calculated depth-integrated rates, to the depth of the deepest sample using a conventional trapezoid procedure. The integration depth varies from 10 to 150 m with a mean of 53 m and a median of 40 m. Since respiration measurements have historically been made in support of gross production measurements, the deepest depth sampled in most cases is the 1% light depth rather than the mixed layer depth. The means for the whole depth integrated dataset ($n = 229$), the open-ocean dataset ($n = 186$), and the coastal/shelf seas dataset ($n = 43$) are 114 ± 5.1, 116 ± 8.5, and 107 ± 11 mmol O_2 m^{-2} d^{-1} (mean $\pm$ standard error; standard deviation $\sim$100 mmol O_2 m^{-2} d^{-1}), respectively.

The respiration data have been assessed with respect to latitudinal and seasonal distribution and covariance with more frequently made measures of plankton biomass and activity. Figure 9.2 indicates the spatial distribution of the respiration data presented on a satellite image of ocean color. Such a portrayal of the data emphasizes the urgent need for more data (Jahnke and Craven 1995; Williams 2000).

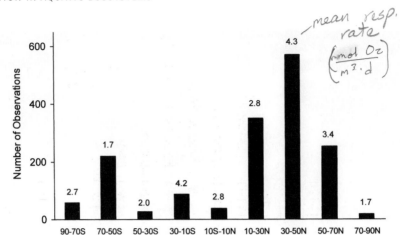

Figure 9.3 Latitudinal distribution of the number of observations of volumetric rates of community respiration. The mean rate for each latitude band (mmol O_2 m^{-3} d^{-1}) is shown above each histogram.

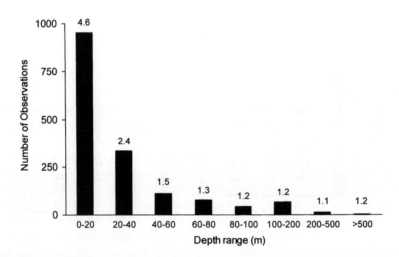

Figure 9.4 Depth distribution of the number of observations of volumetric rates of community respiration. The mean rate for each depth range (mmol O_2 m^{-3} d^{-1}) is shown above each histogram.

The frequency histogram of respiration versus latitude (Fig. 9.3) reveals that the greatest number of samples have been collected within the latitudinal band between 30°N and 50°N where the mean respiration rate is 4.3 mmol O_2 m^{-3} d^{-1}. As mentioned above, the majority of respiration measurements have been made within the euphotic zone. Less than 100 measurements of respiration (~5% of the dataset) were collected below 100 m (Fig. 9.4). Interestingly, of the small number of measurements made, the mean rate is still relatively high (>1 mmol O_2 m^{-3} d^{-1}) and easily measurable with a 24 h oxygen incubation technique. Data are unevenly distributed throughout the year (Fig. 9.5). As might be expected, most of the measurements were collected during the temperate and austral spring. In the Northern Hemisphere the mean respiration rates for April, May, June and July, were 3.5, 6.2, 5.8, and 8.9 mmol O_2 m^{-3} d^{-1}, respectively. The mean respiration rates for data collected during

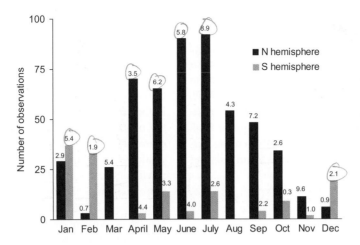

Figure 9.5 Seasonal distribution of the number of observations of volumetric rates of community respiration. The mean rate for each calendar month (mmol O_2 m^{-3} d^{-1}) is shown above each histogram.

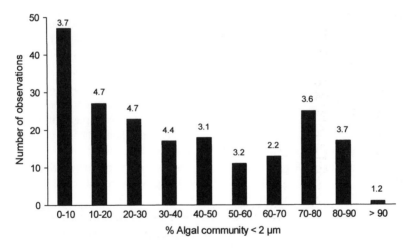

Figure 9.6 Distribution of the number of observations of volumetric rates of respiration with respect to the algal community size structure. The mean rate for each category (mmol O_2 m^{-3} d^{-1}) is shown above each histogram.

December, January, and February in the Southern Hemisphere are 2.1, 5.4, and 1.9 mmol O_2 m^{-3} d^{-1}, respectively. High mean respiration rates measured in September and November in the Northern Hemisphere are biased by coastal studies undertaken at this time of the year.

A focus of research in the late 1980s and 1990s was the study of global carbon cycling and the role of phytoplankton in drawing down atmospheric carbon dioxide (JGOFS Science Plan). Hence, one

expected bias in the respiration data is that the majority of the data were collected during diatom dominated spring blooms (Williams 1998). We have therefore analyzed a frequency histogram of respiration against a broad estimate of phytoplankton community structure—the percentage of total chlorophyll attributable to the smallest cells present (Fig. 9.6). Note that some authors have measured the fraction between 0.2–2 μm as their smallest size band, while others have used <1 μm. There does seem to be a bias towards data collected during larger cell dominated algal populations—with a smaller peak at populations, which have a large proportion (70–80%) of smaller cells. This is possibly due to the inclusion of the three datasets from the Atlantic Meridional Transect program (AMT), which sample the picoautotroph dominated mid-ocean gyres (Gonzalez *et al.* 2002; Robinson *et al.* 2002*a*). There is no clear pattern between the percentage of chlorophyll attributed to <2 μm cells and the magnitude of respiration.

9.3.2   Analysis of the data

The respiration data are biased with respect to time, place, community structure, and depth (see Figs 9.2–9.6). The dataset is particularly deficient in data from deeper water and for periods of the year of low photosynthetic activity. As a consequence the dataset is not comprehensive enough to allow the preparation of maps. In this respect, we are not at the stage that was reached by the ^{14}C community in 1968, when it was possible to produce a realistic map of oceanic productivity (see discussion in Barber and Hilting 2002).

Relationships between respiration and chlorophyll, bacterial abundance, particulate organic carbon, and attenuation
As community respiration is determined by the activity of the algal, bacterial, and microzooplankton community present, one may expect that respiration would be correlated with the abundance or biomass of one or more of the components of the plankton community. Robinson *et al.* (2002*a*) found significant ($p < 0.001$) relationships between respiration and bacterial abundance, chlorophyll *a*, beam attenuation, and particulate organic carbon (POC) of samples collected along a 12 000 km latitudinal transect of the Eastern Atlantic Ocean. Robinson *et al.* (2002*b*) combined these data with those collected during a study of a coccolithophore bloom in the North Sea, and similarly found significant relationships between respiration, and bacterial and microzooplankton biomass, POC, and chlorophyll *a*. Since these latter parameters are less time

and labor intensive in their collection and analysis, these authors drew attention to their value as "predictors" of community respiration. Of the variance in respiration, 40–50% could be accounted for by the variation in bacterial biomass or POC. Using the larger database of respiration measurements collected for the present review we were curious of the predictive power of one or more of these single variable regression equations. Figure 9.7 shows weak relationships between respiration, chlorophyll *a*, POC, bacterial abundance, and attenuation with their associated major reduced axis regressions. Only 20–30% of the variance in respiration can be accounted for by the variation in one of these more routinely measured parameters. Unfortunately, the concurrent data are too sparse to probe this unexpected result. Of the 18 individual studies where respiration and chlorophyll are reported, in only 50% of cases is the r^2 of the relationship between respiration and chlorophyll *a* greater than 0.3. These nine studies tend to be in shelf seas with large ranges in chlorophyll (0.2–50 mg m^{-3}) and respiration (1–53 mmol O$_2$ m^{-3} d^{-1}). The relationship between respiration and chlorophyll *a* does not appear to depend on autotrophic community structure, for example, within the subset of data where less than 50% of the total chlorophyll is attributed to cells <2 μm in diameter (i.e. communities dominated by larger cells) respiration and total chlorophyll were not correlated.

Relationships between respiration and photosynthesis
Assuming that the open ocean is essentially a closed system (and this is debated), then respiration will be principally fueled by photosynthetic production of organic material. The coupling between photosynthesis and respiration will be both physiological (in algae) but also trophic (in heterotrophs). In the former case, the coupling may be tight and on physiological rather than ecological timescales. The flow and recycling through the food web introduce time lags, so that the two processes of community photosynthesis and respiration become out of phase with one another—the latter trailing the former. Temporal studies in pelagic marine ecosystems on timescales that allow the examination of the phasing of the two processes are few (Blight *et al.* 1995; Serret *et al.*

is the ocean a closed system?

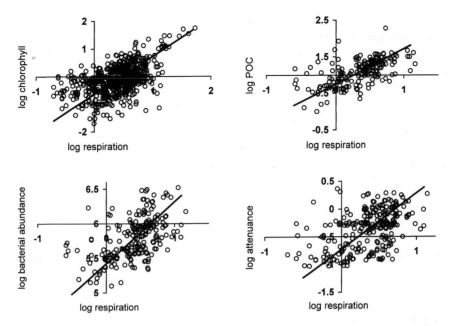

Figure 9.7 Log$_{10}$–log$_{10}$ plots of volumetric rates of respiration against chlorophyll a, POC, bacterial abundance, and attenuance. Respiration rates as mmol O$_2$ m^{-3} d^{-1}, chlorophyll as mg chl a m^{-3}, bacterial abundance as 10^9 cell dm^{-3}, POC as mmol C m^{-3}, attenuance as m^{-1}. The fitted lines are a Model 2 major reduced axis. Chlorophyll: $y = 1.4x + 0.64$; $r^2 = 0.27$, $n = 628$; bacterial abundance: $y = 0.89x + 5.4$; $r^2 = 0.27$, $n = 205$; POC: $y = 0.97x - 0.75$; $r^2 = 0.32$, $n = 170$; and attenuance: $y = 1.01x - 0.72$; $r^2 = 0.22$, $n = 260$.

⤷ om, particles

1999; Williams 2000). In open-water situations, such studies are hindered by advective processes, hence mesocosms offer a more successful environment for such work. Figure 9.8 shows the results of studies in a series of replicate 11 000 dm^3 mesocosms undertaken in a Norwegian Fjord. Figure 9.8(a) shows the simple time series. This illustrates the timescale of the phasing between photosynthesis and respiration. In this instance respiration trails photosynthesis by about 4 days. Figure 9.8(b) and (c) illustrate the swing from autotrophy to heterotrophy. Figure 9.8(d) shows the instantaneous relationship between photosynthesis and respiration. Two lines are fitted to the data—an ordinary least squares (OLS) and major reduced axis (MRA) regression. Both lines have slopes less than unity and this reflects the smaller variance in the respiration dataset as compared with concurrent measurements of photosynthesis. This almost certainly arises as a consequence of time delays stemming from the flow of organic production through the various trophic groups, thus dampening the fluctuations seen in

primary production (see Aristeguí and Harrison 2002). The system being intermittently charged up by pulses of photosynthesis and discharged by a continuous leakage of organic material to respiration (Williams 1995). These concepts are useful in the interpretation of open field data.

The dataset of respiration, when compared to that of ^{14}C-determined photosynthesis is small, about 1% of the photosynthetic observations (Williams and del Giorgio, Chapter 1). An obvious solution to this problem is to seek for relationships between respiration and photosynthesis. If such relationships can be found, then basin wide estimates of respiration could be derived from the massive database of photosynthesis, and potentially from models of photosynthesis.

In Fig. 9.9 we compare the frequency distributions of field observations of photosynthesis and respiration from our database. In the case of the volumetric rates, the respiration rates have a narrower distribution than the photosynthetic rates—consistent with the capacitance effect of the food

OLS - ordinary least sq.
MRA - major reduced axis

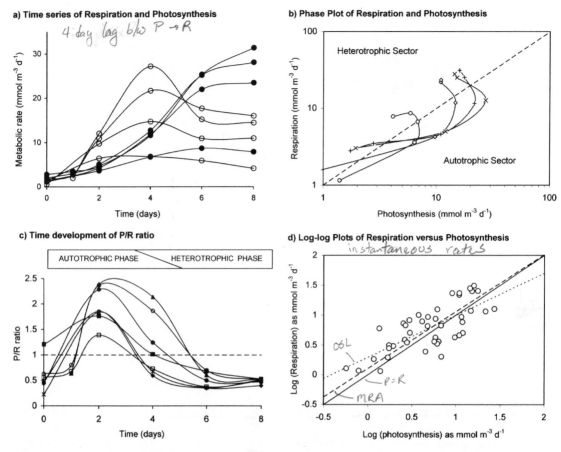

Figure 9.8 Analysis of time series of photosynthesis and respiration followed over 8 days in a series of 11 m³ mesocosms. (a) Time-series plot of photosynthetic rate (open circles) and respiration (filled circles), (b) phase plots of photosynthesis and respiration, (c) time development of P/R ratio, and (d) $\log_{10}-\log_{10}$ plots of photosynthesis and respiration. The dotted line shows an OLS fit, the dashed line an MRA fit and the solid is the $P = R$ line.

web on respiration. The frequency distribution of the depth-integrated rates of respiration is skewed at the upper end—a product of a small number (5–6 profiles) of very high respiration values. For this reason we have worked with log-normalized rates, as most of the line-fitting procedures we use require a normal distribution of values within the dataset.

The first attempt to analyze a wide range of environments for the relationship between photosynthesis and respiration was made by del Giorgio et al.(1997). They found that respiration was scaled to photosynthesis with an exponent less than unity. Duarte and Agusti (1998) using a larger dataset

(280 observations) confirmed the findings of del Giorgio et al. They found that community respiration was scaled as the approximate two-thirds power of photosynthesis, the correlation between the two parameters was high ($r^2 = 0.42$). The slope of the log–log relationship for the open ocean was somewhat lower (0.5). We have repeated this analysis (see Fig. 9.10) on the larger database now available to us (957 paired observations of volumetric rates of oxygen flux derived gross production and respiration). The correlation is much the same ($r^2 = 0.43$), but the slope is reduced to 0.4, this is most likely due to the addition of datasets from coastal regions and low latitudes (Holligan et al.

(a)

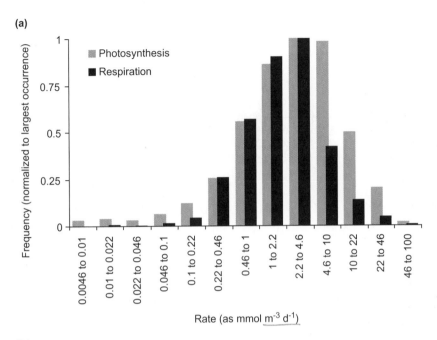

(b)

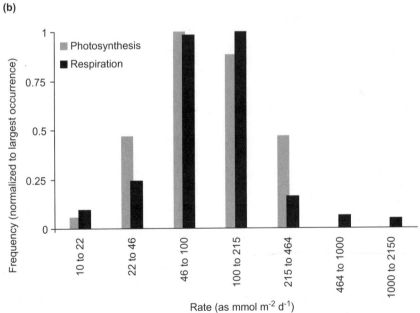

Figure 9.9 Frequency distribution of photosynthetic and respiration rates. The data has been sorted into three bins per decade and are normalized to the largest sample. Frequency distributions of (a) volumetric rates and (b) depth-integrated rates are shown.

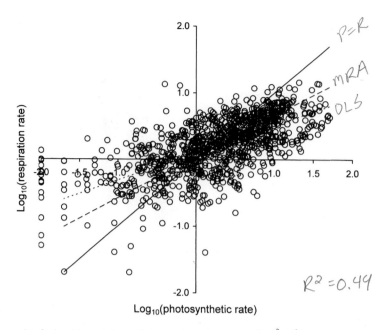

Figure 9.10 Log_{10}–log_{10} plot of volumetric respiration and photosynthesis. Rates as mmol m^{-3} d^{-1}. The dotted line shows an OLS fit ($y = 0.412x + 0.1$), the dashed line an MRA fit ($y = 0.62x + 0.04$), and the solid is the $P = R$ line. The R^2 of the log-normalized data was 0.44.

1984; Arístegui *et al.* 1996; Robinson *et al.* 1999; Robinson unpublished data) where low P/R ratios have been recorded. The variance of photosynthesis (expressed as the standard deviation) is greater than respiration (0.72 versus 0.42), which is the basis of the low slope. Both del Giorgio *et al.* (1997) and Duarte and Agusti (1998) noted that at low photosynthetic rates, the rates of respiration exceeded those of photosynthesis. Based on their derived relationship between photosynthesis and respiration, and the estimated primary production in each biogeochemical province in the ocean, Duarte and Agusti (1998) concluded that "80% of the ocean's surface are expected to be heterotrophic...," supported by a net autotrophy in the remaining 20% of the ocean. This was a controversial conclusion. The two matters of contention were the timescale of the net heterotrophy (whether permanent or transient) and the supply mechanism of the carbon required to support the net heterotrophy.

With volumetric rates one encounters a separation of photosynthesis and respiration, in time (as illustrated in Fig. 9.8) but also in space (depth). This complicates the elucidation of the ecological

relationship between photosynthesis and respiration. Williams (1998) suggested that the depth-integrated rates would overcome the second of these two effects. The penalty is a fivefold reduction in the number of paired photosynthesis and respiration observations from 957 volumetric rates to about 200 depth-integrated rates. The depth-integrated rates are analyzed in Fig. 9.11.

There are important differences between the relationships between photosynthesis and respiration in the volumetric and depth-integrated datasets. The principal difference is that in the log normalized plots, most (72%) of the correlation between photosynthesis and respiration seen in the volumetric rates is lost in the depth-integrated rates. The reason for such a major loss in correlation is not wholly clear and critical in understanding its significance. It could be that the main basis of the relatively high correlation in the volumetric plots came from internal correlation in individual profiles—were this the case, then the relationships obtained from analyses of volumetric plots would have little capability to predict regional differences. A further reason could be sampling bias, either with respect to depth

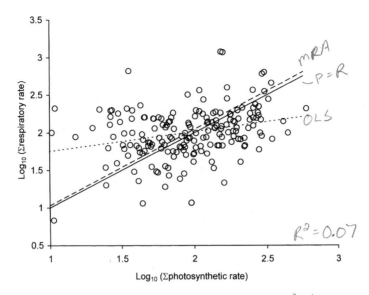

Figure 9.11 Log_{10}–log_{10} plot of <u>depth-integrated respiration and photosynthesis</u>. Rates as mmol m^{-2} d^{-1}. The dotted line shows an OLS fit ($y = 0.26x + 1.5$), the dashed line an MRA fit ($R = 1.0x + 0.01$), and the solid is the $P = R$ line. The R^2 of the log-normalized data was 0.07.

(Fig. 9.4), or community structure (Fig. 9.6) as proposed by Serret *et al.* (2001). Due to the different depth distributions of photosynthesis and respiration, the disparity between the variances in the photosynthetic and respiration rates, seen in the volumetric plots is lost in the depth-integrated plots (standard deviation respiration = 0.34; standard deviation photosynthesis = 0.33). This implies that most of the nonlinear scaling between volumetric rates of photosynthesis and respiration comes from the differences in the depth distribution of these two properties.

The low correlation ($r^2 = 0.13$) between depth-integrated photosynthesis and respiration, suggests that areal photosynthesis is a weak predictor of areal respiration. In many respects it is a disappointing result but was the prediction of Williams and Bowers (1999) and the conclusion of Serret *et al.* (2001, 2002). The conclusion does not sit comfortably with the proposition, at the beginning of the section, that respiration is principally fueled by autochthonous photosynthesis. This may be due to a number of factors: for example, the systems are not closed, or there are temporal offsets between respiration and photosynthesis.

The argument has been made (e.g. Duarte and Agusti 1998; Serret *et al.* 2002) that there are differences between the balance of photosynthesis and respiration in different biogeochemical zones, thus simply working from global averages is inappropriate. Longhurst (1998) identified 55 biogeochemical zones in the oceans, these would represent the most logical framework into which to bin the present data, however with only ~200 depth-integrated observations this degree of resolution is presently far too fine. As an alternative, in order to attempt an analysis, we have grouped the data in 10° of latitude, this gives 17 bins. Even with such a coarse separation the distribution is exceedingly patchy and does not in any way match the areas of ocean within each latitudinal band (Fig. 9.12). The Pacific and the Indian oceans are particularly poorly sampled with no measurements south of 10°N and above 30°N. Most of the measurements lie between 20–50°N in the Atlantic. In the Atlantic the estimates for the latitudes greater than 20°N show some degree of consistency (see Fig. 9.13), with photosynthesis around 120 mmol O_2 m^{-2} d^{-1} and respiration close to 95 mmol O_2 m^{-2} d^{-1}. When the variance in the dataset is considered the difference

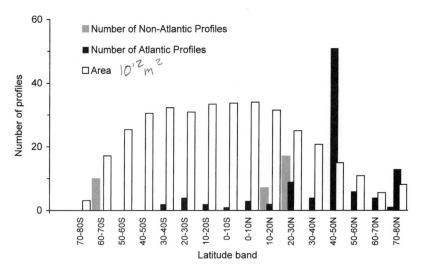

Figure 9.12 Comparison of the latitudinal distribution of depth-integrated rates of community respiration in the Atlantic and other oceans. The data is divided into Atlantic and non-Atlantic Ocean and presented alongside the area (as $10^{12}\,m^2$) of each of the latitudinal bands of ocean.

of $+25\,mmol\,O_2\,m^{-2}\,d^{-1}$ between photosynthesis and respiration is just significant. The rates reported for the latitudes between $20°N$ and $20°S$ are higher (photosynthesis $168\,mmol\,O_2\,m^{-2}\,d^{-1}$; respiration $230\,mmol\,O_2\,m^{-2}d^{-1}$), the number of samples is however low and the variance considerable, leaving the difference ($-62\,mmol\,O_2\,m^{-2}\,d^{-1}$) of borderline significance.

Whereas the differences between respiration and photosynthesis within the current dataset may be on the edge of significance, the pattern of negative net community production on the eastern periphery of the North Atlantic Gyre is a repeated observation (Duarte and Agusti 1998; Duarte et al. 2001; Robinson et al. 2002a; Serret et al. 2001, 2002) and in contradiction with geochemical studies of biogenic (O_2 and CO_2) gas exchange (Jenkins and Goldman 1985; Keeling and Shertz 1992; Emerson et al. 1997, 2002). Several suggestions have been made to account for this anomaly, none of which have as yet been fully proved or refuted. Sampling methodologies could be underestimating gross production, second the area could be fueled by a carbon source from distant previously net productive areas, finally local temporal separation of photosynthesis from respiration could alias the results. The second explanation was put forward by Harrison et al. (2001) to account

for the net heterotrophic areas in the eastern subtropical Atlantic where they argued that the subsidy was derived from filaments advecting offshore from productive areas.

There are very few seasonal studies in low productivity areas where net heterotrophy has been reported—hence one could be observing a net heterotrophic phase following an unobserved net autotrophic one (Serret et al. 1999). This has recently been remedied with a 13-month study at the HOT site (station ALOHA at $22°45'N$ and $158°00'W$) in the subtropical Pacific gyre (Williams et al. 2004). The annual totals (photosynthesis $22\,mol\,O_2\,m^{-2}\,a^{-1}$; respiration $31\,mol\,O_2\,m^{-2}\,a^{-1}$) give a deficit of $9 \pm 1.6\,mol\,O_2\,m^{-2}\,a^{-1}$, which is significant and begs an explanation. In this case the proposition of allochthonous carbon sources is not compelling, as the areas in question are physically isolated. The gradients of dissolved organic carbon have been measured and they are in the opposite direction (Abell et al. 2000). The scale of the difference and the lack of any period of protracted seasonal accumulation of organic material exclude long term out of phase relationships between photosynthesis and respiration. The explanation favored by Williams et al. (2004) is based on intermittency of photosynthesis; the argument (Karl et al. 2003) being that there

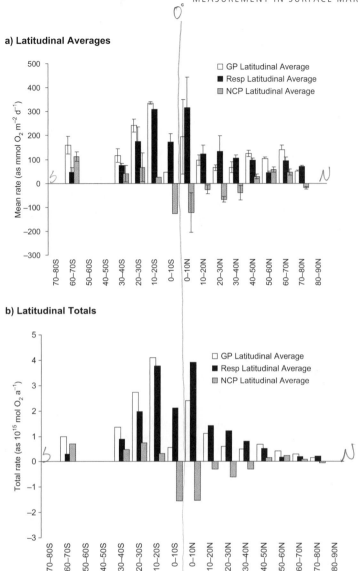

a) Latitudinal Averages

b) Latitudinal Totals

Figure 9.13 Comparison of the latitudinal distribution of depth-integrated rates of community respiration, net community production, and gross production. (a) the geometric means for 10° latitude bands, the error bars represent the standard error for each band; (b) the total rates for each latitude band where data exists, calculated from the data in Fig. 9.12(a) and the oceanic area for each latitude band. The production values are concurrent O_2-derived rates of gross and net community production.

are short intensive bursts of photosynthesis, which charge up the organic reservoir, which is then slowly and steadily discharged by respiration. The evidence for this comes from long-term continuous *in situ* observations of dissolved oxygen in the euphotic zone, which show periods of rapid accumulation of biologically produced oxygen. The explanation

has an interesting and provocative corollary—that because of its better integrating properties, respiration may be a superior determinant of time-integrated production than the measurement of photosynthesis itself. Whereas these explanations are the best we can currently offer; one inevitably asks why our sampling protocols have persistently missed these pulses and always made measurements during the periods of deficit. Without doubt we have much to learn about these systems and a great deal more work to do before we can expect an understanding of how they function.

9.4 Distribution of respiration within the planktonic community

When we consider the distribution of respiration within the community, we implicitly address two quite separate questions. The first is the distribution of respiration between the autotrophic and the heterotrophic communities; autotrophic respiration being that part of primary production (photosynthesis) not available to the rest of the community. The second question is the distribution of respiration within the heterotrophic component of the plankton—this provides insight into ecosystem functioning.

As there is no accepted procedure for blocking the metabolism of different sectors of the plankton with selective inhibitors, alternative ways of apportioning respiration within the planktonic community have had to be devised. They fall into five categories:

1. Calculations of the distribution of respiration from either observations of biomass or from algorithms of biomass distribution.
2. Measurements of the size distribution of respiration from which the respiration of organism groups may be inferred.
3. Derivation of respiration rates from measured or estimated rates of production, and growth efficiencies.
4. Determination of algal respiration as the difference between gross production and ^{14}C derived net primary production.
5. Derivation of respiration rates for various trophic groups from plankton food web models.

9.4.1 Calculations from biomass

Calculations from observations of biomass
Whereas there are a number of datasets for the biomass of particular plankton groups there are few comprehensive studies—in part due to the time and skill required to make a full analysis of biomass. No study, to our knowledge, contains estimates for all the major components of the food web (bacteria, protozoa, algae, and larval and adult zooplankton), so the apportionment of the relative contribution to metabolism will be incomplete. Clearly, if a class is omitted, then the contribution of the groups reported will be overestimated. The problem of ascribing rates to groups across the full size spectrum of the plankton is nontrivial. It is tempting to search for universal allometric relationships. In the case of the algae it seems possible to constrain biomass-related respiration reasonably well (Langdon 1993). Metazoa and protozoa would appear to have a sufficiently stable respiratory metabolism that a single allometric equation may be acceptable (see Fenchel, Chapter 4, Section 4.3), however one runs into major difficulties with the bacteria.

Table 9.2 contains a summary of reported calculations of respiration for various trophic groupings, expressed as a percentage of the sum of the reported rates. By far the most comprehensive study was that of Williams (1982) who used the data from the CEPEX study. This comprised data on bacteria, 3 categories of protozoa, 9 categories of algae, and some 44 categories of zooplankton, spanning the adults and all the larval stages. These observations were made on a large (1300 m^3) mesocosm at 2- or 3-day intervals over a 2-month period. The data in the table show broad agreement between the various estimates of the contribution of different trophic groups to overall respiration, with bacteria comprising 12–58%, the protozoa a further 11–36%, the algae 8–70%, and the larvae and adult zooplankton 3–9%.

The CEPEX dataset analyzed by Williams was sufficiently extensive (it spanned $3\frac{1}{2}$ orders of magnitude in the size of the organisms studied) that a size distribution of metabolism could be constructed. This is shown in Fig. 9.14, along with projections of respiration made from generalizations of biomass distribution. In its broad features it accords very well with observations of size-fractionated respiration and the projections from theoretical biomass profiles, the main discrepancy being in the 1–3 µm size range. No organisms of that size were logged in the CEPEX study, probably not because they

Table 9.2 Estimates of the contribution of various groupings to community respiration

		Bacteria	Protozoa				Phytoplankton				Zooplankton			
		Total	Nanoflagellates	Heterotrophs dinos.	Ciliates	Total	Pico	Nano	Micro	Total	Larval	Adults	Others	Total
Calculation from biomass determinations														
Williams (1981) (*n* = 3; geometric mean)	Mesocosm (Canada)	52	Not included in estimate					40 →		40	3	1	1	5
Holligan *et al.* (1984)	English Channel	← →	← 83 →				7.8 →		7.8	Not included in estimate			9.4
Robinson *et al.* (1999) (*n* = 7; geometric mean)	East Antarctic	12	3	11	1.4	15	69 →		69	Not included in estimate			
Robinson and Williams (1999) (*n* = 6; geometric mean)	Arabian Sea	11				20			13				
Sondegaard *et al.* (2000) (*n* = 3; geometric mean)	Mesocosm (Norway)	51	← 35 →			35	12 →		12	Not include in estimate			
Robinson *et al.* (2002*b*) (*n* = 6, 8; geometric mean)	North Sea	58(18[a])		21	5	28			21				
Summary statistics														
Arithmetic mean of all observations		32	3	18	4	24			34	3	1	1	6
Standard deviation of all observations		22	1	12	4	16			28	1	1	1	3
Total number of observations		26	7	13	13	26			27	3	3	3	4
Outcome from plankton models														
Fasham *et al.* (1999) (inverse model)	Northeast Atlantic	51	← 11 →			11	4.2	22	6.2				6
Nagata (2000) (plankton flow model)	Oligotrophic ocean	35	← 36 →			36	Not inlcuded in estimate						

[a] Two calculation methods for bacterial respiration.

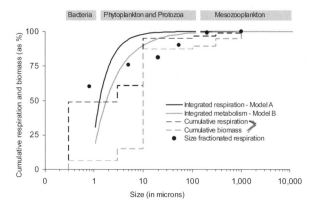

Figure 9.14 Size distribution of observed biomass and calculated and observed cumulative respiration. Integrated metabolism, Model A based on Platt and Denman (1977), Model B based on Sheldon *et al.* (1972). See Box 9.1 for basis of calculation. Cumulative and size-fractionated respiration and biomass taken from Williams (1982).

were absent but that the skills were not available to recognize and enumerate them.

There is one caution with the data derived from field observations generally. As discussed earlier the sampling in many cases is biased to the productive part of the plankton cycle, this would put a bias to the data toward high algal contributions to overall community respiration.

Calculations from generalized biomass profiles
Generalized continuous biomass profiles of the plankton have been derived from field observations (Sheldon *et al.* 1972) and from theory based on an idealized food chain (Platt and Denman 1978). Both of these studies were evolved prior to the modern understanding of the microbial food web however, it would appear that they give a fair account of the modern view of the size of microbial biomass and metabolism.

Sheldon *et al.* (1972) concluded that the biomass profile was flat within logarithmic size classes. Platt and Denman (1978) found a decreasing biomass with increasing size, the slope being −0.22 (the broad structure of Platt and Denman's derivation is given in Box 9.1). These biomass/size distributions can be combined with equations of respiration versus size to produce size distributions of respiration. In general they show the same features, the size classes below 10 μm account for more than 95% of the respiration; the size classes from 30 to 10 000 μm accounting for less than 1% of the total. A similar analysis was made by Platt *et al.* (1984)

who also highlighted the importance of microbial metabolism.

9.4.2 Experimental determination of the size distribution of respiration

In lieu of selective inhibitors to block particular sectors of the plankton, size-fractionation filtration procedures have been used to progressively remove components of the plankton size spectrum—with respiration measurements being made on the various filtrates. This was used initially in studies of the size distribution of photosynthesis and was applied subsequently by Williams (1981) to establish the scale of bacterial respiration in relation to the whole community. The approach has a number of limitations: it cannot, for example, resolve between different functional groupings occupying the same size category. They may be inferred with a degree of certainty in the case of the extreme end-members, that is, bacteria (<1 μm) or the mesozooplankton (>100 μm), as they dominate their class sizes. Intermediate size classes will contain both heterotrophs (mainly protozoa) and autotrophs in variable proportions. There are also problems with attached organisms being included in larger-size classes and the increase in prey consequent upon the removal of its predator. These cautions must be taken as constraints when interpreting the data. Time-series studies of size-fractionated samples and comparison of the size-fractionated respiration rates with those of the whole community can act as indicators of the scale of the constraint. The

---

Box 9.1 Theoretical calculation of biomass and respiration size distribution

A general theory of size distribution within the plankton was derived by Platt and Denman (1977). It divides the size spectrum into a number of logarithmically increasing steps, either decades or octaves, and seeks to relate the biomass within a bandwidth to weight of the size class using an equation of the form:

$$b_w/b_0 = (w/w_0)^\alpha \qquad (1)$$

where:

b_w is the total biomass within a bandwidth;
w is the characteristic weight of the size class;
α is an allometric scaling constant; and
b_0 and w_0 are values for one particular size class, which is used to fix the relationship.

Assuming a unidirectional flow and Fenchel's (1974) empirical equations for the weight dependence of turnover ($\tau_w = Aw^\chi$) and respiration ($R_w = Bw^\gamma$), where $A, B,$

χ and γ are constants, one arrives at a solution for the size distribution for biomass equation (1) as:

$$b_w/b_0 = (w/w_0)^{-0.22} \qquad (2)$$

The value for α of -0.22 in equation (2) comes from the constants $A, B,$ and Fenchel's observation that $\chi + \gamma \approx 1$ Equation (2) may be rearranged as:

$$w/w_0 = (b_w/b_0)^{-1/0.22} \qquad (3)$$

$$w = w_0(b_w/b_0)^{-1/0.22} \qquad (4)$$

Equation (4) can then be embedded into an overall allometric cell mass versus respiration equation ($R_w = Bw^\gamma$), such that:

$$R_w = B\, w_0^\gamma\, (b_w/b_0)^{-\gamma/0.22}, \qquad (5)$$

where characteristically γ has a value in the region of -0.3.

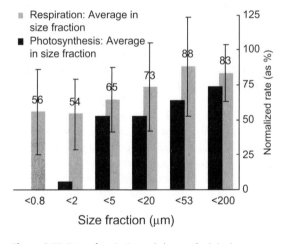

■ Respiration: Average in size fraction
■ Photosynthesis: Average in size fraction

Figure 9.15 Rates of respiration and photosynthesis in size fractions. The rates are normalized against that of the unfractionated sample. Compiled from 62 sets of observations. Error bars are the standard errors.

size distribution of respiration from several studies (Harrison 1986; Griffith *et al.* 1990; Blight *et al.* 1995; Boyd *et al.* 1995; Robinson *et al.* 1999) is summarized in Fig. 9.15, alongside size-fractionated photosynthetic rates.

9.4.3 Derivation from rates of production and growth efficiencies

Several studies (Robinson and Williams, 1999; Robinson *et al.* 1999, 2002*a,b*) have apportioned community respiration to that attributable to planktonic trophic groups using measured rates of bacterial production, microzooplankton herbivory, and/or bacterivory and algal photosynthesis together with an estimate of the respective growth efficiencies.

Bacterial respiration (BR) can be derived from measured bacterial production (BP) if the bacterial growth efficiency (BGE) can be estimated (i.e. BR = BP(1 − BGE)/BGE). If not directly measured, bacterial growth efficiency is either taken from the range of published values, for example, 5–40%, or derived from the equation (BR = 3.7 × BP$^{0.41}$) reported by del Giorgio and Cole (1998) or the temperature relationship (BGE = 0.374 − 0.0104 × t°C) of Rivkin and Legendre (2001). Bacterial growth efficiencies estimated by these two latter equations were found to differ by almost ninefold during a study of a coccolithophore bloom in the North Sea in June 1999 (Robinson *et al.* 2002*b*, also Table 9.2). Microzooplankton respiration (μZR) can be derived

$$BR = \frac{BP(1-BGE)}{BGE}$$

from an estimate of herbivory, assuming a constant growth efficiency of 25% (Straile 1997) and an egestion + excretion rate of 20% (Robinson *et al.* 2002*b*). Algal respiration (AR) can be derived from the $R_{max} : P_{max}$ relationship of Langdon (1993) for the dominant autotroph present, that is, 0.16 ± 0.04 for Prymnesiophyceae and 0.35 ± 0.17 for Dinophyceae, where P_{max} is estimated from ^{14}C primary production or dissolved oxygen gross production measurements.

Given that one can estimate the rates for these various groups, overall community respiration should be equivalent to their sum (BR + μZR + AR). In the studies where this type of accounting exercise has been attempted (Robinson and Williams 1999; Robinson *et al.* 1999, 2002*a,b*), the sum of the estimates of respiration attributable to the component plankton groups lie within 50% of the measured whole community respiration. This is probably as good as can be expected bearing in mind the levels of uncertainty on accepted measures of phyto, micro-zoo, and bacterioplankton activity, not to mention the four to tenfold range in measured and estimated growth efficiencies.

9.4.4 Estimates from ^{14}C determination of net primary production

Since net primary production (NPP) is equivalent to gross production (GP) minus algal respiration (NPP[C] = GP[C] – AR[C]), if one can assume that the ^{14}C technique measures net primary production, that gross production may be derived from oxygen flux (GP[O$_2$]) and that it can be reliably converted to carbon flux GP[C] via a photosynthetic quotient (PQ), then algal respiration can be determined from the difference between GP[O$_2$]/PQ and ^{14}C uptake. Heterotrophic respiration (HR[C]) could then be determined from the difference between measured total community respiration in carbon units (R[C]) and this estimated algal respiration (AR[C]). The approach has not gained much popularity due to the uncertainty over whether or not, and when, the ^{14}C measures gross or net production, as well as the lesser problem of ascribing a value for the photosynthetic quotient however see Marra and Barber (2004).

9.4.5 Inverse analysis

As described in Section 9.2.5 above, several authors have used equations and models of carbon flow through the microbial food web to estimate the respiration attributable to a particular plankton group, for example, bacteria or size class, for example, <5 μm (Fasham *et al.* 1999; Ducklow *et al.* 2000; Anderson and Ducklow 2001). Fasham *et al.* 1999 used a size structured ecosystem model forced with data collected during a 20 day spring bloom study in the NE Atlantic to conclude that 62% of the community respiration came from organisms <5 μm in size. The data extracted from this model are given in Table 9.2, along with that from a model of the food web of an oligotrophic oceanic region derived by Nagata (2000).

9.4.6 Summary

In Fig. 9.16, we have attempted to collate the results from these various approaches to the determination of the size distribution of respiration. Considering the scope for variation and error, the general pattern is remarkably consistent. All the approaches attribute a major proportion (70% or more) of respiration to the smaller-size classes: $42 \pm 10\%$ to organisms less than 1 μm (the size class where bacterial biomass usually dominates), $32 \pm 20\%$ to the size class 1–10 μm (bacteria, protozoans, and phytoplankton). Much smaller proportions are found for the two larger-size classes: $6 \pm 5\%$ to the class 10–100 μm (the larger phytoplankton and protozoan ciliates) and the largest-size class (100–1000 μm—the larval and adult mesozooplankton) accounting for $10 \pm 8.5\%$. Of course one should not expect a fixed pattern, as the composition of the plankton varies in space and with time.

9.5 Synthesis and conclusions

Respiration is one of the two determinants of the balance of organic material in the euphotic zone and the potential to export organic material to the deep ocean. As such it is a major influence on the scale of the biological pump—which sequesters carbon dioxide from the atmosphere.

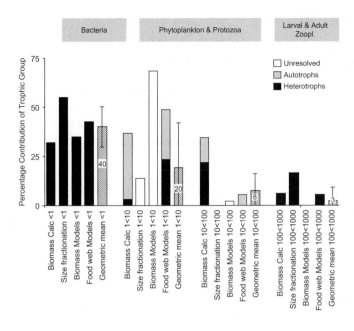

Figure 9.16 Summary of size distribution observations. The various analyses are grouped into categories <1 μm, 1–10 μm, 10–100 μm, and 100–1 000 μ m. Where a distinction was made in the analyses between autotrophs and heterotrophs, this distinction has been retained, otherwise they are grouped as "unresolved".

9.5.1   Estimation of global rates of respiration from direct field observations

The "light/dark" bottle incubation technique is the primary method for measuring respiration. It is not without its errors and limitations, however it is without doubt in our minds as good, if not better, than other methods of measuring plankton metabolism. Acquiring appropriate sized datasets is massively costly in ship and human time. Nonetheless, since the late 1960s we have had reasonably good maps of oceanic photosynthesis based on *in vitro* observations. In the case of photosynthesis the problem is greatly simplified in that one can use measurements of light, chlorophyll and photosynthetic–irradiance relationships to fill in the gaps. Satellite derived ocean color has rendered this as a powerful approach. No analogous approach is available for respiration. The present database of respiration is woefully inadequate—a mere 693 surface measurements covering an area of 320×10^6 km^2. The ideal minimum would be seasonal descriptions of depth-integrated respiration for the 55 biogeochemical zones described by Longhurst (1998). Figure 9.2 makes it clear that we are a long way away from

achieving this aim; a crude separation by 10° of latitude still leaves a number of the latitude bands with no or very little data. This is frustrating as it leaves us uncertain whether or not the trends of respiration and net community production we see with latitude in Fig 9.13 are real or simply reflect inadequate sampling.

The respiration data are biased with respect to time, place, community structure, and depth, and so cannot provide a definitive global estimate of plankton respiration. However, they can be used to derive a maximum and minimum figure for global ocean photic zone respiration, which can then be compared with estimates derived from mass-balance equations. Summing the latitudinal totals of depth-integrated respiration given in Fig. 9.13(b) gives a global (oceanic + coastal) respiration rate of 17.2 Pmol O$_2$ a^{-1} (186 Gt C a^{-1}, using a respiratory quotient of 0.89, see Chapter 14). The mean depth-integrated respiration rate in oceanic waters of 116 $\pm$ 8.5 mmol O$_2$ m^{-2} d^{-1} equates to a global oceanic respiration rate of 13.5 $\pm$ 1 Pmol O$_2$ a^{-1} (146 $\pm$ 12 Gt C a^{-1}). These are undoubtedly maximum values given the bias in sampling toward times and places of high productivity. The lowest values recorded in the database occur in the

Southern Ocean (e.g. Bender *et al.* 2000) and during the autumn and winter (e.g. Serret *et al.* 1999) and give a lower limit to the estimate of oceanic respiration (35 mmol O_2 m^{-2} d^{-1}; 4.1 Pmol O_2 a^{-1}; 44 Gt C a^{-1}). It is worth noting that the lowest estimate which could be derived from the dark oxygen incubation technique is ~10 mmol O_2 m^{-2} d^{-1} or 1.2 Pmol O_2 a^{-1} (13 Gt C a^{-1}) (based on a rate of 0.2 mmol O_2 m^{-3} d^{-1} and a 50 m euphotic zone) and so this lower dataset value is not simply a consequence of the limit of detection of the method. The 13-month study at HOTS provides a crucial seasonally integrated estimate of respiration in an area representative of the open-ocean gyres. The HOTS estimate (31 mol O_2 m^{-2} a^{-1}) would extrapolate to a global ocean estimate of 9.9 Pmol O_2 a^{-1} (119 Gt C a^{-1}). Open-ocean primary production estimates, derived from ^{14}C measurements lie in the range 2.3–4.3 Pmol C a^{-1} (28–52 Gt C a^{-1}, del Giorgio and Duarte 2002). Comparisons between ^{14}C and gross production derived from dissolved oxygen flux suggest that ^{14}C underestimates gross production by 35–65% (Bender *et al.* 1999; Robinson and Williams 1999; Laws *et al.* 2000; Rees *et al.* 2002). Correcting the open-ocean production estimates by a mean of 50% gives a range of 5–9 Pmol C a^{-1} (see also Chapter 14). Comparison with the range in respiration estimates (4–17 Pmol C a^{-1}) could be seen as a major imbalance in the functioning of the oceans but more likely arises from a lack of our understanding in the seasonal and regional variations in photosynthesis and respiration.

9.5.2 Derivation of global rates from correlation analysis

Recognizing the limitations in the current database of respiration, several authors have attempted to derive simple relationships between respiration and indicators of plankton biomass (Robinson *et al.* 2002*a,b*) or photosynthesis (del Giorgio *et al.* 1997; Duarte and Agusti 1998; Serret *et al.* 2001). The analysis we have made in the present chapter (Section 8.3.2) on an enlarged database is not encouraging in this respect. Only 20–30% of the variance in respiration can be accounted for by the variation in routinely measured indicators of plankton

biomass. The low correlation between respiration and photosynthesis precludes the use of this simple property relationship to make useful global predictions of respiration—consistent with the arguments of Williams and Bowers (1999) and Serret *et al.* (2001, 2002). As the major source of carbon for oceanic respiration comes from oceanic photosynthesis (see Box 9.2) there has to be some form of relationship, although clearly it is not a simple one.

9.5.3 Derivation of rates from ocean organic mass-balance calculations

Simple mass-balance calculations enable us to derive a fairly precise estimate of the difference between production and consumption and so in principle provide a constraint to the global estimates of respiration derived from direct measurements. The strength of the mass-balance calculation is that the assumptions are clear and the implication of differing assumptions easy to explore. Their limitation is that they cannot give an independent estimate of both photosynthesis and respiration. The calculation also implicitly assumes steady state over the medium term (10–100 years). This assumption appears to be justifiable as the size of the principal organic reservoir in the upper water column (about 5–8 mol DOC m^{-2}) is about a third or less than that of the annual biological turnover of carbon. Thus if there were persistent imbalances in the upper water column carbon cycle, major changes in the carbon reservoirs would result. We have reasonably reliable observations of DOC concentrations that span 45 years and can be fairly certain that major changes in the DOC pool have not occurred. As estimates of the external organic inputs into the ocean (predominantly river-borne material) exceed those of the outputs (predominantly sedimentation), the ocean as a whole must be net heterotrophic, albeit only very slightly (0.4%, see Box 9.2). This was essentially the argument of Smith and Mackenzie (1987). The difference between production and consumption is small, in relation to the individual processes themselves. Any estimate we obtain for respiration from mass-balance calculations is primarily dependent upon the value we take for oceanic primary production, so in a way we cycle back to the problem

---

Box 9.2 Mass-balance calculations for respiration in the <u>whole oceanic water column</u>

The basic mass-balance equation assumes that as there is no accumulation of organic material, the sum of all the inputs must equal that of all the outputs:

that is, $P + \sum I - (R + \sum O) = 0$,

where:

  P is the annual rate of photosynthesis;
  R is the annual rate of respiration;
  $\sum I$ is the sum of the external inputs;
  $\sum O$ is the sum of the external outputs from the oceanic water column.

The calculation considers annual sums of all the individual sources and sinks, the units are Tmol C a^{-1}.

The individual inputs (see Williams 2000) are:

  I_R = river input of organic material = 34 Tmol C a^{-1};
  I_A = atmospheric deposition of organic material = 2 Tmol C a^{-1}.

The individual outputs (see Williams 2000) are:

  O_S = net sedimentation = 14 Tmol C a^{-1};
  O_A = input of organic material into the atmosphere = 2 Tmol C a^{-1}.

Then, $P + 36 - (R + 16) = 0$, thus $R = P + 20$.

If we take 5500 Tmol C a^{-1} as an estimate of whole ocean gross production (see Box 9.3), then $R = 5500 + 20 = 5520$ Tmol C a^{-1} and <u>$P/R = 0.996$.</u>

⟶ Trophic balance = 0.4% *heterotrophic*.

---

of the uncertainties inherent in this latter estimate. However the one thing we can be certain of from the mass-balance calculation is that there cannot be the massive difference between oceanic production and respiration we derive from analysis of field observations.

Whereas the mass-balance approach is a useful technique on the large scale, it quickly runs into severe limitations when it is attempted for subdivisions of the ocean. In Box 9.3, the calculation is repeated with the simple division of the oceans into the coastal-and the open-ocean zones. The export of organic carbon from the coastal to the open ocean has been part of a major international program (LOICZ—Land–Ocean Interactions in the Costal Zone). The overall spread of values for the total export from the coastal to the open ocean is large but we seem to be approaching some consensus in the estimates. Liu *et al.* (2000*b*) reported a range from 17 to 390 Tmol C a^{-1}. They regard 70 Tmol C a^{-1} to be a representative value. Ducklow and McCallister (in press) adopt a higher figure of 200 Tmol C a^{-1}. These two estimates (70 and 200 Tmol C a^{-1}) probably bracket the likely range and have been used in Box 9.3 to calculate maximum and minimum scenarios. In both situations the open-ocean water column is calculated to be net

heterotrophic, weakly so (1.7%) in the low export case, more significantly so (5%) in the high export case. The higher figure for carbon export to the open ocean calls for a strongly (13%) autotrophic coastal ocean, the lower export figure, a weakly (3%) autotrophic one.

As the focus of this chapter is the epipelagic zone, we have made mass-balance calculations for this zone. It is regarded (Liu *et al.* 2000*a*; Ducklow and McCallister in press) that the export of organic material from the coastal ocean occurs down the continental slope and thus enters the mesopelagic, rather than the epipelagic zone. This leaves the budget for the epipelagic dominated by a single import/export term: the transfer of material to the mesopelagic zone. Recent estimates for the export to the mesopelagic are 667 Tmol C a^{-1} (Liu *et al.* 2000*b*), 782 Tmol C a^{-1} (Ducklow and McCallister, in press), 1700 Tmol C a^{-1} (Arístegui, Chapter 10) and 2292 Tmol C a^{-1} (del Giorgio and Duarte 2002). In Box 9.4 we have calculated a budget for the epipelagic zone using the two extremes, as well as an intermediate figure of 1000 Tmol C a^{-1}. The broad pattern is the same—it requires the epipelagic ocean to be distinctly (16–57%) net autotrophic. The difference between photosynthesis and respiration in the epipelagic (euphotic) zone is set by the

Box 9.3 Mass-balance calculations for respiration in the coastal and open ocean water columns

Separate calculations are made for coastal (area = 0.36×10^{14} m^2) and oceanic (area = 3.2×10^{14} m^2) zones. The sedimentation rates for the coastal and open ocean zones are taken as 12 and 2 Tmol C a^{-1}, respectively (see Box 9.2). The output to atmosphere (O_A = 2 Tmol C a^{-1}, see Box 9.2) is divided *pro rata* by area between the coastal and open ocean giving, respectively, 0.2 and 1.8 Tmol C a^{-1}. The input from the atmosphere (I_A = 2 Tmol C a^{-1}, see Box 9.2) has been apportioned assuming twice the fallout rate in the coastal zone giving 0.36 and 1.64 Tmol C a^{-1}, respectively for the coastal and open ocean zones. Neither of these last two estimates is critical in the calculation. Estimates for the productivity of the coastal ocean vary three to fourfold: Liu *et al.* (2000*b*) gives a range of 340–750 Tmol C a^{-1}. We adopt Ducklow and McCallister's (in press) recent estimate of 1200 Tmol C a^{-1} for net primary production and maintain a ratio of coastal to oceanic production similar to that of Ducklow and McCallister (in press). As we need to base the calculation on gross production, we scale up net primary production by 25% to give gross production so obtaining a figure of 1500 Tmol C a^{-1} for the coastal ocean and 4000 Tmol C a^{-1} for the open ocean. The final terms (O_O and I_C), the output from the coastal production to the open ocean and the import of organic matter produced in coastal waters, must be numerically equal. Two situations have been calculated, a low value (70 Tmol C a^{-1}) for O_O is taken from Liu *et al.* (2000*b*) and a high value (200 Tmol C a^{-1}) from Ducklow and McCallister (in press). The calculations (see Box 9.2 for notation) are as follows:

Coastal ocean: mass-balance equation
$$P + I_A + I_R - (R + O_S + O_A + O_R + O_O) = 0$$

Open ocean: mass-balance equation
$$P + I_A + I_R + I_C - (R + O_S + O_A) = 0$$

Authors vary over whether to separate organic material of planktonic and river origin in their estimates of export from the coastal zone, it is a detail that is unnecessary in the present account and so these export/import terms have been embedded as single values (O_O and I_C). The above equations then simplify to:

Coastal ocean: mass-balance equation
$$P + I_A + I_R - (R + O_S + O_A + O_O) = 0$$

Open ocean: mass-balance equation
$$P + I_A + I_C - (R + O_S + O_A) = 0$$

Coastal ocean: high export calculation
  $P + 0.36 + 34 - (R + 12 + 0.2 + 200) = 0$
Thus, $R = P - 178$ Tmol C a^{-1}.
Assume, $P = 1500$ Tmol C a^{-1}.
Then $R = 1322$ Tmol C a^{-1}, and $P/R = 1.13$.
Coastal ocean would be 12% *autotrophic*.

Open ocean: high export calculation
  $P + 1.6 + 200 - (R + 2 + 1.8) = 0$
Thus, $R = P + 198$ Tmol C a^{-1}.
Assume, $P = 4000$ Tmol C a^{-1}.
Then $R = 4198$ Tmol C a^{-1} and $P/R = 0.95$.
Open ocean would be 5% *heterotrophic*.

Coastal ocean: low export calculation
  $P + 0.36 + 34 - (R + 12 + 0.2 + 70) = 0$
Thus, $R = P - 48$ Tmol C a^{-1}.
Then $R = 1452$ Tmol C a^{-1} and $P/R = 1.03$.
Coastal ocean would be 3% *autotrophic*.

Open ocean: low export calculation
  $P + 1.6 + 70 - (R + 2 + 1.8) = 0$
Thus, $R = P + 68$ Tmol C a^{-1}.
Then, $R = 4068$ Tmol C a^{-1}, and $P/R = 0.98$.
Open ocean would be 1.7% *heterotrophic*.

Box 9.4   Mass-balance calculation for the open ocean epipelagic zone

The prevailing view is that the exported planktonic production from the coastal regions enters the mesopelagic, rather than the epipelagic zone of the open ocean (see Liu *et al.*, 2000a; Arístegui *et al.*, Chapter 10). We make calculations for the epipelagic (0–150 m) zone of the ocean, assuming that coastal production is exported to the mesopelagic zones. We may expect that the material of river origin passes into both zones, however as will be seen below whatever assumption we make of its fate is of little consequence in the overall calculation.

Mass-balance equation,

$$P + I_A + I_R + I_C - (R + O_M + O_A) = 0,$$

where O_M is the export from the epipelagic into the mesopelagic zone.

As $O_M \gg (O_A + I_A + I_R + I_C)$, the equation simplifies to:

$$\boxed{P - (R + O_M) = 0.}$$

Recent estimates for the export to the mesopelagic (O_M) are 667 Tmol C a^{-1} (Liu *et al.* 2000a), 782 Tmol C a^{-1} (Ducklow and McCallister in press), 1700 Tmol C a^{-1} (Arístegui

et al., Chapter 10) and 2292 Tmol C a^{-1} (del Giorgio and Duarte 2002). Here we make the calculation for the extremes of the estimate of O_M (667 and 2292 Tmol C a^{-1}), and what would seem be a median value of 1000 Tmol C a^{-1}.

Epipelagic ocean: low export calculation
$R = P - 667$ Tmol C a^{-1}.
Then $R = 3333$ Tmol C a^{-1},
and $P/R = 1.2$.
Epipelagic ocean: would need to be 16% *autotrophic*.

Epipelagic ocean: high export calculation
$R = P - 2292$ Tmol C a^{-1}.
Then $R = 1708$ Tmol C a^{-1},
and $P/R = 2.3$.
Epipelagic ocean would need to be 57% *autotrophic*.

Epipelagic ocean: intermediate export calculation
$R = P - 1000$ Tmol C a^{-1}.
Then $R = 3000$ Tmol C a^{-1},
and $P/R = 1.33$.
Epipelagic ocean would need to be 25% *autotrophic*.

---

export term, which is reasonably well constrained; if anything it may be an underestimate thus requiring even higher levels of net autotrophy. Whereas the difference between photosynthesis and respiration is determined by a single term, the absolute values are dependent upon the value taken for oceanic photosynthesis. del Giorgio and Duarte (2002) in their review adopted a twofold spread of values—their upper figure for the ocean as a whole (6400 Tmol C a^{-1}) is somewhat higher than the one we have used and it would reduce the calculated relative net autotrophy to a range of 14–49%.

Significantly, the range of epipelagic zone oceanic respiration rates derived from mass-balance considerations (1.7–3 Pmol C a^{-1}) lie below even the lowest estimate of oceanic respiration derived from direct measurements, and are two to eightfold lower than the mid- and highest estimates derived from direct measurements. This does not derive from an error in the mass-balance calculation, but

for the need to supply a figure for the rate of photosynthesis—essentially we return to the problem brought to light in the preceding paragraph. The prediction from mass-balance equations that the epipelagic zone is 25% or so autotrophic is at odds with the direct measurements that imply a balanced or slightly heterotrophic situation. This discrepancy will in part result from the different time and space scales considered by the two approaches. However, they also point to the weaknesses in our understanding of the global balance between organic carbon production and respiration. Mass-balance equations and direct observations indicate (del Giorgio and Duarte 2002; Williams *et al.* in press) that we may be underestimating the rate of organic matter production, by a factor of two or more. In the final chapter (Chapter 14) del Giorgio and Williams discuss this in more detail and achieve some resolution of the problem. Combined with the sampling bias of oceanic respiration measurements, our appreciation of regional and global photosynthesis to respiration balances is rudimentary. If we are to predict the oceans's response to global change, then current and future national and international oceanographic programs will need to incorporate the determination of oceanic respiration and to address the question of whether the ocean biota act as a net source or sink of carbon as a priority research objective.

ocean - atm

Acknowledgments

Many thanks to the following for making published and unpublished data available: Javier Arístegui, Natalia Gonzalez, Dominique Lefèvre, Paul Morris, and Nelson Sherry, and to Alison Fairclough at the British Oceanographic Data Centre (BODC) for her meticulous help in accessing the UK BOFS, PRIME, and Arabesque data. CR was funded on a NERC Advanced Research Fellowship (GT5/96/8/MS) and the Plymouth Marine Laboratory Core Strategic Research Programme. Satellite data in Fig. 9.2 were processed by Gareth Mottram and Peter Miller at the Plymouth Marine Laboratory Remote Sensing Group (www.npm.ac.uk/rsdas/). SeaWiFS data courtesy of the NASA SeaWiFS project and Orbital Sciences Corporation.

References

Abell, J., Emerson, S., and Renaud, P. 2000. Distributions of TOP, TON and TOC in the North Pacific subtropical gyre: implications for nutrient supply in the surface ocean and remineralization in the upper thermocline. *J. Mar. Res.*, **58**: 203–222.

Anderson, T. R. and Ducklow, H. W. 2001. Microbial loop carbon cycling in ocean environments studied using a simple steady-state model. *Aquat. Microb. Ecol.*, **26**: 37–49.

Arístegui, J. and Harrison, W. G. 2002. Decoupling of primary production and community respiration in the ocean: implications for regional carbon studies. *Mar. Ecol. Prog. Ser.*, **29**: 199–209.

Arístegui, J. and Montero, M. F. 1995. The relationship between community respiration and ETS activity in the ocean. *J. Plankton Res.*, **17**: 1563–1571.

Arístegui, J., Montero, M. F., Ballesteros, S., Basterretxea, G., and van Lenning, K. 1996. Planktonic primary production and microbial respiration measured by ^{14}C assimilation and dissolved oxygen changes in coastal waters of the Antarctic Peninsula during the austral summer: implications for carbon flux studies. *Mar. Ecol. Prog. Ser.*, **132**: 191–201.

Arístegui, J., Denis, M., Almunia, J., and Montero, M. F. 2002. Water-column remineralization in the Indian sector of the Southern Ocean during early Spring. *Deep-Sea Res. II*, **49**: 1707–1720.

Barber, R. T. and Hilting, A. K. 2002. History of the study of Plankton Productivity. In P. J. le B. Williams, D. N. Thomas, and C. S. Reynolds (eds) *Phytoplankton Productivity: Carbon Assimilation in Marine and Freshwater Ecosystems*. Blackwell Science Ltd, Oxford, UK, p. 386.

Bender, M., Grande, K., Johnson, K., Marra, J., Williams, P. J. le B., Sieburth, J., Pilson, M., Langdon, C., Hitchcock, G., Orchardo, J., Hunt, C., Donaghay, P., and Heinemann, K. 1987. A comparison of four methods for determining planktonic community production. *Limnol. Oceanogr.*, **32**: 1085–1097.

Bender, M., Orchardo, J., Dickson, M-L., Barber, R., and Lindley, S. 1999. *In vitro* O_2 fluxes compared with ^{14}C production and other rate terms during the JGOFS Equatorial Pacific experiment. *Deep-Sea Res. I*, **46**: 637–654.

Bender, M. L. 2000. Tracer from the sky. *Science*, **288**: 1977–1978.

Bender, M. L., Dickson, M-L., and Orchardo, J. 2000. Net and gross production in the Ross Sea as determined by incubation experiments and dissolved O_2 studies. *Deep-Sea Res. II*, **47**: 3141–3158.

Blanco, J. M., Quinones, R. A., Guerrero, F., and Rodriguez, J. 1998. The use of biomass spectra and allometric relations to estimate respiration of plankton communities. *J. Plankton Res.*, **20**: 887–900.

Blight, S. P., Bentley, T. L., Lefèvre, D., Robinson, C., Rodrigues, R., Rowlands, J., and Williams P. J. le B. 1995. The phasing of autotrophic and heterotrophic plankton metabolism in a temperate coastal ecosystem. *Mar. Ecol. Prog. Ser.*, **128**: 61–74.

Biddanda, B., Opsahl, S., and Benner, R. 1994. Plankton respiration and carbon flux through bacterioplankton on the Louisiana shelf. *Limnol. Oceanogr.*, **39**: 1259–1275.

Boyd, P., Robinson, C., Savidge, G., and Williams, P. J. le B. 1995. Water column and sea ice primary production during austral spring in the Bellingshausen Sea. *Deep-Sea Res. II*, **42**: 1177–1200.

Carlson, C. A., Bates, N. R. T., Ducklow, H. W., and Hansell, D. F. A. 1999. Estimation of bacterial respiration and growth efficiency in the Ross Sea, Antarctica. *Aquat. Microb. Ecol.*, **19**: 229–244.

Chipman, D. W., Marra, J., and Takahashi, T. 1993. Primary production at 47°N 20°W in the North Atlantic Ocean: a comparison between the ^{14}C incubation method and the mixed layer carbon budget. *Deep-Sea Res. II*, **40**: 151–169.

Christensen, J. P., Owens, T. G., Devol, A. H., and Packard, T. T. 1980. Respiration and physiological state in marine bacteria. *Mar. Biol.*, **55**: 267–276.

Cota, G. F., Pomeroy, L. R., Harrison, W. G., Jones, E. P., Peters, F., Sheldon, Jr., W. M., and Weingartner, T. R. 1996. Nutrients, primary production and microbial heterotrophy in the southeastern Chukchi Sea: Arctic summer nutrient depletion and heterotrophy *Mar. Ecol. Prog. Ser.*, **135**: 247–258.

del Giorgio, P. A. 1992. The relationship between ETS (electron transport system) activity and oxygen consumption in lake plankton—a cross system calibration. *J. Plankton Res.*, **14**: 1723–1741.

del Giorgio, P. A. and Cole, J. J. 1998. Bacterioplankton growth efficiency in natural aquatic systems. *Annu. Rev. Ecol. Syst.*, **29**: 503–541.

del Giorgio, P. A. and Duarte, C. M. 2002. Respiration in the open ocean. *Nature*, **420**: 379–384.

del Giorgio, P. A., Cole, J. J., and Cimbleris A. 1997. Respiration rates in bacteria exceed plankton production in unproductive aquatic systems. *Nature*, **385**: 148–151.

Daneri, G. 1992. Comparison between in vitro and *in situ* estimates of primary production within two tracked water bodies. *Arch. Hydrobiol. Beih.*, **37**: 101–109.

Daneri, G., Dellarossa, V., Quinones, R., Jacob, B., Montero, P. and Ulloa, O. 2000. Primary production and community respiration in the Humboldt current system off Chile and associated oceanic areas. *Mar. Ecol. Prog. Ser.*, **197**: 41–49.

Dickson, M.-L. and Orchardo, J. 2001. Oxygen production and respiration in the Antarctic Polar Front region during the austral spring and summer. *Deep-Sea Res. II*, **48**: 4101–4126.

Dickson, M.-L., Orchard, J., Barber, R. T., Marra, J., McCarthy, J. J., and Sambrotto, R. N. 2001. Production and respiration rates in the Arabian Sea during the 1995 Northeast and Southwest Monsoons. *Deep-Sea Res. II*, **48**: 1199–1230.

Duarte, C. M. and Agusti S. 1998. The CO_2 balance of unproductive aquatic ecosystems. *Science*. **281**: 234–236.

Duarte, C. M., Agusti, S., Arístegui, J., Gonzalez, N., and Anadon, R. 2001. Evidence for a heterotrophic subtropical northeast Atlantic. *Limnol. Oceanogr.*, **46**: 425–428.

Ducklow, H. W. and S. L. McCallister. In press. The biogeochemistry of carbon dioxide in the coastal oceans.

Chapter 10 In *The Sea. Volume 13 The Global Coastal Ocean: Multiscale Interdisciplinary Processes*. A. R. Robinson, K. Brink, H. W. Ducklow, R. Jahnke and B. J. Rothschild. eds. Harvard University Press. Cambridge, MA. pp.

Ducklow, H. W., Dickson, M-L., Kirchman, D. L., Steward, G., Orchardo, J., Marra, J., and Azam, F. 2000. Constraining bacterial production, conversion efficiency and respiration in the Ross Sea, Antarctica, January–February, 1997. *Deep-Sea Res. II*, **47**: 3227–3247.

Eissler, Y. and Quinones, R. A. 1999. Microplanktonic respiration off northern Chile during El Nino 1997–1998. *J. Plankton Res.* **21**: 2263–2283.

Emerson, S., Quay, P., Karl, D., Winn, C., Tupas, L., and Landry, M. 1997. Experimental determination of the organic carbon flux from open ocean surface waters. *Nature*, **389**: 951–954.

Emerson, S., Stump, C., Johnson, B., and Karl, D. M. 2002. *In situ* determination of oxygen and nitrogen dynamics in the upper ocean. *Deep-Sea Res. I*, **49**: 941–952.

Fasham, M. J. R., Boyd, P. W., and Savidge, G. 1999. Modeling the relative contributions of autotrophs and heterotrophs to carbon flow at a Lagrangian JGOFS station in the Northeast Atlantic: the importance of DOC. *Limnol. Oceanogr.*, **44**: 80–94.

Gonzalez, N., Anadon, R., Mourino, B., Fernandez, E., Sinha, B., Escanez, J., and de Armas, D. 2001. The metabolic balance of the planktonic community in the north Atlantic subtropical gyre: the role of mesoscale instabilities. *Limnol. Oceanogr.*, **46**: 946–952.

Gonzalez, N., Anadon, R., and Maranon, E. 2002. Large-scale variability of planktonic net community metabolism in the Atlantic ocean: importance of temporal changes in oligotrophic subtropical waters. *Mar. Ecol. Prog. Ser.*, **333**: 21–30.

Grande, K. D., Marra, J., Langdon, C., Heinemann, K., and Bender, M. 1989*a*. Rates of respiration in the light measured in marine phytoplankton using an ^{18}O isotope labelling technique. *J. Exp. Mar. Biol. Ecol.*, **129**: 95–120.

Grande, K. D., Williams, P. J. le B., Marra, J., Purdie, D. A., Heinemann, K., Eppley, R. W., and Bender, M. L. 1989*b*. Primary production in the North Pacific gyre: a comparison of rates determined by the ^{14}C, O_2 concentration and ^{18}O methods. *Deep-Sea Res. I*, **36**: 1621–1634.

Griffith, P. C., Douglas, D. J., and Wainwright, S. C. 1990. Metabolic activity of size-fractionated microbial plankton in estuarine, near shore, and continental shelf waters of Georgia. *Limnol. Oceanogr.*, **59**: 263–270.

Hanson, A. K., Tindale, N. W., and Abdel-Moatia, M. A. R. 2001. An Equatorial Pacific rain event: influence on the distribution of iron and hydrogen peroxide in surface waters. *Mar. Chem.*, **75**: 69–88.

Harrison, W. G. 1986. Respiration and its size-dependence in microplankton populations from surface waters of the Canadian Arctic. *Polar Biol.*, **6**: 145–152.

Harrison, W. G., Arístegui, J., Head, E. J. H., Li, W. K. W., Longhurst, A. R., and Sameoto, D. D. 2001. Basin-scale variability in plankton biomass and community metabolism in the sub-tropical North Atlantic Ocean. *Deep-Sea Res. II*, **48**: 2241–2269.

Herut, B., Shoham-Frider, E., Kress, N., Fiedler, U., and Angel, D. L. 1998. Hydrogen peroxide production rates in clean and polluted coastal marine waters of the Mediterranean, Red and Baltic Seas. *Mar. Pollut. Bull.* **36**: 994–1003.

Hitchcock, G. L., Vargo, G. A., and Dickson, M-L. 2000. Plankton community composition, production and respiration in relation to dissolved inorganic carbon on the West Florida shelf, April 1996. *J. Geophys. Res.*, **105**: 6579–6589.

Holligan, P M, Williams, P. J. le B., Purdie, D. A., and Harris, R. P. 1984. Photosynthesis, respiration and nitrogen supply of plankton populations in stratified, frontal and tidally mixed shelf waters *Mar. Ecol. Prog. Ser.*, **17**: 201–203.

Hopkinson, C. S., Sherr, B., and Wiebe, W. J. 1989. Size fractionated metabolism of coastal microplankton. *Mar. Ecol. Prog. Ser.*, **51**: 155–166.

Jacquet, S., Partensky, F., Lennon, J. F., and Vaulot, D. 2001. Diel patterns of growth and division in marine picoplankton in culture *J. Phycol.*, **37**: 357–369.

Jahnke, R. A. and Craven, D. B. 1995. Quantifying the role of heterotrophic bacteria in the carbon cycle: a need for respiration rate measurements. *Limnol. Oceanogr.*, **40**: 436–441.

Jenkins, W. J. and Goldman, J. C. 1985. Seasonal oxygen cycling and primary production in the Sargasso Sea. *J. Mar. Res.*, **43**: 465–491.

Johnson, K. M., Davis, P. G., and Sieburth, J. M^CN. 1983. Diel variation of TCO_2 in the upper layer of oceanic waters reflects microbial composition, variation and possibly methane cycling. *Mar. Biol.*, **77**: 1–10.

Johnson, K. M., Sieburth, J. M^CN., Williams, P. J. le B., and Brandstrom, L. 1987. Coulometric total carbon dioxide analysis for marine studies: automation and calibration. *Mar. Chem.*, **21**: 117–133.

Karl, D. M., Morris P. J., Williams, P. J. le B., and Emerson, S. 2003. Metabolic balance of the open sea Nature, **426**: 32.

Keeling, R. F. and Shertz, S. R. 1992. Seasonal and interannual variations in atmospheric oxygen and implications for the global carbon cycle Nature, **358**: 723–727.

Kenner, R. A. and Ahmed, S. I. 1975. Measurements of Electron transport activities in marine phytoplankton. *Mar. Biol.*, **33**: 119–127.

Kiddon, J., Bender, M. L., and Marra, J. 1995. Production and respiration in the 1989 North Atlantic Spring bloom—an analysis of irradiance dependent changes *Deep-Sea Res. I*, **42**: 553–576.

Kruse, B. 1993. Measurement of plankton O_2 respiration in gas-tight plastic bags. *Mar. Ecol. Prog. Ser.*, **94**: 155–163.

Langdon, C. 1993. The significance of respiration in production measurements based on oxygen. *ICES Mar. Sci. Symp.*, **197**: 69–78.

Laws, E. A., Landry, M. R., Barber, R. T., Campbell, L., Dickson, M.-L., and Marra, J. 2000. Carbon cycling in primary production incubations: inferences from grazing experiments and photosynthetic studies using ^{14}C and ^{18}O in the Arabian Sea. *Deep-Sea Res. II*, **47**: 1339–1352.

Lefèvre, D., Minas, H. J., Minas, M., Robinson, C., Williams, P. J. le B., and Woodward, E. M. S. 1997. Review of gross community production, primary production, net community production and dark community respiration in the Gulf of Lions *Deep-Sea Res. II*, **44**: 801–832.

Liu, K. K. Atkinson, L., Chen, C. T. A., Gao, S., Hall, J. Macdonald, R. W., McManus, L. T., and Quinones, R. 2000a. Exploring continental margin carbon fluxes on a global scale *EOS*, **81**: 641–642.

Liu K.-K., Iseki, K., and Chao, S.-Y. 2000b. Continental margin carbon fluxes. In R. B. Hanson, H. W. Ducklow, and J. G. Field (eds) *The Changing Ocean Carbon Cycle: A Midterm Synthesis of the Joint Global Ocean Flux Study* IGBP Series 3. Cambridge University, Press, Cambridge, UK, pp. 187–239.

Longhurst, A. 1998. *Ecological Geography of the Sea.* Academic Press, San Diego, p. 398.

Luz, B. and Barkan, E. 2000. Assessment of oceanic productivity with the triple-isotope composition of dissolved oxygen. *Science*, **288**: 2028–2031.

Marra, J. and Barber, R. T. 2004. Phytoplankton and heterotrophic respiration in the surface layer of the ocean. *Geophys. Res. Lett.* **31**: L09314.

Marra, J. and Heinemann, K. 1984. A comparison between non contaminating and conventional incubation procedures in primary production measurements. *Limnol. Oceanogr.*, **29**: 389–392.

Moffett, J. W. and Zafiriou, O. C. 1993. The photochemical decomposition of hydrogen peroxide in surface waters of the eastern Caribbean and Orinoco River. *J. Geophys. Res. Oceans*, **98**: 2307–2313.

Moncoiffe, G., Alvarez-Salgado, X. A., Figueiras, F. G., and Savidge, G. 2000. Seasonal and short-time-scale dynamics of microplankton community production and respiration in an inshore upwelling system. *Mar. Ecol. Prog. Ser.*, **196**: 111–126.

Nagata, T. 2000. Production mechanisms of dissolved organic matter. In D. L. Kirchman (ed.) *Microbial Ecology of the Oceans*. John Wiley and Sons, Inc., New York.

Packard, T. T. 1971. The measurement of respiratory electron transport activity in marine phytoplankton. *J. Mar. Res.*, **29**: 235–244.

Packard, T. T. 1985. The measurement of respiratory electron transport activity in microplankton. In H. W. Jannasch and P. J. le B. Williams (eds) *Advances in Aquatic Microbiology*. Vol. 3 Academic Press, London, pp. 207–261.

Packard, T., Berdalet, E., Blasco, D., Roy, S. O., St-Amand, L., Lagace, B., Lee, K., and Gagne, J.-P. 1996. CO_2 production predicted from isocitrate dehydrogenase activity and bisubstrate enzyme kinetics in the marine bacterium *Pseudomonas nautica*. *Aquat. Microb. Ecol.* **11**: 11–19.

Pamatmat, M. M. 1997. Non-photosynthetic oxygen production and non-respiratory oxygen uptake in the dark: a theory of oxygen dynamics in plankton communities. *Mar. Biol.*, **129**: 735–746.

Platt, T. and Denman, K. 1978. The structure of pelagic marine ecosystems. *Rapp. P.-V. Reun. Cons. Int. Explor. Mer.*, **173**: 60–65.

Platt, T., Lewis, M., and Geider, R. 1984. Thermodynamics of the pelagic ecosystem: elementary closure conditions for biological production in the open ocean. In M. J. R. Fasham (ed.) *Flows of Energy and Materials in Marine Ecosystems*. Plenum Press, New York, pp. 49–84.

Pomeroy, L. R., Sheldon, J. E., and Sheldon, W. M. 1994. Changes in bacterial numbers and leucine assimilation during estimations of microbial respiratory rates in seawater by the precision Winkler method. *Appl. Environ. Microbiol.*, **60**: 328–332.

Pomeroy, L. R., Sheldon, J. E., Sheldon Jr., W. M., and Peters, F. 1995. Limits to growth and respiration of bacterioplankton in the Gulf of Mexico. *Mar. Ecol. Prog. Ser.*, **117**: 259–268.

Rees, A. P., Joint, I., Woodward, E. M. S., and Donald, K. M. 2001. Carbon, nitrogen and phosphorus budgets within a mesoscale eddy: comparison of mass balance with *in vitro* determinations. *Deep-Sea Res. II*, **48**: 859–872.

Rees, A. P., Woodward, E. M. S., Robinson, C., Cummings, D. G., Tarran, G. A., and Joint, I. 2002. Size fractionated nitrogen uptake and carbon fixation during a developing coccolithophore bloom in the North Sea during June 1999. *Deep-Sea Res. II*, **49**: 2905–2927.

Rivkin, R. B. and Legendre, L. 2001. Biogenic carbon cycling in the upper ocean: effects of microbial respiration. *Science*, **291**: 2398–2400.

Robinson, C. 2000. Plankton gross production and respiration in the shallow water hydrothermal systems of Milos, Aegean Sea. *J. Plankton Res.*, **22**: 887–906.

Robinson, C. and Williams, P. J. le B. 1991. Development and assessment of an analytical system for the accurate and continual measurement of total dissolved inorganic carbon *Mar. Chem.*, **34**: 157–175.

Robinson, C. and Williams, P. J. le B. 1993. Temperature and Antarctic plankton community respiration *J. Plankton Res.*, **15**: 1035–1051.

Robinson, C. and Williams, P. J. le B. 1999. Plankton net community production and dark respiration in the Arabian Sea during September 1994. *Deep-Sea Res. II*, **46**: 745–766.

Robinson, C., Archer, S., and Williams, P. J. le B. 1999. Microbial dynamics in coastal waters of East Antarctica: plankton production and respiration. *Mar. Ecol. Prog. Ser.*, **180**: 23–36.

Robinson, C., Serret, P., Tilstone, G., Teira, E., Zubkov, M. V., Rees, A. P., and Woodward, E. M. S. 2002*a*. Plankton respiration in the Eastern Atlantic Ocean *Deep-Sea Res. I*, **49**: 787–813.

Robinson, C., Widdicombe, C. E., Zubkov, M. V., Tarran, G. A., Miller, A. E. J., and Rees, A. P. 2002*b*. Plankton community respiration during a coccolithophore bloom *Deep-Sea Res. II*, **49**: 2929–2950.

Roy, S. O. and Packard, T. T. 2001. CO_2 production rate predicted from isocitrate dehydrogenase activity, intracellular substrate concentrations and kinetic constants in the marine bacterium *Pseudomonas nautical Mar. Biol.* **138**: 1251–1258.

Sampou, P. and Kemp, W. M. 1994. Factors regulating plankton community respiration in Chesapeake Bay *Mar. Ecol. Prog. Ser.*, **110**: 249–258.

Serret, P., Fernandez, E., Sostes, J. A., and Anadon, R. 1999. Seasonal compensation of microbial production and respiration in a temperate sea. *Mar. Ecol. Prog. Ser.*, **187**: 43–57.

Serret, P., Robinson, C., Fernandez, E., Teira, E., and Tilstone, G. 2001. Latitudinal variation of the balance between plankton photosynthesis and respiration in the E. Atlantic Ocean. *Limnol. Oceanogr.*, **46**: 1642–1652.

Serret, P., Fernandez, E. and Robinson, C. 2002. Biogeographic differences in the net ecosystem metabolism of the open ocean. *Ecology*, **83**: 3225–3234.

Sheldon, R. W., Prakash, A. and Sutcliffe W. H. 1972. The size distribution of particles in the Ocean. *Limnol. Oceanogr.*, **17**: 327–340.

Sherr, E. B., Sherr, B. F. and Sigmon, C. T. 1999. Activity of marine bacteria under incubated and in situ conditions. *Aquatic Microbial Ecology*, **20**: 213–223.

Sherry, N. D., Boyd, P. W., Sugimoto, K., and Harrison, P. J. 1999. Seasonal and spatial patterns of heterotrophic bacterial production, respiration, and biomass in the subarctic NE Pacific. *Deep-Sea Res., II*, **46**: 2557–2578.

Smith S. V. and Mackenzie, F. T. C. 1987. The oceans as a net heterotrophic system: implications from the carbon biogeochemical cycle. *Glob. Biogeochem. Cyc.*, **1**: 187–198.

Sondergaard, M., Williams, P. J. le B., Cauwet, G., Rieman, B., Robinson, C., Terzic, S., Woodward, E. M. S., and Worm, J. 2000. Net accumulation and flux of dissolved organic carbon and dissolved organic nitrogen in marine plankton communities. *Limnol. Oceanogr.*, **45**: 1097–1111.

Straile, D. 1997. Gross growth efficiencies of protozoan and metazoan zooplankton and their dependence on food concentration, predator prey weight ratio and taxonomic group. *Limnol. Oceanogr.*, **42**: 1375–1385.

Sullivan, C. W., Arrigo, K. R., McClain, C. A., Comiso, J. C., and Firestone, J. 1993. Distribution of phytoplankton blooms in the Southern Ocean. *Science*, **262**: 1832–1837.

Upstill-Goddard, R. C., Watson, A. J., Wood, J., and Liddicoat, M. I. 1991. Sulfur hexafluoride and ^{3}He as seawater tracers—deployment techniques and continuous underway analysis for sulfur-hexafluoride. *Anal. Chim. Acta*, **249**: 555–562.

Vezina, A. F. and Platt, T. 1988. Food web dynamics in the ocean. I. Best-estimates of flow networks using inverse methods. *Mar. Ecol. Prog. Ser.*, **42**: 269–287.

Williams, P. J. le B. 1981. Microbial contribution to overall marine plankton metabolism: direct measurements of respiration. *Oceanol. Acta*, **4**: 359–364.

Williams, P. J. le B. 1982. Microbial contribution to overall plankton community respiration studies in enclosures, In G. D. Grice and M. R. Reeve (eds) *Marine Microcosms*. Springer-Verlag, Berlin, pp. 305–321.

Williams, P. J. le B. 1995. Evidence for the seasonal accumulation of carbon-rich dissolved organic material, its scale in comparison with changes in particulate material and the consequential effect on net C/N assimilation ratios. *Mar. Chem.*, **51**: 17–29.

Williams, P. J. le B. 1998. The balance of plankton respiration and photosynthesis in the open oceans. *Nature*, **394**: 55–57.

Williams, P. J. le B. 2000. Net production, gross production and respiration: what are the interconnections and what controls what ? In R. B. Hanson, H. W. Ducklow, and J. G. Field (eds) *The Changing Ocean Carbon Cycle: A Midterm Synthesis of the Joint Global Ocean Flux Study*. Cambridge University Press, Oxford, UK, p. 514.

Williams, P. J. le B. and Bowers, D. G. 1999. Regional carbon imbalances in the oceans. *Science*, **284**: 1735b.

Williams, P. J. le B. and Gray, R. W. 1970. Heterotrophic utilization of dissolved organic compounds in the sea. II: observations on the responses of heterotrophic marine populations to abrupt increase in amino acid concentration *J. Mar. Biol. Assoc. UK*, **50**: 871–881.

Williams, P. J. le B. and Purdie, D. A. 1991. *In vitro* and *in situ* derived rates of gross production, net community production and respiration of oxygen in the oligotrophic subtropical gyre of the North Pacific Ocean. *Deep-Sea Res. I*, **38**: 891–910.

Williams, P. J. le B., Morris, P. J., and Karl, D. M. 2004. Net community production and metabolic balance at the oligotrophic ocean site: Station ALOHA. *Deep-Sea Res.*.

Yocis, B. H., Kieber, D. J., and Mopper, K. 2000. Photochemical production of hydrogen peroxide in Antarctic waters *Deep-Sea Res. I*, **47**: 1077–1099.

Yuan, J. C. and Shiller, A. M. 2001. The distribution of hydrogen peroxide in the southern and central Atlantic Ocean *Deep-Sea Res. II*, **48**: 2947–2976.

Zubkov, M. V., Sleigh, M. A., and Burkill, P. H. 2000. Assaying picoplankton distribution by flow cytometry of underway samples collected along a meridional transect across the Atlantic Ocean. *Aquat. Microb. Ecol.*, **21**: 13–20.

Respiration in the mesopelagic and bathypelagic zones of the oceans

Javier Arístegui[1], Susana Agustí[2], Jack J. Middelburg[3], and Carlos M. Duarte[2]

[1] *Facultad de Ciencias del Mar, Universidad de las Palmas de Gran Canaria, Spain*
[2] *IMEDEA (CSIC–UIB), Spain*
[3] *Netherlands Institute of Ecology, The Netherlands*

Outline

In this chapter the mechanisms of transport and remineralization of organic matter in the dark water-column and sediments of the oceans are reviewed. We compare the different approaches to estimate respiration rates, and discuss the discrepancies obtained by the different methodologies. Finally, a respiratory carbon budget is produced for the dark ocean, which includes vertical and lateral fluxes of organic matter. In spite of the uncertainties inherent in the different approaches to estimate carbon fluxes and oxygen consumption in the dark ocean, estimates vary only by a factor of 1.5. Overall, direct measurements of respiration, as well as indirect approaches, converge to suggest a total dark ocean respiration of 1.5–1.7 $Pmol\,C\,a^{-1}$. Carbon mass balances in the dark ocean suggest that the dark ocean receives 1.5–1.6 $Pmol\,C\,a^{-1}$, similar to the estimated respiration, of which >70% is in the form of sinking particles. Almost all the organic matter ($\sim$92%) is remineralized in the water column, the burial in sediments accounts for <1%. Mesopelagic (150–1000 m) respiration accounts for $\sim$70% of dark ocean respiration, with average integrated rates of 3–4 $mol\,C\,m^{-2}\,a^{-1}$, 6–8 times greater than in the bathypelagic zone ($\sim$0.5 $mol\,C\,m^{-2}\,a^{-1}$). The results presented here renders respiration in the dark ocean a major component of the carbon flux in the biosphere, and should promote research in the dark ocean, with the aim of better constraining the role of the biological pump in the removal and storage of atmospheric carbon dioxide.

10.1 Introduction

The ocean supports respiratory activity throughout its entire volume. In contrast to the destruction of organic matter, primary synthesis of organic matter is restricted—except for a marginal contribution of chemosynthesis—to the upper 100–200-m skin of the ocean, where sufficient light penetrates to support photosynthesis. The synthesis processes occurring in the photic layer, which encompasses only about 5% of the water column, have been the object of most of the research effort, and the degradation processes occurring in the dark layers of the ocean remain comparatively poorly studied, despite evidence that the rates involved may be substantial (Williams 2000; del Giorgio and Duarte 2002; Arístegui *et al.* 2003; Andersson *et al.* 2004). Yet, the dark ocean is an important site for the mineralization of organic matter, and is the site of long-term organic carbon storage and burial. Hence, the processes occurring in the dark ocean are essential to understand the functioning of and, in particular, carbon cycling in the biosphere. They, therefore, deserve closer attention than hitherto they have been given.

The reason for the paucity of knowledge on biological processes in the dark layers of the ocean is rooted on an early belief that life in the ocean was confined to its upper, lighted skin. The presence of life in the dark ocean was first demonstrated by collections of deep-sea benthos in the mid-nineteenth century, in the expeditions of Sir John Ross (1817–1818) followed by that of his nephew Sir James Clark Ross aboard the *Erebus* and *Terror* (1839–1843). The Challenger expedition (1888) provided evidence of pelagic life in the dark ocean, but the existence of deep planktonic life remained a controversial topic (e.g. Forbe's azoic ocean hypothesis supported by Agassiz, versus contentions of deep planktonic life by Carl Chun) until directly observed in the first immersion by Beede and Barton in 1934, who reached down to 923 m aboard a bathyscaph. The first evidence of deep-water bacteria was obtained by Portier at the beginning of the twentieth century, during two Prince of Monaco's cruises (1901, 1904), but it was not until the mid-century, when Claude ZoBell and Richard Morita, isolated and counted bacteria from great depths, during the Galathea expedition. Studies on deep-water bacteria regained attention in the 1970s as the result of an accident of the research submersible Alvin (Jannasch and Wirsen 1973), which led to preliminary observations of active microbial metabolism in the dark ocean. These studies were paralleled by studies which implicated the importance of migrating zooplankton (Vinogradov 1970). Pomeroy and Johannes (1968) first measured microplankton respiration in the ocean, by concentrating samples, but the first direct measurements of planktonic respiration at natural densities in the dark ocean were obtained much later, at the turn of the century first by Williams and Purdie (1991) and later by Biddanda and Benner (1997). This body of observations lent support to theoretical speculation on the existence of significant respiration in the dark ocean, necessary to account for deep (300–500 m) oxygen minima in ocean circulation models.

The key role of the dark ocean as the site where the bulk of the excess organic matter produced in the photic layer is mineralized is that it re-fuels subsequent new production. This represents one of the cornerstones of contemporary views on the biogeochemical functioning of the ocean (Libes 1992). Paradoxically, however, estimates of respiration rate—the process responsible for the mineralization of organic matter—in the dark ocean are still few, representing a small percent of all of the estimates of respiration in the sea. Moreover, these few estimates are scattered both in space and time and are based on *in vitro* assessments rather than *in situ* incubations (at depth of sampling). Such estimates are, nevertheless, essential to address critical issues in the cycling of organic carbon in the ocean, which are presently inferred from ocean biology general circulation models rather than from up-scaling from empirical estimates. Yet, even the newest depictions of the global carbon cycle (e.g. Liu *et al.* 2000) ignore respiratory processes in the dark ocean or derive them, in the absence of any independent estimate, as the residual of budgetary exercises (see discussions in Ducklow 1995; Williams 2000; and del Giorgio and Duarte 2002).

In this chapter, we first provide an account of the sources of organic matter to the dark ocean, which constrain the respiratory activity therein, and then examine the available respiration rates—estimated both directly and indirectly—in each of two operationally defined layers:[1] (i) the mesopelagic zone, extending from the bottom of the epipelagic layer (about 100–200 m) down to the base of the permanent thermocline, at about 1000 m depth, and (ii) the bathypelagic zone, defined as the water column and sediments extending from the base of the permanent thermocline to the ocean floor. The separation of the dark ocean in these two layers has been used in the past (e.g. Parsons *et al.* 1984), and is justified first by the very distinct residence times of these waters (100 and 900 years, respectively; Libes 1992). Second, there is exchange of materials between the mesopelagic layer and the epipelagic layer within biologically relevant timescales, whereas the bathypelagic zone can be considered as a reservoir of

---

[1] Terminology adopted from Parsons *et al.* (1984). They use 150 m as the upper limit to the mesopelagic zone; where possible we have adhered to this convention, but this is not always possible when drawing upon other work. We have used the term dark ocean as a collective term to refer to meso- and bathypelagic zones.

materials at biologically relevant timescales. We provide a discussion of the discrepancies found between the different approaches to estimate respiration in the dark ocean, and conclude by using the estimates of respiration in the dark ocean to derive a global estimate, to be compared with current estimates of export and new production. This exercise provides a revised perspective of the role of the dark ocean in the oceanic carbon cycle.

10.2 Supply of organic matter to the dark ocean

The dark ocean receives the excess primary production of the epipelagic zone, as well as additional lateral inputs from the continental margins and other oceanic areas. In addition, migrant zooplankton actively transport organic matter grazed in the epipelagic zone to the dark ocean. There is consensus that the bulk of the organic matter entering the mesopelagic zone is recycled therein, implying that the respiration rates responsible for the recycling must be large. However, there are substantial discrepancies as to the magnitude of organic matter inputs to the dark ocean, so it has been difficult to derive a global estimate of dark ocean respiration from organic matter loading alone.

10.2.1 Overall rates of export of organic carbon from the photic layer

The biological carbon pump exports a fraction of the organic carbon produced by oceanic plankton, both in particulate and dissolved form below the thermocline (Antoine et al. 1996). The export of carbon by the biological pump is conventionally regarded to be constrained by the nitrogen input, which is believed to set an upper limit to the organic carbon export equal to the product of the C/N ratio in the export organic carbon and the nitrogen supplied to the biogenic layer (Ducklow 1995; Williams 1995; Field et al. 1998). The nitrogen input to the photic zone of the ocean is mainly derived from the internal supply of nitrate through vertical mixing, estimated at ~ 50 Tmol $N\,a^{-1}$ in subtropical waters (Jenkins and Wallace 1992; Lewis 2002), atmospheric inputs (Ducklow 1995; Williams 1995; Field et al. 1998; Luz

and Barkan 2000), estimated at about 10 Tmol $N\,a^{-1}$ (Longhurst et al. 1995) and nitrogen fixation for which there is little consensus on the global rate (7–14 Tmol $N\,a^{-1}$; Karl et al. 2002). The assumption that this net production equals the net carbon export by the biological pump is, in turn, based on the assumption that the carbon and nitrogen transport in the upward inorganic and the downward organic fluxes are in similar stoichiometric balance, but the latter assumption is unsupported by current data.

Dissolved organic matter in the photic ocean contains excess carbon relative to nitrogen (mean $\pm$ SE C/N ratio = 15.0 ± 1.1; Table 10.1) compared to the exported particulate organic matter (global mean C/N ratio = 7.8; Takahashi et al. 1985). The twofold

Table 10.1 The distribution of the average atomic C/N ratio in DOM and POM. Values for DOM are near-surface values, averaged when several values were reported

Area	C/N ratio	Reference
	DOM	
NW Mediterranean	8.7	Banoub and Williams (1972)
NW Mediterranean	24.6	Zweifel et al. (1993)
NW Mediterranean	17.0	Souchou et al. (1997)
NW Mediterranean	15.5	Doval et al. (1999)
NW Mediterranean	21.3	C.M.Duarte (unpubl. data)
N Atlantic	20.0	Williams (1995)
NW Atlantic	13.0	Gardner and Stephens (1978)
NW Atlantic	14.0	Hopkinson et al. (1997)
Sargasso Sea	17.1	McCarthy et al. (1996)
Gulf of Mexico	17.0	McCarthy et al. (1996)
N Pacific	15.3	McCarthy et al. (1996)
N Pacific	17.0	Cherrier et al. (1996)
N Pacific	7.9	Christian et al. (1997)
N Pacific	20.0	Karl et al. (1998)
NE Pacific	7.1	Martin et al. (1987)
Equatorial Pacific	9.0	Libby and Wheeler (1997)
Southern California	7.1	Holm-Hansen et al. (1966)
Southern California	14.3	Williams (1986)
Southern California	18.3	Hansell et al. (1993)
Southern Ocean	25.0	Jackson and Williams (1985)
Southern Ocean	8.0	Kähler et al. (1997)
Southern Ocean	14.8	S.Agustí (unpubl. data)
Arctic Ocean	14.8	Wheeler et al. (1997)
Mean $\pm$ SE	15.0 ± 1.1	
	POM	
Global average	7.8	Takahashi et al. (1985)

greater C/N ratio in dissolved organic matter compared to particulate organic matter implies that, for the same nitrogen input to the biogenic layer, the potential total organic carbon export is twofold greater if occurring in dissolved organic carbon (DOC) than in particulate (POC) form. The deviations in the dissolved organic C/N ratio from the Redfield ratio have been noted in the past (e.g. Williams 1995; Najjar and Keeling 2000; Benner 2002), but have yet to be incorporated into evaluations of the role of oceanic biota on atmospheric carbon uptake. Most previous assessments of the potential importance of the DOC flux to the net carbon export have assumed the DOC/DON ratio to be similar to the Redfield ratio (Antoine *et al.* 1996). Because the export of organic carbon possible for a given nitrogen supply is twofold greater when the flux occurs as DOC compared to a similar POC flux, even modest changes in the allocation of net community production to POC or DOC may lead to regional differences in the oceanic carbon supply, and possibly respiration, in the dark ocean.

There is an order of magnitude spread in estimates of the organic carbon export from open ocean surface waters, from 0.3 to 2.3 $Pmol\,C\,a^{-1}$ (Smith and Hollibaugh 1993; Sambrottto *et al.* 1993; Ducklow 1995; Hedges and Keil 1995; Emerson *et al.* 1997; Najjar and Keeling 2000; del Giorgio and Duarte 2002). This variance is attributable, as discussed below, to various degrees of underestimation of the dissolved but, mainly, particulate fluxes.

10.2.2 Delivery of POC

The sinking of particles, resulting from dead organisms or plankton excretion, is accepted to be the principal mechanism by which organic carbon is exported from the epipelagic zone to the dark ocean. Much of this organic matter is carried in aggregates, which result from coagulation of smaller particles (Jackson 2002). On average, more than 80% of the sinking organic matter is remineralized in the upper 1500–2000 m, but there is evidence of regional variations in the remineralization patterns and rates in the mesopelagic zone and thus in the export flux to the dark ocean. Based on the analysis of sediment trap data from 64 open-ocean sites, François *et al.*

(2002) formulated the hypothesis that export ratio (the fraction of primary production that is exported below the surface) and extent of mineralization through the mesopelagic zone tend to have contrary effects on the overall efficacy of the biological pump. Export ratios are higher in productive high-latitude regions dominated by diatoms, but the efficiency of transfer through the mesopelagic zone in these regions is very low. Mesopelagic remineralization in high-latitude regions is comparatively higher than in low-latitude regions, due to the more labile nature of the organic matter exported, which is mainly in the form of phytoplankton aggregates. In contrast, mesopelagic remineralization rates in low-latitude productive regions are lower because the organic matter exported is relatively refractory (consisting primarily of fast-sinking fecal pellets with carbonate minerals), having already been processed extensively by the food web in the mixed layer (François *et al.* 2002). In this context, provided the actual primary production is comparable between high- and low-latitude regions, the most efficient at transferring carbon to the bathypelagic zone would be the productive low-latitude regions, such as the equatorial upwelling zones, or the Arabian Sea (Antia *et al.* 2001; François *et al.* 2002).

The flux of POC into the dark ocean (the export particulate production) can be estimated as the POC collected with floating traps deployed below the surface mixed layer/photic zone. Average annual values obtained from monthly measurements during several years using these traps, both in the subtropical Atlantic (Station BATS; Lohrenz *et al.* 1992) and central Pacific (Station ALOHA; Karl *et al.* 1996), yield similar values (0.7–$0.9\,mol\,C\,m^{-2}\,a^{-1}$), about three to fourfold lower than geochemical estimates of new production (Carlson *et al.* 1994; Emerson *et al.* 1997); that is, the part of the carbon fixed that is not respired until it has been removed from the epipelagic zone to the dark ocean (Platt *et al.* 1992). Several studies using particle-reactive natural radio nuclides (^{234}Th, ^{230}Th, and ^{231}Pa) as a tracer for sinking particles, have convincingly demonstrated that both free-floating and moored sediment traps deployed in the mesopelagic zone severely underestimate the particle flux (Buesseler 1991; Buesseler *et al.* 2000; Scholten *et al.* 2001;

Yu *et al.* 2001). In contrast, traps in the bathypelagic zone seem to intercept the vertical flux of particles more effectively (Yu *et al.* 2001).

POC fluxes measured from deep-sea (bathypelagic) moored sediment traps have been used to derive empirical models which predict the vertical profile of POC flux in the whole water column, by relating it to primary production and depth (Suess 1980; Betzer *et al.* 1984; Berger *et al.* 1987; Pace *et al.* 1987; Lohrenzen *et al.* 1992; Antia *et al.* 2001). In this way, the export particulate production may be estimated from the POC flux at the base of the euphotic zone/mixed layer. The weakness of this approach lies in extrapolating small fluxes measured in the bathypelagic zone to the base of the epipelagic zone, since most of the organic carbon mineralization occurs in the mesopelagic zone. Moreover, these models depend on primary production estimates, which must integrate on the same spatial and temporal scales as sediment traps. Antia *et al.* (2001) compiled a large set of particle flux [230]Th-corrected data from deep sediment traps moored in the Atlantic Ocean and estimated, by applying a derived empirical model, that the POC flux at the base of the euphotic zone was about $0.3 \, \text{Pmol} \, \text{C} \, \text{a}^{-1}$. Extrapolating this to the whole ocean would yield a value in the region of $1 \, \text{Pmol} \, \text{C} \, \text{a}^{-1}$, twice the value of previous estimations using uncorrected sediment trap data (e.g. Martin *et al.* 1987).

10.2.3 Delivery of DOC

The export of DOC to the mesopelagic zone may occur by vertical diffusion through the thermocline (e.g. Emerson *et al.* 1997), or by isopycnal transport and convective mixing of surface rich-DOC waters with deep DOC-impoverished water (e.g. Carlson *et al.* 1994; Hansell 2002). In addition, there is some evidence that DOC can be adsorbed onto deep-water particles, which provides a mechanism whereby DOC is added to the sinking flux of particulate material (Druffel *et al.* 1996). Dissolved organic matter can also aggregate by the action of surface charges and microbial activity to form amorphous particles of marine snow, which supplements the sinking flux (e.g. Barber 1966). While the latter are potentially important processes in the transfer

Table 10.2 The average estimated export DOC flux to the mesopelagic zone by vertical diffusive mixing in different areas of the ocean

Area	DOC diffusive flux ($\text{mmol} \, \text{C} \, \text{m}^{-2} \, \text{d}^{-1}$)	Reference
NW Mediterranean	1.5	Copin-Montégut and Avril (1993)
Middle Atlantic Bight	0.2	Guo *et al.* (1995)
Central Atlantic	0.7	Vidal *et al.* (1999)
Gulf of Mexico	1.2	Guo *et al.* (1995)
N Pacific	2.4	Emerson *et al.* (1997)
Equatorial Pacific	8	Feely *et al.* (1995)
Equatorial Pacific	1	Zhang and Quay (1997)
Equatorial Pacific	0.56	Christian *et al.* (1997)
Equatorial Pacific	6	Archer *et al.* (1997)

of DOC to the dark ocean and sediments (Keil *et al.* 1994), the information available is still insufficient to evaluate their overall impact.

Available average estimates of the oceanic downward DOC flux through vertical diffusive mixing into the water column range from negligible to $8 \, \text{mmol} \, \text{C} \, \text{m}^{-2} \, \text{d}^{-1}$, with a geometric mean of $1.4 \, \text{mmol} \, \text{C} \, \text{m}^{-2} \, \text{d}^{-1}$ (Table 10.2). The available estimates have been mostly derived for the subtropical and tropical ocean, but vertical profiles of DOC concentration in high-latitude seas (e.g. Wiebinga and De Baar 1998) also show a gradient from high concentrations in the mixed layer to lower values in the intermediate waters, suggesting that vertical diffusion of DOC may also occur at these latitudes. These estimates rely, however, on the assumption, or calculation, of vertical diffusion coefficients, which may have an uncertainty by an order of magnitude. Hence, although this mechanism of DOC export is unlikely to be as important on a global scale as the POC flux, it still needs to be accurately evaluated (Hansell 2002). Extrapolation to the global ocean, assuming the mean value of DOC export of $1.4 \, \text{mmol} \, \text{C} \, \text{m}^{-2} \, \text{d}^{-1}$, yields an estimated global DOC export by diffusive mixing of $0.17 \, \text{Pmol} \, \text{C} \, \text{a}^{-1}$.

In addition to a diffusive, gradient-driven flux of DOC, subduction and convective mixing have been shown to be a significant transport process in areas of deep-water formation, and other areas experiencing seasonally intense convective mixing where DOC seasonally accumulates (Hansell

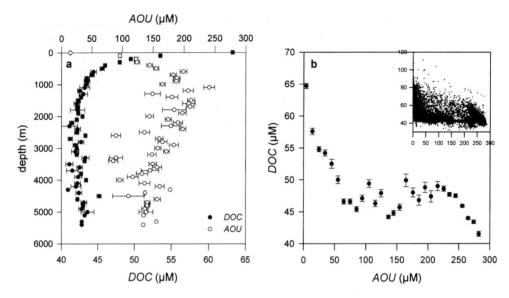

Figure 10.1 DOC and AOU distribution in the dark ocean. (a) The depth distribution of the average (mean ± SE of data grouped by 100 m bins, $n = 9578$) DOC (full circles) concentration and AOU (open circles) in the ocean, and (b) the relationship between DOC and AOU in the ocean. The symbols represent mean ± SE DOC concentrations of data grouped by 10 μM AOU bins ($N = 9823$). The insert in (b) shows the relationship for the raw data ($N = 9823$), which is described in the interval $0 < AOU < 150$, by the fitted regression equation DOC $= 60.3(\pm 0.2) - 0.136(\pm 0.003)$ AOU ($R^2 = 0.28$, $p < 0.0001$, $n = 5541$). The monotonic decrease of DOC when AOU > 225 corresponds to the oxygen minimum zones of the Arabian Sea, Indian Ocean, and Equatorial Pacific (from Arístegui *et al.* 2002a).

2002). Hansell and Carlson (2001) have estimated a global DOC export to the aphotic zone (excluding vertical diffusion) of 0.12 $PmolCa^{-1}$, assuming a global new production of 0.6 $PmolCa^{-1}$ and a contribution of DOC to global export of 20%. Since their calculations are based on the magnitude of new production, evidence for a higher value of new production (e.g. 1.3 $PmolCa^{-1}$; Sambrotto *et al.* 1993; Falkowski *et al.* 1998) would increase the total calculated export by a factor of 2.

Arístegui *et al.* (2002a) compiled a large dataset on the relationship between DOC concentration and apparent oxygen utilization (AOU)—the oxygen anomaly with respect to the dissolved oxygen saturation levels—from various oceans. These data revealed a decline in DOC with increasing AOU in the upper 1000 m, albeit with considerable scatter (Fig. 10.1). In contrast, there is no significant decline in DOC with increasing depth beyond 1000 m (Fig. 10.1), indicating that DOC exported with overturning circulation plays a minor role in supporting respiration in the dark ocean. Assuming a molar

respiratory quotient ($\Delta CO_2/-\Delta O_2$) of 0.69 (Hedges *et al.* 2002), the decline in DOC was estimated to account for about 20% of the AOU within the top 1000 m. This estimate represents, however, an upper limit, since the correlation between DOC and AOU is partly due to mixing of DOC-rich warm surface waters with DOC-poor cold thermocline waters. Removal of this effect by regressing DOC against AOU and water temperature indicated that DOC supports about 10% of the respiration in the mesopelagic waters (Arístegui *et al.* 2002a). Nevertheless, since water temperature and DOC covary in the upper water column, the actual contribution of DOC to AOU probably lies between 10% and 20%.

Considering the above estimates, the overall flux of DOC to the dark ocean can be estimated as the sum of 0.17 $PmolCa^{-1}$ from vertical diffusion and 0.12 $PmolCa^{-1}$ from subduction and convective mixing (cf. 1 $PmolCa^{-1}$ estimate for POC export). Nevertheless, much of the exported DOC by diffusive mixing would be respired in the upper thermocline, contributing little to the ocean-interior carbon reservoir, unless it is transported and respired below the depth of the winter mixed layer.

10.2.4   Transport by migrating zooplankton

The occurrence of active transport of organic matter by migrating zooplankton was proposed decades ago by Vinogradov (1970). Migrating zooplankton contributes to the transport of particulate and DOC by excretion or fecal pellet dissolution. Migratory processes set in motion 0.08 Pmol C (the biomass of the migrant zooplankton) daily between the photic ocean and the dark ocean (Longhurst 1976; Conover 1978). If only a fraction of this organic matter is retained (e.g. excreted) while in the dark zone, this process has the potential to account for a considerable input of organic matter to the dark ocean. Estimates of the actual input involved are, however, still few. Ducklow *et al.* (2001) and Hernández-León and Ikeda (Chapter 5) have summarized the available data of downward transport of carbon by diel migrating zooplankton (the active transport) in the Atlantic and Pacific oceans (including the BATS and HOT time-series stations). The daily average value of active organic matter inputs by zooplankton accounts for 0.54 mmol C m^{-2} (Hernández-León and Ikeda, Chapter 5). Extrapolating this value to the whole ocean the global transport of organic matter from surface to mesopelagic waters mediated by zooplankton would represent about 0.06 Pm C a^{-1}. However, since most of the compiled data by the latter authors correspond to oligotrophic regions, the actual contribution of zooplankton to the downward organic carbon flux could be considerably higher when including estimates from high-productive systems.

10.2.5   Coastal inputs to the dark ocean

Oceanic waters near continental margins may receive inputs of both particulate and DOC from coastal ecosystems, and because these inputs have not been well constrained, this adds additional uncertainty to the mineralization rates in the dark ocean inferred from mass-balance carbon fluxes. The inputs of total organic carbon (TOC) from the coastal ocean to the dark open ocean have been estimated

at 0.17 Pmol C a^{-1} (e.g. Liu *et al.* 2000), representing a significant fraction of the global TOC export from the coastal to the open ocean, which have been estimated to range from 0.23 Pmol C a^{-1} (Gattuso *et al.* 1998) to 0.5 Pmol C a^{-1} (Duarte and Cebrián 1996; Duarte *et al.* 1999).

These estimates of carbon export from the continental margins are based on carbon mass balances, and there are, as yet, no comprehensive measurements of the actual transport processes supporting this global export. In fact, the exchange rate of water between the global coastal and open oceans remains a matter of speculation, supported by a few case studies (Walsh *et al.* 1981; Falkowski *et al.* 1988; Liu *et al.* 2000). For instance, Wollast and Chou (2001) calculated that 15% of the primary production in the Gulf of Biscay was exported to the intermediate and deep waters. Mesoscale instabilities, such as filaments, may accelerate the transport of particulate and DOC from the coastal to the open ocean at upwelling regions (e.g. Gabric *et al.* 1993; Barton *et al.* 1998). Alvarez-Salgado *et al.* (2001) estimated that up to 20% of new production in the NW Spain coastal upwelling region transforms into labile DOC, which is exported offshore to the adjacent open-ocean waters. There is, therefore, a need to directly estimate transport processes to better constrain the input of organic matter from the continental shelves to the dark ocean.

10.2.6   Summary of organic carbon inputs

The total export of biogenic carbon from the photic zone (the sum of the open-ocean sinking POC flux, DOC flux and zooplankton active flux, plus TOC advected from coastal ecosystems) should be equivalent, on an annual basis, for the magnitude of the new production. Our review indicates that the sum of the above four carbon inputs amounts to 1.5–1.6 Pmol C a^{-1}, a value close to some recent estimates of new production (Sambrotto *et al.* 1993; Falkowski *et al.* 1998), but somewhat below the estimates of export production proposed recently by del Giorgio and Duarte (2002; 2.3 Pmol C a^{-1}). Uncertainties in the estimates of the above fluxes could lead to differences between estimates of new production and export biogenic fluxes. Michaels *et al.* (1994) found a large imbalance in the annual carbon cycle at the Bermuda Atlantic Time-series Station (BATS), which they explained by inaccuracies in the determination of the fluxes of POC using shallow sediment traps. This was corroborated by sediment-trap measurements of the ^{234}Th flux during the same period of study (Buesseler 1991). A more detailed interpretation of the particle fluxes at the BATS station during a decade of sampling revealed that surface sediment trap records could be biased by sampling artifacts due to physical dynamics during deep-mixing events (Steinberg *et al.* 2001).

10.3   Pelagic respiration in the dark ocean

The recent evidence that dark ocean respiration must be important is leading to an increase in the number of direct measurements of rates of metabolism in these layers. These low rates of respiration in deep-water stretch conventional direct methods used to measure planktonic respiration in the euphotic zone to and beyond their limits. Indirect estimations of deep-water respiration have been obtained, based on large-scale geochemical mass balances of oxygen and organic matter fields, although these estimates themselves do not always agree in their magnitude (e.g. Michaels *et al.* 1994). In this section, we review the different approaches used to estimate respiration in the dark ocean, and compare their results in order to derive an estimate of the global pelagic environment in the dark ocean.

10.3.1   *In vitro* oxygen consumption measurements

Direct measurements of planktonic respiration may be obtained by the *in vitro* oxygen method (e.g. Williams and Jenkinson 1982). Nevertheless the limited sensitivity of the method has confined generally these measurements to the upper 200 m of the surface ocean, which are thus overwhelmingly restricted to the very upper layers of the aphotic zone (see Robinson and Williams, Chapter 9). Biddanda and Benner (1997) were, however, able to estimate *in vitro* microplankton

respiration rates in intermediate waters in the Gulf of Mexico, because oxygen consumption was rather high (1–2 mmol O_2 m^{-3} d^{-1}) in the upper 500 m. They calculated that the integrated respiration within the thermocline (100–500 m) exceeded the integrated respiration in the top 100 m (456 and 305 mmol O_2 m^{-2} d^{-1}, respectively). More recently, respiration rates in the order of 0.2–0.3 mmol O_2 m^{-3} d^{-1} have been measured down to 1000 m, in the coastal transition zone region between Northwest Africa and the open ocean waters of the North Atlantic subtropical gyre, by J. Arístegui and C. M. Duarte (unpublished data).

The oxygen method, even when applied in the photic zone, often requires long incubations and there are always concerns that these could induce changes in microbial community structure, biomasses, and rates (see Robinson and Williams, Chapter 9, for a review on bottle effects). Nevertheless, time-series measurements indicate that in most occasions, when samples are handled with care, respiration rates remain uniform (Arístegui *et al.* 1996, Robinson and Williams, Chapter 9), even during long (ca. 72 h) deep-water experiments (J. Arístegui and C. M. Duarte, unpublished), in spite of changes in bacterial production and biomasses (J. M. Gasol, unpublished). The contrasting bottle effects on microbial respiration and biomass or bacterial production (as estimated by leucine and thymidine incorporation) must be further assessed in order to test the applicability of the oxygen method to measure microbial respiration rates during relatively long (>24 h) incubation periods in mesopelagic waters (e.g. Pomeroy *et al.* 1994).

10.3.2   Microplankton ETS activity measurements

At depths where respiration rates are too low to be resolved through the direct measurement of oxygen consumption, the ETS (respiratory Electron Transport System) technique (Packard 1971) has proved to be a useful tool to estimate these rates (Packard *et al.* 1988). As the enzymatic analysis can be made after the field work, the ETS approach allows the collection of large datasets during field cruises, facilitating the measurement of oceanic respiration

Table 10.3 Areal estimates of microplankton respiration (R) in the dark ocean. (from Arístegui *et al.* 2003)

Ocean	Depth range (m)	R (mol C m^{-2} a^{-1})	Number of stations	Method
Atlantic (Sargasso Sea)[a]	200–1000	1.5	8	ETS
Atlantic (Sargasso Sea)[a]	1000–bottom	0.7	8	ETS
Atlantic (west Iberia/Morocco)[b]	200–1500	6.9	1	ETS
Atlantic (meddies)[b]	200–1500	6.1	2	ETS
Atlantic (Canary Islands)[c]	200–1000	2.4	11	ETS
Atlantic (Gulf of Mexico)[d]	100–500	118.1	7	Winkler
Mediterranean (west)[b]	200–1500	6.8	2	ETS
Mediterranean (west)[e]	200–3000	1.3	8	ETS
Mediterranean (west)[f]	200–800	2.2	24	ETS
Mediterranean (west)[g]	200–1000	1.2	10	ETS
Mediterranean (east)[h]	200–3000	4.3	>10	ETS
Indian Ocean (Arabian Sea)[i]	200–2400	5.2	6	ETS
Indian Ocean (Bay of Bengal)[i]	200–2400	2.2	6	ETS
Pacific Ocean (Guinea Dome)[a]	200–1000	3.9	>15	ETS
Pacific Ocean (Guinea Dome)[a]	1000–bottom	2.6	>4	ETS
Southern Ocean (Indian sector)[j]	200–1000	1.5	7	ETS

ETS = Respiratory Electron Transport System activity; Winkler = Winkler oxygen determination.
[a]Packard *et al.* 1988; [b]Savenkoff *et al.* 1993a; [c]J. Arístegui, unpublished data; [d]Biddanda and Benner 1997; [e]Christensen *et al.* 1989; [f]Savenkoff *et al.* 1993b; [g]Lefèvre *et al.* 1996; [h]La Ferla and Azzaro, 2001; [i]Naqvi *et al.* 1996; [j]Arístegui *et al.* 2002b.

over large temporal and spatial scales. The ETS method estimates the maximum overall activity of the enzymes associated with the respiratory electron transport system under substrate saturation, in both eukaryotic and prokaryotic organisms. ETS activity measurements represent, therefore, potential respiration rates, and hence calculation of *in situ* respiration rates from ETS activity requires the use of empirically determined algorithms. The R/ETS relationships for deep-water plankton have been derived from monospecific cultures of bacteria and zooplankton (Christensen *et al.* 1980) grown under laboratory conditions. Thus, respiration rates calculated from ETS activity have to be interpreted with a degree of caution.

The available dark-ocean respiration estimates for microplankton (<200 μm) derived from ETS (Table 10.3) have been aggregated into depth-average bins (Arístegui *et al.* 2003). These estimates yield a relationship between respiration rate and depth (Fig. 10.2(a)) best described by an exponential function of the form,

$$R(\mu mol\, O_2\, m^{-3}\, d^{-1}) = 18\, e^{-0.00053\, z},$$

which z is the depth in meters. The equation prescribes that respiration rates decline by about 53% over the mesopelagic layer of the ocean (150–1000 m).

Examination of the residuals (Fig. 10.2(b)) showed an enhanced respiratory activity at 1000–2000-m depth, corresponding to the base of the permanent thermocline, separating the mesopelagic from the bathypelagic layers. Inspection of the individual profiles indicated this enhanced activity to be a consistent feature in various seas and oceans, rather than an anomaly of a particular basin. The origin of these enhancements is unclear, but may derive from mixing of different water masses, or may be produced *in situ* by migrant zooplankton.

Available data are yet too sparse as to allow the elucidation of differences in the magnitude of microplankton respiration across the different

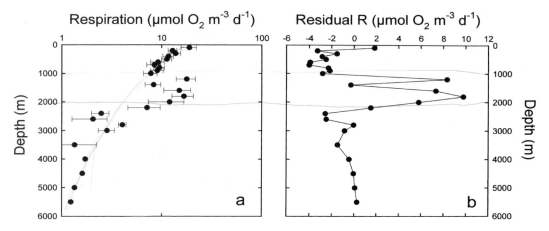

Figure 10.2 The depth distribution of (a) the average ($\pm$ SE) respiration rate in the dark ocean and (b) the depth distribution of the residuals from the fitted exponential equation. Rates compiled from references in Table 10.4 (from Arístegui *et al.* 2003).

oceanic provinces of the dark ocean. There are, however, obvious regularities in the various published reports that help constrain the overall microplankton respiration within the dark layers of the global open ocean. In particular, most studies show that integrated respiration, in the layer extending from the bottom of the photic zone to 800–1000 m, is of a similar magnitude as the integrated microplankton respiration in the photic layer of the open ocean (del Giorgio and Duarte 2002), and sometimes greater.

Integration of respiration rates derived from ETS measurements in the dark ocean yield a global respiration ranging from $1.7\,\mathrm{Pmol\,C\,a^{-1}}$, when derived from the areal rates, to $2.8\,\mathrm{Pmol\,C\,a^{-1}}$, when derived from the depth distribution of the rates weighed for the hypsographic curve of the ocean (Arístegui *et al.* 2003). These rates are comparable to the estimates of the total microplankton respiration in the photic layer, estimated to be $2–4\,\mathrm{Pmol\,C\,a^{-1}}$ (del Giorgio and Duarte 2002; Robinson and Williams, Chapter 9). Examination of the depth-averaged cumulative respiration rate throughout the dark ocean (Fig. 10.3) indicated that 70% of the microplankton respiration within the dark ocean occurs between 200 and 1000 m depth. This represents an estimate of $1.2\,\mathrm{Pmol\,C\,a^{-1}}$ for the mesopelagic zone, and $0.5\,\mathrm{Pmol\,C\,a^{-1}}$ for the ocean interior, when derived from the areal rates, although the bathypelagic rates could be overestimated (see Section 10.5).

10.3.3   Measured rates of bacterial carbon flux

The rates of bacterial production in the dark ocean constrain their contribution to the respiratory rates there. The vertical profiles of bacterial production show a steep exponential decline spanning two to three orders of magnitude from the surface to the dark ocean (Fig. 10.4; Nagata *et al.* 2000; Turley and Stutt 2000). Whereas bacterial biomass in the mesopelagic (150–1000 m) and bathypelagic (1000–4000 m) layers of the water column may exceed that in the photic layer, the integrated bacterial production in each of the two dark layers is much smaller (5–20 fold) than that in the surface ocean (Nagata *et al.* 2000), indicative that bacterial growth in the dark ocean is generally slow. Cell-specific production rates may be considerably high in particle-associated bacteria from mesopelagic waters (Turley and Stutt 2000). Several authors (Smith *et al.* 1992; Cho and Azam 1988; Nagata *et al.* 2000) have proposed that large sinking particles are hydrolyzed to release DOC, which would support production of free-living bacteria both in the mesopelagic and bathypelagic zones.

The total flux of organic carbon through bacteria in the dark ocean (the so-called bacterial carbon flux) can be used to provide a minimum estimate of the long-term, whole-community respiration rate. The bacterial carbon flux (BC) is the sum of bacterial

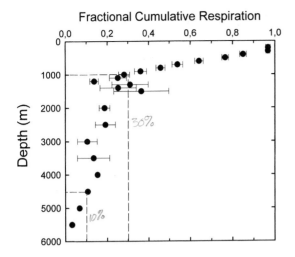

Figure 10.3 The depth distribution of the average (± SE) cumulative respiration rate in the dark ocean. Rates compiled from references in Table 10.4. Dotted lines indicate depths comprising 30% and 10% of the water column respiration in the dark ocean (from Arístegui *et al.* 2003).

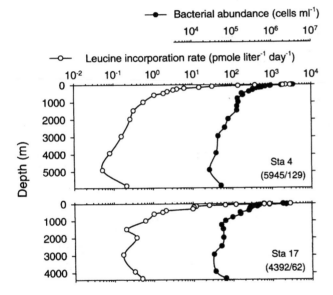

Figure 10.4 Depth profiles of bacterial abundance (closed circles) and leucine incorporation rate (open circles) at two stations from the Subarctic Ocean (from Nagata *et al.* 2000).

production (BP) and bacterial respiration (BR). It may be calculated as the ratio between the bacterial production and the bacterial growth efficiency (BC = BP/BGE = (BP + BR)/BGE). The estimation of BC fluxes relies, however, on assumed factors that are highly uncertain: (i) the ratio of leucine incorporation to bacterial production, which may range from <1 to 20 KgC mol^{-1} leucine (Kirchman *et al.* 1985; Simon *et al.* 1992; Gasol *et al.* 1998; Sherry *et al.* 1999), and (ii) the bacterial growth efficiency (BGE = BP/(BP + BR)), which ranges from 2% to

70% in surface waters of aquatic habitats (del Giorgio and Cole 2000), with no published estimates available for the mesopelagic oceanic waters. Hence the estimates of respiration rates derived from bacterial carbon flux in the dark ocean are highly sensitive to poorly constrained conversion factors and, provided these are not rigorously established, must be considered with some degree of caution.

Nagata *et al.* (2000), using an assumed BGE of 0.2, calculated that integrated BC fluxes for different oceanic regions ranged between 1 and

Table 10.4 Bacterial carbon flux (the total organic carbon consumed by bacteria) in different regions from the dark ocean. Units in mmol C m^{-2} d^{-1} (from Nagata *et al.* 2000)

Location	Mesopelagic zone (100–1000 m)	Bathypelagic zone (>1000 m)	Reference
North Pacific Gyre	6.5		Cho and Azam (1988)
Santa Monica basin	42		Cho and Azam (1988)
Subarctic Pacific	3.5–13		Simon *et al.* (1992)
Indian Ocean	7.5–51		Ducklow (1993)
NE Atlantic	9.8	0.4	Turley and Mackie (1994)
Subarctic Pacific	2.8 ± 0.9	0.5 ± 0.1	Nagata *et al.* (2000)
N subtropical Pacific	1.1	0.1	Nagata *et al.* (2000)

52 mmol C m^{-2} d^{-1} for the mesopelagic zone, and between 0.05 and 0.5 mmol C m^{-2} d^{-1} for the bathypelagic zone (Table 10.4). The geometric mean of the mesopelagic estimates (8 mmol C m^{-2} d^{-1}) represents an annual rate of 2.9 mol C m^{-2}, which coincides with the estimate of oxygen consumption in the mesopelagic zone derived from large-scale tracer balances (Section 10.3.5) and from ETS activity (Table 10.3; geometric mean = 2.9 mol C m^{-2} a^{-1}, excluding the very high value reported by Biddanda and Benner 1997).

10.3.4 Measured mesozooplankton respiration rates

Because of operational reasons the rates of water column respiration often exclude the respiration of metazoan zooplankton, which can have a significant contribution to planktonic respiration. The estimates of the contribution of mesozooplankton (>200 μm) to total respiration in the ocean are highly variable. Joiris *et al.* (1982) estimated that zooplankton contributed less than 1% of the total respiration in productive coastal areas, but estimates in the less productive open ocean seem to be higher. Early studies in open ocean sites suggested a contribution of less than 5% (Pomeroy and Johannes 1968), and subsequent studies report similar ranges (Williams 1981; Holligan *et al.* 1984; Hernández-León *et al.* 2001). On the basis of the reported rates, the contribution of zooplankton to total plankton respiration was estimated by del Giorgio and Duarte (2002) at about 0.22 Pmol C a^{-1} (or about an additional 10% of the pelagic respiration in the dark ocean). This value agrees with the global estimate (0.21 Pmol C a^{-1}) calculated by Hernández-León and Ikeda (Chapter 5) from a major review of zooplankton respiration in the mesopelagic (0.18 Pmol C a^{-1}) and bathypelagic (0.03 Pmol C a^{-1}) zones, based both on biomass and ETS activity.

10.3.5 Inference of organic matter mineralization from oxygen fields

Early insight into the likely importance of respiratory processes in the dark ocean was derived from the examination of the distribution of oxygen and organic matter fields. Evidence of a deep oxygen minimum at depths ranging from 100 m in the Indian Ocean to 1000 m depth in the Pacific Ocean suggested significant oxygen consumption at those depths, although the presence and position of the oxygen minimum may be largely a result of oceanic circulation (cf. Wyrtki 1962). Examination of the AOU in the dark ocean, also provided evidence suggesting an important respiratory activity there (e.g. Broecker and Peng 1982). In fact, the distribution of AOU in the dark ocean parallels the path of dark ocean circulation, indicating that as much as 195 mmol O$_2$ m^{-3} is removed from the oldest dark oceanic waters (Fig. 10.5).

The link between the distribution of AOU and the age of the water masses is so strong that a

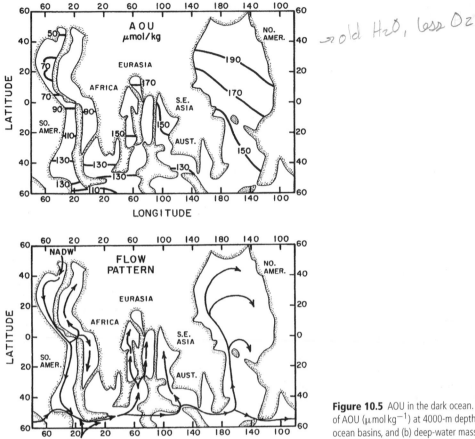

→ old H₂O, less O₂

Figure 10.5 AOU in the dark ocean. (a) The distribution of AOU (μmol kg^{-1}) at 4000-m depth in the world's major ocean basins, and (b) deep-water mass flow pattern at the same depth (from Broecker and Peng 1982).

relationship between AOU and an independent estimate of the age of the water masses can be used to derive long-term average estimates of respiration (the oxygen utilization rate, OUR) in the dark ocean (cf. Jenkins and Wallace 1992). The first calculations of OURs were based on 3D advection–diffusion models based on geostrophic velocity estimates (Riley 1951). More recent approaches include box-model estimates, and advective-diffusive estimates, using tracer mass-balance calculations. Several tracers (e.g. ^{228}Ra, ^{3}H, ^{3}He, CFC) have been satisfactorily applied to estimate OURs, by providing an apparent age for a water parcel with respect to its last exposure to the sea surface (see review by Jenkins and Wallace 1992, and references therein). Despite the rather different approaches, all these

techniques yield similar absolute OUR estimates ($\sim$5–6 mol O$_2$ m^{-2} a^{-1}) for the subtropical North Atlantic Ocean, adding confidence to the results. Unfortunately, these techniques can only be applied where the age of the water mass can be calculated with relative precision and mixing effects can be assumed to be relatively unimportant, which is an uncommon situation. Consequently, these techniques have not yet allowed global estimates of respiration in the dark ocean.

Oxygen consumption rates in the mesopelagic zone have been also deduced from dissolved oxygen profiles and particulate barium stocks. Dehairs *et al.* (1997) used a 1D advection/diffusion inverse model of the steady-state distribution of conservative and non-conservative tracers in

OUR - AOU/time

the water column, with the aim of estimating oxygen consumption in the Southern Ocean. The weakness of this method lies, as well as in other advection/diffusion models, on the difficulty associated with the determination of eddy diffusivity. Their estimated oxygen consumption rates, integrated between 175 and 1000 m, ranged from 1 to 6 mmol O_2 m^{-2} d^{-1}. These rates, extrapolated to a yearly basis (0.4–2.2 mol O_2 m^{-2} a^{-1}) are somewhat lower than those estimated for the mesopelagic zone in the subtropical North Atlantic (Jenkins and Wallace 1992), but broadly agree with ETS estimates in the mesopelagic zone of the Southern Ocean (Arístegui *et al.* 2002*b*).

10.3.6 Inferences of organic matter mineralization from carbon fluxes

Particulate organic carbon is solubilized to yield DOC, which is then partially respired by bacteria, as it sinks through the water column. This results in an exponential decrease in the flux of POC with depth, when sinking across the water column below the euphotic zone. Suess (1980) was the first to propose a model, based on sediment trap data, to predict the flux of particulate organic carbon at a given depth. His model, as well as other subsequent ones (e.g. Betzer *et al.* 1984; Berger *et al.* 1987; Pace *et al.* 1987), predicts the particle flux J_z at a given depth (z in meters), as a function of depth and the surface primary production (P) through the expression (Suess 1980):

$$J_z = P/0.024z$$

Assuming a global primary production rate of 3 Pmol C a^{-1} (Antoine *et al.* 1996), the fraction of POC respired between 150 and 1000 m is estimated to be 0.7 Pmol C a^{-1}. If photosynthesis were closer to 4 Pmol C a^{-1} (del Giorgio and Duarte, 2002), the fraction respired in the 150–1000 m would be ~1 Pmol C a^{-1}. According to the Suess expression:

$$\text{Fraction POC respired} = 1 - \frac{1}{(0.024z + 0.21)}$$

95% of the POC entering the dark ocean is respired within the mesopelagic zone (i.e. above

1000-m depth), a percentage higher than estimated through other approaches. The Suess relationship is based, however, on sediment trap data, which seem to underestimate particulate fluxes (see above), due to hydrodynamics, resuspension, swimmers, and degradation of organic matter between settling and trap collection, as well as to the inefficacy in the trapping performance (Antia *et al.* 2002).

10.3.7 Inference of organic matter mineralization from sediment oxygen consumption

Particulate organic matter fluxes based on sediment traps are often too low to sustain sedimentary demands (Smith *et al.* 1992). Sediments, however, provide the ultimate sediment trap and integrate carbon fluxes over a considerable period of time. Following these considerations, Andersson and coworkers have used sediment oxygen consumption rates to derive an expression for the depth attenuation of the organic carbon flux in the ocean (Andersson *et al.* 2004, see also, Wijsman 2001):

$$\text{Flux} = 31.5e^{-0.018z} + 6.5e^{-0.00046z}$$

where the flux of carbon is given in mmol O_2 m^{-2} d^{-1} and z is water depth in meters. At steady state, the divergence of this organic carbon flux with depth should balance oxygen utilization rate at that depth (Suess 1980). Consequently, differentiation of the above equation with depth gives an expression for the oxygen utilization rate as a function of water depth (mmol O_2 m^{-3} d^{-1}):

$$\text{OUR} = 0.56e^{-0.018z} + 0.0028e^{-0.00046z}$$

The flux equation was combined with ocean bathymetry to derive integrated sediment respiration. Integrated water column respiration was estimated from the difference in organic carbon flux between the top and bottom of the sections considered. Table 10.5 shows the derived respiration rates for the dark water column and sediments. These rates correspond only to particulate organic carbon remineralization, and do not account for the

AOU vs. OUR

Table 10.5 Global estimates of respiration in the dark water column and sediments (Pmol C a^{-1}), derived from oxygen consumption measurements in sediments (from Andersson *et al.* 2004)

	Mesopelagic zone (100–1000 m)	Bathypelagic zone (>1000 m)
Water	0.63	0.23
Sediments	0.04	0.09

Oxygen units converted to carbon units using a respiratory quotient, RQ = 0.69.

dissolved fraction, in contrast to other approaches (AOU/tracers, ETS measurements).

10.3.8 Inference of organic matter mineralization from ocean biology general circulation models

Estimates derived from modeling exercises are rendered uncertain by the influence of mixing processes, because results are highly sensitive to the parameterization of diffusivity, ventilation, and mixing rates, and are stacked to a fixed depth which does not extend to the entire water column (e.g. Bacastow and Meier-Reimer 1991; Najjar and Keeling 2000). Approaches based on general circulation models also require assumptions on the carbon to nutrient stoichiometry, which can be highly variable depending on whether remineralization is based on POC or DOC data (Bacastow and Meier-Reimer 1991). Moreover, the model results are highly dependent on the model formulation (e.g. Ducklow 1995), so that the model output requires independent verification. Because of these limitations, there are, at present no consistent estimates of total respiration in the dark ocean derived from general circulation models.

10.4 Benthic respiration in the dark ocean

Sediments are the ultimate depository of particles in the ocean. Particulate organic matter arriving at the seafloor has been subject to degradation in the photic and aphotic zones of the water column. Consequently, only a very small proportion of the surface water primary production (on the order of a few percent) reaches the seafloor. Moreover, virtual all organic matter delivered to the sediments is respired in the top few decimeters so that little organic matter is buried. Nevertheless it is sufficient to support a large number (3.5×10^{30}) and high biomass (25 Pmol C) of prokaryotic cells in subsurface oceanic sediments (Whitman *et al.* 1998) that appear to be metabolically active (D'Hondt *et al.* 2002).

The global organic carbon burial, recently estimated at 13 Tmol C a^{-1} (range 11–18 Tmol C a^{-1}; Hedges and Keil 1995) represents only a fraction (4–9%) of global sedimentary mineralization (190 Tmol C a^{-1}, Jørgensen 1983; 220 Tmol C a^{-1}, Smith and Hollibaugh, 1993; 140–260 Tmol C a^{-1}, Middelburg *et al.* 1997; 210 Tmol C a^{-1}, Andersson *et al.* 2004). This is often expressed in terms of the burial efficiency: that is, the organic carbon accumulation rate below the diagenetic active surface layer divided by the organic carbon delivery to sediment surface. Burial efficiencies are typically less than 1% and 10% for deep-sea and ocean slope sediments, respectively, but may be much higher for rapidly accumulating shallow-water sediments (Hedges and Keil 1995). Consequently, most organic carbon burial occurs in shallow sediments (>80–90%; Hedges and Keil 1995; Middelburg *et al.* 1997).

Respiration of organic matter in marine sediments has received a fair measure of attention and major advances in our understanding and capabilities to measure and model it have been made. While respiration rates for abyssal waters are usually based on indirect methods and depend on empirical relations and model parameterizations, this is not the case for sediment respiration. Sediment respiration is usually based on sediment oxygen consumption, because these can be measured directly and *in situ* by incubation with autonomous benthic landers (Tengberg *et al.* 1995). Moreover, a simple reaction–diffusion model combined with *in situ* microelectrode data can be used to derive oxygen fluxes in deep-sea sediments (Reimers *et al.* 2001). Oxygen fluxes provide an accurate measure for total sediment respiration because aerobic respiration accounts for almost all organic matter mineralization in deep-sea sediments and most mineralization in slope sediments (Epping *et al.* 2002) and because most products of anaerobic mineralization (ammonium, reduced iron and manganese, sulfide) are efficiently re-oxidized (Soetaert *et al.* 1996, and see Middelburg *et al.* Chapter 11). Oxygen consumption coupled to re-oxidation of reduced components varies from about 20% in deep-sea sediments (oxygen consumption due to nitrification) to more than 60% in shelf sediments (Soetaert *et al.* 1996). Although oxygen consumption does not provide a good measure for aerobic respiration in slope sediments, it gives a reliable measure of total respiration because of efficient, quantitative re-oxidation of anaerobic respiration products (Jørgensen 1982).

Sediment oxygen consumption rates range over 4 orders of magnitude from more than 200 in coastal sediments to about $0.02 \, \mathrm{mmol} \, O_2 \, m^{-2} \, d^{-1}$ in some deep-sea sediments (Jørgensen 1983; Middelburg *et al.* 1997; Andersson *et al.* 2004). A number of researchers have derived empirical relationships from compiled datasets of sediment oxygen consumption. The depth attenuation of sediment oxygen consumption rates has been traditionally described with a power function with exponents varying from -0.36 to $-0.93 \, km^{-1}$ (Devol and Hartnett 2001), although other parameterizations have also been proposed: exponential and double exponential relations. These datasets have also been used to derive global estimates of benthic respiration below 1000 m, which range about tenfold from 0.01 (Jørgensen 1983), 0.05 (Jahnke 1996), 0.08 (Christensen, 2000) to $0.13 \, \mathrm{Pmol} \, O_2 \, a^{-1}$ (Andersson *et al.* 2004). Differences between these estimates result from differences in data availability, data selection criteria (*in situ* only or all available data), and the depth range considered to derive the predictive equations used for upscaling (below 1000 m only or all depths). The latter, most recent estimate, is based on a larger dataset and comprises only *in situ* total sediment community consumption estimates and has therefore been selected. Jørgensen (1983) and Andersson *et al.* (2004) also obtained estimates for total sediment respiration of 0.19 and $0.31 \, \mathrm{Pmol} \, O_2 \, a^{-1}$, respectively, with sediments at intermediate depths (100–1000 m) consuming

$0.05 \, \mathrm{Pmol} \, O_2 \, a^{-1}$. Sedimentary respiration below the photic zone then amounts to $0.18 \, \mathrm{Pmol} \, O_2 \, a^{-1}$ (58% of total sediment respiration).

10.5 Synthesis: budgeting respiration in dark ocean

The pelagic rates of oxygen consumption in the bathypelagic zone (>1000 m) are extremely low and very difficult to measure, resulting in rather few direct measurements of respiration in the ocean's interior. In spite of the scarcity of data, the preceding discussion clearly indicates that the variance in the estimates of respiration rates in the bathypelagic ocean and sediments, where mixing is a minor source of error, is considerablly lower than that in the mesopelagic zone. Fiadeiro and Craig (1978) calculated a rate of respiration for the whole water column below 1000 m of $2 \, \mu l \, O_2 \, l^{-1} \, a^{-1}$ ($0.24 \, \mu mol \, O_2 \, m^{-3} \, d^{-1}$), similar to other estimates available for the ocean interior based on oxygen fields and large-scale models (Riley 1951; Munk 1966; Broecker *et al.* 1991). These rates are comparable in magnitude to the deep-water estimates derived from oxygen consumption in sediments (Table 10.5), but considerablly lower than estimates derived from ETS measurements (Table 10.3). The bathypelagic ETS rates were derived from only 25 measurements, and presumably overestimated the global rate (Arístegui *et al.* 2003). Excluding the global rate derived from ETS, the resulting estimates converge to a total respiration rate in the bathypelagic water column of about $0.1–0.2 \, \mathrm{Pmol} \, C \, a^{-1}$ (cf. $0.1 \, \mathrm{Pmol} \, C \, a^{-1}$ for the sediments).

The global respiration rate in mesopelagic waters (which is thought to represent the bulk of the total dark ocean respiration) is not as well constrained as that of the bathypelagic zone, and there is no accepted global estimate. Nevertheless, in spite of the uncertainties inherent to the different methods used to calculate respiration in the mesopelagic zone, our synthesis shows that the average integrated rates obtained from direct metabolic measurements (oxygen consumption, ETS activity, bacterial carbon flux) vary between 3 and $4 \, \mathrm{mol} \, C \, m^{-2} \, a^{-1}$, and converge to global range of $1–1.3 \, \mathrm{Pmol} \, C^{-1}$, when extrapolated to the whole ocean. These estimates of global dark ocean respiration can be placed within the context of other estimates of respiration based on indirect approaches, which include carbon mass balances, and indirect estimates based on benthic respiration and POC flux. The estimates derived from the different approaches that we have reviewed are synthesized in Table 10.6.

There is a growing number of published oceanic carbon mass balances (i.e. carbon input into the dark ocean minus carbon output) that implicitly contain estimates of the total respiration in the dark ocean. The respiration rate in the dark ocean can be estimated as the difference between the organic carbon inputs to the dark ocean and the organic carbon burial in deep-sea sediments. This mass–balance analysis yields highly divergent respiration rates, ranging from 0.3 to 2.3 $PmolCa^{-1}$ (Sambrottto

et al. 1993; Smith and Hollibaugh 1993; Ducklow 1995; Hedges and Keil 1995; Emerson et al. 1997; Najjar and Keeling 2000; del Giorgio and Duarte 2002). Whereas the negative term of the mass-balance equation, the organic carbon burial, is relatively well-constrained, that of the additive term, the organic carbon inputs, is not. Indeed, estimates of organic carbon inputs to the dark ocean have been increasing from low estimates of about 0.4 $PmolCa^{-1}$, considering only POC inputs from the photic layer (e.g. Martin et al. 1987), to estimates of >2 $PmolCa^{-1}$ when considering the TOC flux, the active flux mediated by zooplankton and inputs from the continental shelf (del Giorgio and Duarte 2002). In contrast, estimates of organic carbon burial in the deep sea have remained relatively stable at about 0.005 $PmolCa^{-1}$. Table 10.6 summarizes the estimates of organic carbon inputs to the dark ocean

Table 10.6 Summary of annual organic carbon fluxes in the dark water-column

	Chapter Section	Areal (mol C m^{-2} a^{-1})	Global (Pmol C a^{-1})
Organic carbon inputs to the mesopelagic zone			
POC sedimentation (corrected for ^{230}Th)	10.2.2	3.1	1
DOC export by diffusive mixing (geomean Table 10.2)	10.2.3	0.5	0.17
DOC export by convection and subduction	10.2.3	0.4	0.12
POC export by migrating zooplankton	10.2.4	0.2	0.06
TOC advection from the coast	10.2.5	0.5	0.17
Mesopelagic water column respiration			
Bacterial carbon flux (geomean Table 10.4)	10.3.3	2.9	1
Microplankton respiration (from ETS)	10.3.2	3.7–6.2	1.2–2
Mesozooplankton respiration (from ETS)	10.3.4	0.56	0.18
Inference from AOU/tracers (North Atlantic)	10.3.5	3.5–4	1.1–1.3
Inference from oxygen fields/barium (Southern Ocean)	10.3.5	0.3–1.5	0.1–0.5
Inference from carbon fluxes	10.3.6	2.2–2.9	0.7–1
Inference from sediment oxygen consumption (Table 10.5)	10.3.7	1.9	0.63
Inference from carbon-mass-balance[a]		5.6–7.1	1.8–2.3
Bathypelagic water column respiration			
Bacterial carbon flux (geomean Table 10.4)	10.3.3	0.27	0.08
Microplankton respiration (from ETS)	10.3.2	1.5–2.5	0.5–0.8
Mesozooplankton respiration (from ETS)	10.3.4	0.09	0.03
Inference from sediment oxygen consumption (Table 10.5)	10.3.7	0.62	0.2
Inference from carbon-mass-balance[a]		0.34–0.4	0.11–0.13
Inference from oxygen fields/circulation models	10.5.2	0.15–0.43	0.05–0.14

[a]del Giorgio and Duarte (2002); mesopelagic area $= 3.29 \times 10^{14}$ m^2; bathypelagic volume $= 8.4 \times 10^{17}$ m^3.

discussed in Section 9.2. The sum of our reviewed estimates yields a global input of organic carbon to the dark ocean of about 1.5–1.6 $PmolCa^{-1}$. This estimate results from the sum of POC sedimentation (1 $PmolCa^{-1}$), DOC delivered by vertical diffusion, isopycnal transport, and convective mixing (0.29 $PmolCa^{-1}$), active transport of POC and DOC by migratory zooplankton (0.06 $PmolCa^{-1}$), and advection of POC and DOC from coastal waters (0.17 $PmolCa^{-1}$).

An alternative approach used to derive global respiration rates in the dark ocean is based on the rates of oxygen consumption in sediments (Wijsman 2001; Andersson *et al.* 2004). The method yields an estimate for the mesopelagic respiration around 0.6 $PmolCa^{-1}$ (Table 10.6), much lower than the 1.2 $PmolCa^{-1}$ that results from the extrapolation of areal rates obtained by direct measurements of microplankton ETS respiratory activity in the mesopelagic zone (Arístegui *et al.* 2003). Andersson's approach is based only on particulate organic matter, while the ETS method measures total microplankton respiration, which includes both dissolved and particulate organic carbon. This difference could account for some, but probably not all, of the difference between the estimates derived from the two methods since, as was discussed in previous sections, DOC seems to account for only 10–20% of the oxygen consumption in the dark ocean (Arístegui *et al.* 2002a). In this regard, the estimates of respiration rates derived from ETS generally agree with oxygen consumption estimates derived from bacterial carbon flux or inferred from AOU/tracers (Table 10.6). Figure 10.6 compares the global respiration rates derived from ETS measurements and oxygen consumption in sediments, with oxygen utilization rates in the subtropical North Atlantic estimated from large-scale tracer balances. If we consider that the latter estimates are typical of the whole ocean, the rates derived from oxygen consumption in sediments would underestimate the mesopelagic respiration by 30%, while the ETS estimates would match the AOU/tracers measurements. Conversely, and as argued above, the ETS—derived rates may severely overestimate respiration in the bathypelagic zone, where the method based on oxygen consumption in sediments

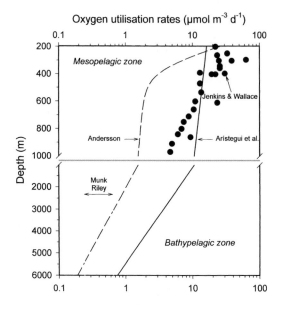

Figure 10.6 Global estimates of OUR versus depth, derived from ETS measurements (Arístegui *et al.* 2003) and oxygen consumption in sediments (Andersson *et al.* 2004). Rates are compared with OUR in the North Atlantic mesopelagic zone (<1000 m), estimated from AOU and large-scale tracers (Jenkins and Wallace 1992), and with oxygen consumption in the bathypelagic zone (>1000 m) derived from oxygen fields in the North Atlantic (Riley 1951) and Pacific Ocean (Munk 1966) (see text for details).

yields more accurate respiratory rates. Hence, it seems preferable to combine both approaches (the AOU/tracers measurements and ETS for the mesopelagic, and the oxygen consumption in sediments for the bathypelagic) to obtain an estimate of respiration in the dark ocean, and this yields a global rate of about 1.2–1.4 $PmolCa^{-1}$. Respiration of zooplankton (0.2 $PmolCa^{-1}$) and sediments (0.13 $PmolCa^{-1}$) added to this estimate, yields a global dark respiration of 1.5–1.7, which is consistent with the calculated global inputs of organic matter to the dark ocean reported above (Table 10.6), but somewhat lower than the total dark ocean respiration estimate of 1.9–2.4 $PmolCa^{-1}$ proposed by del Giorgio and Duarte (2002). Figure 10.7 represents a tentative organic carbon budget for the dark ocean, where organic carbon inputs—according to the reviewed data of Table 10.6—have been slightly rounded to look for a balance.

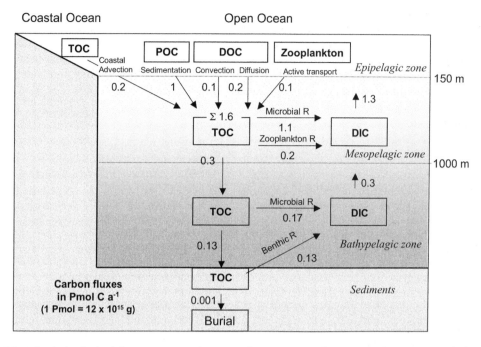

Figure 10.7 <u>Carbon budget for the dark ocean.</u> TOC = Total Organic Carbon; POC = Particulate Organic Carbon; DOC = Dissolved Organic Carbon; DIC = Dissolved Inorganic Carbon; R = Respiration.

These estimates raise an interesting problem, for according to the calculations above, the ratio between exported production and primary production (assuming the presently accepted primary production of 3.3 $PmolCa^{-1}$) would be close to 0.5. This *f*-ratio is much higher than currently accepted values, of about 0.2, implying that either respiration is overestimated here or that primary production is severely underestimated. There is growing evidence that estimates of primary production, derived from extrapolation from ^{14}C-based measurements, underestimate both gross and net primary production (e.g. del Giorgio and Duarte 2002; Marra 2002 and Chapter 14). If so, many models that use primary production as an input to estimate export production to the dark ocean or remineralization of organic carbon in the mesopelagic zone will severely underestimate the global fluxes of carbon and associated respiration rates. In addition, there are allochthonous carbon inputs to the ocean, from exchange with the continental shelves and atmospheric inputs that may further increase the organic matter available to

be exported from the photic layer. Elucidation of the relative importance of these pathways remains critical to achieve a reliable understanding of the carbon cycle in the ocean.

Whereas an improved precision on the estimates of organic carbon inputs to the dark ocean is widely acknowledged (e.g. Ducklow *et al.* 2001), efforts to improve these calculations must continue and proceed in parallel with efforts to increase the very meager empirical base of direct estimates of respiration rate in the dark ocean. The synthesis presented here clearly identified respiration in the dark layers of the ocean to be a major, though largely neglected, component of the carbon flux, and, therefore, a major path in the global biogeochemical cycle of the elements.

Acknowledgments

This research has been supported by the Spanish Plan Nacional de I+D (Project COCA; REN-2000-1471), by a PIONIER grant (833.02.2002) of the Netherlands Organization of Scientific Research, and by the EU projects EVK2-CT-2001-00100, and EVK3-2001-00152 (OASIS). This is publication 3186 of the Netherlands Institute of Ecology. We thank Paul A. del Giorgio and Peter J. le B. Williams for their thoughtful reviews of the manuscript.

References

Alvarez-Salgado, X. A., Gago, J., Míguez, B. M., and Pérez, F. F. 2001. Net ecosystem production of dissolved organic carbon in a coastal upwelling system. The Ría de Vigo, Iberian margin of the North Atlantic. *Limnol. Oceanogr.*, **46**: 135–147.

Andersson, J. H., Wijsman, J. W. M., Herman, P. M. J., Middelburg, J. J., Soetaert, K., and Heip, C. 2004 Respiration patterns in the deep ocean. *Geophys. Res. Lett.*, **31**. doi: 10.1029/2003GL018756.

Antia, A. N., Koeve, W., Fisher, G., Blanz, T., Schulz-Bull, D., Scholten, J., Neuer, S., Kremling, K., Kuss, J., Peinert, R., Hebbeln, D., Bathmann, U., Conte, M., Fehner, U., and Zeitzschel, B. 2001. Basin-wide particulate carbon flux in the Atlantic Ocean: Regional export patterns and potential for atmospheric CO_2 sequestration. *Glob. Biogeochem. Cyc.*, **15**: 845–862.

Antoine, D., André, J.-M., and Morel, A. 1996. Oceanic primary production. 2. Estimation at global scale from satellite (coastal zone color scanner) chlorophyll. *Glob. Biogeochem. Cyc.*, **10**: 57–69.

Archer, D., Peltzer, E., and Kirchman, D. L. 1997. A timescale of DOC production in equatorial Pacific surface waters. *Glob. Biogeochem. Cyc.*, **11**: 435–452.

Arístegui, J., Montero, M. F., Ballesteros, S., Basterretxea, G., and van Lenning, K. 1996. Planktonic primary production and microbial respiration measured by C-14 assimilation and dissolved oxygen changes in coastal waters of the Antarctic Peninsula during austral summer: implications for carbon flux studies. *Mar. Ecol. Prog. Ser.*, **132**: 191–201.

Arístegui, J., Duarte, C. M., Agustí, S., Doval, M., Alvarez-Salgado, X. A., and Hansell, D. A. 2002*a*. Dissolved organic carbon support of respiration in the dark ocean. *Science*, **298**, 1967.

Arístegui, J., Denis, M., Almunia, J., and Montero, M. F. 2002*b*. Water-column remineralization in the Indian sector of the Southern Ocean during early spring. *Deep-Sea Res. II*, **49**: 1707–1720.

Arístegui, J., Agustí, S., and Duarte, C. M. 2003. Respiration in the dark ocean. *Geophys. Res. Lett.*, **30**: 1041. doi:10. 1029/2002GLO16227.

Bacastow, R. and Mainer-Reimer, E. 1991. Dissolved organic carbon in modelling oceanic new production. *Glob. Biogeochem. Cyc.*, **5**: 71–85.

Banoub, M. W. and Williams, P. J. le B. 1972. Measurement of microbial activity and organic material in the western Mediterranean Sea. *Deep-Sea Res. I*, **19**: 433–443.

Barber, R. T. 1966. Interaction of bubbles and bacteria in the formation of organic aggregates in sea water. *Nature*, **211**: 257–258.

Barton, E. D., Arístegui, J., Tett, P., Cantón, M., García-Braun, J., Hernández-León, S., Nykjaer, L., Almeida, C., Almunia, J., Ballesteros, S., Basterretxea, G., Escánez, J., García-Weill, L., Hernández-Guerra, A., López-Laatzen, F., Molina, R., Montrero, M. F., Navarro-Pérez, E., Rodríguez, J. M., van Lenning, K., Vélez, H., and Wild, K. 1998. The transition zone of the Canary Current upwelling region. *Progr. Oceanogr.*, **41**: 455–504.

Berger, W. H., Fisher, K., Lai, C., and Wu, G. 1987. Ocean carbon flux: global maps of primary production and export production. In C. Agegian (ed) *Biogeochemical Cycle and Fluxes between the Deep Euphotic Zone and other Oceanic Realms*, SIO, University of California, pp. 131–176.

Betzer, P. R., Showers, W. J., Laws, E. A., Winn, C. D., di Tullio, G. R., and Kroopnick, P. M. 1984. Primary production and particle fluxes on a transect of the equator at 153°W in the Pacific Ocean. *Deep-Sea Res. I*, **31**: 1–11.

Biddanda, B. and Benner, R. 1997. Major contribution from mesopelagic plankton to heterotrophic metabolism in the upper ocean. *Deep Sea Res. I*, **44**: 2069–2085.

Broecker, W. S., and Peng, T. -H. 1982. *Tracers in the Sea*. Lamont-Doherty Geological Observatory, New York.

Broecker, W. S., Klas, M., Clark, E., Bonani, G., Ivy, S., and Wolffi, W. 1991. The influence of $CaCO_2$ dissolution on core top radiocarbon ages for deep-sea sediments. *Paleoceanography*, **6**: 593–608.

Buesseler, K. O. 1991. Do upper-ocean sediment traps provide an accurate record of particle flux? *Nature*, **353**: 420–423.

Buesseler, K. O., Steinberg, D. K., Michaels, A. F., Johnson, R. J., Andrews, J. E., Valdes, J. R., and Price, J. F. 2000. A comparison of the quantity and composition of material caught in a neutrally buoyant versus surface-tethered sediment trap. *Deep-Sea Res. I*, **47**: 277–294.

Carlson, C. A., Ducklow, H. W., and Michaels, A. F. 1994. Annual flux of dissolved organic carbon from the euphotic zone of the northwestern Sargasso Sea. *Nature*, **371**: 405–408.

Cherrier, J., Bauer, J. E., and Druffel, E. R. M. 1996. Utilization and turnover of labile dissolved organic matter by bacterial heterotrophs in eastern North Pacific surface waters. *Mar. Ecol. Prog. Ser.*, **139**: 267–279.

Cho, B. C. and Azam, F. 1988. Major role of bacteria in biogeochemical fluxes in the ocean's interior. *Nature*, **332**: 441–443.

Christensen, J. P. 2000. A relationship between deep-sea benthic oxygen demand and oceanic primary production. *Oceanol. Acta*, **23**: 65–82.

Christensen, J. P., Owens, T. G., Devol, A. H., and Packard, T. T. 1980. Respiration and physiological state in marine bacteria. *Mar. Biol.*, **55**: 267–276.

Christensen, J. P., Packard, T. T., Dortch, F. Q., Minas, H. J., Gascard, J. C., Richez, C., and Garfield, P. C. 1989. Carbon oxidation in the deep Mediterranean Sea: evidence for dissolved organic carbon source. *Glob. Biogeochem. Cyc.*, **3**: 315–335.

Christian, J. R., Lewis, M. R., and Karl, D. M. 1997. Vertical fluxes of carbon, nitrogen, and phosphorus in the North Pacific Subtropical Gyre near Hawaii. *J. Geophys. Res.*, **102**: 15667–15677.

Conover, R. J. 1978. Transformation of organic matter. In O. Kinne (ed) *A Comprehensive, Integrated Treatise on Life in Oceans and Coastal Waters*, Vol 4, Wiley, New York, pp. 221–499.

Copin-Montégut, G., and Avril, B. 1993. Vertical distribution and temporal variation on dissolved organic carbon in the North-Western Mediterranean Sea. *Deep-Sea Res. I*, **40**: 1963–1972.

Dehairs, F., Shopova, D., Ober, S., Veth, C., and Goeyens, L. 1997. Particulate barium stocks and oxygen consumption in the Southern Ocean mesopelagic water column during spring and early summer: relationship with export production. *Deep-Sea Res. II*, **44**: 497–516.

del Giorgio, P. A. and Cole, J. J. 2000. Bacterial energetics and growth efficiency. In D. L. Kirchman (ed) *Microbial Ecology of the Oceans*. Wiley-Liss, New York, pp. 289–326.

del Giorgio, P. A. and Duarte, C. M. 2002. Total respiration and the organic carbon balance of the open ocean. *Nature*, **420**: 379–384.

Devol, A. H., and Hartnett, H. E. 2001. Role of the oxygen-deficient zone in transfer of organic carbon to the deep ocean. *Limnol. Oceanogr.*, **46**: 1684–1690.

D'Hondt, S., Rutherford, S., and Spivack, A. J. 2002. Metabolic activity of subsurface life in deep-sea sediments. *Science*, **295**: 2067–2070.

Doval, M. D., Pérez, F. F., and Berdalet, E. 1999. Dissolved and particulate organic carbon and nitrogen in the northwestern Mediterranean. *Deep-Sea Res. I*, **46**: 511–527.

Druffel, E. R. M., Bauer, J. E., Williams, P. M., Griffin, S., and Wolgast, D. 1996. Seasonal variability of particulate organic radiocarbon in the northeast Pacific Ocean. *J. Geophys. Res.*, **101**: 20543–20552.

Duarte, C. M., and Cebrián, J. 1996. The fate of marine autotrophic production. *Limnol. Oceanogr.*, **41**: 1758–1766.

Duarte, C. M., Agustí, S., del Giorgio, P. A., and Cole, J. J. 1999. Regional carbon imbalances in the oceans. *Science*, **284**: 1735b.

Ducklow, H. W. 1993. Bacterioplankton distribution and production in the northwestern Indian Ocean and the Gulf of Oman, September 1986. *Deep-Sea Res. II*, **40**: 753–771.

Ducklow, H. W. 1995. Ocean biogeochemical fluxes: New production and export of organic matter from the upper ocean. *Rev. Geophys. Suppl.*, **31**: (Part 2 Suppl.): 1271–1276.

Ducklow, H. W., Steinberg, D. K., and Buesseler, K. O. 2001. Upper ocean carbon export and the biological pump. *Oceanography*, **14**: 50–58.

Emerson, S., Quay, P., Karl, D., Winn, C., Tupas, L., and Landry, M. 1997. Experimental determination of the organic carbon flux from open-ocean surface waters. *Nature*, **389**: 951–954.

Epping, E., van der Zee, C., Soetaert, K., and Helder, W. 2002. On the oxidation and burial of organic carbon in sediments of the Iberian margin and Nazaré Canyon (N. E. Atlantic). *Prog. Oceanogr.*, **52**: 399–431.

Falkowski, P. G., Barber, R. T., and Smetacek, V. 1998. Biogeochemical controls and feedbacks on ocean primary production. *Science*, **281**: 200–206.

Falkowski, P. G., Flagg, C. N., Rowe, G. T., Smith, S. L., Whitledge, T. E., and Wirick, C. D. 1988. The fate of a spring phytoplankton bloom—export or oxidation. *Cont. Shelf Res.*, **8**: 457–484.

Feely, R. A., Wanninkhof, R., Cosca, C. E., Murphy, P. P., Lamb, M. F., and Steckley, M. D. 1995. CO_2 distribution in the equatorial Pacific during the 1991–1992 ENSO event. *Deep-Sea Res. II*, **42**: 365–386.

Fiadeiro, M. E., and Craig, H. 1978. Three-dimensional modeling of tracers in the deep Pacific Ocean: I. salinity and oxygen. *J. Mar. Res.*, **36**: 323–355.

Field, C. B., Behrenfeld, M. J., Randerson, J. T., and Falkowski, P. 1998. Primary production of the biosphere: integrating terrestrial and oceanic components. *Science*, **281**: 237–240.

François, R., Honjo, S., Krishfield, R., and Manganini, S. 2002. Running the gauntlet in the twilight zone: the effect of midwater processes on the biological pump. *U. S. JGOFS News*, **11**: 4–6.

Gabric, A. J., García, L-, van Camp, L., Nykjaer, L., Eifler, W., and Schrimpf, W. 1993. Offshore export of shelf production in the Cape Blanc (Mauritania) giant filament as derived from coastal zone color scanner imagery. *J. Geophys. Res.*, **98**: 4697–4712.

Gardner, W., and Stephens, J. 1978. Stability and composition of terrestrially-derived dissolved organic nitrogen in continental shelf waters. *Mar. Chem.*, **11**: 335–342.

Gasol, J. M., Doval, M. D., Pinhassi, J., Calderón-Paz, J. I., Guixa-Boixareu, N., Vaqué, D., and Pedrós-Alió, C. 1998. Diel variations in bacterial heterotrophic activity and growth in the northwestern Mediterranean Sea. *Mar. Ecol. Prog. Ser.*, **164**: 107–124.

Gattuso, J.-P., Franjignoulle, M., and Wollast, R. 1998. Carbon and carbonate metabolism in coastal aquatic ecosystems. *Ann. Rev. Ecol. Syst.*, **29**: 405–433.

Guo, L., Santschi, P. H., and Warnken, K. W. 1995. Dynamics of dissolved organic carbon (DOC) in oceanic environments. *Limnol. Oceanogr.*, **40**: 1392–1403.

Hansell, D. A. 2002. DOC in the global ocean carbon cycle. In D. A. Hansell, and C. A. Carlson (eds) *Biogeochemistry of Marine Dissolved Organic Matter*. Academic Press, San Diego, pp. 685–715.

Hansell, D. A. and Carlson, C. A. 2001. Marine dissolved organic matter and the carbon cycle. *Oceanography*, **14**: 41–49.

Hansell, D. A., Williams, P. M., and Ward, B. B. 1993. Measurements of DOC and DON in the Southern California Bight using oxidation by high temperature combustion. *Deep-Sea Res. I*, **40**: 219–234.

Hedges, J. I. and Keil, R. G. 1995. Sedimentary organic matter preservation: an assessment and speculative synthesis. *Mar. Chem.*, **49**: 81–115.

Hedges, J. I., Baldock, J. A., Gélinas, Y., Lee, C., Peterson, M. L., and Wakeham, S. G. 2002. The biochemical and elemental compositions of marine plankton: a NMR perspective. *Mar. Chem.*, **78**: 47–63.

Hernández-León, S., Gómez, M., Pagazaurtundua, M., Portillo-Hahnefeld, A., Montero, I., and Almeida, C. 2001. Vertical distribution of zooplankton in Canary Island waters: implications for export flux. *Deep-Sea Res. I*, **48**: 1071–1092.

Holligan, P. M., Williams, P. J. le B., Purdie, D., and Harris, R. P. 1984. Photosynthesis, respiration and nitrogen supply of plankton populations in stratified, frontal and tidally mixed shelf waters. *Mar. Ecol. Prog. Ser.*, **17**: 201–213.

Holm-Hansen, O., Strickland, J. D. H., and Williams, P. M. 1966. A detailed analysis of biologically important substances in a profile off Southern California. *Limnol. Oceanogr.*, **11**: 548–561.

Hopkinson, C. S. Jr, Fry, B., and Nolin, A. L. 1997. Stoichiometry of dissolved organic matter dynamics on the continental shelf of the northeastern USA. *Cont. Shelf Res.*, **17**: 473–489.

Jackson, G. A. 2002. Collecting the garbage of the sea: the role of aggregation in ocean carbon transport. *US JGOFS News*, **11**: 1–3.

Jackson, R. J. and Williams, P. M. 1985. Importance of dissolved organic nitrogen and phosphorus to biological nutrient cycling. *Deep-Sea Res.*, **32**: 225–235.

Jahnke, R. A. 1996. The global ocean flux of particulate organic carbon: areal distribution and magnitude. *Glob. Biogeochem. Cyc.*, **10**: 71–88.

Jannasch, H. W. and Wirsen, C. O. 1973. Deep-sea microorganisms: *In situ* response to nutrient enrichment. *Science*, **180**: 641–643.

Jenkins, W. J. and Wallace, D. W. R. 1992. Tracer based inferences of new primary production in the sea. In P. G. Falkowski, and A. D. Woodhead (eds) *Primary Productivity and Biogeochemical Cycles in the Sea*. Plenum, pp. 299–316.

Joiris, C., Billen, G., Lancelot, C., Daro, M. H., Mommaerts, J. P., Bertels, A., Bossicart, M., Nijs, J., and Hecq, J. H. 1982. A budget of carbon cycling in the Belgian coastal zone: relative roles of zooplankton, bacterioplankton and benthos in the utilization of primary production. *Neth. J. Sea Res.*, **16**: 260–275.

Jørgensen, B. B. 1982. Mineralization of organic matter in the sea bed: role of sulphate reduction. *Nature*, **296**: 643–645.

Jørgensen, B. B. 1983. Processes at the sediment–water interface. In B. Bolin and R. B. Cook (eds) *The Major Biogeochemical Cycles and their Interactions*. John Wiley & Sons, New York, pp. 477–509.

Kähler, P., Bjørnsen, P. K., Lochte, K., and Antia, A. 1997. Dissolved organic matter and its utilization by bacteria during spring in the Southern Ocean. *Deep-Sea Res. II*, **44**: 341–353.

Karl, D. M., Christian, J. R., Dore, J. E., Hebel, D. V., Letelier, R. M., Tupas, L. M., and D. Winn. 1996. Seasonal and interannual variability in primary production and particle flux at Station ALOHA. *Deep-Sea Res. II*, **43**: 539–568.

Karl, D. M., Hebel, D. V., Björkman, K., and Letelier, R. M. 1998. The role of dissolved organic matter release in the productivity of the oligotrophic North Pacific Ocean. *Limnol. Oceanogr.*, **43**: 1270–1286.

Karl, D., Michaels, A., Bergman, B., Capone, D., Carpenter, E., Letelier, R., Lipschutz, F., Paerl, H., Sigman, D., and Stal, L. 2002. Dinitrogen fixation in the world's oceans. *Biogeochemistry*, 57/58, 47–98.

Keil, R. G., Montlucon, D. B., Prahl, F. G., and Hedges, J. I. 1994. Sorptive preservation of labile organic matter in marine sediments. *Nature*, **370**: 549–552.

Kirchman, D. L., K'Nees, E., and Hodson, R. E. 1985. Leucine incorporation and its potential as a measure of protein synthesis by bacteria in natural aquatic systems. *Appl. Environ. Microbiol.*, **49**: 599–607.

La Ferla, R., and Azzaro, M. 2001. Microbial respiration in the Levantine Sea: evolution of the oxidative processes in relation to the main Mediterranean water masses. *Deep-Sea Res. I*, **48**: 2147–2159.

Lefèvre, D., Denis, M., Lambert, C. E., and Miquel, J.- C. 1996. Is DOC the main source of organic matter remineralization in the ocean water column? *J. Mar. Syst.*, **7**: 281–294.

Lewis, M. R. 2002. Variability of plankton and plankton processes on the mesoscale. In P. J. le B. Williams, D. N. Thomas, and C. S. Reynolds (eds) *Phytoplankton Productivity*. Blackwell Science, Oxford, UK, pp. 141–155.

Libby, P. S., and Wheeler, P. A. 1997. Particulate and dissolved organic nitrogen in the central and eastern equatorial Pacific. *Deep-Sea Res. II*, **44**: 345–361.

Libes, S. M. 1992. *An Introduction to Marine Biogeochemistry*. John Wiley & Sons, New York.

Liu, K.-K., Atkinson, L., Chen, C. T. A., Gao, S., Hall, J., Mac-Donald, R. W., Talaue McManus, L., and Quiñones, R. 2000. Exploring continental margin carbon fluxes on a global scale. *EOS*, **81**: 641–644.

Lohrenz, S. E., Knauer, G. A., Asper, V. L., Tuel, M., Michaels, A. F., and Knap, A. H. 1992. Seasonal variability in primary production and particle flux in the northwestern Sargasso Sea: U. S. JGOFS Bermuda Atlantic Time-series Study. *Deep-Sea Res. I*, **39**: 1373–1391.

Longhurst, A. 1976. Vertical migration. In D. H. Cushing and J. J. Walsh (eds) *The ecology of the seas*, Blackwell, pp. 116–137.

Longhurst, A., Sathyendranath, S., Platt, T., and Caverhill, C. 1995. An estimate of global primary production in the ocean from satellite radiometer data. *J. Plankton Res.*, **17**: 1245–1272.

Luz, B., and Barkan, E. 2000. Assessment of oceanic productivity with the triple-isotope composition of dissolved oxygen. *Science*, **288**: 2028–2031.

Martin, J. H., Knauer, G. A., Karl, D. M., and Broenkow, W. W. 1987. VERTEX: carbon cycling in the northeastern Pacific. *Deep-Sea Res. I*, **34**: 267–285.

Marra, J. 2002. Approaches to the measurement of plankton production. In P. J. le B. Williams, D. N. Thomas, and C. S. Reynolds (eds) *Phytoplankton Productivity*. Blackwell Science, Oxford, UK, pp. 78–108.

McCarthy, M. D., Hedges, J. I., and Benner, R. 1996. Major biochemical composition of dissolved high molecular weight organic matter in seawater. *Mar. Chem.*, **55**: 281–297.

Michaels, A. F., Bates, N. R., Buesseler, K. O., Carlson, C. A., and Knap, A. 1994. Carbon-cycle imbalances in the Sargasso Sea. *Nature*, **372**: 537–540.

Middelburg, J. J., Soetaert, K., and Herman, P. M. J. 1997. Empirical relationships for use in global diagenetic models. *Deep-sea Res. I*, **44**: 327–344.

Munk, W. 1966. Abyssal recipes. *Deep-Sea Res. I*, **13**: 707–730.

Nagata, T., Fukuda, H., Fukuda, R., and Koike, I. 2000. Bacterioplankton distribution and production in deep Pacific waters: large-scale geographic variations and possible coupling with sinking particle fluxes. *Limnol. Oceanogr.*, **45**: 419–425.

Najjar, R. G., and Keeling, R. F. 2000. Mean annual cycle of the air-sea oxygen flux: a global view. *Glob. Biogeochem. Cyc.*, **14**: 573–584.

Naqvi, S. W. A., Shailaja, M. S., Dileep Kumar, M., and Sen Gupta, R. 1996. Respiration rates in subsurface waters of the northern Indian Ocean: evidence for low decomposition rates of organic matter within the water column in the Bay of Bengal. *Deep-Sea Res. II*, **43**: 73–81.

Pace, M. L., Knauer, G. A., Karl, D. M., and Martin, J. H. 1987. Primary production, new production and vertical fluxes in the eastern Pacific. *Nature*, **325**: 803–804.

Packard, T. T. 1971. The measurement of respiratory electron transport activity in marine phytoplankton. *J. Mar. Res.*, **29**: 235–244.

Packard, T. T., Denis, M., Rodier, M., and Garfield, P. 1988. Deep-ocean metabolic CO_2 production: calculations from ETS activity. *Deep-Sea Res. I*, **35**: 371–382.

Parsons, T. R., Takahashi, M., and Hargrave, B. 1984. *Biological Oceanographic Processes*, 3rd edition. Pergamon Press, Oxford.

Platt, T., Jauhari, P., and Sathyendranath, S. 1992. The importance and measurement of new production. In P. G. Falkowski, and A. D. Woodhead (eds) *Primary Productivity and Biogeochemical Cycles in the Sea*. Plenum, New York, pp. 273–284.

Pomeroy, L. R., and Johannes, R. E. 1968. Occurrence and respiration of ultraplankton in the upper 500 meters of the ocean. *Deep-Sea Res. I*, **15**: 381–391.

Pomeroy, L. R., Sheldon, J. E., and Sheldon, W. M. 1994. Changes in bacterial numbers and leucine assimilation during estimations of microbial respiratory rates in seawater by the precision Winkler method. *Appl. Environ. Microbiol.*, **60**: 328–332.

Reimers, C. E., Jahnke, R. A., and Thomsen, L. 2001. *In situ* sampling in the benthic boundary layer. In B. P. Boudreau, and B. B. Jørgensen (eds) *The Benthic Boundary Layer*. Oxford University Press, Oxford, pp. 245–268.

Riley, G. A. 1951. Oxygen, phosphate and nitrate in the Atlantic Ocean. *Bull. Bingham Oceanogr. Meteorol.*, **9**: 5–22.

Sambrotto, R. N., Savidge, G., Robinson, C., Boyd, P., Takahashi, T., Karl, D. M., Langdon, C., Chipman, D., Marra, J., and Codispoti, L. 1993. Elevated consumption of carbon relative to nitrogen in the surface ocean. *Nature*, **363**: 248–250.

Savenkoff, C., Lefèvre, D., Denis, M., and Lambert, C. E. 1993*a*. How do microbial communities keep living in the Medieterranean outflow within northeast Atlantic intermediate waters? *Deep-Sea Res. II*, **40**: 627–641.

Savenkoff, C., Prieur, L., Reys, J.-P., Lefèvre, D., Dallot, S., and Denis, M. 1993*b*. Deep microbial communities evidenced in the Liguro-Provençal front by their ETS activity. *Deep-Sea Res. II*, **40**: 709–725.

Scholten, J. C., Fietzke, J., Vogler, S., Rutgers van der Loeff, M. M., Mangini, A., Koeve, W., Waniek, J., Stoffers, P., Antia, A., and Kuss, J. 2001. Trapping efficiencies of sediment traps from the deep Eastern North Atlantic: the ^{230}Th calibration. *Deep Sea Res. II*, **48**: 2383–2408.

Sherry, N. D., Boyd, P. W., Sugimoto, K., and Harrison, P. J. 1999. Seasonal and spatial patterns of heterotrophic bacterial production, respiration, and biomass in the subarctic NE Pacific. *Deep-Sea Res. II*, **46**: 2557–2578.

Simon, M., Welschmeyer, N. A., and Kirchman, D. L. 1992. Bacterial production and the sinking flux of particulate organic matter in the subarctic Pacific. *Deep-Sea Res. I*, **39**: 1997–2008.

Smith, S. V. and Hollibaugh, J. T. 1993. Coastal metabolism and the oceanic organic carbon balance. *Rev. Geophys.*, **31**: 75–89.

Smith, K. L., Baldwin, R. J., and Williams, P. M. 1992. Reconciling particulate organic-carbon flux and sediment community oxygen-consumption in the deep north Pacific. *Nature*, **359**: 313–316.

Soetaert, K., Herman, P. M. J., and Middelburg, J. J. 1996. A model of early diagenetic processes from the shelf to abyssal depths. *Geochim. Cosmochim. Acta*, **60**: 1019–1040.

Sorokin, Y. I. 1978. Decomposition of organic matter and nutrient regeneration. In O. Kine (ed) *Marine Ecology*. John Wiley & Sons, Chichester, pp. 501–506.

Souchou, P., Gasc. A., Cahet, G., Vaquer, A., Colos, Y., and Deslous-Paoli, J. M. 1997. Biogeochemical composition of Mediterranean waters outside the Thau Lagoon. *Estuar. Coast. Shelf Sci.*, **44**: 275–284.

Steinberg, D. K., Carlson, C. A., Bates, N. R., Johnson, R. J., Michaels, A. F., and Knap, A. F. 2001. Overview of the US JGOFS Bermuda Atlantic Time-series Study (BATS): a decade-scale look at ocean biology and biogeochemistry. *Deep-Sea Res. II*, **48**: 1405–1447.

Suess, E. 1980. Particulate organic carbon flux in the oceans-surface productivity and oxygen utilization. *Nature*, **288**: 260–263.

Takahashi, T., Broecker, W. S., and Langer, S. 1985. Redfield ratios based on chemical data from isopycnal surfaces. *J. Geophys. Res.*, **90**: 6907–6924.

Tengberg, A., De Bovee, F., Hall, P., Berelson, W., Chadwick, D., Ciceri, G., Crassous, P., Devol, A., Emerson, S., Gage, J., Glud, R., Graziottini, F., Gundersen, J., Hammond, D., Helder, W., Hinga, K., Holby, O., Jahnke, R., Khripounoff, A., Lieberman, S., Nuppenau, V., Pfannkuche, O., Reimers, C., Rowe, G., Sahami, A., Sayles, F., Schurter, M., Smallman, D., Wehrli, B., and Wilde, P. De. 1995. Benthic chamber and profiling landers in oceanography: a review of design, technical solutions and functioning. *Prog. Oceanogr.*, **35**: 253–294.

Turley, C. M. and Mackie, P. J. 1994. Biogeochemical significance of attached and free-living bacteria and the flux of particles in the NE Atlantic Ocean. *Mar. Ecol. Prog. Ser.*, **115**: 191–203.

Turley, C. M., and Stutt, E. D. 2000. Depth-related cell-specific bacterial leucine incorporation rates on particles and its biogeochemical significance in the northwest Mediterranean. *Limnol. Oceanogr.*, **45**: 408–418.

Vidal, M., Duarte, C. M., and Agustí, S. 1999. Dissolved organic nitrogen and phosphorus pools and fluxes in the Central Atlantic Ocean. *Limnol. Oceanogr.*, **44**: 106–115.

Vinogradov, M. E. 1970. Vertical distribution of the oceanic zooplankton. Israel program for scientific translations.

Walsh, J. J., Rowe, G. T., Iverson, R. L., and McRoy, C. P. 1981. Biological export of shelf carbon: a neglected sink of the global CO_2 cycle. *Nature*, **291**: 196–201.

Wheeler, P. A., Watkins, J. M., and Hansing, R. L. 1997. Nutrients, organic carbon and organic nitrogen in the upper water column of the Arctic Ocean: implications for the sources of dissolved organic carbon. *Deep-Sea Res. II*, **44**: 1571–1592.

Whitman, W. B., Coleman, D. C., and Wiebe, W. J. 1998. Prokaryotes: the unseen majority. *Proc. Natl. Acad. Sci. USA*, **95**: 6578–6583.

Wiebinga, C. J., and de Baar, H. J. W. 1998. Determination of the distribution of dissolved organic carbon in the Indian Sector of the Southern Ocean. *Mar. Chem.*, **61**: 185–201.

Wijsman, J.W.M. 2001. Early Diagenetic Processes in North-western Black Sea Sediments. Ph.D Thesis, University of Groningen, ISBN 90-367-1337-4.

Williams, P. J. le B. 1981. Microbial contribution to overall marine plankton metabolism: Direct measurements of respiration. *Oceanologr. Acta*, **4**: 359–364.

Williams, P. J. le B. 1995. Evidence for the seasonal accumulation of carbon-rich dissolved organic material, its scale in comparison with changes in particulate material and the consequential effect on net C/N assimilation ratios. *Mar. Chem.*, **51**: 17–29.

Williams, P. J. le B. 2000. Heterotrophic bacteria and the dynamics of dissolved organic material. In D. L. Kirchman (ed) *Microbial Ecology of the Oceans*. Wiley-Liss, New York, pp. 153–200.

Williams, P. J. le B. and Jenkinson, N. W. 1982. A transportable microprocessor-controlled preciseWinkler titration suitable for field station and shipboard use. *Limnol. Oceanogr.*, **27**: 576–584.

Williams, P. J. le B. and Purdie, D. A. 1991. *In vitro* and *in situ* derived rates of gross production, net community production and respiration of oxygen in the oligotrophic subtropical gyre of the North Pacific Ocean. *Deep-Sea Res. I*, **38**: 891–910.

Williams, P. M. 1986. *In* R. W. Eppley (ed) *Plankton Dynamics of the Southern California Bight*. Springer Verlag, Berlin, pp. 121–135.

Wollast, R., and Chou, L. 2001. Ocean Margin Exchange in the northern Gulf of Biscay: OMEX I. An introduction. *Deep-Sea Res. II*, **48**: 2971–2978.

Wyrtki, K. 1962. The oxygen minima in relation to ocean circulation. *Deep-Sea Res. I*, **9**: 11–23.

Yu, E. -F., Françoise R., Bacon, M. P.,Honjo, S., Fleer, A. P., Manganini, S. J., Rutgers van der Loeff, M. M., and Ittekot, V. 2001. Trapping efficiency of bottom-tethered sediment traps estimated from the intercepted fluxes of ^{230}Th and ^{231}Pa. *Deep-Sea Res. I*, **48**: 865–889.

Zhang, J. and Quay, P. D. 1997. The total organic carbon export rate based on ^{13}C and ^{12}C of DIC budgets in the equatorial Pacific region. *Deep-Sea Res. II*, **44**: 2163.

Zweifel, U. L., Norrman, B., and Hagström, Å. 1993. Consumption of dissolved organic carbon by marine bacteria and demand for inorganic nutrients. *Mar. Ecol. Prog. Ser.*, **101**: 23–32.

CHAPTER 11

Respiration in coastal benthic communities

Jack J. Middelburg[1], Carlos M. Duarte[2], and Jean-Pierre Gattuso[3]

[1] *Netherlands Institute of Ecology, The Netherlands*
[2] *IMEDEA (CSIC-UIB) Spain*
[3] *Laboratoire d'Océanographie de Villefranche, France*

Outline

This chapter reviews coastal benthic communities with the aim of deriving a global estimate for respiration in these ecosystems. Reefs, mangroves, salt marshes, macroalgae, sea grasses, and unvegetated sediments dominate respiration in the coastal ocean. Estimates of coastal benthic respiration are not well constrained but converge on about 620 Tmol C a^{-1}. In coastal benthic ecosystems autotrophs and multicellular heterotrophs contribute significantly, and in some systems even dominate respiration unlike most other oceanic ecosystems in which bacteria dominate respiration.

11.1 Introduction

The coastal zone is characterized by the presence of an active benthic compartment in close contact and interaction with the pelagic one (Soetaert *et al.* 2000). Benthic coastal communities are highly diverse and include systems in which biological entities are a structuring factor (e.g. coral reefs, mangroves, sea grass beds) as well as those in which physical features and processes determine the landscape (e.g. rocky shores, rippled sandy sediments). Benthic communities differ greatly in structure and their role in ecosystem metabolism depending on whether or not they extend within or below the euphotic zone. Benthic communities receiving sufficient irradiance as to support photosynthesis typically comprise photoautotrophs. These include microalgae, which form highly productive communities on the sediment surface (Cahoon 1999), and macrophytes, including macroalgae and angiosperms (Hemminga and Duarte 2000), which form extensive, highly productive beds and meadows. Similarly, a number of invertebrates (e.g. corals and giant

clams), particularly those in tropical waters, contain photosynthetic symbionts (zooxanthellae), which confer them a significant photosynthetic capacity (Kühl *et al.* 1995). In addition, all structuring macroorganisms present in the illuminated benthic layer support epiphytic microalgae, which add substantially to the primary production of these ecosystems. In contrast, the benthic communities in the aphotic zone rely on organic matter produced elsewhere or by chemoautotrophs to support their metabolic demands.

Benthic communities often dominate ecosystem processes and metabolism, particularly in shallow coastal waters, such as reef lagoons, and are important sites for carbon cycling and bacterial activity, and are all potentially important contributors to ecosystem respiration. Whereas the metabolic balance of the coastal ocean has been subject of much debate (Smith and Hollibaugh 1993; Heip *et al.* 1995; Duarte and Agusti 1998; Gattuso *et al.* 1998), there is, at present, no comprehensive evaluation of the respiration rate of benthic coastal ecosystems, so that their

Table 11.1 Global and specific respiration rate per benthic ecosystem

Benthic ecosystem respiration	Surface area (10^6 km^2)	Respiration (mmol C m^{-2} d^{-1})	Global (Tmol C a^{-1})
Coral Reefs[a]	0.6	359	79 *13%*
Mangroves[b]	0.2	426	28 *15%*
Salt Marshes[b]	0.4	459	67 *45*
Sea grasses[a]	0.6	158	34 *>45%*
Macroalgae[a]	1.4	483	247
Sediments (sublittoral)[a]	24	—[c]	166 *26%*
Sum	27		621

[a] This study.

[b] Gattuso *et al.* (1998).

[c] Benthic respiration (R, mmol m^{-2} d^{-1}) in marine sediments is assumed to depend on water depth (metres) as $R = 32.1\, e^{-0.0077z}$.

contribution to the respiration of the global ocean remains undetermined.

✳ The objective of this chapter is to derive a global estimate for respiration in coastal benthic communities. We will first introduce the methodological approaches used to derive respiration rates in coastal ecosystems before discussing the rates measured in individual coastal benthic communities. The approach to derive a global system-specific respiration estimate differs per habitat because of differences in available data, research traditions, and inherent differences in ecosystem functioning. The respiration rates from the individual components of coastal ecosystems will then be combined to derive a global estimate by bottom–up scaling procedures (Table 11.1). The resulting estimate will be compared with alternative estimates, based on primary producer budgets (Duarte and Cebrián 1996) and coastal ocean carbon budgets (e.g. Smith and Hollibaugh 1993; Wollast 1998). Units of mmol C m^{-2} and per day are used throughout this chapter though some of the data are better expressed per hour (incubation based estimates) or per year (mass balance based estimates) or have been converted from oxygen units using a respiratory quotient (RQ) of 1. Global respiration estimates are given in Tmol a^{-1} (1 Teramol = 10^{12} mol).

11.2 Approaches and methods

Respiration rates can be defined on the basis of (i) the consumption of organic matter, (ii) the use of oxidants, or (iii) the production of inorganic carbon. Quantification of respiration through consumption of organic matter is usually not feasible *i* because changes in stocks are very small (during incubations) compared to natural variability and measurement precision. Moreover, any decrease in organic matter is the result of gross consumption and subsequent resynthesis by secondary producers.

The most straightforward method to measure respiration is the oxygen consumption method. It *ii* is a simple method because oxygen is readily measured using Winkler titration, colorimetry, and electrodes or opt(r)odes. Moreover, the solubility of oxygen in water is rather limited and even low rates of consumption result in detectable changes of the stock. However, oxygen consumption cannot directly be equated with aerobic respiration, because oxygen consumption also results from nitrification and reoxidation of reduced mineralization products (see below). In many coastal environments respiratory activities are so high that oxidant demand exceeds the ambient stock and supply rate of oxygen, and alternative oxidants are used through anoxic metabolic pathways. This is particularly prevalent within sediments, where the limited diffusion rates further constrain the capacity to resupply oxygen from the overlying waters (Soetaert *et al.* 1996, 2000; Glud *et al.* 2003). In coastal sediments, the major alternative electron acceptors are nitrate, nitrite, manganese oxides, iron oxides, and sulfate. These alternative oxidants are utilized not simultaneously but sequentially (Table 11.2), though there is often considerable overlap between the various respiration pathways. Coastal sediments often exhibit a vertical redox zonation of oxidant depletion in the order oxygen > nitrate > manganese oxide > ferric oxides > sulfate (Thamdrup and Canfield 2000 and see King, Chapter 2).

The electron acceptors are not only consumed in organic matter oxidation, but also in the reoxidation of the reduced components produced during anaerobic mineralization. These reduced components include ferrous iron, sulfide, methane, reduced

underestimating rsp?

Table 11.2 Respiration pathways in coastal benthic communities

Aerobic respiration	$[CH_2O]^* + O_2 \rightarrow CO_2 + H_2O$
Denitrification	$5[CH_2O] + 4NO_3^- \rightarrow 4HCO_3^- + 2N_2 + CO_2 + 3H_2O$
Manganese reduction	$[CH_2O] + 2MnO_2 + H_2O \rightarrow HCO_3^- + 2Mn^{2+} + 3OH^-$
Iron reduction	$[CH_2O] + 4Fe(OH)_3 \rightarrow HCO_3^- + 4Fe^{2+} + 7OH^- + 3H_2O$
Sulfate reduction	$2\,[CH_2O] + SO_4^{2-} \rightarrow 2HCO_3^- + H_2S$

*Note: $[CH_2O]$ is used as a notation for organic material.

manganese, and ammonia. Each of these can be oxidized with an electron acceptor higher in the redox sequence and this occurs chemically as well as mediated by microbial activity (chemolithotrophy). Chemolithotrophs thus play a key role in elemental cycles and are essential to fully utilize the energy in organic matter. The widespread distribution of these chemolithotrophs adds considerable complexity to the system and complicates the relation between respiration and electron acceptor use (Jørgensen 2000). For instance, in deep-sea sediments, oxygen consumption is due to aerobic respiration, while the majority of oxygen use in coastal sediments derives from the reoxidation of sulfide, iron, and ammonium and other reduced components (Soetaert *et al.* 1996; Jørgensen 2000; Thamdrup and Canfield 2000). Thus, although oxygen consumption cannot be equated directly with aerobic respiration, it does provide a reliable estimate of total respiration because of the almost complete reoxidation of anaerobic respiration products. However, during incubations the production of reduced species may be decoupled from measured oxygen production. Despite this potential bias, estimates of respiration derived from rates of oxygen consumption encompass both oxic and anoxic pathways. The uncertainty related to reoxidation processes could be avoided if the production of dissolved inorganic carbon ($\Sigma CO_2 \equiv CO_2 + HCO_3^- + CO_3^{2-}$) is measured, because inorganic carbon is the final product of all respiration pathways (Table 11.2). However, ambient concentrations of dissolved inorganic carbon are usually high (>2 mM; that is, about 10 times that of dissolved oxygen), relative to the changes in dissolved inorganic carbon from respiration, so that highly accurate measurements of dissolved inorganic carbon are required to resolve the rates. Moreover, processes other than respiration (e.g. precipitation and dissolution of

calcium carbonate, sulfate reduction, carbon dioxide assimilation by chemoautotrophs, and nitrification through its effect on pH) induce changes in dissolved inorganic carbon concentrations. The most appropriate method to quantify respiration consequently depends on the community and ecosystem.

The use of oxygen (and other oxidants) and production of dissolved inorganic carbon are related via the respiration ratio or RQ (Williams and del Giorgio, Chapter 1). For pure sucrose (CH_2O as in Table 11.2) or the carbohydrate end-member of marine organic matter ($C_6H_{10}O_5$; Hedges *et al.* 2002) there is a $1:1$ molar relationship between oxygen and dissolved inorganic carbon. However, respiration quotients for marine plankton are sometimes higher than 1 because additional oxygen is required for oxidation of hydrogen-rich, nitrogen-, and sulfur-containing material. For instance, use of model Redfield organic matter $C_{106}H_{260}O_{106}N_{16}P_1$ and average marine plankton $C_{106}H_{177}O_{37}N_{17}S_{0.4}P_1$ yield respiration coefficients (for complete oxidation) of 1.3 and 1.45, respectively (Hedges *et al.* 2002). The application of conventional Redfield C:N:P (106:16:1) ratios in coastal benthic ecosystems is not always justified, because riverine organic matter (119:8.9:1; Meybeck 1982) and marine benthic plants (550:30:1; Atkinson and Smith 1983) have stoichiometry ratios significantly differing from that of plankton.

Respiration measurements, whether based on oxygen consumption or inorganic carbon generation, can be based on direct (i.e. rate measurements during incubations) or indirect methods (i.e. chemical gradients combined with transport or mass-balance consideration) and hybrids or combined approaches. The most direct, integrative approach is based on mass-balance budgets for oxygen, carbon dioxide, and nutrients. Nutrient regeneration is linked to respiration and mineralization via Redfield stoichiometry of organic matter (see above), and increases in nutrient concentrations can thus be used to estimate net respiration. However, this approach may fail in systems with nutrient-poor organic matter since heterotrophic bacteria may assimilate inorganic nutrients there to gain enough nutrients for growth (Zehr and Ward 2002), and because many coastal systems are open to allochtonous nutrient inputs (e.g. riverine or atmospheric inputs), so that the use of nutrient budgets to infer respiration rates becomes very uncertain. Moreover, respiration rates for marine macropytes based on Redfield stoichiometry will likely be too low because of low nutrient to carbon ratios of marine benthic plants (Atkinson and Smith 1983). Mass budgets usually attempt to balance the oxygen, carbon dioxide, or nutrient content of inflowing and outflowing water with that of gas exchange across the air–water interface, or they are based on concentration changes in chambers enclosing an area of sediment. The former method is also sometimes referred to as the upstream–downstream method (as it has been developed for unidirectional flowing waters) and it has been applied to riverine, estuarine, reef, and other coastal systems (e.g. Odum and Hoskins 1958). The second, flux-chamber method, is the methodology used most frequently to study non-vegetated, non-reef sediments.

Most respiration measurements are carried out in the dark to exclude oxygen production and carbon dioxide consumption by photosynthesis. However, respiration rates in the light may be higher than those in the dark, both at the level of individual organisms (physiological) as well as the entire community (heterotrophs consuming substrates produced in the light). Light enhanced respiration is widely recognized (see Raven and Beardmore, Chapter 3) and has been reported for benthic microalgal (Epping and Jorgensen 1996) and coral reef communities (Kühl et al. 1995; Langdon et al. 2003). The significance of this for coastal benthic respiration rates is yet to be evaluated.

11.3 Respiration in coastal benthic ecosystems

In this section we discuss the respiration rates of specific types of benthic systems, and roughly follow the communities described in Gattuso et al. (1998): we have somewhat arbitrarily divided coastal benthic communities (Table 11.1) into those dominated by emergent macrophytes (saltmarshes and mangroves), submerged macrophytes (sea grasses and macroalgae), coral reefs, and unvegetated sediments (though in the photic zone this may be covered by benthic algae).

11.3.1 Coastal sediments

In the context of sediment biogeochemistry and sediment community respiration, coastal sediments can be classified into the following typologies (partly based on Aller 1998): (i) steadily accreting silty and muddy sediment, (ii) highly mobile, silty, and muddy sediments (fluidized bed reactors), (iii) non-accumulating or eroding sediments (bypass zones), and (iv) sandy, permeable sediments. A further subdivision can be made between sediments within and below the euphotic zone, although in most sediment respiration studies irradiation conditions are not reported. The relative importance of these sediment typologies depends on the coastal system.

The majority of studies on sediment community respiration have been conducted on steadily accreting sediments below the euphotic zone (e.g. see compilations in Heip et al. 1995; Aller 1998, Thamdrup and Canfield 2000). This bias toward accreting silty sediments limits our understanding of sediment processes in general, and hampers any upscaling endeavors. Respiration in silty coastal sediments is mainly bacterial and occurs via anaerobic pathways because oxygen resupply by diffusion and/or irrigation of animal tubes cannot compensate for oxygen consumption through respiration and reoxidation of reduced products. The contribution of meiofauna to total sediment respiration is rather small (less than a few percent), but macrofauna typically contributes 10–30% of total respiration (Herman et al. 1999). The contribution of fauna to total respiration can be partitioned into direct contributions due to faunal respiration and indirect contributions due to faunal stimulated bacterial activities (Aller and Aller 1998; Glud et al. 2003), but the available data are too limited to allow quantification at the global scale.

Mobile muds are most prominent in deltaic sediments from major rivers such as the Amazon, Orinoco, and Fly rivers (Aller 1998), but do occur in other high-energy coastal systems, provided there is a supply of fine materials (e.g. fluidized muds in the high turbidity zones of tidal estuaries, Abril *et al.* 2000). These mobile beds are thought to function as suboxic, fluidized bed reactors: due to repetitive cycles of erosion and deposition there is an efficient incorporation of fresh organic matter and oxidants from the overlying water column, a steady removal of inhibitory metabolites, and a repetitive recycling among dissolved and particulate phases and among aerobic and anaerobic pathways, which together promote efficient degradation and respiration of

organic materials (Aller 1998). These mobile muds are nevertheless the prime sites for oceanic carbon burial because they account for about 40–50% of all marine organic carbon burial at present (Hedges and Keil 1995). Bacteria dominate respiration in these mobile muds because the biomass of macro and meiofauna is limited (relative to accreting muddy sediments), due to high levels of physical stress that the animals encounter in these sites.

Sandy sediments, which cover about 70% of global continental shelves and most beaches, have conventionally been considered as biogeochemical deserts harboring little life. This misconception is based on the idea that significant reactions, fluxes, and respiration rates require large standing stocks

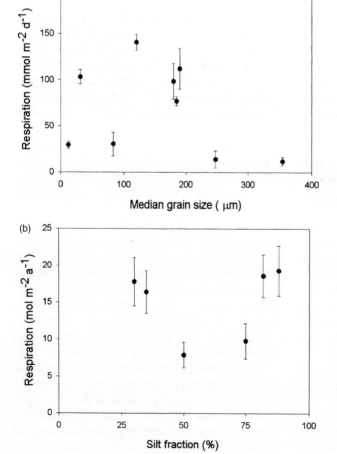

Figure 11.1 Respiration versus measures for sediment grain size. (a) Respiration (mmol m^{-2} d^{-1}) versus median grain size in intertidal flat sediments (Middelburg *et al.* 1996). (b) Respiration (mol m^{-2} a^{-1}) versus percentage silt (i.e. fraction <63 μm) in North Sea sediments (Dauwe *et al.* 2001).

of organic matter (Boudreau *et al.* 2001). Concentrations of organic matter (when expressed on weight basis) are low because of the well-established negative correlation between organic carbon content and (median) grain size (Middelburg *et al.* 1996). However, when organic matter concentrations are expressed on an area or volume basis (i.e. quantity per m^2 or m^3), then organic carbon concentrations in sandy and silty/muddy sediments are much more similar (Middelburg *et al.* 1996). Moreover, the overall degradability of organic carbon in sandy sediments appears to be somewhat higher than that in muddy/silty sediments (Boudreau *et al.* 2001), mainly because accumulating aged, more refractory organic material does not dilute organic carbon in sands. As a consequence, respiration rates in organic-poor sandy sediments may be substantial. Figure 11.1 shows that there are no systematic relationships between sediment community respiration and grain size either expressed as median grain size (Fig. 11.1(a)) or as fraction of silt (less than 63 μm) (Fig. 11.1(b)).

Non-accumulating or bypass zones occur in most coastal environments when physical conditions are not favorable for settling or when deposited material is rapidly resuspended due to current and/or wave activity. Non-accumulating sediments may nevertheless have significant respiration activity. Animals, suspension and interface feeders in particular, may even use eroding sediments as a substrate to settle and live while exploiting the particulate organic carbon in the water column as their main food resource. While high current velocities or turbulence levels may not be favorable for organisms relying on passive settling of organic matter, it may be advantageous for suspension feeders, though too high current velocities may again lower filtration efficiency of suspension feeders (Herman *et al.* 1999). Sediments that show no accumulation or even erosion on the longer term, may nevertheless receive deposition during certain periods; for example, during spring bloom or during slack tides. Organisms can exploit this temporarily deposited material and these sediments can thus support an active community.

Although these sediment typologies differ in community structure and biogeochemical functioning, it appears that sediment oxygen uptake largely depends on water depth. Figure 11.2 shows a compilation of sediment oxygen uptake data for marine sediments shallower than 300 m (Andersson, unpublished results). There is a

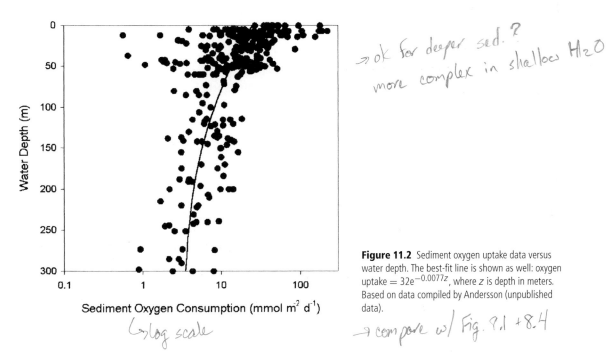

Figure 11.2 Sediment oxygen uptake data versus water depth. The best-fit line is shown as well: oxygen uptake $= 32e^{-0.0077z}$, where z is depth in meters. Based on data compiled by Andersson (unpublished data).

highly significant decrease with depth, consistent with observations in the open ocean (Jørgensen 1983, see also Aristegui *et al.*, Chapter 10) and estuaries (Heip *et al.* 1995 and see Hopkinson and Smith, Chapter 8). These data can be fitted with an exponential equation for sediment respiration versus depth:

$$\text{Respiration} = 32e^{-0.0077z}$$

where z is water depth (m). Combining this equation with the surface area of the ocean shallower than 200 m (23.9×10^6 km^2) results in a total sedimentary respiration of 166 Tmol C a^{-1} (See Table 11.1).

11.3.2 Coral reefs

Estimates of the global surface cover of coral reefs vary quite considerably (reviewed by Spalding and Grenfell 1997) due to the various definitions of coral reefs that have been used. The consensus opinion is that coral reefs cover approximately 0.6×10^6 km^2. Data on community metabolism are numerous in coral reefs relative to other coastal ecosystems. The reason is that reefs are often subject to a unidirectional current from the ocean to the back reef area. Consequently, it is relatively easy to estimate ecosystem metabolism in transects, for example, 1 m wide and several hundreds of meters long, using Lagrangian techniques that measure changes in the water chemistry as the water mass crosses the reef system (Gattuso *et al.* 1999). Some measurements have also been carried out on smaller areas during standing water periods. Another approach, based on ecological stoichiometry and mass balance of carbon and phosphorus (Smith 1991) has provided estimates of community metabolism of entire reef systems, mostly atolls. One limitation of these techniques is that they do not adequately include the portion of the reef that faces the open water (outer reef slope) that is one of the most active zones. Finally, although reef mesocosms do not qualify as exact proxies of coral reefs, they are increasingly being used to address the metabolic response of corals to environmental forcing (e.g. Leclercq *et al.* 1999; Langdon *et al.* 2003).

Although the term "coral reef" is often used generically, this term encompasses various categories, defined according to the community structure or the geomorphology, and these different types of systems exhibit distinct metabolic performances. It is therefore useful to examine each of these categories separately. We use in the present contribution the categories defined by Kinsey (1985): "complete" reef systems, outer reef slope, high activity areas of near total cover by hard substratum (excluding pavement), algal pavement zone, reef-flat coral/algal zones, shallow lagoon environments, algal turfs and algal/sand flats, algal-dominated flats, "sand" areas, and uncertain designation.

The rate of community respiration spans more than one order of magnitude (1.3–910 mol C m^{-2}a^{-1}) and exhibits a rather high variability even within a single category (Fig. 11.3). It is lowest in sandy areas (39 ± 23 mol C m^{-2}a^{-1}) and, as expected, highest in the high activity areas (413 ± 187 mol C m^{-2}a^{-1}). Complete reef systems generally comprise lagoons with large areas covered by sediments. For instance, the southwest of New Caledonia reef complex comprises 95% sediment and 5% of hard-bottoms, covered by coral substratum (Clavier, personal communication). It is therefore not surprising that community respiration of complete reef systems (131 ± 46 mol C m^{-2}a^{-1}; $N = 7$) is much closer to the value of sandy areas to than to the value of high activities areas.

Like in other ecosystems, the rates of reef community respiration have exclusively been measured during the night time and the 24 h values are estimated assuming that respiration is similar in the light and dark. However, it is well established that respiration in the light often exceeds dark respiration in numerous aquatic autotrophs. For example, Kühl *et al.* (1995) estimated that respiration of one coral species investigated in laboratory conditions is more than 6 times higher at a saturating irradiance than in the dark. Despite the strong evidence that the assumption upon which community respiration is based is probably flawed, there is presently no direct estimate of the changes in respiration in any natural reef system. However, a recent breakthrough has been made by Langdon *et al.* (2003) in the Biosphere 2 coral reef mesocosm. They measured gross respiration (light plus dark) with a new ^{14}C isotope dilution method and discovered that light

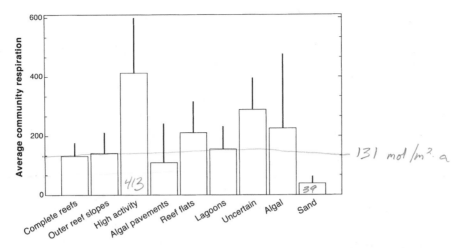

Figure 11.3 Community respiration (mol C m^{-2} a^{-1}) in complete reefs and other reef areas. An RQ of 1 has been assumed to express in units C the rates reported in units O$_2$. Mean ± s.d.

respiration is about twice the dark rate, with the result that the conventional daily respiration rate is underestimated by 40%. Biosphere 2 cannot be considered to represent an "average" reef because it mimics a high latitude, low energy, and algal-dominated system. The results of Langdon *et al.* cannot therefore be extended to all natural reefs but their data do provide, for the first time, a rough estimate of what the difference between light and dark respiration could be in a reef system.

One of the goals of the present book is to provide global estimates of respiration of aquatic organisms and ecosystems. However, estimating the global rate of respiration in coral reefs is extremely difficult for several reasons. First, the fact that community respiration during the day exceeds respiration at night now seems unquestionable. Second, there are only data for seven "complete" reef systems (that do not include the outer reef slope), which makes upscaling to the global scale very uncertain. Third, although there are many estimates for subsections of coral reefs (the categories described above), it is not yet possible to exploit these data to derive an estimate for an "average" reef. Considering both the overall objectives of this book and these three limitations, and without even attempting to provide an error estimate, we suggest the best current estimate of the global respiration of coral reefs at 79 Tmol C a^{-1} (i.e. 131 mol m^{-2} a^{-1} × 6 × 10^{11} m^2).

11.3.3 Emergent macrophytes: mangroves and salt marshes

Mangroves and salt marshes are distinct ecosystems in which primary producers are structuring factors that emerge above the water. Mangroves are intertidal forests that grow above mean sea level. They cover about 0.18 × 10^6 km^2 along sheltered tropical and subtropical shores (31°N to 39°S). Salt marshes are rooted-macrophyte dominated ecosystems in temperate zones that occur within the intertidal zone and are characterized by a zonation of plant species depending on elevation. They cover an area of 0.4 × 10^6 km^2, that is, about twice that of mangroves (Gattuso *et al.* 1998). Although mangroves and salt marsh plants are the dominant producers in their respective ecosystems, there are also significant contributions from other producers (macroalgae, epiphytes, phytoplankton, and benthic microalgae) and from adjacent ecosystems (Middelburg *et al.* 1997). Moreover, a significant part of tree and salt marsh plant production is allocated to below-ground organs and usually not included in community metabolism studies.

Respiration measurements in mangroves and salt marsh systems are difficult to make and even more complex to be analyzed. First, these trees and some macrophytes are so large that compartment and chamber type of measurements for respiration and production are complicated. Moreover, production is normally reported as net primary production based on biomass accumulation derived from allometric approaches. Second, a significant proportion of the respiration occurs in the parts of the tree and plant that are above the water. Third, measurements of respiration in the sediment compartment are complicated by the alternation between emergent and submerged conditions. This complicates not only the measurements, but also respiration rates and respiration pathways may vary with tidal stage. Fourth, sediment respiration in mangrove and salt marsh systems includes respiration from the tree/plant roots and rhizomes as well as that from heterotrophic organisms living on plant litter and allochthonous carbon inputs. Core incubation techniques (the usual technique for assessment of respiration rates in coastal sediments) can not be used because root respiration is excluded, while respiration fueled by root exudates is likely overestimated due to damage to (hair) roots during sampling (Gribsholt and Kristensen 2002). There is definitely a need for novel approaches to quantify respiration *in situ* in mangrove and salt marshes and to partition this respiration between plants and heterotrophs. Perhaps it is possible to apply Keeling-plot type analysis (Flanagan and Eheleringer 1998) to exploit the difference in carbon isotope composition between salt marsh plants and mangroves on the one hand and sedimentary organic matter on the other. In coastal systems with high rates of respiration or large contributions of salt marshes and/or mangroves to coastal (water column) respiration, it may be possible to obtain integrative measures from oxygen and carbon dioxide mass balances (Cai *et al.* 1999).

Given the large uncertainties and the limited amount of new data available since the compilation of Gattuso *et al.* (1998), we have adopted their values in Table 11.1: 426 and 459 mmol m^{-2} d^{-1} and 28 and 67 Tmol a^{-1} for mangroves and salt marshes, respectively.

11.3.4 Submerged macrophytes: sea grasses and macroalgae

Sea grasses and macroalgae are important components of coastal ecosystems. Sea grasses encompass a group of about 50–60 species, which grow on the illuminated (>11% of surface irradiance) sandy and muddy shores of all continents except Antarctica (Hemminga and Duarte 2000; Duarte *et al.* 2002). Sea grass communities form highly productive meadows, which cover an estimated area of about 0.6 × 10^6 km^2 (Charpy-Roubaud and Sournia 1990) developing a global biomass of about 9.2 Tmol C (460 g dry wt m^{-2} on average, Duarte and Chiscano 1999), which results in a net primary production of about 20 Tmol C a^{-1} (92 mmol C m^{-2} d^{-1} on average, Duarte and Chiscano 1999). Macroalgae comprise about 8000 species (Dring 1992), which are the dominant benthic primary producers in rocky shores and also form dense communities in sandy sediments, particularly in subtropical waters, where species of green macroalgae are often important primary producers (Loebb and Harrison 1997). Macroalgae extend over an estimated area of 1.4 × 10^6 km^2 (Gattuso *et al.* 1998), contributing a net primary production of 212 Tmol C a^{-1} (Charpy-Roubaud and Sournia 1990). Macroalgae develop the lushest plant communities in the sea, which are particularly well represented by the kelp forests growing on the rocky shores of cold, nutrient-rich waters (Loebb and Harrison 1997). Although these communities may develop very high biomasses, the average biomass of macroalgal communities (40 g dry weight m^{-2}, in Cebrían and Duarte 1994) is much lower than that of sea grass meadows, which develop an important below-ground biomass. The global macroalgal biomass is about 1.9 Tmol C. Macroalgae are also important components of coral reef communities, but the contribution of coral reefs to the metabolism of the coastal ocean are discussed in Section 11.3.2 above.

In addition to the important macrophyte biomass in sea grass and macroalgal stands, macrophyte-dominated ecosystems support an important biomass of epiphytic microalgae, which are believed to contribute about 20–50% of the total production of macrophyte-dominated benthic systems (Heip *et al.* 1995; Alongi 1998; Hemminga and Duarte 2000). Macrophyte-dominated marine ecosystems are, therefore, highly productive components of the coastal ocean, and are responsible for an important share of the organic carbon production available for export (Duarte and Cebrián 1996).

Marine macrophytes respire, on average, about 51–57% of their gross primary production (Duarte and Cebrián 1996). Accordingly, the direct respiration by the macrophytes themselves can be estimated, using the estimates of their net primary production discussed above, as 220 and 19.7 $TmolCa^{-1}$ for macroalgae and sea grasses, respectively, yielding a total direct respiration of about 240 $TmolCa^{-1}$. Alternatively, the respiration rate of marine macrophytes can be estimated from the specific respiration rate of their tissues, which has been extensively measured in connection to estimates of photosynthetic activity (Enríquez et al. 1995, 1996; Vermaat et al. 1997). The respiration rate of marine macrophytes tends to decline, as that of land plants, with increasing tissue thickness (Enríquez et al. 1995, 1996), and this decline is associated to the decline in growth rate with increasing tissue thickness of autotrophs (Nielsen et al. 1996). The specific respiration rates of the photosynthetic tissues of marine macrophytes averages about 40 $mmolC\,molC^{-1}\,d^{-1}$ (Enríquez et al. 1995, 1996; Vermaat et al. 1997; Touchette and Burkholder 2000), and this average can be used to derive an independent estimate of the respiration rate of marine macrophytes from their global photosynthetic biomass as 28 and 65 $TmolCa^{-1}$ for macroalgae and sea grasses, respectively, yielding a total respiration of marine macrophytes of about 93 $TmolCa^{-1}$. This combined estimate is lower than that derived from the percent respiration of macrophyte gross primary production (240 $TmolCa^{-1}$), probably because it does not account for the respiration contributed by the roots and rhizomes of sea grasses, which is about 15–57% of the total sea grass respiration (Hemminga and Duarte 2000), and because the specific respiration of tropical and subtropical macrophytes is likely to be above the mean value used. The respiration by marine macrophytes has been shown to be highly dependent on temperature (Touchette and Burkholder 2000), with a Q_{10} of about 2.5 (e.g. Marsh et al. 1986), so that tropical and subtropical communities are expected to show higher respiration rates than temperate and polar macrophyte communities of comparable

biomass. This also suggests that global warming is likely to increase significantly the contribution of macrophyte respiration to the respiration of coastal ecosystems.

The metabolism of sea grass communities, including the plants and heterotrophic organisms, was studied since the early introduction of metabolic studies in aquatic ecology by Odum and Hoskins (1958). Since then, the respiration rate of a number of sea grass ecosystems has been studied in a number of sea grass meadows (Table 11.3), largely through the use of benthic chambers, although most of the studies refer to single estimates and annual estimates are still few. The estimates so far available yield a mean respiration rate of 159 ± 19 $mmolC\,m^{-2}\,d^{-1}$, ranging from 14.1 to 596 $mmolC\,m^{-2}\,d^{-1}$, with a log-normal distribution (Fig. 11.4). The mean respiration derived corresponds, when scaled to the surface area covered by sea grasses, to a global respiration of 34.6 $TmolCa^{-1}$, so that the estimated direct contribution of sea grasses that we derived in the previous paragraph (19.7 $TmolCa^{-1}$) would represent about 57% of the community respiration. Indeed, the total respiration of macrophyte-dominated benthic systems is higher than that of the macrophytes alone, because macrophyte-dominated systems are habitats for a rich heterotrophic community (Loebb and Harrison 1997; Hemminga and Duarte 2000; Mann 2000), which also contribute significantly to total system metabolism. This must be particularly important for sea grass marshes, where, in addition to the rich heterotrophic epiphytic community they support, the underlying sediments are an active site for bacterial decomposition of organic matter. Bacterial metabolism is enhanced in sea grass sediments due to the increased inputs of organic matter under the plant canopies, and the direct stimulation of bacterial metabolism by release of organic matter together with oxygen from sea grass roots (Hemminga and Duarte 2000).

In contrast to sea grasses, the total respiration of macroalgae-dominated communities has been examined in only a limited number of studies (Table 11.4). A few additional estimates can be derived from carbon flow analyses (e.g. Newell and Field 1983). The available estimates indicate an average respiration rate of macroalgal beds

Table 11.3 Estimates of respiration rate in sea grass-dominated coastal benthic communities

Species	Location	R (mmol C m^{-2} d^{-1})	Author
Mixed sea grass meadow	The Philippines	213	Barrón (unpublished data)
Enhalus acoroides	The Philippines	45	Barrón (unpublished data)
Cymodocea nodosa	Spanish Mediterranean	48	Barrón (unpublished data)
Posidonia oceanica	Spanish Mediterranean	68	Barrón (unpublished data)
Zostera noltii	South Portugal	182	Barrón (unpublished data)
Cymodocea nodosa	South Portugal	105	Barrón (unpublished data)
Zostera marina	Norway	65	Duarte *et al.* (2000)
Mixed community	Indonesia	238	Erftemeijer *et al.* (1993)
Mixed community	Sri Lanka	72	Johnson and Johnstone (1995)
Halodule uninervis	Indonesia	250	Lindboom and Sandee (1989)
Thalassia hemprichii	Indonesia	267	Lindboom and Sandee (1989)
Thalassia/Enhalus	Indonesia	250	Lindboom and Sandee (1989)
Zostera marina	Netherlands	94	Lindeboom and deBree (1982)
Zostera marina	Northeast USA	225	Murray and Wetzel (1987)
Zostera marina	Northeast USA	191	Nixon and Oviatt (1972)
Thalassia testudinum	Texas	531	Odum and Hoskins (1958)
Mixed	Texas	56	Odum and Hoskins (1958)
Mixed	Texas	447	Odum and Wilson (1962)
Thalassia testudinum	Texas	322	Odum (1963)
Thalassia testudinum	Puerto Rico	488	Odum *et al.* (1959)
Halodule wrightii	Mexico	597	Reyes and Merino (1991)
Thalassia + Halodule	Mexico	541	Reyes and Merino (1991)
Thalassia testudinum	Mexico	478	Reyes and Merino (1991)
Thalassia testudinum	Texas	131	Ziegler and Benner (1998)

Note: Overall mean = 159, median = 94, standard error = 19, N = 64.

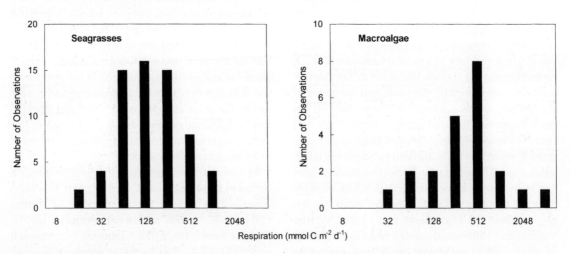

Figure 11.4 Frequency distribution of respiration rate in macrophyte-dominated coastal benthic communities. Note logarithmic x-axis.

Table 11.4 Estimates of respiration rate in macroalgae-dominated coastal benthic communities

Species	Location	R (mmol C m^{-2} d^{-1})	Author
Mixed community	Caribbean	1271	Adey and Steneck (1985)
Caulerpa spp.	The Philippines	103	Barrón (unpublished data)
Halimeda spp.	The Philippines	111	Barrón (unpublished data)
Brown algae	Australia	200	Cheshire *et al.* (1996)
Mixed community	Caribbean	975	Connor and Adey (1977)
Mixed community	French Polynesia	241	Gattuso *et al.* (1997)
Mixed community	Caribbean	833	Griffith *et al.* (1987)
Mixed community	Caribbean	16	Hawkins and Lewis (1982)
Mixed community	Hawaii	288	Kinsey (1979)
Kelp community	South Africa	341	Newell and Field (1983)
Fucus-dominated	Norway	96	Pedersen (1987)
Mixed community	Mexico	384	Reyes and Merino (1991)
Mixed community	Caribbean	255	Rogers and Salesky (1981)
Mixed community	Pacific	501	Smith and Marsh (1973)
Mixed community	Pacific	501	Smith (1973)
Mixed community	Caribbean	58	Vooren (1981)

Note: Overall mean = 483, median = 338, standard error = 29; N = 21.

of 483 mmol C m^{-2} d^{-1}, with a log–normal distribution (Fig. 11.4). This average respiration rate yields, when upscaled to the global area covered by macroalgae, a global rate of 247 Tmol C a^{-1}, which implies that the direct contribution of macroalgae (220 Tmol C a^{-1}) accounts for about 90% of community respiration. This high contribution of macroalgae to community respiration might reflect a bias in the available community respiration data, or it is due to an overestimation of macroalgae respiration based on a fixed fraction of gross primary production.

From the calculations presented above, the total respiration of macrophyte-dominated benthic systems can be conservatively estimated at about 281 Tmol C a^{-1}, of which about 12% corresponds to sea grass ecosystems and the remaining corresponding to macroalgal beds. The total respiration in macrophyte-dominated benthic systems could well exceed the local primary production, in particular in sea grasses because they receive an important input of sestonic organic matter. Sea grass canopies have been shown to be enriched in particles (Duarte *et al.* 1999) and be capable of effectively trapping sestonic particles (Agawin and Duarte 2002), which represent an important input of

organic matter. For example, for a *Posidonia oceanica* meadow where these processes were quantified, the input of seston was comparable to the local production of the sea grass and associated epiphytes (Gacia *et al.* 2002). The majority (70%) of the reports on the metabolism of macrophyte-dominated benthic systems (Tables 11.3 and 11.4) correspond to autotrophic systems in which production is higher than respiration, particularly for macroalgae, for which 95% of the communities studied (Table 11.4) were reported to be autotrophic. Although sea grass-dominated systems can be heterotrophic during certain periods of their seasonal development, all of the systems so far studied have been shown to be autotrophic at an annual basis (Hemminga and Duarte 2000). This is consistent with (i) the significant carbon burial in sea grass meadows (e.g. Mateo *et al.* 1997; Pedersen *et al.* 1997; Gacia *et al.* 2002), which is estimated to average 16% of the net production of the sea grasses (Duarte and Cebrián 1996), and (ii) the important export of sea grass detritus, which averages about 24% of their net primary production (Duarte and Cebrián 1996). Hence, sea grass meadows appear to be autotrophic systems (Gattuso *et al.* 1998; Hemminga and Duarte 2000) despite often large inputs of organic matter they receive. Similarly, seaweed-dominated systems have also been shown to export significant amounts of organic matter (Barrón *et al.* 2003). Although those growing on rocky shores cannot store organic carbon in the sediments, they export substantial amounts of organic matter both as DOC released and detached fragments (Loebb and Harrison 1997; Mann 2000; Barrón *et al.* 2003), which must fuel respiration elsewhere. The export from macroalgal beds has been estimated to average 43% of their net primary production (Duarte and Cebrián 1996), thereby representing an important loss of materials which cannot be, therefore, respired within the system but that will fuel respiration rates in adjacent ecosystems or compartments (e.g. the pelagic compartment).

11.4 Global rate of respiration in coastal benthic communities

The estimation of the global rate of respiration in coastal benthic communities poses a major challenge because of the diversity of systems in question and the overall sparcity of data on certain components and their geographical distribution. Here we will follow the traditional approach based on upscaling from the rates of the contributing components, but we will in addition constrain the global respiration rate from above, that is, by deriving it from global estimates of organic carbon fixation (and inputs) and from coastal carbon budgets. Each of these approaches has severe limitations and requires further study before confidence limits can be established on the global estimates.

11.4.1 Scaling problems

Upscaling refers to the aggregation of information collected at a small scale to a larger scale. The traditional way of upscaling involves the multiplication of habitat area with habitat respiration activity followed by summation over the habitat involved: that is, the so-called bottom–up procedure (Table 11.1). Areal–up or bottom–up scaling is relatively straightforward provided the rates are truly additive, that is, the entities for upscaling can

be considered closed systems and homogeneous, and that there are sufficient data. However, these conditions are very seldom met. First, the diversity of benthic ecosystems in the coastal ocean is very high and it is still unclear how best to partition these habitats into representative elementary units. Each of these ecosystems comprises many biological communities and shows heterogeneity at a variety of spatial and temporal scales. There is a fundamental lack of knowledge on the level of aggregation required to allow effective bottom–up scaling. In this chapter we have adopted a rather coarse resolution of benthic communities: emergent macrophytes (salt marshes, mangroves), submerged macrophytes (sea grasses, macroalgae), coral reefs and nonvegetated sediments (though in the photic zone the latter may be covered by benthic algae). Second, even at this high level of aggregation there is large uncertainty in the respective surface areas because detailed geo-referenced biological datasets are scarce. Third, pelagic and benthic ecosystems in the coastal zone are intimately coupled and it is therefore difficult to partition benthic systems from their surroundings. There is also significant transfer of organic matter and energy among ecosystems, for example, the external subsidies to sustain metabolic imbalances. This open nature of coastal communities implies that conventional additive scaling procedures introduce uncertainties. Fourth, for some of these benthic habitats there are only a few data available (e.g. macroalgae) and most available datasets are incomplete in terms of temporal and spatial coverage. Fifth, each of these coastal benthic ecosystems shows a range of respiration rates as a consequence of variability in light and nutrient availability, carbon input, water depth, climate, community structure patterns, and other factors. Sixth, human perturbations of natural coastal communities may have affected and are still changing coastal ecosystem functioning, hence likely respiration, over decadal timescales (Rabouille *et al.* 2001). It is clear that these complications preclude the derivation of an accurate estimate of coastal benthic respiration, yet the data compiled can be used to derive a first-order approximation of the magnitude of respiration in the benthic compartment of the global coastal ocean, which can help assess the relevance of this compartment in the context of total ocean respiration.

11.4.2 Bottom-up estimate

In the preceding sections we have evaluated and summarized the available data on respiration in the individual benthic communities and have derived ecosystem-specific global respiration rates (Table 11.1). These ecosystem-specific global respiration estimates amount to about 621 $\text{Tmol}\,\text{C}\,\text{a}^{-1}$ with coral reefs, submerged macrophytes, emergent macrophytes, and open sediments accounting for 13%, 45% 15%, and 26%, respectively. Submerged macrophytes, and particularly macroalgae, appear to be the dominant contributors to benthic respiration in the coastal ocean, despite the fact that they cover only about 7% of the surface of the coastal ocean. The large contribution of sediments is due to their large surface area, while the significance of coral reefs, macroalgae, and salt marshes is primarily due to their high area specific rates of respiration. A significant, but unknown, part of mangrove and salt marsh respiration occurs in the air, above the water, and does not contribute to respiration in aquatic ecosystems, the subject of this book.

It appears that autotrophs and multicellular heterotrophs are responsible for a major part of the respiration in coastal benthic systems. Macrofauna in coastal sediments accounts for about 10–30% of total sediment community respiration. Autotrophs and multicellular organisms dominate total reef community respiration, in particular in hard-bottom communities. The high above- and below-ground biomasses of sea grasses, mangroves, and salt marsh plants and high production of macroalgae imply major contributions of these autotrophs to total community respiration. This would suggest that in coastal benthic ecosystems, microbes contribute relatively little to total community respiration and that coastal benthic ecosystems diverge from the majority of other aquatic ecosystems where microbes are the most important group in terms of their contribution to respiration.

Our estimate of global benthic respiration in the coastal zone (621 $\text{Tmol}\,\text{C}\,\text{a}^{-1}$) is larger than that reported by biogeochemists studying sediment

oxygen uptake (e.g. 160 $\text{Tmol}\,\text{C}\,\text{a}^{-1}$, Jørgensen 1983). Those estimates are usually based on a combination of ocean hypsometry and sediment oxygen demand versus depth relationships, and usually exclude macrophyte-covered sediments and coral reef systems that have high area-specific respiration rates.

Gattuso et al. (1998) provided a synthesis of data on carbon metabolism in coastal aquatic ecosystems. They reported a global coastal ocean respiration of 518 $\text{Tmol}\,\text{C}\,\text{a}^{-1}$ based on the sum of global respiration rates of corals (80 $\text{Tmol}\,\text{C}\,\text{a}^{-1}$), saltmarshes (67 $\text{Tmol}\,\text{C}\,\text{a}^{-1}$), mangroves (28 $\text{Tmol}\,\text{C}\,\text{a}^{-1}$), submerged macrophytes (137 $\text{Tmol}\,\text{C}\,\text{a}^{-1}$), and shelf systems (206 $\text{Tmol}\,\text{C}\,\text{a}^{-1}$). Some of these numbers are very similar to those in Table 11.1 because we have either used their data (for salt marshes and mangrove respiration), or because there is significant overlap in the database used (e.g. coral reefs), but the present estimates—based on a larger dataset—significantly increase the respiration of submerged macrophytes (sea grasses and macroalgae). The estimate of Gattuso et al. (1998) for shelf respiration included the pelagic and benthic compartments. We can use our estimate for unvegetated sediment respiration (166 $\text{Tmol}\,\text{C}\,\text{a}^{-1}$) or Jørgensen's (1983) estimate of 160 $\text{Tmol}\,\text{C}\,\text{a}^{-1}$ to derive the pelagic (40–46 $\text{Tmol}\,\text{C}\,\text{a}^{-1}$) contribution to the total shelf respiration estimate of 206 $\text{Tmol}\,\text{C}\,\text{a}^{-1}$ (Gattuso et al. 1998). The estimate of global coastal respiration from these author's data (518 $\text{Tmol}\,\text{C}\,\text{a}^{-1}$) can then be corrected for pelagic shelf respiration and we obtain 472–478 $\text{Tmol}\,\text{C}\,\text{a}^{-1}$ which is still somewhat lower than the estimate we propose here (621 $\text{Tmol}\,\text{C}\,\text{a}^{-1}$).

11.4.3 Coastal carbon budget constraints

A number of geochemists have constructed carbon budgets for the entire coastal ocean and have reported a global rate of respiration for the coastal zone that includes the pelagic as well as the benthic. Smith and Hollibaugh (1993) made a cross-system analysis based on 22 coastal and estuarine systems and they derived a global coastal ocean respiration rate of 507 $\text{Tmol}\,\text{C}\,\text{a}^{-1}$, with about 30% (150 $\text{Tmol}\,\text{C}\,\text{a}^{-1}$) occurring on the bottom. Their estimate did not include coral reefs or mangroves. Wollast (1998) reported a global coastal ocean respiration rate of $300\,\mathrm{Tmol\,C\,a^{-1}}$ (range 95–$548\,\mathrm{Tmol\,C\,a^{-1}}$) based on a small dataset ($N = 10$) comprising temperate and boreal shelf systems from the Northern Hemisphere. This author also provided a separate estimate for coastal oceanic sediments of $150\,\mathrm{Tmol\,C\,a^{-1}}$ (range 65–$333\,\mathrm{Tmol\,C\,a^{-1}}$). Although coastal ocean benthic respiration estimates by Smith and Hollibaugh (1993) and Wollast (1998) are very similar to each other ($150\,\mathrm{Tmol\,C\,a^{-1}}$), and similar to the estimate for sediment respiration proposed by Jørgensen (1983), they likely do not include macrophyte-covered systems and coral reefs. Hence, these geochemical studies have significantly underestimated global coastal benthic respiration.

Rabouille et al. (2001) also investigated the carbon budget of the global coastal ocean but partitioned it into a proximal and a distal zone. The proximal zone includes large bays, estuaries, deltas, inland seas, and salt marshes, has a mean water depth of $20\,\mathrm{m}$ and a surface area of $1.8 \times 10^6\,\mathrm{km^2}$. The distal zone includes the open continental shelves down to $200\,\mathrm{m}$ depth with a surface area of $27 \times 10^6\,\mathrm{km^2}$ and a mean depth of $130\,\mathrm{m}$. They reported a total coastal respiration rate of $348\,\mathrm{Tmol\,C\,a^{-1}}$, with 56.4 and $291.6\,\mathrm{Tmol\,C\,a^{-1}}$ in the proximal and distal coastal zones, respectively. Rabouille et al. (2001) reported benthic respiration rates of 30.4 and $93\,\mathrm{Tmol\,C\,a^{-1}}$ for the global proximal and distal zones, respectively. While their benthic respiration estimate for the distal zone is slightly lower than our estimate for sublittoral sediments ($166\,\mathrm{Tmol\,C\,a^{-1}}$), and those of Jørgensen (1983; $160\,\mathrm{Tmol\,C\,a^{-1}}$) and Wollast (1998; $150\,\mathrm{Tmol\,C\,a^{-1}}$), there is a large discrepancy for the proximal zone. The Rabouille et al. (2001) estimate ($30.4\,\mathrm{Tmol\,C\,a^{-1}}$) is about one order of magnitude smaller than the one from Gattuso et al. (1998; $312\,\mathrm{Tmol\,C\,a^{-1}}$) and our estimate ($455\,\mathrm{Tmol\,C\,a^{-1}}$) for the proximal zone: that is the sum of coral, mangrove, salt marsh, sea grass, and macroalgal respiration.

11.4.4   Coastal primary production constraints

The carbon budgets constructed by geochemists can be complemented by a community ecology approach that explicitly takes into account differences among communities. In this regard, Duarte and Cebrián (1996) compiled extensive datasets on coastal net primary production and the fate of autotrophic production. They reported a total net primary production of the coastal ocean of $836\,\mathrm{Tmol\,C\,a^{-1}}$, with phytoplankton, microphytobenthos, coral reef algae, macroalgae, sea grasses, marsh plants, and mangroves contributing 375, 28, 50, 213, 41, 37, and $92\,\mathrm{Tmol\,C\,a^{-1}}$, respectively. If microphytobenthos, coral reef, macroalgae, sea grass, marsh plant, and mangrove production ($461\,\mathrm{Tmol\,C\,a^{-1}}$), and riverine organic carbon inputs ($34\,\mathrm{Tmol\,C\,a^{-1}}$) are balanced by organic matter burial in sediments (about $13\,\mathrm{Tmol\,C\,a^{-1}}$; Hedges and Keil 1995) and respiration of heterotrophs, then coastal benthic respiration should be at least $482\,\mathrm{Tmol\,C\,a^{-1}}$. This is a minimum estimate since part of the respiration of autotrophs is not included, and part of the production of coastal plankton is also respired in benthic systems. The latter quantity can be estimated from the difference in sediment respiration rates ($166\,\mathrm{Tmol\,C\,a^{-1}}$, Table 1) and sediment primary, that is, microphytobenthos, production ($28\,\mathrm{Tmol\,C\,a^{-1}}$; Duarte and Cebrían 1996): that is, $138\,\mathrm{Tmol\,C\,a^{-1}}$. This implies that the total coastal benthic respiration due to heterotrophs should be about $620\,\mathrm{Tmol\,C\,a^{-1}}$, identical to that derived from the bottom–up approach ($621\,\mathrm{Tmol\,C\,a^{-1}}$). The excellent agreement of the bottom–up and community-production top–down approaches is pleasing, but it might give an impression of false accuracy and precision. The community-production based estimate represents a minimum for coastal respiration because it does not include the respiration of autotrophs, and we have shown before that respiration by autotrophs contributes significantly to, and sometimes even dominates, total ecosystem respiration. For instance, the contribution of submerged macrophytes to coastal benthic respiration alone is $240\,\mathrm{Tmol\,C\,a^{-1}}$ (See Table 11.1).

As an alternative, we can balance total net primary production ($836\,\mathrm{Tmol\,C\,a^{-1}}$; Duarte and Cebrían 1996) or gross primary production ($789\,\mathrm{Tmol\,C\,a^{-1}}$; Gattuso et al. 1998), and riverine carbon input ($34\,\mathrm{Tmol\,C\,a^{-1}}$) with burial (about $13\,\mathrm{Tmol\,C\,a^{-1}}$)

and respiration and export of organic matter from the coastal to open ocean. However, estimates of carbon transfer from the coastal to the open ocean show a large range: 40 Tmol C a^{-1} (Rabouille *et al.* 2001), 183 Tmol C a^{-1} (Wollast 1998), 225 Tmol C a^{-1} (Gattuso *et al.* 1998) and 500 Tmol C a^{-1} (del Giorgio and Duarte 2002) precluding the derivation of respiration rates by difference.

11.5 Conclusions

Respiration in benthic coastal ecosystems constitutes a major part of total coastal ocean respiration. Our estimate based on the sum of individual ecosystem contributions (621 Tmol C a^{-1}) is well above that based on published global coastal carbon budgets (~150 Tmol C a^{-1}), but similar to estimates of coastal respiration by heterotrophs based on the balance between net primary production and river inputs on the one hand, and sediment burial on the other (~620 Tmol C a^{-1}). Our estimate of global benthic respiration is highly uncertain because of the diversity of benthic communities, aggregation problems necessary to upscaling, limited availability of data, and large variability within and among ecosystems. We anticipate that improvement of georeferenced datasets on coastal habitats (typology) and other products of the Land–Ocean Interaction in the Coastal Zone program will ultimately result in more accurate estimates on this significant and changing term in the global oceanic respiration budget.

Human activities have already impacted and are increasingly influencing coastal benthic communities. The effects of human impacts on coastal benthic respiration are multiple and involve changes in the magnitude of respiration of existing communities as well as changes in the community composition and ecosystem functioning. Accelerated sea level rise will result in changing distribution patterns of littoral and shallow benthic communities such as salt marshes, mangroves, sea grasses, and coral reefs. Eutrophication of coastal systems due to excess nutrient inputs may not only result in changes in the magnitude of primary production, hence benthic respiration, but it may also induce alterations in the community composition of dominant primary producers. For instance, sea grass communities may be replaced by macroalgae or phytoplankton. Eutrophication also affects macrobenthos community composition and biomass with consequences for respiration pathways. Land use change and related alterations in riverine organic matter delivery to the coastal zone will have a direct impact on respiration by heterotrophs and indirectly as well on autotrophs. Rising temperatures will result in enhanced rates of respiration by autotrophs, in particular marine macrophytes, and their contribution to respiration in the coastal ocean will likely increase. Rising temperature will, at least temporarily, result in enhanced rates of respiration by heterotrophs, but the effect on the longer term is difficult to predict because respiration by heterotrophs is ultimately governed by organic matter supply (i.e. net production). The combined effect of altered community specific respiration rates and changes in benthic community distribution patterns is difficult to predict. Before we can assess these changes, let alone predict the effects of human impact, we need more knowledge on the factors governing respiration at the community level.

Acknowledgments

This research was partly supported by the EU (EVK3-CT-2000-00040; EVK3-CT-2002-00076), the Spanish Plan Nacional de I+D (REN-2000-1471) and by a PIONIER grant (833.02.002) from the Netherlands Organisation of Scientific Research. We thank our co-workers for use of their unpublished results and the editors Paul del Giorgio and Peter Williams for constructive remarks. This is publication 3185 of the Netherlands Institute of Ecology.

References

Abril, G. W., Riou, S. A., Etcheber, H., Frankignoulle, M., De Wit, R., and Middelburg, J. J. 2000. Transient nitrogen transformations in an estuarine turbidity maximum—fluid mud system (the Gironde, S. W. France). *Estuar. Coast. Shelf Sci.*, **50**: 703–715.

Adey, W. H. and Steneck, R. S. 1985. Highly productive Eastern Caribbean reefs: Synergistic effects of biological, chemical, physical, and geological factors. *NOAA Symp. Undersea Res.*, **3**: 163–188.

Agawin, N. S. R. and Duarte, C. M. 2002. Evidence of direct particle trapping by a tropical sea grass meadow. *Estuaries*, **25**: 1205–1209.

Aller, R. C. 1998. Mobile deltaic and continental shelf muds as suboxic, fluidized bed reactors. *Mar. Chem.*, **61**: 143–155.

Aller, R. C. and Aller, J. Y. 1998. The effect of biogenic irrigation intensity and solute exchange on diagenetic reaction rates in marine sediments. *J. Mar. Res.*, **56**: 905–936.

Alongi, D. M. 1998. *Coastal Ecosystem Processes*. Boca Raton FL, CRC, 419 pp.

Atkinson, M. J. and Smith, S. V. 1983. C:N:P ratios of benthic marine plants. *Limnol. Oceanogr.*, **28**: 568–574.

Barrón, C., Marbá, N., Duarte, C. M., Pedersen, M. F., Lindblad, C., Kersting, K., Moy, F., and Bokn, T. 2003. High organic carbon export precludes eutrophication responses in experimental rocky shore communities. *Ecosystems*, **6**: 144–153.

Boudreau, B. P., Huettel, M., Forster, S., Jahnke R. A., McLachlan, A., Middelburg, J. J., Nielsen, P., Sansone, F., Taghon, G., van Raaphorst, W., Webster, I., Weslawski, J. M., Wiberg, P., and Sundby, B. 2001. Permeable marine sediments: overturning an old paradigm. *EOS,* **82**: 133–136.

Cahoon, L. B. 1999. The role of benthic microalgae in neritic ecosystems. *Oceanogr. Mar. Biol. Ann. Rev.*, **37**: 47–86.

Cai, W.-J., Pomeroy, L. R., Moran, M. A., and Wang, Y. 1999. Oxygen and carbon dioxide mass balance for the estuarine-intertidal marsh complex of five rivers in the southeastern U. S. *Limnol. Oceanogr.*, **44**: 639–649.

Cebrián, J. and Duarte, C. M. 1994. The dependency of herbivory on growth rate in natural plant communities. Functional Ecology **8**: 518–525.

Charpy-Roubaud, C. and Sournia, A. 1990. The comparative estimation of phytoplanktonic microphytobenthic production in the oceans. *Mar. Micr. Food Webs*, **4**: 31–57.

Cheshire, A. C., Westphalen, G., Wenden, A., Scriven, L. J., and Rowland, B. C. 1996. Photosynthesis and respiration of phaeophycean-dominated macroalgal communities in summer and winter. *Aquat. Bot.*, **55**: 159–70.

Connor, J. L. and Adey, W. H. 1977. The benthic algal composition, standing crop, and productivity of a Caribbean algal ridge. *Atoll Res. Bull.*, **211**: 1–40.

Dauwe, B., Middelburg, J. J., and Herman, P. M. J. 2001. The effect of oxygen on the degradability of organic matter in subtidal and intertidal sediments of the North Sea area. *Mar. Ecol. Prog. Ser.*, **215**: 13–22.

del Giorgio, P. A. and Duarte, C. M. 2002. Total respiration and the organic carbon balance of the open ocean. *Nature*, **420**: 379–384.

Dring, M. J. 1992. *The Biology of Marine Plants*. Cambridge University Press, Cambridge.

Duarte, C. M. and Agustí, S. 1998. The CO_2 balance of unproductive aquatic ecosystems. *Science*, **282**: 234–236.

Duarte, C. M. and Cebrián, J. 1996. The fate of marine autotrophic production. *Limnol. Oceanogr.*, **41**: 1758–1766.

Duarte, C. M. and Chiscano, C. L. 1999. Sea grass biomass and production: a reassessment. *Aquat. Bot.*, **65**: 159–174.

Duarte, C. M. Benavent, E., and Sánchez, M. C. 1999. The microcosm of particles within seagrass (*Posidonia oceanica*) canopies. *Mar. Ecol. Prog. Ser.*, **181**: 289–295.

Duarte, C. M., Martínez, R., and Barrón, C. 2002. Biomass, production and rhizome growth near the northern limit of sea grass (*Zostera marina* L.) distribution. *Aquat. Bot.*, **72**: 183–189.

Enríquez, S., Duarte, C. M., and Sand-Jensen, K. 1995. Patterns in the photosynthetic metabolism of Mediterranean macrophytes. *Mar. Ecol. Prog. Ser.*, **119**: 243–252.

Enríquez, S., Nielsen, S. L., Duarte, C. M., and Sand-Jensen, K. 1996. Broad-scale comparison of photosynthetic rates across phototrophic organisms. *Oecologia (Berlin)*, **108**: 197–206.

Epping, E. H. G. and Jørgensen, B. B. 1996. Light-enhanced oxygen respiration in benthic phototrophic communities. *Mar. Ecol. Prog. Ser.*, **139**: 193–203.

Erftemeijer, P. L. A., Osinga, R., and Mars, A. E. 1993. Primary production of sea grass beds in South Sulawesi (Indonesia): A comparison of habitats, methods and species. *Aquat. Bot.*, **46**: 67–90.

Flanagan, L. B. and Ehleringer, A. R. 1998. Ecosystem-atmosphere CO_2 exchange: interpreting signals of change using stable isotope ratios. *Trends Ecol. Evol.*, **13**: 10–14.

Gacia, E., Duarte, C. M., and Middelburg, J. J. 2002. Carbon and nutrient deposition in a Mediterranean sea grass (Posidonia oceanica) meadow. *Limnol. Oceanogr.*, **47**: 23–32.

Gattuso, J.-P., Paai, C. E., Pichon, M., Delesalle, B., and Frankignoulle, M. 1997. Primary production, calcification and air-sea CO_2 fluxes of a macroalgal-dominated coral reef community (Moorea, French Polynesia). *J. Phycol.*, **33**: 729–738.

Gattuso, J.-P., Frankignoulle, M., and Wollast, R. 1998. Carbon and carbonate metabolism in coastal aquatic ecosystems. *Annu. Rev. Ecol. Syst.*, **29**: 405–434.

Gattuso, J.-P., Frankignoulle, M., and Smith, S. V. 1999. Measurement of community metabolism and significance of coral reefs in the CO_2 source-sink debate. *Proc. Nat. Acad. Sci. USA*, **96**: 13017–13022.

Glud, R. N., Gundersen, J. K., Røy, H., and Jørgensen, B. B. 2003. Seasonal dynamics of benthic O_2 uptake in a semi enclosed bay: importance of diffusion and fauna activity. *Limnol. Oceanogr.*, **48**: 1265–1276.

Griffith, P. C., Cubit, J. D., Adey, W. H., and Norris, J. N. 1987. Computer-automated flow respirometry: metabolism measurements on a Caribbean reef flat and in a microcosm. *Limnol. Oceanogr.*, **32**: 442–451.

Gribsholt, B. and Kristensen, E. 2002. Impact of sampling methods on sulfate reduction rates and dissolved organic carbon (DOC) concentrations in vegetated salt marsh sediments. *Wetlands Ecol. Manage.*, **10**: 371–379.

Hawkins, C. M. and Lewis, J. B. 1982. Benthic primary production on a fringing coral reef in Barbados, West Indies. *Aquat. Bot.*, **12**: 355–364.

Hedges, J. I. and Keil, R. G. 1995. Sedimentary organic matter preservation: an assessment and speculative synthesis. *Mar. Chem.*, **49**: 81–116.

Hedges, J. I., Baldock, J. A., Gélinas, Y., Lee, C., Peterson, M. L., and Wakeham, S. G. 2002. The biochemical and elemental compositions of marine plankton: a NMR perspective. *Mar. Chem.*, **78**: 47–63.

Heip, C. H. R., Goosen, N. K., Herman, P. M. J., Kromkamp, J., Middelburg, J. J., and Soetaert, K. 1995. Production and consumption of biological particles in temperate tidal estuaries. *Oceanogr. Mar. Biol. Ann. Reviews*, **33**: 1–150.

Hemminga, M. A., and Duarte, C. M. 2000. *Sea grass Ecology*. Cambridge University Press, Cambridge.

Herman, P. M. J., Middelburg, J. J., van de Koppel, J., and Heip, C. H. R. 1999. Ecology of estuarine macrobenthos. *Advances Ecol. Res.*, **29**: 195–240.

Johnson, P. and Johnstone, R. 1995. Productivity and nutrient dynamics of tropical sea-grass communities in Puttalam Lagoon, Sri Lanka. *Ambio*, **24**: 411–417.

Jørgensen, B. B. 1983. Processes at the sediment-water interface. In B. Bolin, and R. B. Cook (eds) *The Major Biogeochemical Cycles and Their Interactions*. SCOPE, pp. 477–515.

Jørgensen, B. B. 2000. Bacteria and marine biogeochemistry. In H. D. Shulz, and M. Zabel (eds) *Marine Geochemistry*. Springer-Verlag, Berlin, pp. 173–207.

Kinsey, D. W. 1979. Carbon Turnover and Accumulation by Coral Reefs. PhD Thesis, University of Hawaii.

Kinsey, D. W. 1985. Metabolism, calcification and carbon production. I. System level studies. *Proceedings of the 5th International Coral Reef Congress*, **4**: 505–526.

Kühl, M., Cohen, Y., Dalsgaard, T., Jørgensen, B. B., and Revsbech, N. P. 1995. Microenvironment and photosynthesis of zooxanthellae in scleractinian corals studied with microsensors for O_2, pH and light. *Mar. Ecol. Prog. Ser.*, **117**: 159–172.

Langdon, C., Aceves, H., Barnett, H., Takahashi, T., Chipman, D., Goddard, J., Sweeney, C., and Atkinson, M. 2003. Effect of elevated CO_2 on the community metabolism of an experimental coral reef. *Glob. Biogeochem. Cycl.*, **17**: DOI 10. 1029/2002GB001941.

Leclercq, N., Gattuso, J.-P., and Jaubert, J. 1999. Measurement of primary production and respiration in open-top mesocosms: application to a coral reef mesocosm. *Mar. Ecol. Prog. Ser.*, **177**: 299–304.

Lindeboom, H. J. and de Bree, B. H. H. 1982. Daily production and consumption in an eelgrass (*Zostera marina*) community in saline Lake Grevlingen: discrepancies between the O_2 and ^{14}C method. *Neth. J. Sea Res.* **16**: 362–379.

Lindeboom, H. J. and Sandee, A. J. J. 1989. Production and consumption of tropical sea grass fields in eastern Indonesia measured with bell jars and microelectrodes. *Neth. J. Sea Res.*, **23**: 181–190.

Loebb, C. S. and Harrison, P. J. 1997. *Seaweed Ecology Physiology*. Cambridge University Press, Cambridge.

Mann, K. H., 2000. *Ecology of Coastal Waters*. Blackwell, Malden.

Marsh, J. A. Jr., Dennison, W. C., and Alberte, R. A. 1986. Effects of temperature on photosynthesis and respiration in eelgrass (*Zostera marina* L.). *J. Exp. Mar. Biol. Ecol.* **101**: 257–267.

Mateo, M. A., Romero, J., Pérez, M., Littler, M. M., and D. S. Littler. 1997. Dynamics of millenary organic deposits resulting from the growth of the Mediterranean sea grass *Posidonia oceanica*. *Estuar. Coast. Shelf Sci.*, **14**: 103–110.

Meybeck, M. 1982. Carbon, nitrogen, and phosphorus transport by world rivers. *Am. J. Sci.*, **282**: 401–450.

Middelburg, J. J., Klaver, G., Nieuwenhuize, J., Wielemaker, A., de Haas, W., and van der Nat, J. F. W. A. 1996. Organic matter mineralization in intertidal sediments along an estuarine gradient. *Mar. Ecol. Prog. Ser.* **132**: 157–168.

Middelburg, J. J., Nieuwenhuize, J., Lubberts, R. K., and van de Plassche, O. 1997. Organic carbon isotope systematics of coastal marshes. *Est. Coast. Shelf. Sci.*, **45**: 681–687.

Murray, L. and Wetzel, R. L. 1987. Oxygen production and consumption associated with major autotrophic components in two temperate sea grass communities. *Mar. Ecol. Prog. Ser.*, **38**: 231–239.

Newell, R. C. and Field, J. G. 1983. The contribution of bacteria and detritus to carbon and nitrogen flow in a benthic community. *Mar. Biol. Lett.*, **4**: 23–28.

Nielsen, S. L., Enríquez, S., Duarte, C. M., and Sand-Jensen, K. 1996. Scaling of maximum growth rates across photosynthetic organisms. *Funct. Ecol.*, **10**: 167–175.

Nixon, S. W. and Oviatt, C. A. 1972. Preliminary measurements of midsummer metabolism in beds of eelgrass, *Zostera marina. Ecology*, **53**: 150–153.

Odum, H. T. 1963. Productivity measurements in Texas turtle grass and the effects of dredging an intracoastal channel. *Pub. Inst. Mar. Sci. Univ. Tex.*, **9**: 45–58.

Odum, H. T. and Hoskins, C. M. 1958. Comparative studies of the metabolism of marine waters. *Pub. Inst. Mar. Sci. Univ. Tex.*, **5**: 16–46.

Odum, H. T. and Wilson, R. F. 1962. Further studies on reareation and metabolism of Texas Bays, 1958–1960. *Publ. Inst. Mar. Sci. Tex.*, **8**: 23–55.

Odum, H. T., Burkholder, P. R., and Rivero, J. 1959. Measurements of productivity of turtle grass flats, reefs, and the bahia Fosforescente of southern Puerto Rico. *Pub. Inst. Mar. Sci. Univ. Tex.*, **6**: 159–170.

Pedersen, A. 1987. Community metabolism on rocky-shore assemblages in a mesocosm: A. Fluctuations in production, respiration, chlorophyll *a* content and C:N ratios of grazed and non-grazed assemblages. *Hydrobiologia*, **151/152**: 267–275.

Pedersen, M. F., Duarte, C. M., and Cebrián, J. 1997. Rate of changes in organic matter and nutrient stocks during sea grass (*Cymodocea nodosa*) colonization and stand development. *Mar. Ecol. Prog. Ser.* **159**: 29–36.

Rabouille, C., Mackenzie, F. T., and Ver, L. M. 2001. Influence of the human perturbation on carbon, nitrogen, and oxygen biogeochemical cycles in the global coastal ocean. *Geochim. Cosmochim. Acta*, **65**: 3615–3641.

Reyes, E. and Merino, M. 1991. Diel dissolved oxygen dynamics and eutrophication in a shallow well mixed tropical lagoon (Cancún, México). *Estuaries*, **14**: 372–381.

Rogers, C. S. and Salesky, N. H. 1981. Productivity of *Acropora palmata* (Lamarck), macroscopic algae and algal turf from Tague Bay reef, St. Croix, U. S. Virgin Island. *J. Exp. Mar. Biol. Ecol.*, **49**: 179–187.

Smith, S. V. 1973. Carbon dioxide dynamics: a record of organic carbon production, respiration and calcification in the Eniwetok reef flat community. *Limnol. Oceanogr.*, **18**: 106–120.

Smith, S. V. 1991. Stoichiometry of C: N: P fluxes in shallow-water marine ecosystems. In J. Cole, G. Lovett, and S. Findlay, (eds) *Analyses of Ecosytems. Patterns, Mechanisms and Theory*. Springer-Verlag, pp. 259–286.

Smith, S. V. and Marsh, J. A. Jr. 1973. Organic carbon production on the windward reef flat of Eniwetok atoll. *Limnol. Oceanogr.*, **18**: 953–961.

Smith, S. V., and Hollibaugh, J. T. 1993. Coastal metabolism and the oceanic organic carbon balance. *Rev. Geophys.*, **31**: 75–89.

Soetaert, K., Herman, P. M. J., and Middelburg, J. J. 1996. A model of early diagenetic processes from the shelf to abyssal depths. *Geochim. Cosmochim. Acta*, **60**: 1019–1040.

Soetaert, K., Middelburg, J. J., Herman, P. M. J., and Buis, K. 2000. On the coupling of benthic and pelagic biogeochemical models. *Earth Sci. Rev.*, **51**: 173–201.

Spalding, M. D., and Grenfell, A. M. 1997. New estimates of global and regional coral reef areas. *Coral Reefs*, **16**: 225–230.

Thamdrup, B., and Canfield, D. E. 2000. Benthic respiration in aquatic sediments. In O. E. Sala, R. B. Jackson, H. A. Mooney, and R. W. Howarth (eds) *Methods in Ecosystem Science*. Springer, New York, pp. 86–103.

Touchette, B. W., and Burkholder, J. A. 2000. Overview of the physiological ecology of carbon metabolism in sea grasses. *J. Exp. Mar. Biol. Ecol.*, **250**: 169–205.

Vermaat, J. E., Agawin, N. S. R., Fortes, M. D., Uri, J. S., Duarte, C. M., Marbá, N., Enríquez, S., and van Vierssen, W. 1997. The capacity of sea grasses to survive increased turbidity and siltation: The significance of growth form and light use. *AMBIO*, **26**: 499–504.

Vooren, C. M. 1981. Photosynthetic rates of benthic algae from the deep coral reef of Curacao. *Aquat. Bot.*, **10**: 143–161.

Wollast, R. 1998. Evaluation and comparison of the global carbon cycle in the coastal zone and in the open ocean. In K. H. Brink, and A. R. Robinson (eds) *The sea*. Wiley and Sons, New York, pp. 213–252.

Zehr, J. P., and Ward, B. B. 2002. Nitrogen cycling in the ocean: new perspectives on processes and paradigms. *Appl. Environ. Micr.*, **68**: 1015–1024.

Ziegler, S., and Benner, R. 1998. Ecosystem metabolism in a subtropical, sea grass-dominated lagoon. *Mar. Ecol. Prog. Ser.*, **173**: 1–12.

CHAPTER 12

Suboxic respiration in the oceanic water column

Louis A. Codispoti,[1] Tadashi Yoshinari[2] and Allan H. Devol[3]

[1] *University of Maryland Center for Environmental Science, USA*
[2] *Wadsworth Center, New York State Department of Health, USA*
[3] *School of Oceanography, University of Washington, USA*

Outline

Upon oxygen depletion, a suite of alternate oxidants supports microbial respiration and inhibits the onset of sulfate (SO_4^{2-}) reduction. Waters in this intermediate state are referred to as "oxygen deficient" or "suboxic." This chapter discusses types of suboxic water column respiration, their occurrence, variability, and significance. Methodological problems are also discussed. Nitrate (NO_3^-) is the most abundant suboxic electron acceptor in oceanic water, but the suite of alternate oxidants includes nitrite (NO_2^-), nitric oxide (NO), nitrous oxide (N_2O), iodate (IO_3^-), manganese (Mn III and IV), iron (Fe III), and several other oxidants present at low concentrations. Although canonical denitrification, which involves reduction of nitrate to N_2O and N_2 is probably the single most important suboxic respiratory pathway in the water column, important additional respiratory pathways for dinitrogen (N_2) and N_2O production are also considered. Canonical estimates of water column conversion of fixed-N to dinitrogen and of the nitrous oxide flux from the ocean to the atmosphere are too low, because of underappreciated pathways, incorrect stoichiometries, and undersampling. Suboxic water column respiration is sensitive to change and it impacts global budgets for fixed-N and some trace elements even though suboxic waters comprise only $\sim$0.1–0.2% of the oceanic volume. More effort needs to be directed at including this suite of processes in our attempts to understand global change. We also need to improve the available methodologies.

12.1 Introduction

12.1.1 Definition of suboxia

When dissolved oxygen concentrations become vanishingly small and before the onset of sulfate (SO_4^{2-}) reduction, there exists a "suboxic" or "oxygen deficient" (Richards 1965) condition in which respiration is supported by oxidants that provide more energy than SO_4^{2-} respiration. The suite of demonstrated and potential alternate electron acceptors includes oxidized forms of inorganic N (NO_3^-, NO_2^-, NO, N_2O), Mn (III and IV), IO_3^-, Fe (III), molybdenum (Mo VI), uranium (U VI), chromium (Cr VI),

and several other trace constituents. Oxidized inorganic N (primarily nitrate) is, by far, the most abundant suboxic electron acceptor in the oceanic water column (Fig. 12.1; Farrenkopf *et al.* 1997*a,b*; Measures and Vink 1999; Lewis and Luther 2000; Witter *et al.* 2000).

Due primarily to analytical difficulties, working definitions of suboxic conditions vary with upper limit dissolved oxygen concentrations ranging from $\sim$2 to 10 μmol dm^{-3} (e.g. Codispoti *et al.* 1991, 2001; Murray *et al.* 1995). The best available data suggest that suboxic respiration does not become prominent until oxygen concentrations fall below $\sim$2–4 μmol dm^{-3}. Similarly, sulfide concentrations

$$\frac{\mu mol}{dm^3} \rightarrow \frac{nmol}{dL}$$

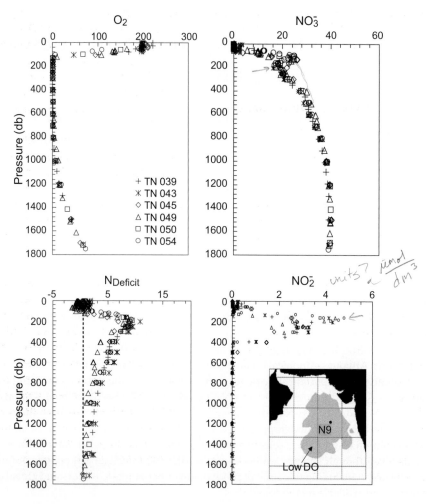

Figure 12.1 Data from station N9 located in the portion of the Arabian Sea that contains suboxic water at depths between ~100 and 1000 db (1 db ~1m). Cruises, TN039, 43, 45, 49, 50, and 54 comprise the US Joint Global Ocean Flux Process Study of the Arabian Sea and cover a complete annual cycle beginning in September 1994 and ending in December 1995. A nitrate (NO_3^-) minimum and nitrite (NO_2^-) maximum within the suboxic zone ($O_2 < 5$ μmol dm^{-3}) signals the presence of enhanced denitrification. The parameter, $N_{deficit}$ estimates the conversion of nitrate to N_2 (see equation 11). The shaded area in the inset indicates the extent of the regions where, on average, the maximum concentration of nitrite (NO_2^-) in the oxygen minimum zone (the secondary nitrite maximum) exceeds 1 μmol dm^{-3} as described by Naqvi (1994). All concentrations are in μ mol dm^{-3}. From (Codispoti *et al.* 2001). These data are available at http://usjgofs.whoi.edu/arabianobjects.html.

in suboxic waters are likely to be lower than some quoted values because of rapid oxidation by suboxic electron acceptors and by the oxygen flux from adjacent aerated zones (Murray *et al.* 1995).

12.1.2 Additional terminology

For convenience, we sometimes use MnO_2 and MnO to represent Mn (IV) and Mn (II), although other forms are more likely abundant in seawater. We use HS^- or "sulfide" to denote $H_2S + HS^- + S^{2-}$. Similarly, we sometimes use NH_3, HNO_2, and HNO_3 to represent ammonium, nitrite, and nitrate, respectively even though the ionized forms (NH_4^+, NO_2^-, NO_3^-) are more abundant in ocean water. Fixed nitrogen or fixed-N denotes all forms of nitrogen other than elemental nitrogen (mainly dinitrogen on our planet). Also note that the reactive phosphorus determined by routine nutrient analyses is sometimes called "phosphate" and represented by PO_4^{3-} and that for the data presented here a decibar (db) is $\sim$0.99 m.

12.2 Suboxic zones and water masses

12.2.1 Occurrence and volume

During the early stages of life on earth, anoxic respiration and suboxic respiration supported by Fe (III) may have been dominant (Lovley 1991, 1993), but this is not the case in the modern ocean's water column where oxygen-supported respiration dominates. Waters with oxygen concentrations $<\sim$5 μmol dm^{-3} occupy only $\sim$0.1–0.2% of the total volume, although the exact amount is difficult to quantify because of the poor quality of historical oxygen data, temporal variability, and undersampling. The largest suboxic water masses are embedded in oxygen minimum zones (OMZs) and occur within the $\sim$100–1000 m depth range on a quasi-permanent basis north and south of the equator in the eastern tropical Pacific Ocean and in the northern Arabian Sea (e.g. Codispoti and Packard 1980; Naqvi 1994; Codispoti et al. 2001; Deutsch et al. 2001; Fig. 12.1). A relatively small quasi-permanent suboxic zone may also exist off SW Africa (Calvert and Price 1971). Despite the low oxygen concentrations, turnover times for these waters can be as short as several years (Codispoti and Packard 1980; Naqvi 1994; Howell et al. 1997). Transient suboxia occurs in the vicinity of the quasi-permanent zones (e.g. Codispoti and Packard 1980; Codispoti et al. 1986). Seasonal or transient suboxic zones also occur in estuaries such as Chesapeake Bay (Hobbie 2000), and over continental shelves such as the western Indian Shelf (Naqvi et al. 2000). Estuarine and shelf occurrences may be on the increase due to anthropogenic influences (e.g. Hobbie 2000; Rabalais et al. 2000; Codispoti et al. 2001). Although relatively small volumes are involved, respiration rates in these shallow coastal waters can be high compared to most of the open ocean. Small volumes of suboxic water are also sandwiched between oxic and anoxic layers that occur in basins such as the Black Sea (e.g. Murray et al. 1995), the Cariaco

Trench (Richards 1975; Scranton et al. 1987), and in anoxic fjords (Anderson and Devol 1987). While the global significance of these "sandwich layers" may be small, they have proven useful as natural laboratories for unravelling the complex of respiratory processes that occur within the oxic/suboxic/anoxic transition zone (Nealson and Safferini 1994; Murray et al. 1995).

Environments that may include suboxic zones, but which have received little study include microenvironments within particles/aggregates. Field data, for example, sometimes display suboxic signals in particle rich waters with low (but not suboxic) oxygen levels (Alldredge and Cohen 1987; Wolgast et al. 1998). Laboratory and modeling studies (Kaplan and Wofsy 1985; Alldredge and Cohen 1987) also suggest that suboxia is possible in settling or suspended aggregates, their existence being dependent on factors such as ambient dissolved oxygen concentration, aggregate size, and sinking rates. Suboxic environments are also likely in the vicinity of deep-sea thermal vents and cold seeps and in brines (Van Cappellen et al. 1998) but have received little attention vis à vis suboxic respiration. Such environments could be important: for example, the entire volume of the oceans may be processed through deep-sea hydrothermal plumes every few thousand years (German 2002).

12.2.2 Chemical features of suboxic water masses

In addition to vanishingly small oxygen concentrations, chemical features of water column suboxia typically include nitrite maxima and nitrate minima (Fig. 12.1). Suspended particle, iodide (I^-), and Mn maxima may also be common (Pak et al. 1980; Garfield et al. 1983; Naqvi 1994; Farrenkopf et al. 1997a, b; Lewis and Luther 2000). Extreme nitrous oxide (N_2O) gradients are also typical with extremely low concentrations found in the "cores" of suboxic zones and extremely high concentrations occurring at the boundaries (Fig. 12.2).

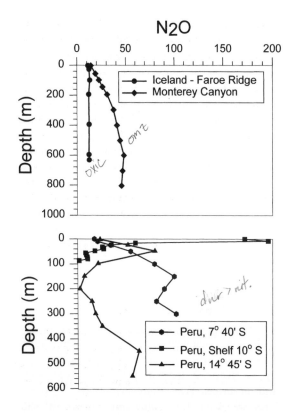

dnr-nit coupled?
denitrification - dnr

nitrification - nit

Figure 12.2 Nitrous oxide (N₂O) concentrations versus depth from different oceanic regions. The North Atlantic data (circles, upper panel) are from a well-oxygenated water column (Hahn 1981). The Monterey Bay region (diamonds, upper panel) is productive and has a well-developed OMZ (minimum oxygen ~10 μmol dm⁻³), but the waters do not become suboxic. Pierotti and Rasmussen's (1980) data (hexagons, lower panel, Peru, 7° 40′S) show high values in low oxygen waters north of the quasi-permanent suboxic zone off Peru. The triangles in the lower panel represent a vertical profile from the quasi-permanent suboxic zone off Peru showing high N₂O values above and below the suboxic layer and extremely low values in its core. These values were taken during the NITROP Expedition (Codispoti *et al.*, 1986; Ward *et al.* 1989; Lipschultz *et al.* 1990). The squares in the lower panel are from a shelf station off Peru taken during the upwelling of suboxic and low oxygen waters (also from the NITROP Expedition). Concentrations are as nmol dm⁻³.

12.2.3 Concentrations of suboxic electron acceptors

Nitrate (NO₃⁻) with an average oceanic concentration of ~31 μmol dm⁻³ is the most abundant suboxic respiratory electron acceptor in the oceanic water column (Fig. 12.1). Nitrite (NO₂⁻) concentrations in the oceanic water column are generally less than 1 μmol dm⁻³ but much higher concentrations occur in the "secondary nitrite maxima" that typify suboxic water columns (Fig. 12.1). Concentrations more than 10 μmol dm⁻³ are fairly common off Peru with a maximum observed value of 23 μmol dm⁻³ (Codispoti *et al.* 1986). Nitric oxide (NO) is essentially absent from well-oxygenated water (O₂ more than 100 μmol dm⁻³), but concentrations approaching 0.5 nmol dm⁻³ occur in low oxygen waters (Ward and Zafiriou 1988). N₂O concentrations as high as ~0.5 μmol dm⁻³ have been observed in association with transient suboxia over the Indian Shelf (Naqvi *et al.* 2000) and have

been attributed to "stop and go" denitrification (see Codispoti *et al.* 2001), and large gradients occur in regions influenced by suboxic waters. The data in Fig. 12.2 support the hypothesis that N₂O turnover is rapid in and at the boundaries of suboxic regions (Codispoti and Christensen 1985) and that the highest and lowest oceanic N₂O concentration often occur in close proximity in association with suboxic waters. Most oceanic N₂O concentrations would fall between the lines defining the North Atlantic and Monterey Bay data in the upper panel, but we see a much larger range of concentrations in the data taken in and near suboxic waters. Typical IO₃⁻ concentrations in oxygenated seawater are ~0.4–0.5 μmol dm⁻³. In suboxic waters, most may be reduced to I⁻ (Farrenkopf *et al.* 1997a,b). Iron and manganese are present in only nanomolar (nmol dm⁻³) concentrations in open-ocean waters with maximum manganese concentrations in suboxic Arabian Sea waters of ~10 nmol dm⁻³, and iron concentrations even lower (Lewis and Luther 2000; Witter *et al.* 2000). Manganese and iron are more abundant in sediments, and within and near suboxic/anoxic interfaces with several hundred $nmol\,dm^{-3}$ concentrations of dissolved iron and more than $10\,\mu mol\,dm^{-3}$ dissolved manganese found near the suboxic/anoxic boundary in the Black Sea (e.g. Murray *et al.* 1995). Much higher manganese and iron concentrations are found in the oxic/anoxic transition region in Orca Basin brines (Van Cappellen *et al.* 1998). Of the other potential and demonstrated suboxic respiratory electron acceptors, molybdenum (Mo (VI)) is most abundant with a total Mo concentration of $\sim$100 $nmol\,dm^{-3}$. Others include uranium (U (VI)) with a concentration of $\sim$10 $nmol\,dm^{-3}$, and chromium (Cr (VI)) with a total Cr concentration of $\sim$5 $nmol\,dm^{-3}$. There are several other possibilities present at even lower concentrations (Lovley 1993; Anbar and Knoll 2002).

12.3 Sequence and types of suboxic respiration

12.3.1 Sequence

Although, there is some variance due to the conditions chosen, most calculations (e.g. Froelich *et al.* 1979; Murray *et al.* 1995; Hulth *et al.* 1999; King, Chapter 2) suggest that respiration employing "Redfieldian" (Redfield *et al.* 1963; Anderson 1995) organic matter as the electron donor and O_2, NO_3^-, and Mn (IV) as electron acceptors yield similar amounts of free energy. Fe-supported respiration yields significantly less, and is usually followed by respiration supported by sulfate (SO_4^{2-}). For example, Froelich *et al.* (1979) suggest that the yield (kJ/mole of glucose) is -3190 for oxygen supported respiration, -3030 for nitrate, -3090 to -2920 for manganese (depending on the form of manganese), -1410 to -1330 for iron (depending on the form of Fe III), and -380 for sulfate (SO_4^{2-}). Farrenkopf *et al.* (1997*b*) suggest that IO_3^- supported oxidation is energetically similar to oxidation with nitrate. These calculations are generally in concert with observations that suggest co-occurrence of I^- maxima, nitrate minima, and nitrite maxima in the

Arabian Sea (Farrenkopf *et al.* 1997*a*,*b*), and with nitrate, manganese, iron, and HS^- gradients in and at the boundaries of the suboxic zone in the Black Sea where we see the following sequence with depth: oxygen exhaustion, nitrate exhaustion, increase in Mn (II), increase in Fe (II), increase in HS^- (Nealson and Saffarini 1994; Codispoti *et al.* 2001). Given the high proportion of IO_3^- reduced to I^- in the Arabian Sea, we suggest that under oceanic conditions one can speculate that the energy yield sequence is: O_2 > IO_3^- > NO_3^- > Mn (IV) > Fe (III) > SO_4^{2-}. Note that the microbial growth yields for denitrification appear to be significantly lower than the yields supported by oxic respiration (Koike and Hattori 1975; Stouthamer *et al.* 1982; King, Chapter 2) even though the energy yields are similar.

12.3.2 Canonical denitrification

Because of the relative abundance of nitrate in the oceanic water column, denitrification is a major suboxic respiratory process. Knowles (1996) provides the following definition of canonical denitrification.

'Some aerobic microorganisms, mainly bacteria have the ability when oxygen becomes limiting of switching over to use the nitrogen oxides, nitrate (NO_3^-), nitrite (NO_2^-), nitric oxide (NO), and nitrous oxide (N_2O) as terminal acceptors for electrons in their metabolism. This process, known as denitrification, permits organisms to continue what is essentially a form of respiration in which the end product is dinitrogen. However, intermediates sometimes accumulate.'

Traditional definitions of denitrification assume gaseous end-products (N_2O and N_2), and a distinction is generally made between dissimilatory nitrate reduction ($NO_3^- \rightarrow NO_2^-$) and denitrification (e.g. King, Chapter 2).

During canonical denitrification, the pathway for N reduction involves four distinct enzyme systems (nitrate reductase, nitrite reductase, nitric oxide reductase, and nitrous oxide reductase; Zumft and Körner 1997). Each step is associated with energy conserving oxidative phosphorylation (e.g. Ferguson 1994). The process can be summarized as:

$$NO_3^- \rightarrow NO_2^- \rightarrow NO \rightarrow N_2O \rightarrow N_2 \qquad (1)$$

Nitrite (NO_2^-), NO, and N_2O are obligatory intermediates that can escape the cell and then be reassimilated and reduced or exported. Data from the oceanic water column and from the laboratory suggest, that both NO_2^- and N_2O accumulate during the initial stages of denitrification and that there is net N_2O consumption during well-established denitrification (e.g. Körner and Zumft 1989; Codispoti *et al.* 1992; Naqvi *et al.* 2000). Codispoti and Packard (1980) suggest that in suboxic waters off Peru, there is a positive correlation between nitrite concentrations and denitrification rates. Whatever the causes, the resulting gradients suggest that well-established water column denitrification zones export nitrite and import nitrous oxide. Anderson *et al.* (1982) and Codispoti and Christensen (1985) discuss some of the implications of these exports and imports for estimating the amount of dinitrogen produced by denitrification. Because of its lability, less is known about the distribution of nitric oxide (NO). Ferguson (1994) and Zumft and Körner (1997) summarize data suggesting that NO should be present only in nanomolar concentrations during steady-state denitrification. As already noted, NO concentrations in low oxygen and suboxic waters are less than 0.5 nmol dm^{-3} with NO essentially absent from well-oxygenated ($O_2 > 100$ μmol dm^{-3}) water. NO is highly toxic, and denitrifying enzyme systems restrict its concentration to nanomolar levels (Ferguson 1994).

Denitrification can be carried out by a wide variety of bacteria in biochemically diverse groups, and by some archaea and fungi (e. g. Knowles 1996; Zumft and Körner 1997). An even larger number are capable of dissimilatory nitrate reduction to nitrite (Zumft and Cárdenas 1979) including at least one ciliated protist (Fenchel, Chapter 4). Knowles (1996) suggests that most denitrifying organisms of significance in nature, are heterotrophic bacteria such as *Pseudomonas* and *Alcaligenes* spp. having an aerobic type of electron transport system. He further suggests that denitrifying bacteria are widely distributed. In accord with these suggestions, we observe canonical denitrification in the water column whenever suboxia, oxidized inorganic N, and labile organic matter cooccur. In open ocean suboxic zones, the activity of canonical denitrifiers generally appears to be limited by the supply of labile organic matter (Lipschultz *et al.* 1990).

Although denitrification is not poisoned by O_2, laboratory experiments and high quality oceanic nitrite and oxygen data suggest that denitrification is not the dominant respiratory mode until oxygen is below 2–4 μmol dm^{-3} (Devol 1978). For example, in Fig. 12.3 we see that the bulk of the elevated NO_2^- concentrations in Arabian Sea waters from depths ≥ 100 db occur at oxygen concentrations of <2.5 μmol dm^{-3} as determined using a high-precision automated Winkler titration system that had been compared with the low concentration method of Broenkow and Cline (1969). If we take elevated nitrite as indicative of suboxia and active denitrification (see Fig. 12.1), it seems likely that the transition to suboxia occurs at oxygen concentrations much lower than suggested by some working definitions that may be influenced by the difficulty of obtaining accurate low concentration oxygen data. Although oxygen generally represses synthesis of denitrifying enzymes and inhibits the activity of preformed enzymes, some denitrifying systems are relatively insensitive to oxygen (Lloyd 1993; Ferguson 1994). Denitrification in the presence of oxygen has been observed in the laboratory on several occasions (e.g. Bell and Ferguson 1991; Yoshinari and Koike 1994). A low level of denitrification is, therefore, a possibility in well-aerated water. Ferguson (1994) suggests that aerobic denitrification might be advantageous when an organism is growing at near suboxic concentrations or in an environment rapidly switching between oxic and anoxic. Even miniscule denitrification rates in oxic waters could be globally significant given that such waters comprise more than 99.8% of the oceanic volume (Codispoti *et al.* 2001).

Stoichiometric equations assume constancy in the ratios of change of carbon, oxygen, nitrogen, and phosphorus during biological processing and are frequently employed for assessing the impact of denitrification on the fixed-N content of water masses. Using these "Redfield ratios", Richards (1965) suggested two possible stoichiometric equations for canonical denitrification:

$$(CH_2O)_{106}(NH_3)_{16}H_3PO_4 + 84.8\,HNO_3 \rightarrow 106\,CO_2$$

$$+ 42.4\,N_2 + 148.4\,H_2O + 16\,NH_3 + H_3PO_4 \qquad (2)$$

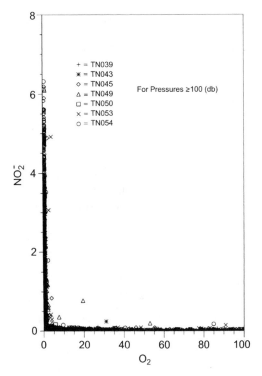

Figure 12.3 Nitrite (NO_2^-) versus dissolved oxygen (O_2) concentrations from depths greater than or equal to 100 db and below in the Arabian Sea. Concentrations are in micromolar and the stations are from all cruises mentioned in Fig. 12.1. This figure is from Morrison *et al.* (1999). Concentrations are as μmol dm^{-3}.

and

$$(CH_2O)_{106}(NH_3)_{16}H_3PO_4 + 94.4HNO_3 \rightarrow 106\,CO_2$$
$$+ 55.2\,N_2 + 177.2H_2O + H_3PO_4 \qquad (3)$$

Equation (3) yields more free energy than equation (2) (e.g. Luther *et al.* 1997), and an accumulation of evidence suggests that organic N is oxidized to dinitrogen during denitrification (e.g. Luther *et al.* 1997; Codispoti *et al.* 2001). Changes in the suggested composition of "Redfieldian" organic matter led Gruber and Sarmiento (1997) to suggest the following stoichiometric equation for canonical denitrification:

$$C_{106}H_{175}O_{42}N_{16}P + 104NO_3^- \rightarrow 4CO_2 + 102HCO_3^-$$
$$+ 60N_2 + 36H_2O + HPO_4^{2-} \qquad (4)$$

VanMooy *et al.* (2002) suggest that suboxic heterotrophs preferentially attack amino acids with C/N ratios of ~ 4 (by atoms) versus the ratio of ~ 6.6 for "Redfieldian" organic matter. They assumed that the organic substrate consisted of an ideal oceanic protein (Anderson 1995) and suggested the following stoichiometric equation for denitrification:

$$C_{61}H_{97}O_{20}N_{16} + 60.2NO_3^- \rightarrow 0.8CO_2 + 60.2HCO_3^-$$
$$+ 38.1N_2 + 18.4H_2O \qquad (5)$$

An important facet of the changes in stoichiometric equations is an increasing acceptance of the idea that a considerable amount of organic N is oxidized to dinitrogen during denitrification. Most estimates of oceanic denitrification are based mainly on nitrate "disappearance" (nitrate deficits) and therefore underestimate dinitrogen production (Codispoti *et al.* 2001). Equation (5) suggests that the total dinitrogen produced during denitrification exceeds that produced from nitrate reduction by $\sim 27\%$, but even this excess does not account for recent estimates of the excess dinitrogen arising from biological processes in the Arabian Sea (Fig. 12.4; Codispoti *et al.* 2001).

12.3.3 Alternate modes of suboxic respiration that produce dinitrogen

Discrepancies between estimates of dinitrogen produced versus nitrate reduced (Fig. 12.4) suggest the importance of pathways other than canonical denitrification for producing dinitrogen. These processes could include the direct oxidation of NH_4^+ by NO_3^- or NO_2^- in suboxic waters with all three being converted to dinitrogen. Such reactions are thermodynamically permissible (e.g. Richards 1965), and several recent studies suggest that such reactions occur in nature. For example, continuous nutrient profiles from the Black Sea suggest a net diffusion of ammonium, nitrite, and nitrate into the suboxic zone (Fig. 12.5) thereby implying that all three species are consumed and presumably converted into dinitrogen (e.g. Murray *et al.* 1995). Both sulfide and dissolved oxygen are essentially zero in the suboxic zone of the Black Sea (hatched region in Fig. 12.5), suggesting that respiration is mainly

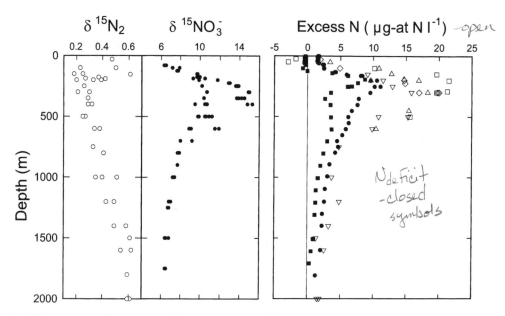

Figure 12.4 $\delta^{15}N_2$ (left panel), $\delta^{15}NO_3^-$ (middle panel), excess-N_2 (open symbols in right panel), and $N_{deficit}$ (closed symbols, right panel) versus depth in the Arabian Sea. These data were collected at different times and locations within the quasi-permanent suboxic zone, and they suggest that the burden of excess N_2 arising from all sources is about twice the nitrate deficit that accounts only for the nitrate converted to N_2 as discussed by Codispoti et al. (2001). They also show the depletion in N_2, ^{15}N, and the enrichment of ^{15}N in nitrate associated with the suboxic zone (see Fig. 12.1). The units of μg-atoms l^{-1} are used in the last panel instead of $\mu mol\,dm^{-3}$ to indicate more clearly that we are comparing atoms of N in the $N_{deficit}$ values with atoms of N in excess N_2.

supported by the suboxic electron acceptors discussed in this chapter although in these data there is one depth within the suboxic zone where dissolved oxygen concentrations were $\sim$7 $\mu mol\,dm^{-3}$. The detectable nitrite (NO_2^-) concentrations ($\sim$0.01–0.02 $\mu mol\,dm^{-3}$) in the anoxic zone are probably artifacts. The nitrite peak near the bottom of the suboxic zone was often more pronounced than in this pump profile. The shape of the reactive P (PO_4^{3-}) profile could be explained by regeneration of reactive-P below the photic zone, oxidation of Mn (II) and Fe (II) by nitrate in the suboxic zone, formation of oxidized manganese and iron rich particles that sequester reactive-P, and by dissolution of these particles in the upper portion of the anoxic zone due to reduction of manganese and iron.

A microbe that can perform the so-called "anammox" (anaerobic ammonium oxidation) reaction (equation (6)), in which nitrite is the oxidant, has been isolated thereby helping to solve the problem of a "lithotrope missing in nature" (Strous et al.

1999; Damste et al. 2002). Studies of the influence of the anammox reaction in sediments (Dalsgaard and Thamdrup 2002; Thamdrup and Dalsgaard 2002), suggest that this reaction which may be summarized as,

$$NH_4^+ + NO_2^- \rightarrow N_2 + 2H_2O \qquad (6)$$

accounted for 24% and 67% of the total dinitrogen production at two typical continental shelf sites. Because of the rapidity of nitrate reduction (presumably via dissimilatory nitrate reduction), it made little difference whether nitrate or nitrite was the initial oxidant. This reaction appeared to be less important in nearshore sediments with a higher supply of organic matter. In 2003, the first papers documenting the existence of the "anammox" reaction in suboxic waters were published (Dalsgaard et al. 2003; Devol 2003; Kuypers et al. 2003).

Luther et al. (1997) suggest a cycle in which Mn (II) can be oxidized by nitrate and Mn (IV) reduced by

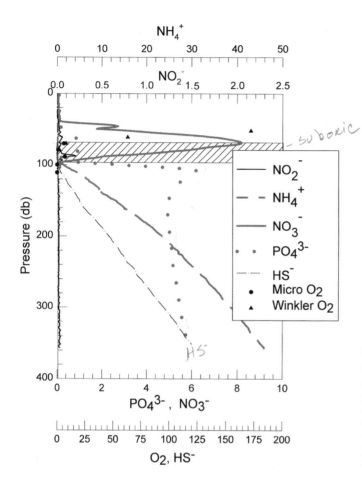

Figure 12.5 Continuous vertical profiles of NO_3^-, NO_2^-, NH_3 ($NH_3 + NH_4^+$), PO_4^{3-}, and sulfide ($H_2S + HS^- + S^{2-}$) in oxic, suboxic, and anoxic portions of the Black Sea. Discrete Winkler and colorimetric dissolved oxygen concentrations taken with the low-concentration method of Broenkow and Cline (1969) are also included. The shaded region approximates the suboxic zone. From Codispoti *et al.* (2001). All concentrations as $\mu mol\,dm^{-3}$.

ammonium according to the following equations:

$$15MnO + 6HNO_3 \rightarrow$$
$$15MnO_2 + 3N_2 + 3H_2O \tag{7}$$

and

$$15MnO_2 + 10NH_3 \rightarrow$$
$$15MnO + 5N_2 + 15H_2O \tag{8}$$

In this pair of reactions, we see that Mn (II) and Mn (IV) can be recycled. Summing equations (7) and (8) gives:

$$6HNO_3 + 10NH_3 \rightarrow 8N_2 + 18H_2O \tag{9}$$

Equation (9) is identical to a thermodynamically permissible equation considered by Richards (1965). It is similar to the "anammox" reaction, but nitrite has been replaced by nitrate, a perhaps insignificant difference given the abundance of bacteria capable of dissimilatory reduction of nitrate to nitrite (Zumft and Cárdenas 1979) and the presence of nitrite maxima in suboxic waters (Fig. 12.1). Thamdrup and Dalsgaard (2000) found that during Mn (IV) oxidation of organic matter in sediments from the Skagerrak, ammonium was produced instead of dinitrogen, so the importance of the manganese mediated reactions in producing dinitrogen in natural systems is uncertain. The manganese mediated reactions have, however, been demonstrated in the laboratory, as has the production of dinitrogen from the oxidation of Fe (II) by nitrate (Straub *et al.* 1996). What is not in question is the existence of biologically mediated lithotrophic reactions that can yield dinitrogen from the oxidation of ammonium with nitrite or nitrate and the reduction of nitrate by Fe (II) and Mn (II). Codispoti *et al.* (2001) discussed how equations (6) or (9) might be coupled with the ammonium flux from sediments underlying suboxic waters. They noted that this ammonium flux could be enhanced by nitrate oxidation of HS^- resulting from bacterial transport of nitrate to underlying sulfidic sediments (e.g. Fossing *et al.* 1995), a process that produces mainly ammonium via "nitrate fermentation" (Jorgensen and Gallardo 1999). Respiratory oxidation of organic matter by Mn (III and IV), IO_3^-, and Fe (III) may also produce dinitrogen (Nealson and Saffarini 1994; Farrenkopf *et al.* 1997*a*, *b*; Luther *et al.* 1997). Thus, the production of dinitrogen in suboxic waters may exceed the yield (relative to nitrate disappearance) suggested by the equations for canonical denitrification (equations (2)–(5)) including those that assume complete conversion of organic N to dinitrogen. At mid-depths in the open ocean low manganese and iron concentrations should reduce the importance of reactions involving these elements, but relatively high concentrations may exist when suboxic waters contact sediments, in suboxic brines and in suboxic layers that exist between oxic and anoxic interfaces in basins such as the Black Sea.

12.3.4 Nitrification at low oxygen tensions

Heterotrophic nitrification is not thought to be important in seawater, and oceanic nitrification is traditionally described as a bacterial process that occurs in two steps: *Nitroso* bacteria oxidize ammonium to nitrite, and *Nitro* bacteria oxidize NO_2^- to NO_3^-. It is also generally understood that ammonium oxidizers, in particular, appear to thrive at low oxygen concentrations and produce NO and N_2O as side-products with the N_2O yield increasing with decreasing oxygen (Kaplan and Wofsy 1985; Anderson *et al.* 1993). This "textbook" description requires modification because of increasing evidence for the ability of

ammonium oxidizing nitrifiers to "denitrify," and for nitrification to proceed under suboxic or near-suboxic conditions. For example, *Nitroso* bacteria produce NO and N_2O during oxidation of ammonium, but they may also "denitrify" NO_2^- to NO, N_2O and perhaps to dinitrogen (Fig. 12.6) with the latter pathway enhanced at low oxygen and high nitrite concentrations (Poth and Focht 1985; Ostrom *et al.* 2000). Although counterintuitive, the pathway of ammonium oxidation to nitrite followed by reduction to N_2O helps explain ^{15}N depletion in some N_2O samples (Yoshida 1988). The ability of nitrifiers to function at low oxygen concentrations and to release nitrate and intermediates (NO_2^-, NO, N_2O) that can be employed by denitrifiers produces a potential for coupling of nitrification and denitrification in sediments and in the water column that has been summarized in the so-called "leaky pipe" model (e.g. Zafiriou 1990; Box 12.1), but the traditional model requires revision to account for the denitrification pathway associated with nitrification (Schmidt *et al.* 2002; Fig. 12.6) and for some of the other reactions discussed herein. There is also increasing evidence for suboxic nitrification. For example, the anammox reaction described above may cease at oxygen concentrations above 2 $\mu mol\,dm^{-3}$, yet laboratory experiments suggest that some nitrate is produced via the following (approximately balanced) chemoautotrophic reaction (Schmidt *et al.* 2002):

$$NH_4^+ + 1.32\,NO_2^- + 0.066\,HCO_3^- + 0.13\,H^+ \rightarrow 0.26\,NO_3^-$$
$$+ 1.02\,N_2 + 0.066\,CH_2O_{0.5}N_{0.15} + 2.03\,H_2O \quad (10)$$

In addition, Hulth *et al.* (1999) suggest the possibility of suboxic nitrification in marine sediments supported by MnO_2 (Hulth *et al.* 1999), although the already cited study of Thamdrup and Dalsgaard (2000) points otherwise.

12.3.5 Nitrous oxide cycling in and near suboxic portions of the water column

Although, N_2O can be produced and consumed by a wide variety of biological processes (Kaplan and Wofsy 1985; Ostrom *et al.* 2000; Codispoti *et al.* 2001), the co-occurrence of suboxia and

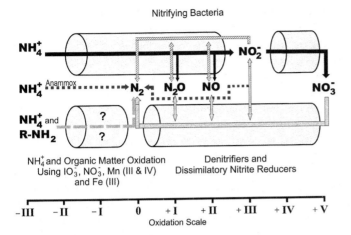

Nitrifying Bacteria

Figure 12.6 An update of Zafiriou's (1990) version of the "leaky pipe" model. See also text Box 12.1.

Box 12.1 The "leaky pipe" concept

Figure 12.6 updates Zafiriou's (1990) "leaky pipe" model that was an elaboration of M. K. Firestone's original conceptual model. We now know that *nitroso* (NH_4^+ to NO_2^-) nitrifiers can denitrify, some in the presence of oxygen (Schmidt *et al.* 2002), and the "anammox" reaction has been demonstrated to exist in suboxic sediments and seawater as described in the text. There are also data to suggest that oxidation of NH_4^+ and organic matter (R-NH$_2$) by suboxic electron acceptors produces N_2, but the question marks indicate that this remains to be confirmed in oceanic suboxic waters for all of the oxidants except nitrite. Recent studies (e.g. Hulth *et al.* 1999; Schmidt *et al.* 2002) also suggest that nitrification can be supported by suboxic electron acceptors as well as by O_2. Our knowledge of these processes is

changing rapidly, so it is likely that this figure will soon need additional revisions. The main point of Fig. 12.6 is to show the multiple pathways that can lead to N_2 production. We have opted not to indicate intermediates that are not well documented in seawater. These include hydroxylamine (NH_2OH) which is known to be an intermediate in nitrification and hydroxylamine, hydrazine(N_2H_2), nitric oxide (NO), nitrogen dioxide (NO_2), and N_2O_2, which are known or potential intermediates in the anammox reaction. The minor fraction of nitrate that may be produced by the "anammox" reaction is also omitted. Finally, note that although canonical denitrification is usually described as a heterotrophic process, the oxidation of inorganic compounds such as Fe (II) by nitrate may also lead to N_2 production.

---

the highest and lowest N_2O concentrations in the ocean (Fig. 12.2) can be explained by the interplay of nitrification and denitrification and the dependence of these processes on oxygen concentrations. These relationships may be summarized as follows: (i) adapted denitrifiers in the "core" of oceanic suboxic zones are net consumers of N_2O; (ii) there is often a buildup of N_2O during the initial stages of denitrification (e.g. Codispoti *et al.* 1992; Naqvi *et al.* 2000); (iii) nitrification at the low oxygen

concentrations associated with the boundaries of suboxic zones produces relatively large amounts of N_2O; and (iv) the N_2O/N_2 yield ratio for denitrifiers may increase as oxygen concentrations increase (Betlach and Tiedje 1981). It is also possible that the high nitrite concentrations associated with suboxic waters also enhance N_2O production (e.g. Anderson *et al.* 1993).

On balance, the extremely high concentrations of N_2O found in proximity to the suboxic zones tend to dominate over the zones of extremely low N_2O found in the "cores" of the suboxic water masses, and the waters bordering the suboxic zones in the eastern tropical South Pacific and the Arabian Sea are strong N_2O sources (Naqvi 1994; Bange *et al.* 2001; Fig. 12.2).

12.4 Methodology

12.4.1 Sampling

Because the transition from oxic to suboxic to anoxic occurs over dissolved oxygen concentrations of only a few micromolar, care must be taken to ensure that sampling bottles are airtight and do not leak or contain "dead-spaces." Regrettably, oceanographic sampling bottles are often "tripped" before they have sufficiently "flushed" with ambient water. This problem is exacerbated by increasingly large CTD/Rosette systems that entrain water. If incubations are planned, sampling bottles must usually be refitted with nontoxic O-rings and springs. Another problem that is often ignored is the possibility of settling of particles/aggregates within the sampling bottles.

12.4.2 Determination of dissolved oxygen and dinitrogen

Dissolved oxygen and excess N_2 concentrations are fundamental properties of suboxic water, but precise oxygen data that properly account for interference from IO_3^-, NO_2^-, and other variations in the seawater blank are rare, and accurate values for the excess-N_2 produced by *in situ* biological processes are rarer still. Although the Winkler titration can be highly precise, as late as the 1960s massive systematic errors existed between institutions (Carritt and Carpenter 1966). Widespread adoption of the modification of the Winkler titration suggested by Carpenter (1965), and the development of automated titration systems (e.g. Williams and Jenkinson 1982), have made it possible to obtain highly precise dissolved oxygen data on a routine basis, but care and skill are still required when drawing samples in an atmosphere that is $\sim20\%$ O_2. Experiments have shown that $\sim3\times$ the volume of the sampling

flask should be employed for flushing (by overflowing) a dissolved oxygen flask when sampling low oxygen water (Horibe *et al.* 1972). In suboxic waters, Winkler reagents should be modified to destroy nitrite. Even today, few investigators check oxygen reagent blanks for the presence of oxidants or reductants in shipboard dionized water or perform blanks on seawater samples. When skilled analysts use contemporary (precision $\sim\pm0.1\%$) Winkler techniques to analyze suboxic samples, good agreement with the low concentration colorimetric method (Broenkow and Cline 1969) can be obtained (Morrison *et al.* 1999). Such data from the Arabian Sea OMZ suggest not only good agreement between the methods but also that the nitrite maxima associated with open ocean suboxia do not occur until oxygen concentrations are less than ~2 μmol dm^{-3} (Fig. 12.3), but even this data set lacks high quality seawater blanks. Improved sampling and analytical techniques, have also cast doubt (e.g. Murray *et al.* 1995) on the existence of an interface layer in the Black Sea where previous investigations suggested that HS$^-$ and oxygen overlap.

Since fixed-N can be converted to dinitrogen via multiple pathways, the best way to determine this signal would be direct measurement of this "excess N_2." The problem is that the background dinitrogen concentration arising from dissolution from the atmosphere is approximately 100 times higher than the maximum biological inputs. Although we can calculate the atmospheric fraction of the dinitrogen that should be dissolved in a water sample based on solubility tables, minor deviations from solubility are of the same magnitude as the denitrification signal. Such deviations can arise from bubble injection, varying barometric pressures, and mixing of water masses (Hamme and Emerson 2002). This problem is partially alleviated by employing N/Ar ratios determined by mass spectrometry to estimate "excess N_2" reasoning that deviations from ideality in the concentration of atmospheric dinitrogen could be computed by knowing the deviation in the inert gas Ar. Recent improvements in mass spectrometry have made the determination of N/Ar ratios more convenient (Emerson *et al.* 1999), and there has been a significant increase in the number of such observations (e.g. Fig. 12.4). Nevertheless, the computation of excess N_2 is still not straightforward. For example, the solubility kinetics of the two gases differ and this can cause their saturation values to differ under various "bubble injection" scenarios (Hamme and Emerson 2002). Thus, N/Ar ratios must be examined in concert with other oceanographic data to define initial N/Ar ratios before the onset of denitrification.

12.4.3 Nutrient determinations

Most studies of oceanic suboxic zones rely heavily on examination of nitrite and nitrate distributions, and many employ reactive phosphorus (PO_4^{3-}) concentrations to estimate the dinitrogen produced from canonical denitrification using the Redfield ratio concept (e.g. Gruber and Sarmiento 1997; Codispoti *et al.* 2001, equations (2)–(5)). Under ideal conditions, the best "routine" methods for NO_2^-, NO_3^-, and PO_4^{3-} may be capable of accuracies of better than 1%, but under normal seagoing conditions, accuracies of ~2% are probably more the norm even for experienced research teams. The typical ammonium methods are even less accurate, but ammonium concentrations are generally low in suboxic waters (e.g. Fig. 12.5). Because of even greater inaccuracies in much of the extant data, they can be useless when attempting to combine NO_2^-, NO_3^-, and PO_4^{3-} data to nitrate deficits. For example, consider the following equation for estimating the nitrate removed by denitrification in the Arabian Sea water column (Codispoti *et al.* 2001),

$$N_{deficit} = (14.89(P - 0.28) - N)\,\mu M \times 0.86, \quad (11)$$

where $N = NO_3^- + NO_2^- + NH_4^+$, $N_{deficit}$ is the estimate of the nitrate converted to dinitrogen, P = reactive phosphorus, and the factor 0.86 accounts for the P released during canonical denitrification. As indicated by this equation, an error of only 0.07 μmol dm^{-3} in reactive phosphorus can cause a 1 μmol dm^{-3} error in $N_{deficit}$: a significant difference (Figs 12.1 and 12.4). Thus, N/P based methods for calculating $N_{deficits}$ and similar parameters such as N* (Gruber and Sarmiento 1997) can be employed only when high quality phosphate and inorganic nitrogen data are both available. Comparing nutrient data from slowly varying abyssal waters can sometimes be employed to reduce systematic errors (e.g Gruber and Sarmiento 1997), but the development and acceptance of distributable nutrient standards is also needed.

The commonly used phenol-hypochlorite ammonium method is troublesome and accuracies are probably no better than ~5%. Fortunately, ammonium concentrations in open-ocean suboxic water are generally <1 μmol dm^{-3}, and a new fluorometric method (Holmes *et al.* 1999) shows promise.

12.4.4 Microbial identification

Genetic sequencing techniques have helped to describe populations of denitrifying microbes (Ward 1996; Braker *et al.* 2001), and immunofluorescent staining combined with autoradiography has provided insight into the distribution of nitrifiers and nitrification rates (Ward *et al.* 1989). Given the recent discoveries regarding alternate metabolic pathways in suboxic waters and recent advances in biotechnology, it is, however, fair to say that we are in the infancy of describing microbial populations in suboxic waters.

12.4.5 Enzyme based techniques

Determining the activities of the enzyme systems involved in the respiratory electron transport system (ETS) or in the reduction steps of the denitrification pathway (nitrate reductase, nitrite reductase, nitric oxide reductase, and nitrous oxide reductase) all present possibilities for providing insight into suboxic respiration. General problems with all such methods include translating enzyme activities into estimates of *in situ* rates, quantitative filtering of the ambient microbes, and quantitative recovery of the enzyme activity. The variety of enzymes involved in the steps of denitrification may also be a complication as is the multiplicity of pathways for NO, N_2O, and dinitrogen production. For example, there are periplasmic and membrane-bound types of nitrate reductase and nitrous oxide reductase, at least two types of nitrite reductase, a type of nitrous oxide reductase that is not directly inhibited by oxygen, etc. In addition, most investigations have concentrated only on Gram-negative bacteria (Bell *et al.* 1990; Bell and Ferguson 1991; Ferguson 1994; Knowles 1996; Zumft 1997).

Despite the above-mentioned problems, denitrification rates determined from ETS activities have in several cases agreed well with estimates based on nitrate deficit transports (e.g. Codispoti and Packard 1980; Naqvi 1994). Both types of estimates relied on earlier stoichiometries, (Codispoti *et al.* 2001), but applying appropriate corrections should produce roughly similar values for the two approaches. Past applications of the ETS method have assumed that the bacteria are in stationary phase and produced reasonable denitrification rates while assuming log-phase growth for the bacteria would yield much larger rates (e.g. Codispoti and Packard 1980; Naqvi 1994). It is likely, however, that some fraction of the microbial population is growing rapidly, and we need to learn more about the distribution of bacteria in these zones in relation to particles/aggregates and organic rich "trails" left by sinking particles/aggregates (Kiørboe 2001). Also, the recent studies summarized above, suggest that we can no longer assume that canonical denitrifiers are the only significant respirers in open-ocean suboxic waters. Attempts to use nitrate reductase and nitrite reductase to elucidate processes in suboxic waters have met with less success, but Shailaja (2001) has made some progress. Also assay of the key autotrophic enzyme, ribuolose 1,5-*bis*-phosphate carboxylase (RUBISCO) has been employed with some success for estimating the activity of nitrifiers below the photic zone in oxic and suboxic waters (Ward, *et al.* 1989).

12.4.6 Incubation techniques

A generic problem in applying incubation techniques to suboxic waters is that sample manipulations may cause oxygen to deviate significantly from *in situ* levels since only small additions of dissolved oxygen are required for the suboxic to oxic transition. Additional artifacts can be introduced by the breakup of aggregates, and by changes in pressure when samples are collected at depths of several hundred meters. Since very small changes in dissolved oxygen concentration can have an impact on suboxic and near-suboxic metabolic pathways, handling and maintaining incubation samples so as to maintain *in situ* dissolved oxygen levels, are desirable but daunting. A less satisfactory alternative might be to maintain ~ 0 dissolved oxygen levels coupled with short incubations that produce linear time-course results. A near ideal incubation might be described as taking place *in situ*, and lasting <1 h (to minimize alterations in the initial population).

Two basic incubation techniques have been used for assessing canonical rates of denitrification in the ocean, ^{15}N tracer technology and the acetylene block technique. The basis of the acetylene block technique arises from the blocking effect of acetylene on nitrous oxide reductase. Thus under appropriate conditions, the addition of acetylene prevents the reduction of N_2O to N_2, and the resulting accumulation of N_2O against the generally low background values can be easily detected by gas chromatography. Unfortunately, acetylene also blocks some types of nitrification (Schmidt *et al.* 2002), and we now know that nitrifiers can be important in suboxic zones.

The addition of ^{15}N enriched nitrate to a water sample and the analysis of the ^{15}N enrichment in dinitrogen permits calculation of rates of canonical ($NO_3^- \rightarrow N_2$) denitrification. Capone and Montoya (2001) describe these incubation techniques in more detail and Seitzinger *et al.* (1993) and Joye *et al.* (1996) have compared these methods *inter alia* and with other approaches. Since nitrate concentrations in the major oceanic suboxic water masses are generally well above 10 μmol dm^{-3} (Fig. 12.1), the addition of [15]NO_3^- tracer does not cause a problem. This is not the case, however, if one wants to trace the fate of ammonium.

Canonical nitrification rates can be determined by the addition of ^{15}N-enriched ammonium or nitrite and separation, extraction, and determination of the ^{15}N enrichment in the nitrite and nitrate produced (e.g. Ward *et al.* 1989).

12.4.7 Natural isotope ratios

Most biological transformations favor uptake of lighter isotopes. Knowledge of the fractionations that occur can provide valuable insight into rates and pathways of oceanic nitrogen transformations. For example, N_2O enriched in heavy isotopes is present in the core of suboxic water masses (Naqvi *et al.* 1998) presumably due to the preferential reduction of lighter N_2O to dinitrogen during denitrification.

The $^{15}N/^{14}N$ ratio in atmospheric N_2, expressed as $\delta^{15}N$ is 0‰, but during denitrification there is a large fractionation that favors ^{14}N utilization. Early work by Cline and Kaplan (1975) in the suboxic zone of the eastern tropical North Pacific (ETNP) suggested that the *in situ* community preferentially reduced the light N isotope in nitrate with a fractionation of 20–40‰. Later work by Brandes *et al.* (1998) and Altabet *et al.* (1999) in the ETNP and the Arabian Sea narrowed this range to between 24‰ and 28‰. In both regions, there is a clear signal in the isotopic composition of nitrate with $\delta^{15}N$ values rising from ~6‰ in the deep waters to a maximum of ~15‰ in the core of the suboxic layer and then decreasing toward the surface (Fig. 12.4). Concomitant with this is a decrease in the isotopic value of dinitrogen in the suboxic zone from about 0.6‰ to 0.2‰. As is the case for nitrate deficits, these isotopic anomalies extend beyond the secondary nitrite maximum (Figs 12.1 and 12.4). These broader features can be explained either by mixing or by a reduction in nitrite accumulation in the deeper portion of the suboxic zone due to more stable conditions and slower rates (Codispoti *et al.* 2001).

12.5 Significance of suboxic and near-suboxic water column respiration

12.5.1 General comments

Rate estimates for denitrification, biological dinitrogen production, and net nitrous oxide production in suboxic and adjacent waters vary widely (e.g. Gruber and Sarmiento 1997; Codispoti *et al.* 2001), but there is general agreement that water column denitrification is a sink for oceanic fixed-N, second only to sedimentary denitrification. It is also becoming clear that the small portions of the ocean adjoining suboxic waters are a globally significant N_2O source (Codispoti and Christensen 1985; Codispoti *et al.* 1992; Naqvi *et al.* 2000; Bange *et al.* 2001).

Low oxygen estuaries are also strong sources of N_2O (De Wilde and De Bie 2000; Bange *et al.* 2002).

12.5.2 Maintaining a habitable planet *—point?*

It has often been pointed out that at chemical equilibrium most of the oxygen in the atmosphere would be combined with N in the +V oxidation state, and the sea would be a dilute solution of nitric acid with a low pH (e.g. Hutchinson 1957, 1975). Fortunately, the timescale for this process to be significant is on the order of several million years because of the large atmospheric oxygen reservoir. Further, the conversion of fixed-N to dinitrogen by the processes outlined herein is a factor in preventing this unhappy outcome for life as we know it.

12.5.3 Barrier to anoxia and sulfide production

Nitrate (NO_3^-) is, by far, the most abundant suboxic electron acceptor in oxygenated seawater with an average concentration of ~31 $\mu mol\ dm^{-3}$. Since the reduction of nitrate to dinitrogen represents a change from the +V to 0 oxidation state, this represents a suboxic electron acceptor capacity in seawater of 155 microequivalents. The minor contributors to the suite of suboxic respiratory electron acceptors that we have mentioned previously add an additional ~3 microequivalents (mainly from IO_3^-). Fully oxygenated "average" seawater (T ~4°C, S ~35, NO_3^- ~31 $\mu mol\ dm^{-3}$) contains enough oxygen to accept ~1300 microequivalents. Thus, the suite of suboxic electron acceptors provides an electron acceptance capacity equivalent to ~12% of that provided by the oxygen present when "average" seawater leaves the surface. Because suboxic waters tend to occur in the upper 1000 m at low latitudes, however, they are warmer than average. In addition, oceanic suboxia tends to occur in regions of high nitrate concentration (Fig. 12.1). As a consequence, the suboxic electron accepting capacity can, in some cases exceed 20% of that provided by dissolved oxygen. Thus, suboxic electron acceptors provide a significant barrier to the onset of anoxia and sulfide production as noted by Redfield *et al.* (1963).

12.5.4 Constraint on sedimentary denitrification

Determining the water column denitrification rate is important in its own right but takes on additional importance because the water column conversion of nitrate to dinitrogen places constraints on the sedimentary denitrification rate via isotope budgets as described by Brandes and Devol (2002). Basically, we must balance the $\delta^{15}N$ of ~0‰ associated with fixed-N arising from nitrogen fixation, the main source of oceanic fixed-N (Table 12.1), with the $\delta^{15}Ns$ arising from water column and sedimentary denitrification. These differ greatly, in part because nitrate is seldom depleted in the major suboxic water masses (e.g. Fig. 12.1), and the $\delta^{15}N$ of oceanic nitrate can only be rationalized if the sedimentary conversion of nitrate to dinitrogen is several-fold greater than the water column rate (Brandes and Devol 2002).

12.5.5 Control on oceanic fixed-N pool

Whether or not the oceanic fixed-N budget is balanced within the rather large error bounds of the source and sink terms (e.g. Gruber and Sarmiento 1997; Codispoti *et al.* 2001; Table 12.1) is a matter of debate. There is general agreement that the overall rates of nitrogen fixation, denitrification, and N_2O turnover require significant upward revision (Table 12.1) and that biological conversion of fixed-N to dinitrogen (listed as denitrification in Table 12.1) is the dominant sink. Correlative with these increased rates is a decrease in the turnover time for fixed-N in the ocean from ~10 000 to 1500 years (Codispoti *et al.* 2001) with implications for the role of the oceanic-N cycle in climate change (Broecker and Henderson 1998). Given the differing requirements for nitrogen fixation which is a (the) major source for oceanic fixed-N and for water column and sedimentary denitrification which are the main sinks, it is unlikely that the oceanic budget is ever in exact balance (Falkowski 1997; Codispoti *et al.* 2001). Over ocean-circulation timescales (~100 to ~1000 years), however, an excess of denitrification will tend to increase N-limitation of primary production and could cause a general increase in

subsurface dissolved oxygen, thereby reducing denitrification (e.g. Broecker and Peng 1982; Codispoti 1989). Thus, decreases in oceanic fixed-N could only go so far in the face of this negative feedback, but how far depends, in part, on just how large denitrification can be at any particular time, and estimates of the total oceanic denitrification rate have been increasing.

The overall nitrate deficit in today's ocean is only a few micromolar below the concentration necessary to produce the canonical "Redfield ratio" of 16/1 (N/P by atoms) for phytoplankton uptake. This fact undergirds the notion that while a small nitrate deficit is required to stimulate nitrogen fixation, phosphate (PO_4^{3-}) is actually the "master" nutrient for oceanic primary production with fixed-N tracking the waxing and waning of the oceanic PO_4^{3-} inventory. This canonical geochemical view contrasts with the opinion of many biologists (see Codispoti 1989 and Falkowski 1997), and it has to confront the following factors:

1. A fixed-N deficit is not a sufficient condition for nitrogen fixation (Codispoti, 1989) which requires relatively large amounts of iron (Falkowski 1997), may be inhibited by turbulence (Paerl 1990), and in the oceanic water column, appears to be largely restricted to waters with temperatures more than 20°C (Carpenter and Romans 1991).
2. Although the overall deficit of about 3 μmol dm^{-3} of fixed inorganic N seems modest compared to the oceanic average nitrate concentration of ~ 31 μmol dm^{-3} (Gruber and Sarmiento 1997), concentrations in waters supplied to the euphotic zone are much lower, on the order of 10 μmol dm^{-3}, suggesting a deficit of ~30% relative to the canonical Redfield ratio.
3. If we look at phytoplankton uptake on the genera level, we see significant deviations from the canonical Redfield ratio. *Phaeocystis* spp. in the Ross Sea (Antarctica), for example, appear to have an N/P uptake ratio of ~19 (Arrigo *et al.* 1999).
4. N/P ratios during water column nitrogen fixation are high ranging from ~57 to 140 (Karl *et al.* 1992; Gruber and Sarmiento 1997), and ratios for the entire communities associated with nitrogen fixation are often more than 20 (Naqvi *et al.* 1986; Karl *et al.* 1992).

Table 12.1 Gruber and Sarmiento's fixed-N budget for the present-day ocean and the changes suggested by Codispoti *et al.* (2001)[a]. All rates as Tg N a^{-1}

Process	Gruber and Sarmiento (1997)	Codispoti *et al.*'s (2001) updated values
Sources		
Pelagic N$_2$ fixation	110 ±40	
Benthic N$_2$ fixation	15 ±10	
River input (DN)	34 ±10	
River input (PON)	42 ±10	
Atmospheric deposition (Net)	30 ±5	
Atmospheric deposition (DON)		56[b]
Total sources	231 ±44	287
Sinks		
Organic N export		1[c]
Benthic denitrification[c]	95 ±20	300
Water column denitrification[c]	80 ±20	150
Sedimentation	25 ±10	
N$_2$O loss	4 ±2	6
Total sinks	204 ±30	482

[a] Tg = 10^{12} g.

[b] Average atmospheric organic N (mostly anthropogenic) deposition value suggested by Cornell *et al.* (1995).

[c] Includes all biological conversions of fixed-N to N$_2$ shown in Fig. 12.6.

5. Thus, the question of whether or not an N-replete ocean would have an N/P ratio similar to today's value or a value above 20 remains open.

12.5.6   A control on oceanic primary production

The conversions of fixed-N to dinitrogen in suboxic waters are major sinks (Table 12.1) for oceanic fixed-N, the major limiting nutrient for oceanic primary production. By removing fixed-N these processes create a need for nitrogen fixation with its high iron requirement (e.g. Falkowski 1997; Codispoti *et al.* 2001), and thus, also contribute to iron limitation of oceanic primary production. If we take the value of 150 Tg N a^{-1} suggested by Codispoti *et al.* (2001) for water column denitrification and multiply it by the 6 C/N (by wt.) Redfield ratio for phytoplankton uptake, we see that water column denitrification removes an amount of fixed-N sufficient to reduce oceanic "new" primary production by ~10^{15} g C a^{-1}, a globally significant amount.

12.5.7   Enhancing the nitrous oxide source

As suggested by the occurrence of extremely high concentrations (Fig. 12.2), respiration at the boundaries of suboxic portions of the water column is a globally significant source of N$_2$O (e.g. Codispoti *et al.* 2001). This is important because N$_2$O is 200–300 times (depending on the selected time interval) more powerful than CO$_2$ as a greenhouse gas (Manne and Richels 2001), and it influences the planetary ozone distribution (Nevision and Holland 1997). The more we study regions with suboxic waters, the more we see cause to raise estimates of the net flux of N$_2$O from the ocean to the atmosphere from ~4 to 6 Tg N a^{-1} (Table 12.1) with further increases a possibility. For example, Naqvi *et al.* (2000) recently discovered a globally significant N$_2$O source over the western Indian Shelf during the development of seasonal suboxia and anoxia.

12.5.8   A potential agent of climate change

Several authors have suggested that the interplay between suboxic dinitrogen production, nitrogen fixation and the iron supply can contribute to climate change by changing the ability of the oceanic "biological pump" to sequester atmospheric carbon dioxide (e.g. Christensen *et al.* 1987; Ganeshram *et al.* 1995; Falkowski 1997; Gruber and Sarmiento 1997; Broecker and Henderson 1998; Codispoti *et al.* 2001; Altabet *et al.* 2002; Devol 2002). Since suboxic dinitrogen production in the water column via denitrification, anammox, etc. is a major sink for oceanic fixed-N, (Table 12.1), suboxic water column respiration may play a role in climate change via changes in the oceanic fixed-N inventory that, in turn, can alter oceanic primary production. In addition, the increased N_2O fluxes to the atmosphere associated with suboxic water column respiration (see 12.5.7) also have the potential to play a role in climate change.

12.5.9 Controlling the distribution of some elements and contaminants

Changes in oxidation state undergone by trace metals and metalloids involved in suboxic respiration can influence their solubilities. For example, Senn and Hemond (2002) conducted a study suggesting the potential influence of nitrate oxidation on arsenic in a polluted lake. Another example is provided by uranium chemistry. The change in oxidation state from U (VI) to U (IV) greatly reduces the solubility of uranium providing the major sink for oceanic uranium as well as a tool for remediation of contaminated solutions. Lovley (1993) reviews this topic for a variety of trace metals and for metalloids.

12.6 Synthesis

12.6.1 Present state of knowledge

This review suggests that we are in the infancy of understanding suboxic water column respiration. We have been limited by methodology and concepts. Estimates of the overall rates of denitrification, nitrogen fixation, and nitrous oxide have undergone drastic upward revision in recent years. We are only beginning to understand the relative importance of all of the pathways that can lead to

dinitrogen production from fixed-N and the roles of oxidants such as Mn (III and IV), and IO_3^-. We still lack adequate descriptions of the bacterial flora in suboxic zones. We do not know whether or not significant denitrification takes place in oxygenated waters or in association with thermal vents, and seeps, etc. We do know enough, however, to suggest that some of the processes that occur in the small portion of ocean space that is suboxic are globally significant, and that the "more we look, the more we are likely to find." This is likely to be true for nitrogen fixation as well (Fuhrman and Capone 2001).

12.6.2 Temporal variability

Given that suboxic conditions exist in only 0.1–0.2% of the oceanic volume, and that the respiratory pathways are sensitive to minute changes in dissolved oxygen concentrations it should be no surprise that even the larger suboxic water masses regions exhibit globally significant variability on a variety of scales ranges ranging from at least the glacial–interglacial time frame to the ~decadal timescale (e.g. Codispoti *et al.* 1986; Altabet *et al.* 2002). Observations of interannual variability in the suboxic water masses off Peru (Codispoti and Packard 1980; Codispoti *et al.* 1986) confirm the rapidity with which these regions can change. Naqvi *et al.* (1990) have discussed seasonal changes in the quasi-permanent suboxic water mass found in the Arabian Sea, and turnover times for the major suboxic water masses found in the Arabian Sea and in the eastern tropical South Pacific are less than ~10 years. (Codispoti and Packard 1980; Howell *et al.* 1997). Calculations also suggest that only small changes in the oxygen and carbon supply to suboxic regions are required to cause such changes (Codispoti 1989). One reason why small changes in the carbon flux can have a large effect on suboxic water masses is that the ratio of C/N (by atoms) during phytoplankton growth is ~6.6 (Redfield *et al.* 1963) whereas the C/N ratio during denitrification is <1 (see equations (2)–(5) and Codispoti *et al.* 2001).

12.6.3 Anthropogenic effects

Human activities have seriously disturbed terrestrial and oceanic ecosystems, and there appears to be no respite in these changes. These alterations include massive intrusions on the global nitrogen cycle (e.g. Smil 1997; Vitousek *et al.* 1997) and on oceanic ecosystem structure (e.g. Watling and Norse 1998). Thus, our observations of oceanic suboxia may well be occurring during a time when the ocean is going through a major transition and they have to be interpreted in this context (Codispoti *et al.* 2001). With respect to anthropogenic effects, we note that globally significant increases in denitrification were observed in the waters off Peru after the collapse of the Peruvian anchoveta fishery (Codispoti and Packard 1980) and after the major 1982–3 el Niño (Codispoti *et al.* 1986). The collapse of the fishery followed a major increase in fishing pressure, and anthropogenic activities may be increasing the frequency of el Niños. Anthropogenic impingement also lowers oxygen concentrations in coastal waters (e.g. Rabalais *et al.* 2000; Hopkinson and Smith, Chapter 7) and may be contributing significantly to the recent reports of high denitrification rates and N_2O concentrations over the Indian continental shelf (Naqvi *et al.* 2000). Overall, it seems likely that anthropogenic effects are significantly altering the rate and distribution of suboxic water column respiration.

Acknowledgments

Over the years, our research on oceanic suboxia has received financial support from several government agencies including EPA, NOAA, NSF, ONR, and the NY State Department of Health. We gratefully acknowledge this financial support and also thank the officers and crews of the research vessels who facilitated our observations, often under difficult conditions. A host of colleagues have been generous with advice, stimulating conversation, and constructive criticism, and we are most grateful for their unselfish assistance. Finally, we thank our families for their forbearance and support.

References

Alldredge, A. L. and Cohen., Y. 1987. Can microscale chemical patches persist in the sea? Microelectrode study of marine snow, fecal pellets. *Science*, **235**: 689–693.

Altabet, M. A., Pilskaln, C., Thunell, C., Pride, C., Sigman, D., Chavez, F., and Francois, R. 1999. The nitrogen isotope biogeochemistry of sinking particles from the margin of the Eastern North Pacific. *Deep-Sea Res. I*, **46**: 655–679.

Altabet, M. A., Higginson, M. J., and Murray, D. W. 2002. The effect of millennial-scale changes in Arabian Sea denitrification on atmospheric CO_2. *Nature*, **415**: 159–162.

Anbar, A. D. and Knoll, A. H. 2002. Proterozoic ocean chemistry and evolution: a bioinorganic bridge? *Science*, **297**: 1137–1142.

Anderson, J. J. and Devol, A. H. 1987. Extent and intensity of the anoxic zone in basins and fjords. *Deep-Sea Res.*, **34**: 927–944.

Anderson, J. J., Okubo, A., Robbins, A. S., and Richards, F. A. 1982. A model for nitrite and nitrate distributions in oceanic oxygen minimum zones. *Deep-Sea Res.*, **29**: 1113–1140.

Anderson, I. C., Poth, M., Homstead, J., and Burdige, D. 1993. A comparison of NO and N_2O production by the autotrophic nitrifier *Nitrosomonas europaea* and the heterotrohpic nitrifier *Alcaligenes faecalis*. *Appl. Environ. Microbiol.*, **59**: 3525–3533.

Anderson, L. A. 1995. On the hydrogen and oxygen content of marine phytoplankton. *Deep-Sea Res. I*, **42**: 1675–1680.

Arrigo, K. R., Robinson, D. H., Worthen, D. L., Dunbar, R. B., DiTullio, G. R., VanWoert, M., and Lizotte, M. P. 1999. Phytoplankton community structure and the drawdown of nutrients and CO_2 in the Southern Ocean. *Science*, **283**: 365–367.

Bange, H. W., Andreae, M. O., Lal, S., Law, C. S., Naqvi, S. W. A., Patra, P. K., Rixen, T., and Upstill-Goddard, R. C. 2001. Nitrous oxide emissions from the Arabian Sea: A synthesis. *Atmos. Chem. Phys.*, **1**: 61–71.

Bell, L. C. and Ferguson, S. J. 1991. Nitric and nitrous oxide reductases are active under aerobic conditions in cells of *Thiosphaera pantotropha*. *J. Biochem.*, **273**: 423–427.

Bell, L. C., Richardson, D. J., and Ferguson, S. J. 1990. Periplasmic and membrane-bound respiratory nitrate reductases in *Thiosphaera pantotropha*. The periplasmic enzyme catalyzes the first step in aerobic denitrification. *FEBS Lett.*, **265**: 85–87.

Betlach, M. and Tiedje, J. M. 1981. Kinetic explanation for accumulation of nitrite, nitric oxide and nitrous oxide during bacterial denitrification. *Appl. Environ. Microbiol.*, **42**: 1074–1084.

Braker, G., Ayala-del-Rio, H. L., Devol, A. H., Fesefeldt, A., and Tiedje, J. M. 2001. Community structure of denitrifiers, bacteria, and archaea along redox gradients in Pacific Northwest marine sediments by terminal restriction fragment length polymorphism analysis of amplified nitrite reductase (nirS) and 16s rRNA genes. *Appl. Environ. Microbiol.*, **67**: 1893–1901.

Brandes, J. A. and Devol, A. H. 2002. A global marine fixed nitrogen isotopic budget: implications for Holocene nitrogen cycling. *Glob. Biogeochem. Cyc.*, **16**: 67.1–67.14.

Brandes, J. A., Devol, A. H., Yoshinari, T., Jayakumar, D. A., and Naqvi, S. W. A. 1998. Isotopic composition of nitrate in the central Arabian Sea and eastern tropical North Pacific: a tracer for mixing and nitrogen cycles. *Limnol. Oceanogr.*, **43**: 1680–1689.

Broecker, W. S. and Henderson, G. M. 1998. The sequence of events surrounding Termination II and their implications for the cause of glacial-interglacial CO_2 changes. *Palaeogeogr. Palaeoclimatol. Palaeoecol.*, **142**: 217–230.

Broecker, W. S. and Peng, T. H. 1982. *Tracers in the sea*. Eldigio, Lamont-Doherty Geology Observatory, New York.

Broenkow, W. W. and Cline, J. D. 1969. Colorimetric determination of dissolved oxygen at low concentration. *Limnol. Oceanogr.*, **14**: 450–454.

Calvert, S. E. and Price, N. B. 1971. Upwelling and nutrient regeneration in the Benguela Current, October, 1968. *Deep-Sea Res.*, **18**: 505–523.

Capone, D. G. and Montoya, J. P. 2001. Nitrogen fixation and denitrification. In J. H. Paul (ed.) *Methods in Microbiology*, Vol. 30. Academic Press, London, pp. 501–514.

Carpenter, J. H. 1965. The Chesapeake Bay Institute Technique for the Winkler dissolved oxygen method. *Limnol. Oceanogr.*, **10**: 141–143.

Carpenter, E. J. and Romans, K. 1991. Major role of the cyanobacterium *Trichodesmium* in nutrient cycling in the North Atlantic Ocean. *Science*, **254**: 1356–1358.

Carritt, D. E. and Carpenter, J. H. 1966. Comparison and evaluation of currently employed modifications of the Winkler method for determining dissolved oxygen in seawater, a NASCO report. *J. Mar. Res.*, **24**: 286–318.

Christensen, J. P., Murray, J. W., Devol, A. H., and Codispoti, L. A. 1987. Denitrification in continental shelf sediments has major impact on the oceanic nitrogen budget. *Glob. Biogeochem. Cyc.*, **1**: 97–116.

Cline, J. D. and Kaplan, I. R. 1975. Isotopic fractionation of dissolved nitrate during denitrification in the eastern tropical North Pacific Ocean. *Mar. Chem.*, **3**: 271–299.

Codispoti, L. A. 1989. Phosphorus vs. nitrogen limitation of new and export production. In W. Berger, V. Smetacek, and G. Wefer (eds) *Productivity of the Ocean: Present and Past*. John Wiley and Sons, Chichester, pp. 377–394.

Codispoti, L. A. and Christensen, J. P. 1985. Nitrification, denitrification and nitrous oxide cycling in the eastern tropical South Pacific Ocean. *Mar. Chem.*, **16**: 277–300.

Codispoti, L. A. and Packard, T. T. 1980. Denitrification rates in the eastern tropical South Pacific. *J. Mar. Res.*, **38**: 453–477.

Codispoti, L. A., Friederich, G. E., Packard, T. T., Glover, H. T., Kelly, P. J., Spinrad, R. W., Barber, R. T., Elkins, J. W., Ward, B. B., Lipschultz, F., and Lostanau, N. 1986. High nitrate levels off of the coast of Peru: a signal of instability in the marine denitrification rate. *Science*, **233**: 1200–1202.

Codispoti, L. A., Friederich, G. E., Murray, J. W., and Sakamoto, C. M. 1991. Chemical variability in the Black Sea: Implications of continuous vertical profiles that penetrated the oxic/anoxic interface. *Deep-Sea Res.*, **38**: S691–S710.

Codispoti, L. A., Elkins, J. W., Yoshinari, T., Friederich, G. E., Sakamoto, C. M., and Packard, T. T. 1992. On the nitrous oxide flux from productive regions that contain low oxygen waters. In B. Desai (ed.) *Oceanography of the Indian Ocean*. Oxford and IBH Publishing Co., New Delhi, pp. 271–284.

Codispoti, L. A., Brandes, J. A., Christensen, J. P., Devol, A. H., Naqvi, S. W. A., Paerl, H. W., and Yoshinari, T. 2001. The oceanic fixed nitrogen and nitrous oxide budgets: Moving targets as we enter the anthropocene? *Sci. Mar.*, **65**: 85–105.

Cornell, S., Rendell, A., and Jickells, T. 1995. Atmospheric inputs of dissolved organic nitrogen to the oceans. *Nature*, **376**: 243–246.

Dalsgaard, T. and Thamdrup, B. 2002. Factors controlling anaerobic ammonium oxidation with nitrite in marine sediments. *Appl. Environ. Microbiol.*, **68**: 3802–3808.

Dalsgaard, T., Canfield, D. E., Petersen J., Thamdrup, B., and Acuña-González, J. 2003. N_2 production by the anammox reaction in the anoxic water column of Golfo Dulce, Costa Rica. *Nature*, **422**: 606–608.

Damsté, J. S. S., Strous, M., Rijpstra, W. I. C., Hopmans, E. C., Geenevasen, J. A. J., van Duin, A. C. T., van Niftrik, L. A., and Jetten, M. S. M. 2002. Linearly concatenated cyclobutane lipids form a dense bacterial membrane. *Nature*, **419**: 708–712.

Deutsch, C., Gruber, N., Key, R. M., and Sarmiento, J. L. 2001. Denitrification and N_2 fixation in the Pacific Ocean. *Glob. Biogeochem. Cyc.*, **15**: 483–506.

Devol, A. H. 1978. Bacterial oxygen uptake kinetics as related to biological processes in oxygen deficient zones of the oceans. *Deep-Sea Res.*, **25**: 137–146.

Devol, A. H. 2002. Getting cool with nitrogen. *Nature*, **415**: 131–132.

Devol, A. H. 2003. Solution to a marine mystery. *Nature*, **422**: 575–576.

de Wilde, H. P. J. and de Bie, M. J. M. 2000. Nitrous oxide in the Schelde estuary: production by nitrification and emission to the atmosphere. *Mar. Chem.*, **69**: 203–216.

Emerson, S., Stump, C., Wilbur, D., and Quay, P. 1999. Accurate measurement of O_2, N_2, and Ar gases in water and the solubility of N_2. *Marine Chem.*, **64**: 337–347.

Falkowski, P. G. 1997. Evolution of the nitrogen cycle and its influence on the biological sequestration of CO_2 in the ocean. *Nature*, **387**: 272–275.

Farrenkopf, A. M., Dollhopf, M. E., Chadhain, S. N., Luther III, G. W., and Nealson, K. H. 1997a. Reduction of iodate in seawater during Arabian Sea shipboard incubations and in laboratory cultures of the marine bacterium *Shewanella putrefaciens* strain MR-4. *Mar. Chem.*, **57**: 347–354.

Farrenkopf, A. M., Luther III, G. W., Truesdale, V. W., and Van der Weijden, C. H. 1997b. Sub-surface iodide maxima: evidence for biologically catalyzed redox cycling in Arabian Sea OMZ during the SW intermonsoon. *Deep-Sea Res. II*, **44**: 1391–1409.

Ferguson, S. J. 1994. Denitrification and its control. *Anton. van Leeuwen.*, **66**: 89–110.

Fossing, H., Gallardo, V. A., Joergensen, B. B., Huettel, M., Nielsen, L. P., Schulz, H., Canfield, D. E., Forster, S., Glud, R. N., Gundersen, J. K., Kuever, J., Ramsing, N. B., Teske, A., Thamdrup, B., and Ulloa, O. 1995. Concentration and transport of nitrate by the mat-forming sulphur bacterium *Thioploca*. *Nature*, **374**: 713–715.

Froelich, P. N., Klinkhammer, G. P., Bender, M. L., Luedtke, N. A., Heath, G. R., Cullen, D., Dauphin, P., Hammond, D., Hartman, B., and Maynard, V. 1979. Early oxidation of organic matter in pelagic sediments of the eastern equatorial Atlantic: suboxic diagenesis. *Geochim. Cosmochim. Acta*, **43**: 1075–1090.

Fuhrman, J. A. and Capone, D. G. 2001. Nifty nanoplankton. *Nature*, **412**: 593–594.

Ganeshram, R. S., Pedersen, T. F., Calvert, S. W., and Murray, J. W. 1995. Large changes in the oceanic nutrient inventories from glacial to interglacial periods. *Nature*, **376**: 755–757.

Garfield, P. C., Packard, T. T., Friederich, G. E., and Codispoti, L. A. 1983. A subsurface particle maximum layer and enhanced microbial activity in the secondary nitrite maximum of the northeastern tropical Pacific Ocean. *J. Mar. Res.*, **41**: 747–768.

German, C. 2002. Bubbling under. *Nature*, **415**: 124–125.

Gruber, N. and Sarmiento, J. L. 1997. Global patterns of marine nitrogen fixation and denitrification. *Glob. Biogeochem. Cyc.*, **11**: 235–266.

Hahn, J. 1981. Nitrous oxide in the oceans. In C. C. Delwiche (ed.) *Denitrification, Nitrification and Atmospheric Nitrous Oxide*. Wiley Interscience, New York, pp. 191–240.

Hamme, R. C. and Emerson, J. R. 2002. Mechanisms controlling the global oceanic distribution of the inert gases, argon, nitrogen and neon. *Geophys. Res. Lett.*, **29**(23): 2120. doi: 10. 1029/2002G1015273.

Hobbie, J. E. (ed.) 2000. Estuarine Science: A Synthetic Approach to Research and Practice. Island Press, Washington, DC.

Holmes, R. M., Aminot, A., Kerouel, R., Hooker, B. A., and Peterson, B. J. 1999. A simple and precise method for measuring ammonium in marine and freshwater ecosystems. *Can. J. Fish. Aquat. Sci.*, **56**: 1801–1808.

Horibe, Y., Kodama, Y., and Shigehara, K. 1972. Errors in sampling procedure for the determination of dissolved oxygen by Winkler method. *J. Oceanogr. Soc. Jpn.*, **28**: 203–206.

Howell, E. A., Doney, S. C., Fine, R. A., and Olson, D. B. 1997. Geochemical estimates of denitrification in the Arabian Sea and the Bay of Bengal during WOCE. *Geophys. Res. Lett.*, **24**: 2549–2552.

Hulth, S., Aller, R. C., and Gilbert, F. 1999. Coupled anoxic nitrification/manganese reduction in marine sediments. *Geochim. Cosmochim. Acta*, **63**: 49–66.

Hutchinson, G. E. 1957 and 1975 (reissue). *A Treatise on Limnology*, Vol. I. John Wiley and Sons, New York.

Jorgensen, B. B. and Gallardo, V. A. 1999. *Thioploca* spp: Filamentous sulfur bacteria with nitrate vacuoles. *FEMS Microbiol.*, **28**: 301–313.

Joye, S. B., Smith, S. V., Hollibaugh, J. T., and Paerl, H. W. 1996. Estimating denitrification rates in estuarine sediments: a comparison of stoichiometric and acetylene based methods. *Biogeochemistry*, **33**: 197–215.

Kaplan, W. A. and Wofsy, S. C. 1985. The biochemistry of nitrous oxide: a review. In H. W. Jannasch and P. J. le B. Williams (eds) *Advances in Aquatic Microbiology 3*. Academic Press, London, pp. 181–206.

Karl, D. M., Letelier, R., Hebel, D. V., Bird, D. F., and Winn, C. D. 1992. *Trichodesmium* blooms and new production in the North Pacific gyre. In E. J. Carpenter and D. G. Capone (ed.) *Marine and pelagic cyanobacteria: Trichodesmium and other Diazotrophs* NATO Science Series Vol. 362. Kluwer, Dordrecht, pp. 219–237.

Kiørboe, T. 2001. Formation and fate of marine snow: small-scale processes with large-scale implications. *Sci Mar.*, **65**: 57–71.

Koike, I. and Hattori, A. 1975. Growth yield of a denitrifying bacterium, *Pseudomonas denitrificans*, under aerobic and denitrifying conditions. *J. Gen. Microbiol.*, **88**: 1–10.

Knowles, R. 1996. Denitrification: microbiology and ecology. *Life Support Biosphere Sci.*, **3**: 31–34.

Körner, H. and Zumft, W. G. 1989. Expression of denitrification enzymes in response to the dissolved oxygen level and respiratory substrate in continous culture of *Pseudomonas stutzeri*. *Appl. Environ. Microbiol.*, **55**: 1670–1676.

Kuypers, M. M. M., Sliekers, A. O., Lavik, G., Schmid, M., Jørgensen, B. B., Kuenen, J. G., Damsté, J. S. S., Strous, M., and Jetten, M. S. M. 2003. Anaerobic ammonium oxidation by anammox bacteria in the Black Sea. *Nature*, **422**: 608–611.

Lewis, B. L. and Luther III, G. W. 2000. Processes controlling the distribution and cycling of manganese in the oxygen minimum zone of the Arabian Sea. *Deep-Sea Res. II*, **47**: 1541–1561.

Lipschultz, F., Wofsy, S. C., Ward, B. B., Codispoti, L. A., Friederich, G., and Elkins, J. W. 1990. Bacterial transformations of inorganic nitrogen in the oxygen-deficient waters of the Eastern Tropical South Pacific Ocean. *Deep-Sea Res.*, **37**: 1513–1541.

Lloyd, D. L. 1993. Aerobic denitrification in soil and sediments: From fallacies to trends. *Ecol. Evol.*, **8**: 352–356.

Lovley, D. R. 1991. Dissimilatory Fe(III) and Mn(IV) reduction. *Microbiol Rev.*, **55**: 259–287.

Lovley, D. R. 1993. Dissimilatory metal reduction. *Annu. Rev. Microbiol.*, **47**: 263–290.

Luther III, G. W., Sundby, B., Lewis, B. L., Brendel, P. J., and Silverberg, N. 1997. The interaction of manganese with the nitrogen cycle in continental margin sediments: alternative pathways for dinitrogen formation. *Geochim. Cosmochim. Acta*, **61**: 4043–4052.

Manne, A. S. and Richels, R. G. 2001. An alternative approach to establishing trade-offs among greenhouse gases. *Nature*, **410**: 675–677.

Measures, C. I. and Vink, S. 1999. Seasonal variations in the distribution of Fe and Al in the surface waters of the Arabian Sea. *Deep-Sea Res. II*, **46**: 1597–1622.

Morrison, J. M., Codispoti, L. A., Smith, S. L., Wishner, K., Flagg, C., Gardner, W. D., Gaurin S., Naqvi, S. W. A., Manghnani, V., Prosperie, L. and Gundersen, J. S. 1999. The oxygen minimum zone in the Arabian Sea during 1995. *Deep Sea Res. II*, **46**: 1903–1931.

Murray, J. W., Codispoti, L. A., and Friederich, G. E. 1995. Oxidation-reduction environments: The suboxic zone in the Black Sea. In C. P. Huang, C. R. O'Melia, and J. J. Morgan (eds) *Aquatic Chemistry: Interfacial and Interspecies Processes*. American Chemical Society, Washington DC, pp. 157–176.

Naqvi, S. W. A. 1994. Denitrification processes in the Arabian Sea. In D. Lal, (ed) *Biogeochemistry of the Arabian Sea*. Indian Academy of Sciences, Bangalore, India, pp. 279–300.

Naqvi, S. W. A., Hansen, H. P., and Kureishy, T. W. 1986. Nutrient uptake and regeneration ratios in the Red Sea with reference to the nutrient budgets. *Oceanol. Acta*, **93**: 271–275.

Naqvi, S. W. A., Noronha, R. J., Somasundar, K., and Sen Gupta, R. 1990. Seasonal changes in the denitrification regime of the Arabian Sea. *Deep-Sea Res.*, **37**: 593–611.

Naqvi, S. W. A., Yoshinari, T., Brandes, J. A., Devol, A. H., Jayakumar, D. A., Narvekar, P. V., Altabet, M. A., and Codispoti, L. A. 1998. Nitrogen isotopic studies in the Arabian Sea. *Proc. Indian. Acad. Sci.*, **107**: 367–378.

Naqvi, S. W. A., Jayakumar, D. A., Narvekar, P. V., Nalk, H., Sarma, V. V. S. S., D'Souza, W., Joseph, S., and George, M. D. 2000. Increased marine production of N_2O due to intensifying anoxia on the Indian continental shelf. *Nature*, **408**: 346–349.

Nealson, K. H. and Saffarini, D. 1994. Iron and manganese in anaerobic respiration: environmental significance, physiology, and regulation. *Annu. Rev. Microbiol.*, **48**: 311–343.

Nevison, C. and Holland, E. 1997. A reexamination of the impact of anthropogenical fixed nitrogen on atmospheric N_2O and the stratosphere O_3 layer. *J. Geophys. Res.*, **102**: 22519–22536.

Ostrom, N. E., Russ, M. E., Popp, B., Rust, T. M., and Karl, D. M. 2000. Mechanisms of nitrous oxide production in the subtropical North Pacific based on determinations of the isotopic abundances of nitrous oxide and di-oxygen. *Chemosphere Glob. Change Sci.*, **2**: 281–290.

Paerl, H. W. 1990. Physiological ecology and regulation of N_2 fixation in natural waters. *Adv. Microb. Ecol.*, **8**: 305–344.

Pak, H., Codispoti, L. A., and Zaneveld, J. R. V. 1980. On the intermediate particle maxima associated with oxygen-poor water off western South America. *Deep-Sea Res.*, **27A**: 783–797.

Pierotti, D. and Rasmussen, R. A. 1980. Nitrous oxide measurements in the eastern tropical Pacific Ocean. *Tellus*, **32**: 56–72.

Poth, M. and Focht, D. D. 1985. [15]N kinetic analysis of N_2O production by *Nitrosomonas europaea*: an examination of nitrifier denitrification. *Appl. Environ. Microbiol.*, **49**: 1134–1141.

Rabalais, N. N., Turner, R. E., Justic, D., Dortch, Q., Wiseman, W. J., and Sen Gupta, B. K. 2000. Gulf of Mexico biological system responses to nutrient changes in the Mississippi River. In J. W. Hobbie (ed) *Estuarine Science: A Synthesis Approach*. Island Press, Washington DC, pp. 241–268.

Redfield, A., Ketchum, B. H., and Richards, F. A. 1963. The influence of organisms on the composition of seawater. In M. N. Hill (ed) *The Sea*. Academic Press, New York, pp. 26–77.

classic Richards, F. A. 1965. Anoxic basins and fjords. In J. P. Riley, and G. Skirrow, (eds) *Chemical Oceanography*. Academic Press, New York, pp. 611–645.

Richards, F. A. 1975. The Cariaco Basin (trench). pp. 11–67. In H. Barns (ed) *Oceanography and Marine Biology: An Annual Review* Vol. 13. George Allen and Unwin Ltd, London.

Schmidt, I., Sliekers, O., Schmid, M., Cirpus, I., Strous, M., Bock, E., Kuenen, J. G., and Jetten, M. S. M. 2002. Aerobic and anaerobic ammonia oxidizing bacteria—competitors or natural partners? *FEMS Microbiol. Ecol.*, **39**: 175–181.

Scranton, M. I., Sayles, F. L., Bacon, M. P., and Brewer, P. G. 1987. Temporal changes in the hydrography and chemistry of the Cariaco Trench. *Deep-Sea Res.*, **34**: 945–963.

Seitzinger, S. P., Nielsen, L. P., Caffrey, J. M., and Christensen, P. B. 1993. Denitrification measurements in aquatic sediments: a comparison of three methods. *Biogeochemistry*, **23**: 147–167.

Senn, D. B. and Hemond, H. F. 2002. Nitrate controls on iron and arsenic in an urban lake. *Science*, **296**: 2373–2376.

Shailaja, M. S. 2001. Respiratory enzyme activities in the oxygen-deficient waters of the Arabian Sea. *Curr. Sci.*, **81**: 684–689.

Smil, V. 1997. Global population and the nitrogen cycle. *Sci. Am.*, **277**: 76–81.

Stouthamer, A. H., Boogerd, F. C., and van Verseveld, H. W. 1982. The bioenergetics of denitrification. *Anton. van Leeuwen.*, **48**: 545–553.

Straub, K. L., Benz, M., Schink, B., and Widdel, F. 1996. Anaerobic, nitrate-dependent microbial oxidation of ferrous iron. *Appl. Environ. Microbiol.*, **62**: 1458–1460.

Strous, M., Fuerst, J. A., Kramer, E. H. M., Logemann, S., Muyzer, G., van de Pas-Schoonen, K. T., Webb, R., Kuenen, J. G., and Jetten, M. S. M. 1999. Missing lithotroph identified as new planctomycete. *Nature*, **400**: 446–449.

Thamdrup, B. and Dalsgaard, T. 2000. The fate of ammonium in anoxic manganese oxide-rich marine sediment. *Geochim. Cosmochim. Acta*, **64**: 4157–4164.

Thamdrup, B. and Dalsgaard, T. 2002. Production of N_2 through anaerobic ammonium oxidation coupled to nitrate reduction in marine sediments. *Appl. Environ. Microbiol.*, **68**: 1312–1318.

Van Cappellen, P., Viollier, E., and Roychoudhury, A. 1998. Biogeochemical cycles of manganese and iron at the oxic-anoxic transition of a stratified marine basin (Orca Basin, Gulf of Mexico). *Environ. Sci. Technol.*, **32**: 2931–2939.

Van Mooy, B. A. S., Keil, R. G., and Devol, A. H. 2002. Impact of suboxia on sinking particulate organic carbon: enhanced carbon flux and preferential degradation of amino acids via denitrification. *Geochim. Cosmochim. Acta*, **66**: 457–465.

Vitousek, P. M., Aber, J. D., Howarth, R. W., Likens, G. E., and Matson, P. A. 1997. Human alteration of the global nitrogen cycle: Sources and consequences. *Ecol. Appl.*, **7**: 737–750.

Ward, B. B. 1996. Nitrification and denitrification: probing the nitrogen cycle in aquatic environments. *Microb. Ecol.*, **32**: 247–261.

Ward, B. B. and Zafiriou, O. C. 1988. Nitrification and nitric oxide in the oxygen minimum of the eastern tropical North Pacific. *Deep-Sea Res.*, **35**: 1127–1142.

Ward, B. B., Glover, H. E., and Lipschultz, F. 1989. Chemoautotrophic activity and nitrification in the oxygen minimum zone off Peru. *Deep-Sea Res.*, **36**: 1031–1057.

Watling, L. and Norse, E. A. 1998. Disturbance of the seabed by mobile fishing gear: a comparison to forest clearcutting. *Conserv. Biol.*, **12**: 1180–1197.

Williams, P. J. le B. and Jenkinson, N. W. 1982. A transportable microprocessor-controlled precise Winkler titration suitable for field station and shipboard use. *Limnol. Oceanogr.*, **27**: 576–584.

Witter, A. E., Lewis, B. L., and Luther III, G. W. 2000. Iron speciation in the Arabian Sea. *Deep-Sea Res. II*, **47**: 1517–1539.

Wolgast, D. M., Carlucci, A. F., and Bauer, J. E. 1998. Nitrate respiration associated with detrital aggregates in aerobic bottom waters of the abyssal NE Pacific. *Deep Sea Res.*, **45**: 4–5.

Yoshida, N. 1988. ^{15}N-depleted N_2O as a product of nitrification. *Nature*, **2335**: 528–529.

Yoshinari, T. and Koike, I. 1994. The use of stable isotopes for the studies of gaseous nitrogen species in the marine environment. In K. Lajtha, and R. H. Michener (eds) *Stable Isotopes in Ecology and Environmental Science*. Blackwell Scientific, Oxford, pp. 114–137.

Zafiriou, O. C. 1990. Laughing gas from leaky pipes. *Nature*, **347**: 15–16.

Zumft, W. G. 1997. Cell biology and molecular basis of denitrification. *Microbiol. Mol. Biol. Rev.*, **61**: 533–616.

Zumft, W. G. and Cardenas, J. 1979. The inorganic biochemistry of nitrogen bioenergetic processes. *Naturwissenschaften*, **66**: 81–88.

Zumft, W. G. and Körner, H. 1997. Enzyme diversity and mosaic gene organization in denitrification. *Anton. van Leeuwen.*, **71**: 43–58.

CHAPTER 13

Incorporating plankton respiration in models of aquatic ecosystem function

Kevin J. Flynn

Ecology Research Unit, University of Wales, Swansea, UK

Outline

This chapter considers the needs, ways and means to describe respiratory activities within ordinary differential equation based dynamic models of planktonic organisms. An overview of the relationship between respiration, regeneration, and biogeochemical cycling precedes an introduction to ordinary differential equation modeling and to types of model. Individual subsections considers modeling respiration in bacteria, phytoplankton and zooplankton (both micro' and meso' forms). The conclusions include a consideration on future directions, noting that a critical limiting factor is the lack of experimental data from ecologically important organisms that are suitable for model development and parameterization.

13.1 Introduction

13.1.1 Why and when model

Models are constructed for various purposes but typically they are made to provide a dynamic framework based on current knowledge (both conceptual and in terms of numeric values) to enable scientists to judge how complete the knowledge base is and/or to conduct "what if" tests. If the model is constructed using a good knowledge base then the model is more likely to be of value for testing. The attempt at constructing a model is itself a worthy activity, indicating where knowledge is weak and hence where experimentation is needed. Experimentation and data collection, and the construction and use of models, should thus be an iterative process.

At the outset, one needs to consider the level of interest; specifically in the context of this work which atomic constituent(s) will be considered, whether the subject is considered as an individual (e.g. a copepod) or a population (e.g. bacteria per liter of seawater), together with the setting (organism as a detached entity, such as in a test tube, or within an ecosystem).

Is it necessary to model respiratory activity explicitly? The modeling of respiration, at the simplest level, just provides a "loss" term. If net growth (equals gross growth minus respiration) is the required primary output of the model, then one may argue that it is not necessary to explicitly describe respiration at all. Indeed many models of oceanographic processes do not even describe C (e.g. Evans and Garçon 1997). Accounting for the regeneration of nutrients, which is inextricably linked to respiratory activity, is given prominence but may be attained by assigning a proportion of intake (e.g. of N) that is retained versus that which is lost (which in reality is regenerated through various processes, including respiration). Flows of carbon may then be calculated assuming fixed stoichiometric (Redfield-like) ratios of $C:N:P(:Si)$. However, the assumption of fixed ratios is flawed and can lead to significant errors in model output (as discussed by Flynn 2003 for phytoplankton models). The corollary is that respiration should itself be modeled explicitly and not included within a general loss term for C.

13.1.2 Types of model

Models may be divided into empirical and mechanistic. The former describe relationships in mathematical terms with no specific attempt to explain the underlying biological relationship. These may include allometric terms, for example, relating respiration rate to body size and temperature (though such relationships also turn up in the most complex of models, for example, Carlotti and Hirche 1997). Often, especially in older works, the term "model" may be used to describe a single linear, exponential, or polynomial equation for such an allometric relationship. The "test" for such models is thus a statistic such as regression.

The term model as used, in this chapter, will refer to a series of equations with a dynamic, time-based component. These models describe the flow of materials between compartments, be they abiotic and/or biotic, in processes such as trophic webs.

All models, but especially simple models, make assumptions as to which processes are most important and how (if at all) the implications of processes that are not explicitly included are to be acknowledged. For respiration, one could describe separately basal (resting) respiration, that associated with biosynthesis of biomass from acquired nutrients ("food"), and that which supports behavior such as predatory (hunting) activity. Or one could lump these three together and relate total respiration to the growth rate, arguing that the higher the growth rate, the greater the synthesis and feeding activities, etc., and hence higher the overall respiration burden. The more complex the model, the greater the scope for detail, but also the greater the problem is for parameterization (assigning values to constants and finding real data for comparison with model output).

Mechanistic models, or perhaps more realistically pseudo-mechanistic models, have a construction that is reasoned upon biological knowledge (e.g. biochemical or physiological basis). They may contain various state variables between which material flows. These have a history, such as gut contents, biomass, and eggs in a model of zooplankton, for example, Carlotti and Hirche 1997. Typically these flows are subject to interactions with feedback loops, a critical feature of real biological systems. The detail of such models may be extensive (akin to a biochemical pathway) but more likely specific activities, such as nutrient transport or prey capture, are described in more detail than others, such as growth at the expense of gut contents.

An important aspect of respiration is that it is a rate. If the topic of this work was biomass or nutrient concentration then our treatment within experiments and models could be very different. In modeling terms, biomass and concentrations are assigned as state variables; these have histories. A sample can be taken and (provided steps are taken to prevent deterioration during storage) such components remain unchanged in the future. Rate processes are auxiliary variables within models; they change with time with no direct history. You cannot take a rate and put it in a bottle for future analysis. Indeed, and of critical importance, the process rate may change rapidly in response to sampling during our estimation of the process during incubations. While dynamic models are obvious platforms for exploring the implications of rate processes, parameterizing them is difficult. In the absence of experimentally derived rate estimates, respiration rates in models have to depend on the tuning (optimization) of the whole model against values for state variables. The trouble is that the net rate of growth of the state variable for biomass is in reality the result of several (perhaps very many) rate processes, none of which necessarily have robust (safe) experimentally derived estimates over a good time course. An incubation of 12–24 h duration for determining respiration may well equate to a generation time (or so) for a bacterium, but only a small fraction of a life cycle of a copepod.

Some rate processes are more likely to be computed within models as differences in a balancing act between inputs and outputs. It is easier to define "loss" of C (as the difference between gross and net growth) than to separate it into respiration plus excretion or defecation. Because of the great intrinsic variability of respiration (especially with behavior) simulating changes in the respiration rate with time is not often included; establishing an "average" rate estimate is difficult enough and such values are easiest to include as "constants" in models. By the same token, photosynthesis is often integrated over the daily light–dark cycle, although there is an accompanying loss of fidelity that can be important (Flynn and Fasham 2003).

Models may be constructed to describe individual organisms (individual based models, IBMs) or populations in terms of total biomass (usually as a concentration). The larger the organism, the more likely an IBM is used. Larger organisms are also more complex, with multiple stage life cycles, in contrast with the binary fission of microbes. Given sufficient computing power, a population could be simulated as an array of IBMs (e.g. Woods and Onken 1982).

13.1.3 Modeling platforms

Most biological models make use of ordinary differential equations (ODEs). Traditionally the use of mathematics at this level, together with the programming skills required to code the mathematics, acted to prevent the use of modeling by biologists. Over the last decade, however, there has been the development of a several object orientated modeling platforms for use on desktop computers. These include Stella, iThink, and Powersim Constructor. Now the biologist can concentrate on the science of getting the structure correct rather than worrying so much about the mechanics. That said, it is still very easy for the novice to make basic blunders. Although the software will often identify programming errors it will not locate structural or integration errors that may, for example, result in a model failing to give an output that provides a mass balance (i.e. the total amount of carbon or nitrogen from all sources is not constant as it should be). It is important that submodels be tested to ensure that mass balance is attained, that all carbon supplied as prey can be accounted for as predator biomass, eggs, CO_2, faeces, and dissolved organic C, for example. The work of Haefner (1996) gives a good starting point for the biologist who is interested in this type of biological modeling.

13.2 Functions and interactions of respiration

Respiratory activity is associated with several functions that we may divide between basal (r_{bas}), biosynthetic (r_{syn}), and behavioral (r_{beh}). We can thus describe net growth, μ_{net} according to equation (1), noting that the rate of respiration associated with biosynthetic activity and (here) behavior are functions of the gross growth rate, μ_{gross}, while the basal respiration rate is scaled to the maximum net growth rate, μ_{max}. In this instance there is actually no justification in separating metabolic and behavioral respiration, while if behavioral activity were constant (perhaps for an organism that swims constantly) then r_{beh} should be scaled to μ_{max} and hence combined with r_{bas}. I shall return to behavioral activity but first will focus on basal and metabolic respiration.

$$\mu_{net} = \mu_{gross} - (\mu_{max} \cdot r_{bas} + \mu_{gross} \cdot r_{syn}$$
$$+ \mu_{gross} \cdot r_{beh}) \tag{1}$$

13.2.1 Basal and metabolic respiration

A level of basal respiration is needed by an organism just to stay alive, to maintain pH and ion gradients, and repair damage/decay. The work of Pirt (1982) indicates that the level of basal respiration depends on the limiting substrate in bacteria. P-limitation in particular can result in high levels (Fig. 13.1). One may question whether this is really basal respiration *per se* or if P-limitation coupled with excess carbon leads to poorly controlled (cyclic) metabolic respiration.

Most low molecular weight metabolites are derived at one level or another from products of respiratory cycles, such as the tricarboxylic acid cycle. The operation of these biochemical cycles for anaplerotic activity, aside from the operation of pathways to supply adenosine triphosphate and reductant, inevitably results in respiration. One may make the assumption that biochemical activities are not very different between organisms and thus that the cost of assimilation of 1 g of nitrogen

runs which form intermediates of the TCA cycle

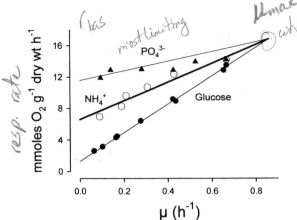

resp. rate (handwritten)

Handwritten annotations: r_{bas} · *most limiting* · *μmax* · ① *why do these converge?* · → *scaling r_{bas} to $μmax$*

Figure 13.1 Relationship between growth rate limited by glucose, ammonium, or phosphate, and respiration rate in *Klebsiella aerogenes* growing at steady-state in a chemostat. Data of Neijssel and Tempest (1976) with lines from the model of Pirt (1982).

into bacterial biomass is similar to that into copepod biomass. It may or may not be safe to do so, but provided one discriminates between gross differences such as anaerobic fermentation versus aerobic processes, more often than not we have little option. Usefully, through the use of mechanistic models, we may explore how sensitive output is to variations in factors such as metabolic costs. The repair of damage and decay is a continual process, a proportion of which is associated with turnover of proteins such as enzymes as a function of cellular regulation. A proportion of "basal" respiration is thus linked closely to biosynthetic activity.

Pirt (1982) argued that the form of equation (1) (ignoring the behavioral component) was incorrect and that two types of basal respiration should be considered, one of which (r_{bas1}) declined as a function of increasing growth rate (referenced to the maximum gross growth rate, $μ_{gmax}$), while the other (r_{bas2}) is constant, as in equation (2). The value of maximum gross growth rate, $μ_{gmax}$, is given by $μ_{max} \cdot (1 + r_{syn} + r_{bas2})$, as it must provide sufficient growth to counter the maximum rates of synthesis-linked and basal respiration (note that at this maximum level of growth $r_{bas}1$ is no longer of consequence). The unit for all parameters here are element specific (e.g. $mol\,C\,mol^{-1}\,C\,d^{-1}$, that is, the net dimensions d^{-1} are the same as for growth rate).

$$\mu_{net} = \mu_{gross} - \left[\mu_{max} \cdot \left(\left(1 - \frac{\mu_{gross}}{\mu_{gmax}} \right) \cdot r_{bas1} \right. \right.$$
$$\left. \left. + r_{bas2} \right) + \mu_{gross} \cdot r_{syn} \right] \qquad (2)$$

Basal respiration r_{bas1} is a function of the limiting nutrient(s), perhaps associated with metabolic processes that the cell has to undertake in order to circumvent that particular stress. Thus r_{bas1} appears to be higher in P-limited bacteria than in glucose-limited bacteria (Fig. 13.1). As the stress is alleviated, so this type of respiration is minimized and eventually disappears. The second type of basal respiration, r_{bas2}, is present throughout (associated with maintaining cellular homeostasis), while the respiration associated with biosynthetic metabolism is as before. We need to question what r_{bas2} really is and whether we can actually discriminate in experiments between respiration associated purely with biosynthesis and that associated with protein recycling, for example. One may expect that the rate of recycling is not likely to be a constant but to vary with the overall growth rate. So, if r_{syn} is associated solely with new synthesis and r_{bas2} is associated with repair and decay, then r_{bas2} should be a function of $μ_{gross}$. Unless we have data upon which to identify these forms of respiration, then r_{bas2} in equation (2) may be deleted and the respiration described by it distributed between r_{bas1} and r_{syn}, with r_{syn} now describing both new and recycling metabolic activity, equation (3).

$$\mu_{net} = \mu_{gross} - \left[\mu_{max} \cdot \left(1 - \frac{\mu_{gross}}{\mu_{gmax}} \right) \cdot r_{bas1} \right.$$
$$\left. + \mu_{gross} \cdot r_{syn} \right] \qquad (3)$$

The relationship between $μ$ and respiration is linear for both types of relationship. The reason why one may wish to use equation (2) or (3) rather than equation (1), is to simulate the type of steady-state response shown in Fig. 13.1, where respiration at growth rates near maximal are similar irrespective of the limiting nutrient, but they are different at low growth rates. Hence r_{syn}

Handwritten at bottom: $μ_{gmax} = μ_{max} \cdot (1 + r_{syn} + r_{bas2})$

is constant while r_{bas1} differs depending on the limiting nutrient. Equation (1) can only give this experimentally derived response with a constant r_{bas} and a substrate-specific r_{syn} that is used at all growth rates; this argument is not easy to accept, especially if the rate limitation of growth is switched between nutrients of widely different r_{syn}. In fact, however, r_{syn} is not necessarily a constant even for a single nutrient type, but may depend on the form in which that nutrient is supplied; we shall return to this when considering a model of bacterial growth.

Another problem is that of deciding which value of r_{bas1} to employ in a situation where the most limiting nutrient changes. Figure 13.1 describes a situation where only one nutrient is limiting while in nature several nutrients are likely to be rate limiting at any one instant. If the usual threshold method is used (where the most limiting nutrient sets the growth rate in the model) then the operational value of r_{bas1} will suddenly change. This can be most easily overcome by the use of a proportionating equation to weight the operational value of r_{bas1} depending on the relative levels of nutrient stress.

13.2.2 Behavior

We now return to respiratory activity associated with behavioral functions, typically associated with motility, some of which may be for food capture, predator avoidance, or reproduction. An organism surrounded with feed will be able to assimilate nutrients more readily while not expending energy on hunting. Clearly there is a level of cost–benefit in movement for hunting but to what extent one models it depends on the aims of the modeler. For some organisms, specifically microbes, it has been argued that the cost of motility is minor (e.g. Raven and Richardson 1984) and as (except at death) many of these organisms are constantly moving then any cost may be included with r_{syn} in equation (3), or with r_{bas2} in equation (2). If behavioral activity is inversely related to growth rate (a starving organizer hunts for food, while a food-replete organism surrounded by food does not need to), then we could approximate this relationship to that for r_{bas1}. The behavioral cost is then added to the basal

respiratory activity (r_{bas1}) and is lessened as growth rate improves with nutritional status (i.e. as μ_{gross} approaches μ_{gmax}). A more complex relationship would require inclusion of a specific term related to growth rate, to nutritional status, or perhaps to some feature of the external environment (food type or physicochemical feature such as turbulence).

13.2.3 Temperature

Temperature is a critical factor affecting growth and other activities and hence respiration. In models, that link may or may not be included. If it is then often an Arrhenius equation (e.g. Collins and Boylen 1982; see also Pace and Prairie, Chapter 7; Hernández-León and Ikeda, Chapter 5), reference to Q_{10} (e.g. Robinson and Williams 1993 report a Q_{10} close to 2 for Antarctic bacteria), or perhaps to a full temperature-activity curve, may be employed. However, temperature does not just affect respiration but rather it affects growth rates in general terms. For example, Rivkin and Legendre (2001) report that bacterial growth efficiency is inversely related to temperature, implying a differential effect on growth and respiration. Temperature effects on respiration in models can probably for the most part relate *pro rata* to effects on growth and nutrient/feed acquisition processes. This can be achieved simply by making μ_{max} a function of temperature, assuming that the derivation of μ_{gross} and μ_{gmax} are (as one would expect) also functions of μ_{max}.

13.2.4 Death

In reality, how long an organism can survive in the absence of a positive value of μ_{net} depends on the availability of previously accumulated carbon. Within a model, however, the implications of a negative μ_{net} will depend on whether the simulation is an IBM or for a population. If the model is an IBM then there is a finite amount of reserve carbon (e.g. oil within a copepod) and once that is exhausted the organism being simulated "dies." The model of Fiksen and Carlotti (1998) for the vertically migrating copepod *Calanus finmarchicus* stresses the sensitivity of the model outcome on the respiratory rate during dormancy. For a population, however, the model could simply display a decline in total biomass; that is the mathematical consequence but most likely not the biological consequence. The implication is that the total biomass is an energy reserve for the individual, which is clearly nonsense. One could justify this type of response where cannibalism occurs or for a bacterial population where individuals survive by degrading the cells of others. Otherwise in reality members of the population would either all die over a short period of time, or they would form a resting stage with near-zero or zero respiration. It is worth noting that in ecosystem models "phytoplankton," "bacteria," and "zooplankton" represent cohorts of individual species that are likely to exhibit different activities (including respiration rates) and in some instances (especially zooplankton) may consume each other.

For ecosystem models of annual events (such as oceanic production), and of course in reality, it is vital that the population does not die during the winter else it would never restart. To avoid having to construct a detailed mathematical description of such events, it is easiest simply to ignore basal respiration in such models. Respiration then only occurs when gross growth occurs. Other death terms may be included (sinking, predation, etc.), but the unstated assumption is that basal respiration only occurs concurrently with other activity and hence can be effectively included as a loss term with biosynthetic and behavioral terms.

The desire for simplification that pervades most modeling activities usually prevents a simulation of processes such as encystment that would "turn off" basal respiration when μ_{net} falls below 0. The formation of resting stages, and recovery from them, is a behavioral activity, and only detailed models consider the implications of r_{beh}. Rather rapidly, then, equation (3) collapses to equation (4), and we are left with essentially a loss term associated with growth activities.

$$\mu_{net} = \mu_{gross} \cdot (1 - r_{syn}). \tag{4}$$

This assumes a linear relationship between respiratory loss and growth rate. One may argue that this should not be linear, but making r_{syn} a power or sigmoidal function of μ_{gross} does not alter the final

relationship between μ_{gross} and μ_{net} greatly over reasonable values of r_{syn} and generates equations that would be better resolved by inclusion of the biological detail implicit with the use of r_{bas} and r_{beh}. Often a general "loss" term for mortality will then be included.

Frequently then, as a mathematical modeling expedient, basal respiration is ignored. Yet if the response seen in the data of Pirt (1982) is in anyway representative of bacteria, and other organisms, in general then this basal respiration is extremely important. One may argue that the growth efficiency of marine bacteria (Connolly et al. 1992) is so much lower than one may expect from the efficiency of organisms such as Escherichia coli (computed from data in Ingraham et al. 1983), at least in part, because their growth is rate limited by substrate availability and hence basal respiration is relatively much more important. An empirical model will overcome this problem by making the value of r_{syn} in equation (4) much greater to compensate, or by using a value of growth efficiency to drive the model. In a mechanistic model the growth efficiency will be an output, not an input, variable. There is no right or wrong way to model (assuming the model is not dysfunctional within its operating environment—see Flynn 2003), but it is important to appreciate the implications of the simplifications; while a biologist may do so, a mathematician may not.

13.3 Models of autotrophic versus heterotrophic organisms

There are some fundamental differences between phytoplankton and heterotrophs such as bacteria and zooplankton. Aside from the obvious difference in energy capture, autotrophs typically obtain nutrients as discrete packages, such as carbon dioxide, nitrate, and phosphate. The organisms, and hence our models of them, can regulate resource capture to match current demands. Heterotrophs typically obtain a proportion (bacteria) if not all (phagotrophs, etc.) of their nutrients in combined form. Thus they consume food with $C:N:P$ ratios that are not necessarily in keeping with their metabolic demands. This is a critical difference that affects respiration and nutrient regeneration and forms the basis of modeling the same. A good starting point for a consideration of the implications of heterotrophic nutrition in this context is Anderson (1992), who works through the issues stepwise generating steady-state solutions for the fate of ingested food through respiration (carbon as CO_2), regeneration (here nitrogen as NH_4^+, but could include phosphorus as PO_4^{3-}), voiding (combined organic material), and into new biomass. The voiding of carbon in this treatment is the balance of ingestion minus respiration and incorporation. Respiration as such is, then, assumed to vary as a function of growth rate, with the basal rate added on.

Models of microbes (bacteria, phytoplankton, and to some extent also microzooplankton) often have common components, based on Monod (1949) and/or quota (Droop 1968; Caperon and Meyer 1972) equations. The Monod equation makes growth a function solely of external nutrient supply. Thus in equation (5), S is the substrate concentration, K_s is the half-saturation constant for growth and μ_{max} is the maximum growth rate, and μ the resultant growth rate as a rectangular hyperbolic function of S. For a predator model, S is prey concentration. A threshold term may be included that prevents nutrient (or prey) acquisition unless the external concentration is above a set level.

$$\mu = \mu_{max} \cdot \frac{S}{S + K_s}. \qquad (5)$$

In the quota model (the Caperon and Meyer 1972 version is given in equation (6)), growth is made a rectangular hyperbolic function of the internal nutrient concentration described here as a ratio of nutrient X to carbon biomass, with XC_0 the minimum ratio and K_{qx} a constant for altering the curve form. For a whole model, the quota model is coupled with a Monod type equation controlling the entrance of nutrient or prey into the organism as a function of external availability.

$$\mu = \mu_{max} \cdot \frac{XC - XC_0}{XC - XC_0 + K_{qx}}. \qquad (6)$$

Note that neither of these extensively used equations make any reference to respiration; μ describes net growth. The simplest route to simulating the removal of a resource that is subsequently respired

is not even to consider its entry into the organism but just to make an additional loss term on the resource equivalent to $\mu \cdot r \cdot B$, where r is the biomass or nutrient specific respiration rate, μ is (as before) the net biomass or nutrient specific growth rate, and B is the biomass concentration (e.g. $molC\ dm^{-3}$).

13.4   Phytoplankton models

Of all components of the aquatic ecosystem phytoplankton, the primary producers, have been the subject of the greatest modeling effort. Typically in ecosystem models, however, growth is simulated as net photosynthesis integrated over 24 h; the actual processes of respiration are not simulated. That may seem particularly peculiar given the importance of the critical depth (in comparison with the mixing depth) in defining the start of phytoplankton growth with all its implications for ecosystem productivity (Cushing 1995). Respiratory activity is typically only incorporated into detailed models of phytoplankton, specifically those considering the cost implications of assimilating oxidized versus reduced forms of N, diel processes (especially dark nutrient assimilation) and the implications of Fe-stress.

Banse (1976) considers the cell-size relationship with respiration and whether this could be employed at a rather superficial level to compute the net growth efficiency (NGE) for algal growth. One may argue that, in general terms, smaller cells should have lower carbon-specific respiration rates and hence higher NGEs. However, Banse could not find substantive evidence that NGE was actually different in algal cells of different size (conflicting with Laws 1975) This is convenient because in ecosystem models, while some workers do use size-specific growth differences, cell size is not usually a driving force. Rather, organism-specific (diatom versus non-diatom) features are used, as these are more important as keystone factors in trophic and biogeochemical processes. At a logical minimum then, respiration in these organisms may be described using a method like that in Collos (1997), making use of a basal term (as a function of the maximum photosynthetic rate, typically 5–10%) and a metabolic respiration term relating respiration to the current growth (or photosynthesis rate). This is the form of equation (1). Laws and Bannister (1980) suggest that respiration during darkness amounts to 10–30% of day-light carbon-fixation, but that depends on organism types (dinoflagellates respire more than diatoms) and on the level of dark-N assimilation.

13.4.1 Respiration and photoreductant

Much respiratory activity in phytoplankton occurs concurrently with photosynthesis, including the provision of photoreductant for the reduction of nitrate through to ammonium prior to nitrogen-assimilation. The interaction between carbon, nitrogen, and phosphorus in respiration is very complex; fortunately we do not need to model these interactions mathematically at anything like the level shown in the conceptual (biochemical) model in Gauthier and Turpin (1997). The way that this provision is most easily simulated (e.g. Flynn et al. 1997; Geider et al. 1998) is to assume that actually photosynthesis gives rise to organic carbon that is then respired to support the assimilatory activity. This has the advantage that one does not need to be concerned over the phasing of photo-generated versus respiration-generated support for nutrient assimilation during exposure of the organism to changes in irradiance over the course of the day. Flynn et al. (1997) assumed these costs would be similar in phytoplankton to those in bacteria and yeast, obtaining 2 and 1.75 mol C mol^{-1} N for the reduction of nitrate to ammonium and for incorporation of ammonium into cell-nitrogen, respectively.

Another assumption made is that protein recycling is non-significant. This is unlikely to be the case. Inclusion of a recycling route within the model, as used by John and Flynn (2002) for the simulation of the synthesis of toxins using recycled-nitrogen, leads to an increased respiratory cost. In this instance, the organism in question, dinoflagellate, typically do have higher respiratory rates (Harris 1978). A major problem with modeling recycling is simply providing a half-life to use for protein turnover. In the absence of detailed information, John and Flynn (2002) set a value that resulted in a final respiration rate in keeping with the

expected rate. Thus respiration accounts for around 30% of gross photosynthesis in their model.

13.4.2 Dark nutrient assimilation and diel vertical migration

On average, half the day is spent in darkness but assimilation of nitrogen can continue in darkness at the expense of previously accumulated carbon. The extent of that assimilation depends on how efficient the cell is at controlling respiratory costs and using ammonium rather than nitrate. At first sight the extra cost in reducing nitrate would appear to be of great significance. However, when one includes the drain on carbon for the synthesis of amino acids, the difference is lessened to 1.31 rather than 2.14 (Clark and Flynn 2002). One then needs to ask whether it is necessary to model the light–dark cycle of carbon and nitrogen assimilations. There can be an important degree of decoupling (e.g. Eppley and Sharp 1975) so that if simulations are to be run over time frames of days (rather than months) then it is important that the model can simulate the dynamics correctly.

In a series of modeling studies, Flynn et al. (2002) and Flynn and Fasham (2002, 2003) showed that the differences in assimilating nitrate or ammonium for diel vertical migrating (DVM) species were lessened further by the fact that only cells that were significantly N-stressed (large surplus of carbon) would engage in DVM. In these cells the lack of nitrogen was of greater importance than the efficiency of the use of the surplus carbon. This type of study is an example where models can contribute to our understanding of biology through enabling the running of "virtual" experiments that would be difficult or even impossible to conduct in reality. Such studies can, however, only be conducted using models with a sound mechanistic basis.

13.4.3 Fe-stress and respiration

Iron is required primarily in support of photosynthesis, respiration, and nitrate/nitrite reduction (and also for nitrogenase in N_2-fixers). The balance of distribution depends on the irradiance, growth rate, and N-source (Fig. 13.2). The simulation of

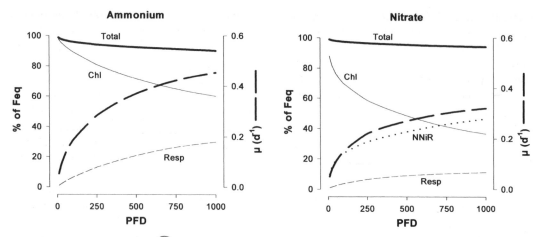

Figure 13.2 Distribution of the Fe quota (Feq) within an Fe-limited phytoplankton cell growing on ammonium or nitrate at different photo flux densities (PFD) μmol photons m^{-2} s^{-1}). The labels denote: total, total Fe accounted for; Chl, Fe associated with photosystems; Resp, Fe associated with respiration; NNiR, Fe associated with the enzymes of nitrate and nitrite reduction. Note that growth using nitrate is slower, and hence the respiration rates and allocation of Fe to respiration are less. Output generated by the model of Flynn and Hipkin (1999).

the division of cellular iron between the activities assumes certain costs (Raven 1990). While the synthesis of photosystems (through photoacclimation) and perhaps the synthesis of the enzymes of nitrate reduction can be made functions of the availability of unaccounted cellular-iron, respiratory activity does not have a state variable in the model of Flynn and Hipkin (1999). The iron-cost of respiration (cytochromes, etc.) was computed according to the current respiration rate and an assumed cost of that process. The synthesis of respiratory systems, such as mitochondria, were thus considered to have priority over other demands for iron.

13.5   Bacterial models

13.5.1   Respiration and regeneration during use of inorganic nutrients versus organics

Many models of aquatic system do not even include bacteria as a state variable. If they are described then their primary role is in converting organic material (decay processes) back to inorganic nutrients for phytoplankton use. Respiration, then, equals nutrient regeneration and the gross growth efficiency (GGE), or its equivalent, is usually fixed (though there is doubt as to whether bacterial GGE is indeed

invariant: Goldman *et al.* 1987; del Giorgio and Cole 2000; Goldman and Dennett 2000). The division of bacterial respiration into different basal (maintenance) and biosynthetic (metabolic) costs associated with growth (Pirt 1982) has already been discussed. Placing these concepts into a model, Thingstad (1987) explored the meaning of the term "maintenance" respiration, finding the simulation of such metabolism easier when using a quota-based model (making reference to surplus cellular-carbon) than using a Monod structure. Thingstad (1987) also indicated the importance of phosphorous-stress in affecting respiration in bacteria. Alas, most experimental studies of marine bacteria concentrate on carbon and/or nitrogen limitations.

For those who wish to model bacterial respiration in greater detail there are considerable data available, albeit for *E. coli* (Ingraham *et al.* 1983). From such information it is possible to derive growth and respiration rates when using different amino acid substrates. More realistically for aquatic systems, studies of the activity of bacteria using different mixed substrates are required. Bacteria, though to a lesser extent than phytoplankton, regulate their nutrient intake to match their needs. Phagotrophic organisms are very different in this respect; they cannot select parts of an organism, specifically the $C:N:P$, prior to engulfing the feed. Bacteria can also retrieve particular parts of nutrient molecules making use of extracellular digestion: a good example being the use of alkaline phosphatase to liberate inorganic-phosphorus for uptake from organic-phosphorus. This is important because it affects the resultant fate of the "surplus" part of the molecule and hence how we model it. The issue is this: given a bacteria with a set $C:N:P$ and nutrients with a different ratio (often a higher carbon content) what is the fate of the surplus nutrient? Is it rejected or respired/regenerated? For a bacterium, it is more likely that surplus carbon entering the cell will be respired. It is quite easy to construct a stochiometric model to simulate either form of the "loss" of carbon, but that loss could actually be a rejection associated with extracellular digestion (in which instance the carbon, dissolved organic carbon (DOC) is still available as an energy source for other microbes) or respiration (carbon is lost from organics and thus no longer of use to heterotrophs). Under anaerobic conditions (in particular) DOC is also likely to be lost as fermentation products. For the bacterium in question it does not matter, but for future generations of cells that may consume these products it is important that this aspect of a simulation is correctly modeled. Suffice to say that modeling bacterial ecophysiology probably presents a greater challenge than that for phytoplankton or zooplankton.

Growth of bacteria using organic substrates that are readily assimilated (primary metabolites such as amino acids and simple sugars) should be quicker than that supported by the use of inorganic nitrogen and phosphorus and carbohydrates. At least that is the extrapolation from laboratory cultures growing on nutrient broth versus a simple medium. Partly this is because fewer resources need be directed at synthesis and operation of the associated biochemical machinery. The decreased use of respiratory cycles for metabolite synthesis may also represent a considerable saving. However, if the organic substrates are relatively refractory, not readily being incorporated into bacterial biomass, the gains are lost because the substrates need to be degraded to simpler components (consuming resources) prior to synthesis. Goldman and Dennett (2000) found no evidence of a suppression of ammonium use when

amino acids were present, though any such link will depend on the types and concentration of the amino acids available, as well as on the availability of a carbon source to enable ammonium utilization. The problem for marine bacteria, or more specifically for us, is that we do not know what proportion of bacterial growth these three groups of substrates (simple primary metabolites, inorganic nutrients, and more complex organics) support, and hence the respiratory and regenerative activities are less easy to model.

While the intimacy of the relationship between phytoplankton and bacteria (competition for common nutrients while bacteria use organics leaked/released by the algae) is known to be important, few have attempted to model it in any detail (e.g. Ietswaart and Flynn 1995). To do the job properly we need a model that can switch between the nutrient sources; that is a nontrivial task. Beyond the level of dissolved inorganic-nitrogen and organic-nitrogen (DIN versus DON), going into the detail of DON and DOC composition would be unwarranted, not least because of the lack of experimental and field data to support it. One approach for describing DOC–DON–DIN interactions is developed below.

13.5.2 A model of bacterial growth and respiration

The following gives a brief description of a model describing growth of a bacterium on dissolved inorganic N (DIN), organic-nitrogen (DON), and organic-carbon (DOC). It is one of a series a models developed by Flynn and Anderson (in preparation). A dynamic version of this type of model may be constructed using one of the readily available object-orientated modeling programs available for PCs; a steady-state version is easily constructed within a spreadsheet. The definitions of state and auxiliary variables, constants, and equations are given in Tables 13.1 and 13.2 and Box 13.1, respectively.

The basis of the model is a single state variable (bacterial carbon) with a fixed bacterial $C:N:P$ ratio. It handles nutrients in the form of DOC (defined here as labile material containing only carbohydrate), ᶜᴴᴼ DOM (i.e. carbon coupled with nitrogen and/or *sv vs. av* (handwritten)

Table 13.1 State variables (sv) and auxiliary variables (av) for bacteria model

Parameter	Units	Type	Explanation
BC	molC dm^{-3}	sv	Bacterial C-biomass
BR	molC mol^{-1} C d^{-1}	av	Basal respiration rate
C_{res}	molC mol^{-1} C d^{-1}	av	Total respiration rate
DIX	molX dm^{-3}	sv	Concentration of inorganic X substrate
DIX^{vel}	dl	av	Index for relative use of inorganic X
DIX$_{use}$	molX mol^{-1}C d^{-1}	av	Use of inorganic X
DOC	molC dm^{-3}	sv	Concentration of simple labile DOC (containing no X)
DOC^{vel}	dl	av	Index for relative use of simple labile DOC
DOC$_{use}$	molC mol^{-1}C d^{-1}	av	Use of simple labile DOC
DOM	molC dm^{-3}	sv	Concentration of DOM (i.e. DOC containing X)
DOMC^{vel}	dl	av	Index for relative use of DOM-C
DOM$_{use}$	molC mol^{-1}C d^{-1}	av	Use of DOM-C
DOMX	molX dm^{-3}	sv	Concentration of X associated with DOM
DOMX^{vel}	dl	av	Index for relative use of DOM-X
DOMX$_{use}$	molX mol^{-1}C d^{-1}	av	Use of DOM-X
GE$_C$	dl	av	Growth efficiency for C
GE$_X$	dl	av	Growth efficiency for X
$^{gross}C_{use}$	molC mol^{-1}C d^{-1}	av	Gross use of C
$^{gross}X_{use}$	molX mol^{-1}C d^{-1}	av	Gross use of X
N$_{rsyn}$	molC mol^{-1}N	av	Average respiratory expenditure for assimilation of N
r_{bas}	molC mol^{-1}C	av	Basal respiratory expenditure
relX	dl	av	Index of nutrient sufficiency; value>1 indicates C rather than X limits
totC^{vel}	molC mol^{-1}C d^{-1}	av	Total potential C acquisition rate
totX^{vel}	molX mol^{-1}C d^{-1}	av	Total X acquisition rate
V$_{max}$	d^{-1}	av	Maximum gross C growth rate to attain μ_{max} using DIN
μ	d^{-1}	av	Realized net growth rate

[a] Note: dl—dimensionless.

(handwritten annotations:) state – static, measurable #; auxiliary – rate or ratio (variable)

Table 13.2 Constants for bacteria model

Constant	Units	Value	Explanation
BXC	molX mol^{-1} C	X=N, 0.17; X=P, 0.008	Stoichiometric bacterial ratio of $X:C$
DIN$_{rsyn}$	molC mol^{-1}N	1.75	Respiratory cost for assimilation of DIN-N
DON$_{rsyn}$	molC mol^{-1}N	1	Respiratory cost for assimilation of amino-N
K$_{DOC}$	μmolC dm^{-3}	0.167	Half-saturation constant for DOC acquisition
K$_{DOM}$	μmolC dm^{-3}	1.67	Half-saturation constant for DOM-C acquisition
K$_X$	μmolX dm^{-3}	X=N, 0.0036; X=P, 0.003	Half-saturation constant for X acquisition
r_{bas}C	molC mol^{-1} C	0.05	Basal respiratory cost when C-limited
r_{bas}N	molC mol^{-1} C	0.1	Basal respiratory cost when N-limited
r_{bas}P	molC mol^{-1} C	0.2	Basal respiratory cost when P-limited
relCpref	dl	1	Scaling factor for preference of C from DOM
relXpref	dl	1	Scaling factor for preference of C from DOMX
μ_{max}	d^{-1}	4.158	Maximum growth rate ($\equiv$6 divisions d^{-1})

[a] Note: dl—dimensionless.

Box 13.1 Equations for bacterial model

1. Establishing the maximum uptake rate (DON, DIN and DIP by analogy)

$$DOC^{vel} = \frac{DOC}{DOC + K_{DOC}} \tag{7}$$

$$DOMC^{vel} = \frac{relCpref \cdot DOM}{DOM + K_{DOM}} \tag{8}$$

$$V_{max} = \mu_{max} \cdot (1 + DIN_{rsyn} \cdot BNC \tag{9}$$
$$+ r_{bas}C)$$

2. Establishing rate of C and X acquisition

$$totC^{vel} = MIN[V_{max}, V_{max} \cdot (DOC^{vel} \tag{10}$$
$$+ DOMC^{vel})]$$

$$totX^{vel} = V_{max} \cdot BXC \cdot MIN[DIX^{vel} \tag{11}$$
$$+ DOMX^{vel}, 1]$$

3. Establishing index of nutrient sufficiency
→ determining limiting nut

$$relX = \frac{totX^{vel}/totC^{vel}}{BXC} \tag{12}$$

$$gross C_{use} = MIN[relN, relP, 1] \cdot totC^{vel} \tag{13}$$

4. Partitioning OC use between DOC and DOM

$$DOC_{use} = \frac{gross C_{use} \cdot DOC^{vel}}{DOC^{vel} + DOMC^{vel}} \tag{14}$$

$$DOM_{use} = gross C_{use} - DOC_{use} \tag{15}$$

5. Establishing the respiratory cost of N assimilation

$$N_{rsyn} = \frac{DIN_{rsyn} \cdot DIN^{vel} + DON_{rsyn} \cdot DOMN^{vel}}{DIN^{vel} + DON^{vel}} \tag{16}$$

cost of assimilation

6. Determination of the gross rates of N and P assimilation and, for N, its form (DIN or DON)

$$gross N_{use} = \frac{gross C_{use}}{1/BNC + N_{rsyn}} \tag{17}$$

$$gross P_{use} = gross N_{use} \cdot \frac{BPC}{BNC} \tag{18}$$

$$DOMX_{use} = \frac{gross X_{use} \cdot DOMX^{vel}}{DOMX^{vel} + DIX^{vel}} \tag{19}$$

$$DIX_{use} = gross X_{use} - DOMX_{use} \tag{20}$$

like ①

7. Computing the basal metabolic expenditure (as molC molC^{-1})

$$r_{bas} = r_{bas}C + MAX[(1 - relN) \cdot (r_{bas}N - r_{bas}C)$$
$$\cdot (relN \leq 1), (1 - relP) \cdot (r_{bas}P - r_{bas}C)$$
$$\cdot (relP \leq 1)] \tag{21}$$

8. Calculation of basal, then total metabolic rate (as molC molC^{-1} d^{-1})

$$BR = \mu_{max} \cdot r_{bas} \tag{22}$$

$$BR = \mu_{max} \cdot r_{bas} \cdot \left(1 - \frac{gross C_{use}}{V_{max}}\right) \tag{23}$$ Pirt eq.

choose 1

like eq. 2+3

$$C_{res} = DOMN_{use} \cdot DON_{rsyn} + DIN_{use} \cdot DIN_{rsyn}$$
$$+ BR \tag{24}$$

9. Calculation of resultant growth rates and carbon and nutrient growth efficiencies

$$\mu = gross C_{use} - C_{res} \tag{25}$$

$$GE_C = \frac{gross C_{use} - C_{res}}{gross C_{use}} \cdot 100 \tag{26}$$

$$GE_X = \frac{gross X_{use} - BR \cdot BXC}{gross X_{use}} \cdot 100 \tag{27}$$

phosphorous), DIN, DIP, DOMN (nitrogen in DOM) and DOMP (phosphorous in DOM).

It is assumed that nitrogen and phosphorus associated with DOM are handled as separate elements. DOMP is universally hydrolyzed by the action of external phosphatases and 5′ nucleotidases releasing DIP for uptake, while nitrogen is

either deaminated or taken in as amino/nucleic acids. Accordingly carbon is sourced from DOC and/or DOM, nitrogen from DIN and/or DOMN, and phosphorus from DIP and/or DOMP. From hereon X is read for nitrogen or phosphorous.

Each nutrient acquisition is associated with hyperbolic functions with respective half saturation constants and may also be scaled relative to each other (allowing a simulation of the implications of surge transport kinetics). The result of this is the derivation of a potential rate for acquisition for each nutrient source, giving DOC^{vel}, DIX^{vel}, etc. (e.g. equation (7)). Equation (8) shows the use of a relative preference scaling index (relCpref); this must be $\geqslant 1$; the preferred element source carries the higher index. The value for K_{DOM}, or relCpref, could be made a function of the C:N or C:P of DOM to make the preference lessen as the material becomes more refractory. In order to attain the required maximum net growth rate, μ_{max}, when using the more resource expensive DIN, the required maximum rate of (gross) carbon acquisition is defined by equation (9). Reference to the basal respiration rate when limited by carbon (or more specifically when not limited by X), $r_{bas}C$, is omitted if the Pirt (1982) respiration method is employed as at $\mu = \mu_{max}$ there is no basal respiration. The values of potential velocity for nutrient source acquisition are used to compute the potential velocity for carbon acquisition, thus giving $totC^{vel}$ (equation (10)).

The rate-limiting element is established in equation (12), as the element having the lowest relX (if relX > 1 then carbon rather than X is limiting), with the gross need for carbon then defined by equation (13). Provision of this carbon is partitioned between DOC and DOM according to equations (14) and (15), by reference to DOC^{vel} and $DOMC^{vel}$ (derived as per, equations (7) and (8)).

The cost of assimilating nitrogen provided as DIN and/or DON is given by equation (16). This, together with the stoichiometric bacterial N:C, BNC, is used to derive the acquisition of nitrogen equation (17), with the acquisition of phosphorus being given by proportion (equation (18)). The use of these nutrients from organic and/or inorganic sources are described by equations (19) and (20), respectively.

Basal respiratory costs are defined by equation (21), with reference to $r_{bas}C$, $r_{bas}N$, and $r_{bas}P$, the respiratory costs when carbon-limited, or when limited by nitrogen or P. This is used to obtain the basal respiration rate using either equation (22) or, for the Pirt approach, equation (23). With the latter, basal respiration approaches zero as the growth rate approaches its

maximum. Total carbon-specific respiration cost, accounting for metabolic costs and basal costs, is given by equation (24) and the net growth rate, μ, by equation (25). Growth efficiencies for the use of carbon and of X are given by equations (26) and (27). In this model the values of GE_N and GE_P are the same, locked as they are by the stoichiometric ratio both for acquisition and for excretion (the latter occurring with basal respiration only).

Equation (24) is analogous to equation (3) (Section 13.2.1) except that we do not have a single r_{syn}. Account is made for differences in the assimilatory costs of using DIN and DON. The constant values used here for these costs (DIN_{rsyn} and DON_{rsyn}) were derived from data for E. coli (Ingraham et al. 1983). More often than not species-specific data are unavailable so we have to make assumptions that common biochemical processes are not too dissimilar between sometimes very different organisms. The DIN source is assumed to be ammonium; the use of nitrate would require an additional respiratory expense and also a mechanism within the model to switch between DIN sources, perhaps along the lines described in Evans and Garçon (1997) or in more detail by Flynn (2001). The assimilation cost for DON (DON_{rsyn}) assumes the source is suite of amino acids that are readily metabolized. It also assumes that the carbon-source supporting its assimilation is a simple sugar. The sources of carbon here are DOC and DOM; whether the carbon-source is thus readily used in support of metabolism is not assured. The sources of DOC and DON for a naturally growing organism are most unlikely to be single simple compounds. More likely they may represent a complex mixture, perhaps actually the products of extracellular digestion, with some components more refractory than others. Thus glutamine and arginine fit more readily into metabolic pathways than does histidine. In addition, the concentration of these individual compounds is most likely very low and potentially rate limiting at the point of transport into the cell.

What one can do, however, with a model is to subject it to a sensitivity analysis, to determine how sensitive the output (here the growth rate) is to changes in the values of constants describing substrate affinity and respiratory costs. If the model is shown to be particularly sensitive to certain constants, and importantly assuming that we are happy that this is not purely a function of the structure of the model itself, then those facets of biological activity should be made focal points for future research. If one were to argue that only a few forms of DON in marine waters could be readily incorporated directly into bacterial-N, then it may be safer to assume that DON is deaminated after uptake and that the respiratory cost of assimilation would thus be the same as that for DIN entering as NH_4^+.

An example of the output of the model is shown in Fig. 13.3, using different forms of the respiration function and different costs associated with the assimilation of nitrogen. It can be seen that actually there appears to be little justification for involving the complexity of the Pirt equation (equation (23), Box 13.1); the differences in the absolute values of respiration rates may be altered by changing the respiration constants. What can be seen, however, is the significance of the growth limitation by carbon, nitrogen, or phosphorous. In addition, altering the N-assimilation costs ("high cost" plots in Fig. 13.3) cause growth deficiencies to decline rapidly with substrate limitation. It appears difficult to believe that growth efficiencies for bacteria growing in a nutrient impoverished environment should be lower than those for bacteria growing in nutrient rich systems (such as *E. coli* within the gut) because one would expect natural selection to achieve the converse. However, substrate limitation can explain such a loss in efficiency if as a consequence the basal respiration rate becomes a significant burden, as suggested by the model.

13.6 Zooplankton models

In most ecosystem models zooplankton are treated in a very rudimentary fashion, sometimes they are not described explicitly at all, with predation on phytoplankton being just another loss term (e.g. Evans and Garçon 1997). There are some arguments for this low-level representation, namely the poor and contrasting quality of field data for this diverse group of organisms. Where zooplankton are specified, either as "zooplankton" or divided into microbial predators (ciliates and other protists, the microzooplankton) or metazoa (copepods etc.,

the mesozooplankton), the main inputs and outputs are predation, feeding efficiency (as a global term), and loss (mainly through higher trophic activity and mortality). Respiration, *per se*, is usually not mentioned. Only when we consider more mechanistic models, with multiple currencies (usually carbon and N) is respiration specified as a unique loss term.

Zooplankton, of whatever type, are very different to bacteria and phytoplankton in that their food comes as packages that are taken in whole. A common basis of multinutrient models of zooplankton is stoichiometry (Anderson 1992; Anderson and Pond 2000). The C:N value of the predator is considered to be fixed while that of the food may be lower (bacteria), equal (N-sufficient phytoplankton), or higher (nitrogen-deprived phytoplankton). While feeding on bacteria will clearly result in an excess of nitrogen (regenerated as ammonium), that situation is exacerbated by the inevitable respiration of some of the bacterial-carbon. Thus respiration varies with not only the C:N of the predator and prey, but with growth efficiency of the zooplankton. It also varies with the excretion rate. However, more typically, the respiration rate is assumed to account for a constant fraction of ingested material and hence the stoichiometric equation is arranged to report the excretion rate. Both the excretion rate and the growth efficiency will also vary with the effect of prey quality (in this context, C:N) as feeding activity changes with prey density and palatability. As Davidson (1996) points out, the dynamics of predator-prey interaction are complex with inadequate experimental knowledge to fully support model development.

13.6.1 Microzooplankton models

Models of microzooplankton are reviewed by Davidson (1996). There is scant mention of respiration in models of these organisms and little has changed since that date. Thus respiration is typically a constant fraction of carbon entering the organism and/or of growth rate. Thingstad and Pengerud (1985) simulated respiration rate as a fixed quantity per cells added to the population at each time point for bacteria, but used a fixed fraction of ingested carbon to describe respiration in the predator (with

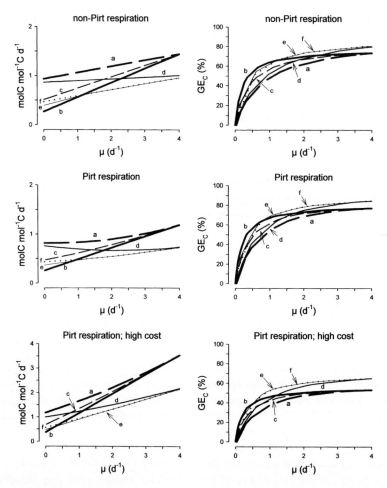

Figure 13.3 Comparison of respiration rates and GE_C versus growth rates for the bacteria model described in Tables 13.1, 13.2, and Box 13.1, with different configurations of the respiratory function. The non-Pirt model used equation (22) (Box 13.1) while the Pirt model used equation (23). For the "high cost" simulations the values of DIN_{rsyn} and DON_{rsyn} (Table13. 2) were tripled. The keys denote: a, P-limited + DOC and DIN saturated; b, DOC-limited + DIN and P saturated; c, DIN-limited + DOC and P saturated; d, P-limited + DOC and DON saturated; e, DON-limited + P saturated; f, DON-limited + P and DOC saturated.

a stoichiometric loss of nitrogen). Davidson *et al.* (1995) combined respiration and voiding of carbon together as "carbon-loss." The safest assumption is to include a basal level of respiration (that also supports motility for hunting) plus a rate associated with biochemical processes for biomass synthesis (see Section 13.2.2 on behavioral respiration). As an approximation we could use the bacterial cost for assimilation of ammonium into biomass (ca. 1.75 molC mol^{-1} N). Because components such as amino acids will actually be gained from the food, rather

than being synthesized *de novo* from ammonium and keto acids, this will actually be an overestimate but then using the rate for bacterial growth on amino acids would most likely be an underestimate because of the (perceived) greater inherent efficiency of bacterial metabolic processes.

13.6.2 Mesozooplankton models

Models of mesozooplankton range from the extremely simplistic (e.g. Evans and Garçon 1997) to the highly detailed (Carlotti and Nival 1992; Caparroy and Carlotti 1996; Carlotti and Hirche 1997). The former do not mention respiration specifically, while the latter includes it at several levels (basal, metabolic/synthesis, and behavioral). It is also far more likely that the detailed models will be of individuals (IBMs), and indeed of particular life stages (such as egg-producing adults).

It is only for IBMs that inclusion of a detailed description of respiration is really warranted; at simpler levels equation (4) is adequate. Respiration rates are, however, of particular importance during periods of relative inactivity, such as with diapause in over wintering species (such as *Calanus* in the Atlantic) and over the day-light hours in species undertaking DVM (Fisken and Carlotti 1998). For diapausal species the issue is one of survival at low temperature for many months making use of previously accumulated ("stored") lipid. Modeling of this process requires a knowledge of the rate of respiration at the *in situ* temperature. Active swimming consumes more energy and it is difficult to measure respiration rates because of the stress caused to the animals in attempting to bring them up from (often great) depth for experiments. It is possible that overwintering animals do eat (although primary production may be minimal, other particulate organics are present), and thus contribute to nutrient regeneration through respiration. Models of such activities will help in the design of the very experiments required for their parameterization.

Diel vertical migration activity is rather different. Here animals migrate considerable distances and spend time at depth in lower temperatures. There is some debate over whether most metabolic activity is concurrent with feeding, and during the descent phase (both most time at depth being inactive), or whether metabolic activity is maintained. Feeding activity associated with DVM may provide a mechanism for the modification of the vertical flux of nitrogen (Hays *et al.* 1997); modeling of this activity presents an interesting and useful challenge in which respiration plays an important part.

The models of Carlotti and Hirche (1997) and Carlotti and Nival (1992) give descriptions for facets of copepod biology; the construction and operation of models of such detail can be instructive. These models include specific reference to respiration. Caparroy and Carlotti (1996) gives a description of different predatory modes, either actively hunting or ambushing prey, enabling the cost–benefit of these types of activity within a calm or turbulent water column to be considered.

13.7 Ecosystem models

While models of individual organisms may be reasonably detailed, the typical ecosystem model resorts to using extremely simple submodels (often Monod-type constructions, like equation (5)) for each organism type (if indeed the organisms themselves are represented). Often, then, respiration is not mentioned as such or if it is then it is held as a constant even in relatively complex models (e.g. Moloney and Field 1991; Baretta-Becker *et al.* 1994; Roelke 2000). In fairness, variations in respiration rates in ecosystem models are the least of the worries, for there are serious problems simply in obtaining biomass data for many organisms, while the submodels themselves may be criticized as being inadequate on grounds that that they contain more profound errors than in simulating respiration (e.g. for phytoplankton, Flynn 2003).

13.8 Conclusions

Models fulfill various functions. Detailed inclusion of respiratory activity in models of aquatic systems is not warranted for current ecosystem models; fixed rates, or at most respiratory terms scaled to growth rates, for groups of organisms suffice. One may argue that only for mechanistic models of individual processes, or of processes within the individual, is real detail required. To a large extent this state of affairs reflects the view (rightly or wrongly) that simulating growth that is net of respiratory activity is sufficient. In simulation scenarios where behavioral responses become important a variable respiration rate requires a more explicit simulation of growth as being the balance between energy input and output. In this context behavior may include selection of alternative feed as well as motility, but these respiratory costs are most likely to be important for larger organisms with discontinuous feeding in a patchy ecosystem. From a biogeochemical perspective, the respiratory activity of bacteria is the most important. More work is required to develop and test models of bacterial growth in response to different nutrient availabilities, especially with respect to multinutrient limitations.

A good test of a model is whether output from a simulation could be passed as real by a biologist. However, unless the biologist has access to an adequate data bank, that judgment may be nontrivial. In this capacity, models may be useful in directing research by identifying areas of particular weakness. Models also form dynamic reviews of knowledge. Each of the values of respiration given elsewhere in this book translate to an estimate encompassing a range of possible values within a steady-state or dynamic environment. The state of knowledge of each system may be gauged by our ability, or more likely inability, to construct models that reproduce those respiration rates.

References

Anderson, T. R. 1992. Modelling the influence of food C : N ratio, and respiration on growth and nitrogen excretion in marine zooplankton and bacteria. *J. Plankton Res.*, **14**: 1645–1671.

Anderson, T. R. and Pond, D. 2000. Stoichiometric theory extended to micronutrients: comparison of the roles of essential fatty acids, carbon and nitrogen in the nutrition of marine copepods. *Limnol. Oceanogr.*, **45**: 1162–1167.

Banse, K. 1976. Rates of growth, respiration and photosynthesis of unicellular algae as related to cell size—a review. *J. Phycol.*, **12**: 135–140.

Baretta-Bekker, J. G., Riemann, B., Baretta, J. W., and Rasmussen, E. K. 1994. Testing the microbial loop concept by comparing mesocosm data with results from a dynamical simulation model. *Mar. Ecol. Prog. Ser.*, **106**: 187–198.

Caparroy, P. and Carlotti, F. 1996. A model for *Acartia tonsa*: effect of turbulence and consequences for the related physiological processes. *J. Plankton Res.*, **18**: 2139–2177.

Caperon, J. and Meyer, J. 1972. Nitrogen-limited growth of marine phytoplankton. I. Changes in population characteristics with steady-state growth. *Deep-Sea Res.*, **19**: 601–618.

Carlotti, F. and Hirche, H. J. 1997. Growth and egg production of female *Calanus finmarchicus*: an individual-based physiological model and experimental validation. *Mar. Ecol. Prog. Ser.*, **149**: 91–104.

Carlotti, F. and Nival, P. 1992. Model of copepod growth and development: moulting and mortality in relation to physiological processes during an individual moult cycle. *Mar. Ecol. Prog. Ser.*, **84**: 219–233.

Clark, D. R. and Flynn, K. J. 2002. N-assimilation in the noxious flagellate *Heterosigma carterae* (Raphidophyceae); dependence on light, N-source and physiological state. *J. Phycol.*, **38**: 503–512.

Collins, C. D. and Boylen, C. W. 1982. Physiological responses of *Anabaena variabilis* (Cyanophyceae) to instantaneous exposure to various combinations of light intensity and temperature. *J. Phycol.*, **18**: 206–211.

Collos, Y. 1997. A physiological basis for estimating inorganic carbon release during photosynthesis by natural phytoplankton. *Ecol. Model.*, **96**: 285–292.

Connolly, J. P., Coffin, R. B., and Landeck, R. E. 1992. Carbon utilization by bacteria in natural water systems. In C. Hurst (ed.) *Modelling the Metabolic and Physiologic Activities of Microorganisms*. John Wiley and Sons, New York, pp. 249–276.

Cushing, D. H. 1995. *Population Production and Regulation in the sea: A Fisheries Perspective*. Cambridge University Press, Cambridge, UK. p. 354.

Davidson, K. 1996. Modelling microbial food webs. *Mar. Ecol. Prog. Ser.*, **145**: 279–296.

Davidson, K., Cunningham, A., and Flynn, K. J. 1995. Predator–prey interactions between *Isochrysis galbana* and *Oxyrrhis marina*. III. Mathematical modelling of predation and nutrient regeneration. *J. Plankton Res.*, **17**: 465–492.

del Giorgio, P. A. and Cole, J. J. 2000. Bacterial Energetics and growth efficiency. In D. L. Kirchman (ed.) *Microbial Ecology of the Oceans*. Wiley, New York, pp. 289–325.

Droop, M. R. 1968. Vitamin B_{12} and marine ecology. IV. The kinetics of uptake, growth, and inhibition in *Monochrysis lutheri*. *J. Mar. Biol. Assoc. UK*, **48**: 689–733.

Eppley, R. W. and Sharp, J. H. 1975. Photosynthetic measurements in the central North Pacific: the dark-loss of carbon in 24-h incubations. *Limnol. Oceanogr.*, **20**: 981–987.

Evans, G. T. and Garçon, V. C. (eds). 1997. One-dimensional models of water column biogeochemistry. *JGOFS Report*, 23/97, JGOFS Bergen, Norway.

Fiksen, O. and Carlotti, F. 1998. A model of optimal life history and diel vertical migration in *Calanus finmarchicus*. *Sarsia*, **83**: 129–147.

Flynn, K. J. 2001. A mechanistic model for describing dynamic multi-nutrient, light, temperature interactions in phytoplankton. *J. Plankton. Res.*, **23**: 977–997.

Flynn, K. J. 2003. Modelling multi-nutrient interactions in phytoplankton; balancing simplicity and realism. *Progr. Oceanogr.*, **56**: 249–279.

Flynn, K. J. and Fasham, M. J. R. 2002. A modelling exploration of vertical migration by phytoplankton. *J. Theor. Biol.*, **218**: 471–484.

Flynn, K. J. and Fasham, M. J. R. 2003. Operation of light-dark cycles within simple ecosystem models of primary production and the consequences of using phytoplankton models with different abilities to assimilate N in darkness. *J. Plankton Res.*, **25**: 83–92.

Flynn, K. J. and Hipkin, C. R. 1999. Interactions between iron, light, ammonium, and nitrate: insights from the construction of a dynamic model of algal physiology. *J. Phycol.*, **35**: 1171–1190.

Flynn, K. J., Fasham, M. J. R., and Hipkin, C. R. 1997. Modelling the interaction between ammonium and nitrate uptake in marine phytoplankton. *Philos. Trans. R. Soc. Lond.*, **352**: 1625–1645.

Flynn, K. J., Clark, D. R., and Owens, N. J. P. 2002. The dark assimilation of nitrogen by phytoplankton II; a modelling study of different strategies. *New Phytol.*, **155**: 109–119.

Gauthier, D. A. and Turpin, D. H. 1997. Interactions between inorganic phosphate (P_i) assimilation, photosynthesis and respiration in the P_i-limited green alga *Selenastrum minutum*. *Plant Cell Environ.*, **20**: 12–24.

Geider, R. J., MacIntyre, H. L., and Kana, T. M. 1998. A dynamic regulatory model of phytoplankton acclimation to light, nutrients and temperature. *Limnol. Oceanogr.*, **43**: 679–694.

Goldman, J. C., Caron, D. A., and Dennett, M. R. 1987. Regulation of gross growth efficiency and ammonium regeneration in bacteria by substrate C:N ratio. *Limnol. Oceanogr.*, **32**: 1239–1252.

Goldman, J. C. and Dennett, M. R. 2000. Growth of marine bacteria in batch and continuous culture under carbon and nitrogen limitation. *Limnol. Oceanogr.*, **45**: 789–800.

Haefner, J. W. 1996. *Modeling Biological Systems*. Chapman and Hall, New York.

Harris, G. P. 1978. Photosynthesis, productivity and growth: the physiological ecology of phytoplankton. *Arch. Hydrobiol. Beih. Ergeb. Limnol.*, **10**: 1–171.

Hays, G. C., Harris, R. P., Head, R. N., and Kennedy, H. 1997. A technique for the *in situ* assessment of the vertical nitrogen flux caused by the diel vertical migration of zooplankton. *Deep-Sea Res. I*, **44**: 1085–1089.

Ietswaart, T. and Flynn, K. J. 1995. Modelling interactions between phytoplankton and bacteria under nutrient-regenerating conditions. *J. Plankton Res.*, **17**: 729–744.

Ingraham, J. L., Maaloe, O., and Neidhardt, F. C. 1983. *Growth of the Bacterial Cell*. Sinuar, Sunderland, MA, 435pp.

John, H. J. and Flynn, K. J. 2002. Modelling changes in paralytic shellfish toxin content of dinoflagellates in response to nitrogen and phosphorus supply. *Mar. Ecol. Prog. Ser.*, **225**: 147–160.

Laws, E. A. 1975. The importance of respiration losses in controlling the size distribution of marine phytoplankton. *Ecology*, **56**: 419–426.

Laws, E. A. and Bannister, T. T. 1980. Nutrient- and light-limited growth of *Thalassiosira fluviatilis* in continuous culture, with implications for phytoplankton growth in the ocean. *Limnol. Oceanogr.*, **25**: 457–473.

Moloney, C. L. and Field J. G. 1991. The size-based dynamics of plankton food webs. I. A simulation model of carbon and nitrogen flows. *J. Plankton Res.*, **13**: 1003–1038.

Monod, J. 1949. The growth of bacterial cultures. *Annu. Rev. Microbiol.*, **3**: 371–394.

Neijssel, O. M. and Tempest, D. W. 1976. Bioenergetic aspects of aerobic growth of *Kelbsiella aerogenes* NCTC 418 in carbon-limited and carbon-sufficient chemostat culture. *Arch. Microbiol.*, **107**: 215–221.

Pirt, S. J. 1982. Maintenance energy: a general model for energy-limited and energy-sufficient growth. *Arch Microbiol.*, **133**: 300–302.

Rivkin, R. B. and Legendre. L. 2001 Biogenic carbon cycling in the upper ocean: effects of microbial respiration. *Science*, **291**: 2398–2400.

Robinson, C. and Williams, P. J. le B. 1993. Temperature response of Antarctic plankton respiration. *J. Plank. Res.*, **15**: 1035–1051.

Raven, J. R. 1990. Predictions of Mn and Fe use efficiencies of phototrophic growth as a function of light availability for growth and C assimilation pathway. *New Phytol.*, **117**: 1–18.

Raven, J. A. and Richardson, K. 1984. Dinophyte flagella: a cost–benefit analysis. *New Phytol.*, **98**: 259–276.

Roelke, D. L. 2000. Copepod food-quality threshold as a mechanism influencing phytoplankton succession and accumulation of biomass, and secondary productivity: a modeling study with management implications. *Ecol. Model.*, **134**: 245–274.

Thingstad, T. F. 1987. Utilization of N, P and organic C by heterotrophic bacteria. I. Outline of a chemostat theory with a consistent concept of "maintenance" metabolism. *Mar. Ecol. Prog. Ser.*, **35**: 99–109.

Thingstad, T. F. and Pengerud, B. 1985. Fate and effect of allochthonous organic material in aquatic microbial ecosystems. An analysis based on chemostat theory. *Mar. Ecol. Prog. Ser.*, **21**: 47–62.

Woods, J. D. and Onken, R. 1982. Diurnal variation and primary production in the ocean—preliminary results of a Lagrangian Ensemble method. *J. Plankton Res.*, **4**: 735–756.

CHAPTER 14

The global significance of respiration in aquatic ecosystems: from single cells to the biosphere

Paul A. del Giorgio[1] and Peter J. le B. Williams[2]

[1] *Département des sciences biologiques, Université du Québec à Montréal, Canada*
[2] *School of Ocean Sciences, University of Wales, Bangor, UK*

Outline

The respiration of aquatic ecosystems represents the largest gap in our understanding of the global carbon cycle. In this chapter, we synthesize the existing information regarding the magnitude and regulation of respiration in the major aquatic ecosystems of the biosphere, and draw from the main conclusions of the different chapters of this book. We also provide a contextual background to place this information, and to identify the major gaps in our current knowledge and future areas of research.

The largest and also most uncertain component of this global estimate is the respiration in the surface layers of the open oceans, which we can proscribe no better than a range from 9 to 12 $Pmol\,C\,a^{-1}$. Globally, the rounded off estimates for respiration in aquatic ecosystems appear to be in the range of 14–17 $Pmol\,C\,a^{-1}$, higher than current estimates of global terrestrial respiration. The global estimates of respiration suggest that primary production, carbon export, and total carbon flux in aquatic ecosystems, particularly in the open oceans, have been severely underestimated. We suggest that respiration provides the most robust estimates of aquatic primary production, and because it integrates so many aspects of the functioning of the ecosystem, long-term shifts in respiration may potentially provide the best warning system for global change.

14.1 Introduction

The most common and classical view of respiration is as the major sink of organic matter in the biosphere, and in this view respiration is a loss term that has significance only in as much as it is compared to the gain, that is, production. This view of respiration as completely tied to production has had both practical and conceptual implications: from a practical standpoint, the assumption of a close coupling and direct dependence of respiration on primary production has resulted in that respiration is most often derived from production, rather than being measured directly, particularly in models of carbon flux. At the conceptual level much of our current understanding of respiration at the ecosystem level rests on the assumptions that aquatic respiration is both regulated and bound by primary production at the scales that are of ecological and biogeochemical interest. The result is that there are extremely few studies that have explicitly assessed the magnitude and regulation of respiration *per se*, and our current understanding of respiration rests very much on the current perceptions concerning production. The shortcoming of the common view that respiration is simply determined by photosynthesis is most clearly seen in the open oceans, where external sources of nutrients are extremely low, and almost all photosynthesis is run on nutrients mineralized at some point in the ocean during respiration. An explicit analysis of aquatic respiration is the main objective of this book, and in this final chapter we attempt to synthesize the findings and conclusions of the different sections of the book, and in turn, integrate these into a global perspective.

We begin by exploring the various roles that respiration may play in aquatic ecosystems, other than the widely accepted role as the main carbon sink. As we will see below, although the regulation of aquatic respiration may ultimately rest on primary production over the entire biosphere and over long periods, at the temporal and spatial scales that are most frequently addressed in ecological and biogeochemical studies, the coupling between respiration and production is often weak and there are immediate factors, other than photosynthesis, that influence respiration. The various chapters of this book have explored the physiological basis of respiration at the cellular and organism level, as well as quantitatively synthesized the existing information on the magnitude and regulation of respiration in the major aquatic ecosystems of the world. Here we attempt to integrate the information provided by these chapters and analyze the factors that may influence respiration at different levels in aquatic ecosystems. We discuss the partitioning of respiration into different biochemical processes and among different biological components, and we attempt to place respiration in aquatic systems in the context of global productivity and carbon cycling. Without doubt, respiration represents the single major area of ignorance in our understanding of the global carbon cycle, and we finally discuss some of the major gaps in our knowledge.

14.2 A conceptual framework for aquatic respiration

At the cellular level, respiration involves a set of pathways that lead to energy production, and is at the core of the functioning of the cell. There is a large diversity of pathways, particularly in the microbial world, but the essential elements of cellular respiration have been the focus of research for decades and

are well understood (see King, Chapter 2; Raven and Beardall, Chapter 3; Fenchel, Chapter 4 this volume). At the organism level, the magnitude and regulation of respiration are not just the simple summation of that of individual cells. For example, Hernández-León and Ikeda (Chapter 5) point out that in zooplankton, there is a wide range of factors that influence cell-specific respiration, some of which act directly at the cellular level, such as temperature, others such as behavior or feeding, acting at the organism level. At first sight, respiration at the ecosystem level, seems a fairly straightforward matter, compared to the complexities of biochemical pathways of respiration at the cellular level: at this level, respiration represents the summation of the respiratory activity of all the organisms that inhabit the ecosystem, which accounts for the total amount of carbon mineralized, oxygen consumed, or carbon dioxide produced. But the simplicity of this concept is misleading, because the factors that control the rate of respiration of individual organisms or even groups of organisms are not necessarily those that control respiration at the community or ecosystem level. In particular, there are feedbacks and loops at this level that are virtually impossible to anticipate from our knowledge of the physiology and autoecology of the organisms involved.

What is respiration and what does it represent from the ecosystem point of view? In the sections below we attempt to provide a conceptual framework that will allow us to interpret better the observed patterns of respiration that exist within and among aquatic ecosystem, which we present in the final sections of the chapter. As we proceed we shall address the questions raised in the introductory chapter.

14.2.1 Respiration at the single cell and ecosystem level

The first four chapters of this book have described respiration from a strictly physiological and bioenergetic point of view. At the cellular and organism level, respiration is the process of generation of chemical energy used to fuel all cellular processes. Flynn in Chapter 13 discusses the concept of maintenance metabolism as it pertains to the bioenergetics of individual cells or organisms, and its importance to the organisms to withstand the fluctuations of their environment. In practice this maintenance metabolism is difficult to measure and separate from the respiration that is linked to cell growth, but the concept is very useful because it emphasizes the fact that even at the cellular level, not all respiration is directly and proportionally related to production and growth. Fenchel in Chapter 4 further discusses the importance of this baseline metabolism in the capacity of protists to both withstand starvation, and to resume growth thereafter. It would appear that in unicellular organisms, maintaining a minimum level of energy flux is essential to their capacity to respond to favorable conditions. In general terms, at an organism level, there can be no growth without respiration, but there can certainly be respiration without growth.

At the cellular level, it is clear that respiration is not one single, homogenous process, but rather a collection of pathways that are related to energy production and cycling within the cells. For example, Williams and del Giorgio in Chapter 1 and Raven and Beardall in Chapter 3 discuss the various types of respiration that occur in planktonic autotrophs. Some, like mitochondrial respiration, are directly coupled to organic matter oxidation and energy generation. Others, such as the various forms of photorespiration, can increase energy dissipation when the processes of reductive assimilation of inorganic carbon and of oxidised nitrogen and sulfur sources are restricted relative to the input of light energy. The subset of photorespiratory reactions involving glycolate is a means of recouping some of the relatively oxidised (i.e. low grade) organic carbon in glycolate through the oxygenase activity of the carbon dioxide-fixing enzyme RUBISCO, especially in the minority of marine phytoplankton organisms which lack an inorganic carbon concentrating mechanism and particularly under conditions where the $pO_2:pCO_2$ ratio is high enough to favour the oxygenase activity of RUBISCO (see Raven and Beardall, Chapter 3).

Perhaps the greatest difference between respiration at the individual and ecosystem level is that organisms can regulate respiration within certain bounds, whereas at the ecosystem level, it would seem that respiration occurs at the maximum rate allowed by the physical and chemical constraints of the system. For example, an algal cell will not respire all the organic matter available in the cell, but rather will accumulate carbohydrate reserves that can be used as energy sources during periods of darkness. At the ecosystem level, there is also storage of organic matter that was not respired, but the regulation of this storage seems the result of physical, chemical (molecular composition, nutrient limitation), or physiological constraints to utilization of the organic matter (such as temperature), rather than as some form of regulation of ecosystem metabolism.

It is interesting to pose the question: is there an analogy between the respiration of a cell or an organism and the respiration of an entire ecosystem? For example, we can ask whether there is a parallel to the idea of maintenance metabolism at the ecosystem level. At the ecosystem level, respiration has multiple other facets, but some of the concepts that have been developed for unicellular organisms may still apply. In the following sections we further explore some of the implications of these conceptual analogies.

14.2.2 The components of aquatic respiration at the ecosystem level

As discussed in the previous section, the total rate of respiration of an organism may be separated into different components, some linked to the maintenance of basic cell functions, some linked macromolecular turnover, and some linked to energy generation to fuel biosynthesis and growth. In this section, we propose to explore the possibility that respiration at the ecosystem level may also be broken down into components that are somewhat analogous to those described for single organisms.

We can make a primary separation of overall respiration into two concrete categories—autotrophic (R_a) and heterotrophic (R_h). This separation of course begs the question whether these two trophic groupings are in themselves discrete—the debate on respiration is probably not at a stage of sophistication where the blurred separation of between autotrophs and heterotrophs is a major hindrance to our discussions. Although the separation into these two categories is conceptually valuable and valid (but noting the caveat in the preceding sentence) it is unfortunately far from easy to accomplish with any precision (see Chapter 9, Section 9.4.5).

_Ra_

Autotrophic respiration simply represents the respiration by algae and other primary producers (macroalgae and sea grasses in the case of benthic communities, for example). It is the component of total respiration that is most tightly linked to primary production, although the form of this relationship is probably neither simple nor linear. For example, increases in the rate of photosynthesis at the community level may be accompanied by a range of increases in algal respiration depending on the level of nutrient availability and other factors. In this regard, Raven and Beardall (Chapter 3) point out that there are systematic differences in the specific respiration rates among major algal groups, so algal community composition will play a major role in determining the shape of the relationship between algal respiration and primary production.

_Rh_

Heterotrophic respiration represents that portion of heterotrophic respiration supported primarily by organic matter derived from _in situ_ primary production, and secondarily by other sources. For example, the respiration rate of zooplankton that results from grazing on phytoplankton would contribute to heterotrophic respiration. The portion of bacterial metabolism that is directly supported by primary production of dissolved organic carbon (DOC) by algae, or indirectly through zooplankton sloppy feeding and excretion, would also be considered in this category.

It is useful to explore whether within one or both of these two concrete categories (R_a and R_h) we can envisage a further category (or categories as there is no third respiring entity) that parallels the maintenance respiration we recognize at the organism level. In the thermodynamic sense it would be the work required to maintain organization of the system, as opposed to the organism. This concept has been discussed by theoretical ecologists and has given rise to the concept of exergy, or negentropy (see Reynolds 2002 for a discussion of exergy).[1] The concept of the need to invest energy (gained by the respiration of some of the organic introduced into

is this
_Rb ?_
or related?

the system) to maintain the system has been discussed by a number of workers. Bjørnsen (1988), when addressing why apparently healthy algal cells released organic material, brought forward the useful analogy of property tax (the service related component) and income tax (the growth related component). Williams (1990) used this broad concept to argue the apparently wasteful release of DOC from the algae into the community at large facilitated the growth of heterotrophic bacteria thus sustaining an active bacterial population. Such a population would ensure the rapid colonization and mineralization of organic detrital particles, so preventing the loss from the euphotic zone of the nutrients associated with them. The penalty to the algae, of feeding their competitor for nutrients, can be argued to be minimal, as the turnover and release in inorganic nutrients back into the environment by the microbial community is rapid. If we take the analogy of the investment into goods or services one step further, we note that the currency is identical in both cases—such an argument by analogy would lead to the conclusion that there will be no fundamental distinction between the sources of energy for these different ecological functions. Whether we can tease out the energy invested in sustaining community infrastructure, from that used by individuals for their own growth and maintenance, remains to be seen.

symbiosis?

This third component of total respiration will be termed baseline respiration (R_b), and this component is also due mostly to heterotrophic organisms (although certain algae in resting stages respiring reserves may also contribute), and this is the component that is most distantly related to _in situ_ primary production, if it responds at all. This component represents the slow but steady consumption of organic matter, that is relatively independent of the contemporary primary production, and the size, sources, and nature of this pool will vary across ecosystems. The organic matter from autochthonous origin that is either not consumed, or consumed and transformed into less labile products, may enter the pool of organic matter that will eventually fuel the baseline respiration. In lakes and estuaries, this may be a mixture of more recalcitrant organic matter derived from phytoplankton production that persists in the system, organic matter resuspended from sediments, and organic matter of terrestrial

_Rb_

---

[1] The concept of negative entropy in biological systems dates back at least as far as Schrödinger (1944).

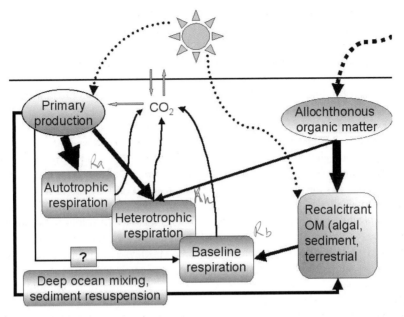

Figure 14.1 A schematic diagram of the main components of aquatic respiration, and their links to other aspects of ecosystem function. The size of the arrows connecting the different components suggest the relative strength of the link.

origin that enters aquatic from the drainage basin. In the oceans, it is likely that the overwhelming majority of the organic matter that supports baseline respiration is of marine origin, but the time lag between the time of synthesis of organic matter to its final oxidation through respiration could be in the order of hundreds to thousands of years, so that in practice, there is little connection to contemporary production, and in this sense, lakes, estuaries, and oceans may not differ substantially. Regardless of the origin of the organic material, it is likely that photochemical reactions play a major role in rendering old, recalcitrant organic matter available for biological consumption (Mopper *et al.* 1991).

The common denominator of baseline respiration is that it is not directly linked to contemporary production, but rather represents an integration of past autochthonous production, or the supply of allochthonous organic matter, that is itself an integration of the primary production of another aquatic or terrestrial ecosystem. For example, Cole and Caraco (2001) suggest that the net heterotrophy (ecosystem respiration in excess of authochthonous primary production) that characterizes the Hudson

River is fueled in part by extremely old (>1000 years) organic matter of terrestrial origin, which finds its way from the soils to the river, where it becomes available for biological decomposition. In support of these observations, McCallister (2002) found that the ^{14}C signature of bacterial biomass in the Hudson River is old, suggesting that at least a portion of bacterial respiration must have been supported by this ancient organic matter as well. Likewise, Cherrier *et al.* (1999) found that bacterial biomass in open-ocean samples also has an old ^{14}C signature, suggesting that in the open ocean, a proportion of bacterial production, and therefore most likely respiration as well, is based on carbon that was produced hundreds of years ago and somehow persisted in the system. The low and relatively constant levels of winter respiration in ice-covered northern temperate lakes (Welch and Bergman 1984) are also probably a manifestation of lake baseline respiration.

Figure 14.1 summarizes the interrelation between the hypothetical components of aquatic respiration and other aspects of the functioning of the aquatic ecosystem. The distinction between the three basic forms of respiration that we propose here is completely operational, and in reality it will often be difficult if impossible to establish, especially between the baseline heterotrophic respiration and the heterotrophic respiration linked more directly to primary production, because although the pools of organic matter used may be different, the organisms responsible are often the same. Even the autotrophic component will be difficult to determine because there are a number of confounding issues. For example, it is now widely accepted that mixotrophic organisms are important components of virtually all pelagic microbial food webs (Stickney *et al.* 1999), and there is little question that these organisms may contribute significantly to total community respiration. The extent to which the respiration of mixotrophs is fueled by their own photosynthetic activity, or by the organic matter they ingest is not known, but clearly this has an impact on our view of autotrophic versus heterotrophc respiration as well as on more general issues of net primary and ecosystem production, which we discuss in sections below. We believe, however, that the partition of respiration in its major components does provide a useful conceptual framework to place some of the patterns of variation that are observed in aquatic ecosystems, as we discuss in the sections below.

14.2.3 Variation and regulation of the components of respiration

These three components of ecosystem respiration (R_a, R_h, and R_b) differ particularly in terms of the timing and strength of response time relative to the growth part of primary production. Autotrophic respiration will be linked to primary production at the scale of minutes to hours. Heterotrophic respiration will be linked to primary production in the scales of hours to days, (bacterial respiration of DOC excreted by algae), days (zooplankton respiration of consumed algal carbon, bacterial respiration of DOC excreted from zooplankton, protozoan respiration of both algal and bacterial carbon), and weeks (respiration based on particulate organic matter derived from algae, zooplankton fecal products, respiration of polymeric DOC that requires extensive enzymatic degradation) (see Robinson and Williams, Chapter 9 Fig. 9.8, and also Blight *et al.* 1995). The baseline

respiration will have either a much longer time-lag, measured in months and years (lakes) and even hundreds of years in the case of open oceans, or no relation at all if most of the baseline respiration is supported by external organic matter inputs.

These components of respiration are in addition associated to different aspects of the function of the aquatic ecosystem, and their regulation might be exerted by factors that are common to all, and others that are unique to each component. Temperature, for example, will affect all components of respiration, although we do not know if the temperature dependence is the same in all cases. Other factors will influence specific components. For example, autotrophic respiration will be affected by the absolute rate of primary production, the level of inorganic nutrient stress, and the composition of the phytoplankton community (see Raven and Beardall, Chapter 3). Heterotrophic respiration will be affected by not only the supply but also the quality of the organic matter available for short-term consumption (see Flynn, Chapter 13), the nutrient stoichiometry of the resources, and by factors that affect the relative distribution of different heterotrophic components (i.e. bacteria, micro and macrozooplankton, see Fenchel, Chapter 4 and Hernández-León and Ikeda, Chapter 5). The baseline respiration, on the other hand, will be regulated by factors that influence the input, persistence and degradation of refractory organic matter in the ecosystem (i.e. Benner *et al.* 1993). The components that make up the total respiration may thus have different responses to the same environmental gradients and forcing factors, and the consequence is that the sum of the components may not be a simple function of any single factor.

One of the most consistent observations that emerge from different types of aquatic ecosystems is that respiration tends to be much less variable, both spatially and temporally, than other key aspects of the ecosystem, such as primary production. This is true for lakes (Pace and Prairie, Chapter 7), estuaries (Hopkinson and Smith, Chapter 8), open-ocean surface waters (Robinson and Williams, Chapter 9), and even coastal benthic communities (Middelburg *et al.* Chapter 11), suggesting that in most ecosystems, the component of respiration that is directly coupled to primary production may be small, and that the other components may often dominate total system metabolism.

What is important in this scheme is that aquatic ecosystems may differ not only in the total magnitude of respiration, but also in the relative composition of this respiration in terms of (i) the relative contribution of the three components to total respiration, and (ii) the sources of organic matter that fuel the heterotrophic respiration, particularly the baseline respiration. In this scenario, aquatic ecosystems may have similar levels of total respiration, but drastically different patterns of respiration relative to production and other environmental factors, if either the relative importance of each component differs, or the sources that fuel the different components differ. Respiration at the ecosystem level is usually modeled as a relatively simple, first-order function of primary production (see Flynn in Chapter 13), but we suggest here that this vision simply fails to capture the complexity of the process.

14.2.4   Links between respiration and the organic matter pools in aquatic ecosystems

Respiration differs fundamentally from photosynthesis in that the latter is driven by a single energy source—photons. Very importantly, there is no potential for storage of photons, or their immediate products—they are either used or permanently lost. By contrast the energy sources for respiration are manifold, with greatly varying reactivities. This concept of varying reactivities of the substrates for respiration has been with us for some time. Water quality engineers early on adopted multiple first-order decay constants to model the time course of BOD, Ogura (1973) used a similar approach to describe the progress curve of DOC decomposition. It is generally recognized that there is a continuum of reactivity within complex DOC mixtures, but it is common to find in the literature DOC categorized into discrete groups—labile, semilabile, and refractory, (Kirchman *et al.* 1993; Middelburg *et al.* 1993). Such categorizations have been useful in describing decomposition processes in the short- and medium-terms (Middelburg *et al.* 1993), but on the timescale of deep ocean circulation (1500 years) the order of the first-order rate constants that would be required

to describe the decomposition of the DOC in deep ocean (10^{-3} to 10^{-4} a^{-1}; Williams 2000) are beyond what most microbiologists would be prepared to accept, as they are way below maintenance rates. To square this circle Williams (2000) argued that in the deep ocean it is necessary to consider the process of decomposition as discontinuous. Thus, although it is tempting, it is probably unwise to search for a "theory of everything" to account for heterotrophic respiration in natural systems.

It is clear that the different components of respiration are linked conceptually, and perhaps practically as well, to the idea of coexisting pools of organic matter that are characterized by different turn over rates in aquatic ecosystems (Ogura 1973; Benner *et al.* 1992). It is difficult, however, to establish the functional link between the patterns of *in vitro* oxygen and DOC consumption, in part because the experimental timescales involved in the two types of measurements are often very different. DOC bioassays typically involve incubations of days to weeks (Hansell *et al.* 1995; del Giorgio and Davis 2002) because even the most precise analytical techniques can only detect changes larger than 1–2 μmol C dm^{-3}. Because of this, DOC bioassays tend to ignore the initial portion of the organic matter consumption that occurs within hours or a few days after the start of the incubation. In contrast, respiration measurements usually involve incubations of less than 2 days, typically 24 h or less (see Hopkinson and Smith, Chapter 8; Robinson and Williams, Chapter 9). These two approaches have rarely carried out in parallel (for exceptions, see Carlson *et al.* 1999), so it is still unclear how they compare. We have developed robust ways of generalizing views of productivity, for example, based on the P versus I curve and the nutrient load versus biomass curve, but we have no comparable general models that link respiration to organic matter composition, lability, and consumption.

14.2.5   The concepts of net primary production and net community production and their significance

If we bundle together all the forms of algal respiration (see Chapter 3), then the difference between this and true gross photosynthesis (GPP)—net primary

$$NPP = GPP - R$$

NEP= P-R

↓ trophic status

production (NPP)—is that part of autotrophic production currently available for growth. As such it is an ecological property of major importance. However, it is questionable how well we are able to measure it. It is difficult, although now possible, to measure a property close to true GPP with the ^{18}O technique, but as we are neither able to separate algal from heterotroph respiration, nor measure all forms of algal respiration, we cannot compute NPP. Despite some 50 plus years of use of the ^{14}C technique, we still have not resolved the debate over to what extent the technique measures gross organic production or NPP. We know it cannot measure true GPP (see Robinson and Williams, Chapter 9), and below the compensation irradiance, when it is negative, we know it cannot measure NPP (i.e. algal respiration exceeds algal organic production) as the ^{14}C technique cannot sensibly return negative values. Most likely (see Williams and Lefèvre 1996) it will occupy a variable position between these two extremes, being closer to NPP at high photosynthetic rates, when recycling of respiratory ^{12}C-rich carbon dioxide will be favored (note—it is almost certainly the cycling, not the respiratory carbon dioxide nor the respiration of ^{14}C-labeled photosynthetic products, that is the major determinant what the technique measures, Williams 1993) and being closer to gross organic production when photosynthetic rates are low. It has been recognized (see Robinson and Williams, Chapter 9) that if we were able to solve this problem of what the ^{14}C technique measures, then it would be a way to use the difference between net community production (NCP) as measured by the oxygen technique and NPP when measured by the ^{14}C technique, along with a RQ to separate out heterotrophic respiration. Presently the unknowns surrounding the ^{14}C techniques are regarded to be too great to make this a viable approach.

 huh

Respiration has been traditionally considered a variable of importance in the context of its relationship to primary production. There are several reasons for this. From the point of view of its regulation, it has traditionally been assumed that respiration is the simple consequence, a result or by-product, of photosynthesis, so that by understanding the former we would understand what regulates the latter. From the point of view of its magnitude, it has also been traditionally considered

that it is the difference between respiration and photosynthesis, and not the absolute magnitude of respiration that is of ecological relevance. This balance has been expressed variously as the P/R ratio, or as the difference between production and respiration, which is usually termed net ecosystem production (NEP) or NCP. Both NEP, and P/R, are considered ecologically important because they are thought to represent the extent to which an ecosystem is a source or a sink of carbon, the degree to which organic matter can be sustainably harvested, and whether the system is a net importer or exporter of organic matter (Odum 1956; Quiñones and Platt 1991). Therefore, there was great interest, and there still is, in determining whether a given ecosystem is autotrophic ($P > R$) or heterotrophic ($P < R$). Much of the focus of research has thus not been on the magnitude of respiration *per se*, but the magnitude of respiration relative to photosynthesis, and the coupling between these two factors. It had generally been expected that such coupling would exist and should be fairly strong.

A common misconception is that in aquatic ecosystems, a negative NEP or a $P/R < 1$ implies that there is little or no organic matter available for higher consumers, for storage in sediment or for export to other ecosystems. These misconceptions arise because the basic assumptions that underlie most studies of $P–R$ coupling in aquatic systems are rarely, if ever met: (i) all aquatic ecosystems are open to various inputs, including those of organic matter, and (ii) the measurement scales are commensurate to the real time lags between photosynthesis and respiration. The only scenario where P–R would provide an index of potential carbon storage or export would be either in a perfectly closed system, or in an open system where all the organic inputs could be accurately determined. There are no closed aquatic ecosystems in the biosphere, and the fact is that in most aquatic ecosystems it is difficult, if not impossible, to determine what proportion of respiration is supported by contemporary primary production and what proportion is supported by other sources, including past autochthonous primary production stored in the system. For example, estuaries often receive large amounts of allochthonous materials, part of which are processed within the system and other parts simply flow through and are discharged to the sea. As Hopkinson and Smith (Chapter 8) point out, many temperate estuaries are net heterotrophic, in the sense that respiration often exceeds photosynthesis. The fact that there may be a negative NEP (or a $P/R < 1$) does not imply that there is little or no organic matter available for higher consumers, for storage in sediment or for export to other ecosystems. Quite the opposite, most estuaries support extremely high rates of secondary, including fish and shellfish, production (Nixon 1988), store large amounts of organic matter in the sediments, and also export organic matter to the coastal ocean.

Pace and Prairie discuss a further example in Chapter 7. Lakes act as traps of materials that derive from the basin, most notably P, N, Fe, particulate and dissolved organic matter. Part of this terrestrial organic matter is decomposed and respired within the lakes, supporting the net heterotrophy that is often observed, particularly in oligotrophic temperate lakes, which not surprisingly, systematically efflux carbon dioxide to the atmosphere. But a significant portion of the organic matter that is imported from the drainage basin also sediments and is then stored. Thus lakes can function simultaneously as a net source of carbon dioxide to the atmosphere, a net sink of terrestrial organic matter, and a source of organic matter to aquatic ecosystems downstream, they often support relatively high levels of secondary production, all the while often having P/R ratios often much lower than 1.

14.3 Aquatic respiration at the ecosystem level

In the following sections, we synthesize the major patterns in aquatic respiration that have emerged from the analysis of published data and information on respiration in the major aquatic ecosystems of the biosphere. When interpreting these data and patterns, it is important to bear in mind the concept of respiration not as a single, uniform process, but rather of respiration as the sum of several processes, each of which having a distinct regulation, a different level of coupling to ecosystem productivity, and different biogeochemical and functional roles in the ecosystem.

14.3.1 The measurement of respiration in aquatic ecosystems

Robinson and Williams (Chapter 9), Hopkinson and Smith (Chapter 8), and Pace and Prairie (Chapter 6) discuss at some length the limitations of bottle measurements based on changes in oxygen and carbon dioxide, and other approaches used to measure pelagic respiration. As Robinson and Williams (Chapter 9) argue, current methods approach the limit of resolution of volumetric techniques, and will be difficult to surpass unless new developments occur. For example, it is currently possible to obtain rates of oxygen consumption or carbon dioxide production in enclosed samples, including benthic chambers and cores, bottles, bags, and flow-through incubation systems, in incubations that are no longer than 24 h, from a wide variety of aquatic ecosystems. But there are still vast regions of the deeper parts of the oceans and even some lakes where incubation would have to extend for days and even weeks before a change could be detected, with all the uncertainties involved in such long incubations.

The rates of oxygen consumption in enclosed samples are often initially linear but inevitably decline, and the point of inflexion varies greatly among samples. The transition is in some cases abrupt and in others continuous—the difference between the two forms of transition raises important questions surrounding the controlling substrates and kinetics. Robinson and Williams (Chapter 9) suggest that in most marine samples, rates of oxygen consumption remain linear for over one to as long as 5 days, but in estuarine and lake samples, inflection often occurs before that. Whether the oxygen decrease during incubations remains linear or not is most often seen as an issue of practical consequence (Robinson and Williams, Chapter 9; Hopkinson and Smith, Chapter 8; Pace and Prairie, Chapter 7), because it influences the calculated rates of respiration, and generally explained as a progressive depletion of substrates, but has seldom been interpreted as reflection of the different components of the total respiration.

Where respiration rates are too low even to be detected with the best volumetric techniques available, other methods have had to be employed.

Plankton respiration rate in the mesopelagic layers of the ocean has been studied using indirect enzymatic approaches, such as ETS activity, as Hernández-León and Ikeda (Chapter 5) and Arístegui *et al.* (Chapter 10) describe in detail. The biochemical approaches are regaining popularity but are not exempt of problems, the most critical being the calibration between the enzymatic rates and the actual rates of plankton respiration. Isotope approaches based on the biological fractionation of isotopes of oxygen have been successfully employed in both lakes and oceans. These approaches may be more effective to determine the net balance between plankton production and respiration, than to derive absolute rates of respiration.

Finally, it is possible to derive whole-system metabolism by following the changes in oxygen and carbon dioxide in the water or in the overlying atmosphere, generally during day/night cycles. The advantage of the whole-system approach is that it provides an integrated estimate of system metabolism, the disadvantage being that physical factors, such as water and wind turbulence and wind speed, greatly influence the estimates and these factors are not always easy to either model or constrain. Variations of this approach have been used in lakes (Pace and Prairie, Chapter 7), peat bogs and swamps (i.e. eddy covariance, Roehm in Chapter 6), estuaries (Hopkinson and Smith, Chapter 8), and even coastal benthic communities (i.e. unidirectional flow studies in coral reefs, Middelburg *et al.* Chapter 11).

In the past, and precisely because of some of the limitations described above, respiration has not been measured directly but rather derived from other processes, such as primary production, nutrient recycling, and carbon loading. There other indirect approaches, for example, using community structure combined with allometry (see Hernández-León and Ikeda, Chapter 5, and Robinson and Williams, Chapter 9), and organic mass balances based on biogeochemical information (see, for example, Robinson and Williams, Chapter 9; Ducklow and McCallister in press). These indirect approaches are necessarily reliant on the assumptions underlying the calculations. In some cases, these indirect approaches yield estimates that are

in fairly good agreement with actual measurements of respiration. For example, Arístegui *et al.* (Chapter 10), show that there is a convergence of estimates of mesopelagic respiration from the extrapolation of measurements of respiration, bacterial carbon flux estimates, apparent oxygen utilization, and organic mass balances.

14.3.2 Rates of respiration in the major aquatic ecosystems

Table 14.1 summarizes the global surface area covered by each aquatic ecosystem, the average (and ranges when provided by the authors) volumetric and depth-integrated rates of respiration for the major components of aquatic ecosystems in the biosphere, as well as their estimated global contribution to respiration. These data are those reported in the chapters that dealt with respiration in individual ecosystems (Chapters 6–11), except for some missing values that were drawn from the literature. It is important to bear in mind some intrinsic problems with these analyses. Almost without exception, the data sets involved are modest, relative to the size and complexity of the ecosystems in question, and are biased toward certain regions and seasons. For example, the overwhelming majority of lake data come from the growing season in northern regions, and there are very few data for the large lakes of world, which contain over half of the freshwaters (Pace and Prairie, Chapter 7). Likewise, the majority of the estuarine data come from temperate systems (Hopkinson and Smith, Chapter 8). As Robinson and Williams point out in Chapter 9, the average values they provide are based on measurements that are biased geographically to the Northern Atlantic as well as seasonally, to the spring and summer, and that probably do not represent, for example, the annual respiration in the ultraoligotrophic gyres. Roehm in Chapter 6 points out the scarcity of data on tropical bogs and swamps, which nevertheless cover huge areas.

There have often been major discrepancies between whole-system metabolic measurements, based on air–water gas flux or whole-water gas concentrations, with bottle or chamber measurements,

Table 14.1 Summary of estimates of respiration for various aquatic ecosystems. The primary data are extracted from the appropriate chapters—listed in the table. There are apparent discrepancies in the table due to different calculation, averaging procedures, and rounding off.

Ecosystem	Dimensions			Mean daily respiration rates					Global annual rates [note 1]	
	Source Chapter	Volume	Surface area	Volumetric	Volumetric	Areal	Areal	Areal	Respiration	Overall totals
				mean	range	mean	totals	range	rates	totals
		as 10^{12} m^3	as 10^{12} m^2	as mmols C m^{-3}		as mmols C m^{-2}			as Tmol C	
Freshwater peat bogs	6	13	6.7			1240			790	[note 2]
Freshwater marshes and swamps	6	7	3.5	–		530			180	[note 2]
Freshwater wetlands total	6									970
Lake water column	7	177	2.7	15.7 (0.6–161)		71			58	
Lake sediments	7	–	2.7	–		10.7		(1.6-33)	4	
Lakes total	7						82			62
Rivers	[note 3]								33	33
Estuarine water column	8	9	1.4	17.8		114			58	
Estuarine sediments	8	–	1.4	–		34			17	
Estuarine total	8						148			75
Coastal margins [note 4] – benthic	11	–	3.2			318	318		460	460
Coastal ocean pelagic	9	29,300	31	7.4		109			1200	
Coastal ocean sediments	11	–		–		19 [note 5]			170	
Coastal ocean total	9						128			1,400
Coastal/Continental margin total	9									1,800
Open ocean epipelagic (note 6)	9	47,600	317	2.9		105			9,000–12,000	[note 7]
Open ocean mesopelagic	10	262,000	308	0.014		12		(1.9–17)	1,400	
Open ocean bathypelagic	10	817,000	268	0.0005		1.6			170	
Open ocean sediments	11	–	308	–		1.2		(0.02–200)	130	
Open ocean total							120		11,000 -	14,000
Totals										
Freshwater/Brackish total									1,100 -	1,100
Marine total									12,500 -	15,500
Aquatic total									13,600 -	16,600
Aquatic benthic									1,800 -	1,800
Aquatic pelagic									12,000 -	15,000
Terrestrial total									9,500 -	10,500
Global total		1,160,000	329						24,000 -	28,000

[handwritten] * conservative estimates *
mostly derived from enclosed measurements

Notes:
1. Values rounded off.
2. These values derive from the sum of three means for separate areas, the areal and global values are calculated independently and not using the sum of the areal rates and total global areas.
3. Data from Caraco and Cole (1999).
4. Sum of estimates for coral reefs, salt marshes, mangroves, and benthic algae. [handwritten] coastal benthic communities
5. The value is back calculated from Middelburg et al. Chapter 11 using their figure of 23.9 × 10^{12} m^2 for area of the coastal zone.
6. The values have been converted from O_2 to C units using a RQ = 0.9, this assumes 80% recycling of mineralized nitogen, that is, f-ratio = 0.2.
7. The upper value of epipelagic ocean respiration is derived using the mean areal rate of respiration for the global surface ocean; the lower value is derived from the extrapolation of the mean annual areal respiration measured at the HOTS station (Williams et al. 2004).

[handwritten] areal rates
vs. volumetric rates

in wetlands (Roehm, Chapter 6), lakes (Pace and Prairie, Chapter 7), and estuaries (Chapter 8). The use of the apparent oxygen utilization (AOU) to derive pelagic respiration in the mesopelagic ocean layers (Arístegui *et al.* Chapter 10) is another example of whole-system measurement. Whole-system estimates are usually higher than enclosed measurements, and this difference has been attributed to the inclusion of components in the whole-system estimates that are excluded from bottle or small enclosure approaches, such as larger animals or sediments. For example, the sum of the average rates of pelagic and benthic respiration in estuaries is approximately 144 mmol C m^{-2} d^{-1}, much lower than the average whole-system estimate of 289 mmol C m^{-2} d^{-1}. Similar discrepancies for lakes are reported by Pace and Prairie (Chapter 7). In the ocean, where the boundaries are well defined, when whole-system and discrete measurements have been compared, the agreement is good (25% or so, see Robinson and Williams, Chapter 9), perhaps because the influence of the benthos is minimal. This is important to bear in mind in the subsequent sections, because the estimated average rates of respiration for each ecosystem that appear in Table 14.1 are mostly derived enclosed measurements, therefore the global rates of respiration that are derived from these should be viewed in general as conservative estimates ecosystem respiration.

Respiration rates in wetlands

By far the highest rates of areal respiration measured in aquatic ecosystems have been recorded in freshwater wetlands, particularly northern peatbogs. Roehm in Chapter 6 reports a range of measured respiration for peatbogs between 17 and 2700 mmol C m^{-2} d^{-1} with a global average of 1240 mmol C m^{-2} d^{-1} (Table 14.1). Swamps and freshwater marshes seem to have lower rates of respiration, with a total range of 17 to 916 mmol C m^{-2} d^{-1} and a global average of 530 mmol C m^{-2} d^{-1}. It should be pointed out that a large fraction of measurements used in the analysis of wetland respiration, particularly those of northern peatbogs, are whole-system estimates based on eddy covariance techniques, which tend to yield higher estimates relative to enclosure approaches, as we discussed above. These whole-system estimates do not allow a discrimination of the respiration due to autotrophs and heterotrophs, but Roehm suggests that total peatbog respiration is on average equally split between plant and microbial respiration in the water column. What is clear, as Roehm points out, is that in addition to the production of carbon dioxide, these freshwater wetlands are a major source of respiratory methane, as we discuss in sections below, so their global role should not be judged only on the basis of carbon dioxide flux.

Respiration rates in pelagic systems

Without doubt, the lowest rates of water column respiration occur in the bathypelagic regions of the open oceans, below 1000 m. There are no direct measurements of respiration in the bathypelagic water column, but the indirect measurements based on mass balances of oxygen or DOC converge to rates in the region of 0.0005 mmol C m^{-3} d^{-1} (Arístegui *et al.* Chapter 10), which are approximately four orders of magnitude lower than the rates found in the surface ocean layers (see below). The large volume of the bathypelagic zone nevertheless results in depth-integrated rates (approximately 1.6 mmol C m^{-2} d^{-1}) within 2 orders of magnitude of those in the surface layers. There are a few direct measurements of respiration in the mesopelagic layers of the oceans (between 150 and 1000 m), most based on ETS (Arístegui *et al.* Chapter 10), and these measurements converge to volumetric rates in the order of 0.014 mmol C m^{-3} d^{-1}, two orders of magnitude lower than those found in the surface layers (average 2.9 mmol C m^{-3} d^{-1}), and which translate into depth-integrated rates in the order of 12 mmol C m^{-2} d^{-1}, 10-fold lower than the average surface areal rates of 105 mmol C m^{-2} d^{-1}.

Thus, there is a four-order of magnitude range in the volumetric rates of respiration along a vertical profile in the open ocean, which is approximately the same range that is found in the surface waters across all aquatic ecosystems (Fig. 14.2). The individual measurements of volumetric rates of surface water respiration reported in the different chapters range from 0.02 (in the surface waters of the Southern Ocean) to over 300 mmol C m^{-3} d^{-1}

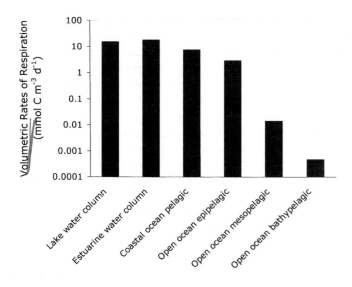

Note log scales!

Figure 14.2 <u>Mean volumetric rates of respiration in</u> the water column of the major aquatic ecosystems. Source of data, Table 14.1.

→ surface waters
vol. rates
→ several orders of mag. variation

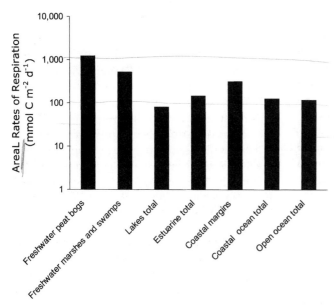

areal rates
→ one order of mag. variation

Figure 14.3 <u>Mean areal rates of respiration</u> in the water column of the major aquatic ecosystems. Source of data, Table 14.1.

(in estuaries), but the average values for each major system are much closer, ranging from 2.9 mmol C m^{-3} d^{-1} in the surface, euphotic waters of the open oceans, 7.4 mmol C m^{-3} d^{-1} for coastal oceans, 16 C m^{-3} d^{-1} in lakes, to 17 mmol C m^{-3} d^{-1} in estuaries (Table 14.1). The <u>average depth-integrated rates of respiration in the upper layers are even less variable than the volumetric rates, ranging from 71 mmol C m^{-2} d^{-1}</u> <u>in lakes, 114 mmol C m^{-2} d^{-1} in estuaries and 116 and 107 mmol C m^{-2} d^{-1} in coastal and open ocean regions, respectively</u> (Table 14.1 and Fig. 14.3). It is interesting to note this convergence of areal pelagic respiration rates across ecosystems that often vary greatly in the importance of external sources of organic matter, the nature of the nutrient limitation and the structure of the food web.

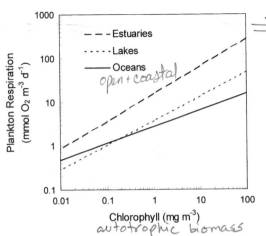

→ chld α PP?
No

Figure 14.4 The relationship between plankton respiration (PR) and chlorophyll (Chl) concentration for lakes (open circles, full line), estuaries (open squares, dashed line), and oceans (open triangles, dotted line). The regression equations are the following: Lake PR = 3.71 × Chl$^{0.56}$ (from Pace and Prairie, Chapter 7); Estuarine PR = 15.49 × Chl$^{0.63}$ (from Hopkinson and Smith, Chapter 8); Ocean PR = 2.74 Chl$^{0.38}$ (from Robinson and Williams, Chapter 9).

log-slope

The available data suggest roughly similar patterns of respiration along productivity gradients in the surface waters of the different major ecosystems. Figure 14.4 shows the relationship between volumetric plankton respiration and chlorophyll concentration, based on the equations derived by Pace and Prairie for lakes (Chapter 7), Hopkinson and Smith for estuaries (Chapter 8), and Robinson and Williams for open and coastal oceans, (Chapter 9). The three relationships have in common log-slopes that are substantially lower than 1 (0.56, 0.63, and 0.38, lakes, estuaries, and oceans, respectively). The pattern of respiration and chlorophyll is remarkably similar in oceans and lakes, despite the obvious differences in the functioning of these systems. The general pattern is similar in estuaries, but respiration in estuaries appears to be on average twice that of lakes and oceans at similar levels of chlorophyll. Whether this reflects the greater influence of allochthonous organic matter in estuaries, or a fundamental differences in the cycling of primary production remains a matter for future research, but these findings do agree with previous reports of extremely elevated levels of secondary production in estuaries relative to other systems (Nixon

1988). The important point here is that systems with similar levels of autotrophic biomass have roughly similar levels of planktonic respiration, which is not necessarily the case with primary production, for example (Baines *et al.* 1994).   *↳rate*

Respiration rates in benthic systems

As respiration in the surface waters average sediment respiration rates in coastal oceans, estuaries and lakes differ only modestly, with a total range of individual measurements that goes from 3 mmol C m^{-2} d^{-1} in coastal sediments to over 140 mmol C m^{-2} d^{-1} in estuarine sediments (Table 14.2). Overall, the average sediment respiration rate varies only threefold, from 11 mmol C m^{-2} d^{-1} in lakes, 19 mmol C m^{-2} d^{-1} in coastal oceans, and 34 mmol C m^{-2} d^{-1} in estuaries (Table 14.2). Deep ocean sediments have rates of respiration that are two orders lower than those recorded for coastal sediments (Arístegui *et al.* Chapter 10). Hopkinson and Smith (Chapter 8) and Pace and Prairie (Chapter 7) argue that sediment respiration is a direct function of the rate of supply of organic matter to the sediments from the upper water column layers. If the average rates of sedimentation discussed above are representative of *in situ* conditions, then the rate of supply in coastal oceans, estuaries, and lakes would appear to also be roughly similar, which is intriguing considering that there are large differences in physical forcing, hydrology, and food web structure among these systems.

Many aquatic ecosystems have rich benthic communities of plants and metazoans that also contribute to total system respiration, in addition to the respiration strictly due to sediments. Examples of such benthic communities are macrophyte beds in lakes, estuaries, and coastal oceans; mollusc reefs in estuaries and coral reefs in coastal oceans. The contribution of these benthic communities to total system respiration is often difficult to establish. Middelburg *et al.* (Chapter 11) have carried out a detailed analysis of respiration in coastal benthic communities and conclude that in these shallow coastal marine ecosystems, total respiration may be dominated by the contribution of benthic communities, as we discuss in the next section.

Table 14.2 Comparison of the rates of benthic and pelagic respiration in the major aquatic ecosystems. Rates as mmol Cm^{-2} d^{-1}

	Pelagic respiration	Benthic respiration	Benthic as % pelagic respiration
Lakes	71	11	15
Estuaries	114	34	30
Coastal ocean	109	19	17
Open ocean	105	1.6	1.5

(handwritten annotation: mmol C/m². d)

14.3.3 The total contribution of benthos versus pelagic respiration

The proportion of benthic versus pelagic respiration varies widely across aquatic ecosystems (Table 14.2). The lowest relative contribution of benthic respiration is in the open oceans, where benthic respiration represents 1.5% or less of the total system respiration (Arístegui *et al.* Chapter 10). On average, benthic respiration represents approximately 15% of the total lake respiration (Pace and Prairie, Chapter 7), but this contribution varies greatly with lake morphometry, and is much higher in the smaller, shallower lakes and is negligible in the largest lakes. The contribution of benthic respiration tends to be higher in estuaries, due to a combination of high productivity and low mean depth, and Hopkinson and Smith suggest that it is in the order of 25% of the total system respiration. Likewise, in shallow rivers, it would appear than much of the respiration occurs in the sediments and not water column (Caraco and Cole 1999), although in deeper rivers such as the Hudson River in New York (USA), water column metabolism also dominates (Caraco and Cole 1999). By far, the greatest contribution of benthic communities to total system respiration, similar to that in the shallowest lakes, occurs in coastal marine regions, where benthic communities, including macroalgal and seaweed beds and coral reefs, may contribute over half of all system respiration (Middelburg *et al.* Chapter 11).

14.3.4 The link between community structure and respiration

The contribution of different functional groups or size classes to total respiration is not easy to determine, because there is no method that can achieve a complete separation between groups. Sequential filtration is the most commonly used method to assess the contribution of different planktonic groups but it seldom achieves an effective separation between groups. For example, bacterial respiration is usually determined as oxygen consumption in water filtered through 0.8–3 μm filters, but this filtration procedure allows passage of small picoautotrophs and flagellates, and more importantly, a variable and often large fraction of the bacterial activity is retained by the filter (del Giorgio Pers. com.).

In spite of these technical problems, most reports converge to suggest a pattern where bacteria (and small picoplankton) tend to dominate community respiration in the most oligotrophic waters, whereas the contribution of algae and metazoans tends to increase with system productivity (Biddanda *et al.* 2001; Smith and Kemp 2001; Robinson and Williams, Chapter 9). The only systems that seem to depart from this general pattern of dominance by microbes is the coastal benthic ecosystems. Middelbug *et al.* (Chapter 11) argue that plants (macroalgae and sea grasses) and macrofauna (corals and benthic fauna) are responsible for the bulk of respiration in these systems, in spite of the fact that there are active microbial communities is all coastal sediments.

In lakes, del Giorgio *et al.* (1999) found a good correlation between the total heterotrophic biomass in the plankton, and community respiration. In addition, there were consistent patterns between the partition of biomass into heterotrophic and autotrophic components, and the P/R ratio of the lake. Others have also reported that the apparent coupling between photosynthesis and respiration in lakes (Schindler *et al.* 1997), as well as in oceans (Serret *et al.* 2001; Smith and Kemp 2001), seems to be strongly influenced by community structure, particularly in terms of the relative importance of the microbial component.

Hernandez-León and Ikeda (Chapter 5) estimate that metazooplankton contribute approximately 1.08 Pmol C a^{-1} in the global ocean, with 0.87 Pmol C contributed in the surface layers alone. The significance of this potential zooplankton respiration depends greatly on the estimate of global ocean respiration, but could represent around 5–10%

of the total surface respiration. For the mesopelagic, Hernadez-León and Ikeda estimate a global zooplankton respiration in the order of $0.18\,\mathrm{Pmol\,C\,a^{-1}}$, which would represent approximately 12% of the total respiration estimated in this layer by Arístegui et al. in Chapter 10 ($1.4\,\mathrm{Pmol\,C\,a^{-1}}$, and see Table 14.1). For the bathypelagic region, the estimate of zooplankton respiration is $0.03\,\mathrm{Pmol\,C\,a^{-1}}$, which again is around 10% of the total respiration estimated by Arístegui et al. ($0.3\,\mathrm{Pmol\,C\,a^{-1}}$, Chapter 10). It would thus appear that zooplankton contribute in the order of 5–10% to total pelagic respiration in the oceans. One practical consequence of this conclusion is that since the standard bottle incubations rarely capture the respiration of metazooplankton, water column respiration in oceans and other aquatic ecosystems may be underestimated by at least this percentage, although the analysis made by Robinson and Williams (Chapter 9) suggests a lower percentage error.

In addition to their direct contribution, zooplankton and other components of the community may contribute indirectly to community respiration through their grazing activity and other biological interactions. For example, Pace and Cole (1999) found that lakes with similar nutrient loading but different food web structure had significantly different plankton community respiration rates. Another example of how changes in the food web can affect system-level respiration is the invasion of the Hudson River by the exotic Zebra mussel (Caraco et al. 2000). Before the invasion, the system was characterized by high chlorophyll levels and extremely high water column respiration rates, mostly due to algae and secondarily to heterotrophic bacteria. A few years following the invasion, a significant portion of system respiration had shifted to the benthos, and the mussels themselves were responsible for a large fraction of the total respiration. There is also evidence that in addition to the invasions of exotic species, the dramatic depletion of key invertebrate and fish populations through overharvesting, may be leading to strong shifts in the ocean's food web (Jackson et al. 2001), but it is impossible at present to predict how coastal- or open-ocean respiration might change following major changes in the food web structure.

14.3.5 The coupling between respiration and production

The term coupling is used very loosely in ecology, often implying that a strong coupling will be reflected in tight patterns of covariation between photosynthesis and respiration, whereas uncoupling is evidenced by weak relationships between the two. But inferring the strength of the coupling between respiration and production simply from the strength of the correlation between the two may be misleading. For example, Pace and Prairie (Chapter 7) show that plankton respiration in lakes is strongly positively correlated to measures of lake productivity, such as chlorophyll or total photosynthesis concentration. This could suggest that in lakes, plankton respiration depends almost entirely on algal production and that this coupling is very dynamic so that variations in lake production are quickly followed by a response in terms of respiration. But it is interesting to note that Pace and Prairie also conclude, as others have done (Wetzel 1992; del Giorgio et al. 1999; Jonsson et al. 2003), that a substantial portion (potentially >50%) of lake respiration could be supported by organic matter inputs from terrestrial ecosystems. If this conclusion is correct, then this will weaken the coupling between lake respiration and primary production.

The strong correlation between lake respiration and lake chlorophyll or total phosphorous seems to be at odds with the observation that much of lake respiration can be supported by non-algal carbon, but the above observations can be reconciled if we consider that total lake respiration is the sum of the autotrophic, heterotrophic, and baseline components discussed in the previous section, each of which is differently and uniquely related to primary production and other forcing factors. For northern temperate lakes in particular, del Giorgio et al. (1999) have suggested that the baseline respiration may be relatively constant along nutrient enrichment gradients. The consequence is that at low levels of autotrophic biomass and primary production, this baseline respiration may represent a large fraction of the total respiration, but its contribution to total respiration in more productive lakes is minimal (Fig. 14.5). The relative constancy of baseline respiration among lakes allows for the

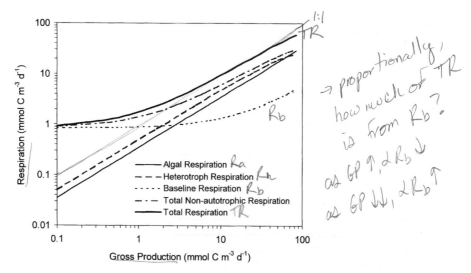

Handwritten annotations on figure:

1:1

TR

→ proportionally, how much of TR is from Rb?

as GP ↑, dRb ↓

as GP ↓↓, dRb ↑

Rb

Ra
Rh
Rb

Figure 14.5 Hypothetical model of the evolution of the main components of aquatic respiration along a lake productivity gradient. Total respiration (TR) is divided into the respiration of autotrophs (AR), respiration of heterotrophs supported directly by primary production (HR), and baseline respiration (BLR), supported indirectly by previous primary production as well as external inputs of organic matter. The model assumes that an increasing fraction (from 10 to 30%) of PP either sediments or is exported and is thus not respired, along a gradient of increasing PP. Of the PP that stays in the system, the model further assumes that small fraction of PP (5%) enters the recalcitrant organic matter pool, which eventually fuels of portion of the baseline respiration (BLR). The remainder of the PP is respired in the short term by both autotrophs or heterotrophs. In addition, the model assumes that a constant amount of allochthonous organic matter (0.9 mmol m^{-3} d^{-1}) is respired in all lakes. The baseline respiration is the sum of this allochthonous carbon respiration plus the fraction of PP that goes into the recalcitrant organic matter pool. Autotrophic respiration (AR) is assumed to be a constant fraction of PP (35%); Heterotrophic respiration (HR) is the difference between primary production and the sum of algal respiration, carbon export, and primary production that enters the long-term pool. The main point of this simple model is that even a small — *how is* amount of baseline respiration can have profound effects on the relationship between respiration and production among aquatic ecosystems. *that shown?*

existence of a strong correlation between total respiration and primary production among lakes, even though in the oligotrophic side of the trophic spectrum total respiration may not be supported primarily by primary production. The main point of the simple model presented in Fig. 14.5 is that even a small amount of baseline respiration can have profound effects on the shape of the relationship between total respiration and production across aquatic ecosystems.

At the other extreme of the trophic continuum, there have been some suggestions that even in the open-ocean gyres the plankton community may consume small amounts of allochthonous organic matter derived from terrestrial or other oceanic regions (del Giorgio and Duarte 2002; Hansell *et al.* 2004). There is little question that in these open-ocean regions, however, phytoplankton production is overwhelmingly the major source of organic matter to fuel ecosystem respiration, and

so it is in these regions where we would expect the strongest coupling between the two. And yet Robinson and Williams (Chapter 9) and Aristegui and Harrison (2002) show that in the surface waters of the open oceans, plankton respiration is often poorly correlated or not correlated at all to primary production and other indices of system productivity. We have a great deal to learn about the connection between photosynthesis and respiration, but what is clear is that statistical relationships—or lack of— between photosynthesis and respiration should be used with caution to infer coupling between the two.

14.4 Aquatic respiration in the biosphere

14.4.1 Linking the aquatic systems

All aquatic systems are open. Thus as well as the internal production and consumption of organic material by photosynthesis and respiration, ecosystems may import and export organic material from/to adjacent ecosystems. The immediate consequence of this is that their internal organic budgets need not balance. Further, all ecosystems contain potential reservoirs of organic material, in the water column or in the sediments, which may also serve as sources and sinks. These, for example, serve to buffer the community against diel and seasonal fluctuations in production. They also confound the analysis of the immediate coupling between photosynthesis and respiration (see Sections 14.2.5 and 14.3.5). Ignoring these short- to medium-term imbalances, we may write the following mass-balance equation, in which the sums of the internal (I) and external (E) sources minus that of the sinks (respiration and export) equal zero:

$$(P + I) - (R + E) = 0. \tag{1}$$

If we envisage a simple series of "cascading" ecosystems (see e.g. Fig. 14.7) we may rewrite the above equation as

$$(P_n + I_n) - (R_n + E_n) = 0, \tag{2}$$

$$(P_{n+1} + I_{n+1}) - (R_{n+1} + E_{n+1}) = 0. \tag{3}$$

Assuming a simple linear flow, and that the upstream ecosystem is the major source of imported organic material (probably a reasonable assumption for most major aquatic ecosystems other than estuaries), the import term for one ecosystem will be the export term for the upstream one that is, $I_{n+1} = E_n$, thus we may substitute the import term and replace it by the upstream export term so simplifying the equation:

$$(P_{n+1} + E_n) - (R_{n+1} + E_{n+1}) = 0. \tag{4}$$

As the import term dictates the potential for the internal processes of photosynthesis and respiration to be systematically out of balance, equation (4) makes the point that the upstream controls the potential balance of metabolism in the immediate (and possibly subsequent) downstream system(s).

Replacing the import term by the upstream export term raised the question of what ecological factors determine the three remaining terms. The export term is the difference between the internal and external sources of organic material and the internal

removal processes. This leads onto the question of what controls this balance. We understand the basic physical, chemical, and biological controls on photosynthesis and there are a number of whole ecosystem models of production. We have a much poorer understanding of the broad controls on the extent of decomposition due to respiration (as opposed to its rate) or as yet any general rules. Two factors might be expected to play important roles: (i) the resistance to decay of the products of photosynthesis and the imported organic material, (ii) the residence time of water in the system, as this sets the period available for decomposition. Whereas it is hard to generalize about the former term, we can make some broad generalizations about the latter. Rivers and estuaries have variable and comparatively short residence times (characteristically days to months), at the other of the scale in the oceans the resident times are quite well prescribed ranging from 1 to 10 years for the coastal ocean, a hundred years or so for the epipelagic (0–150 m) and mesopelagic (150–1000 m) regions of the oceans to values of the order of a millennium or so for the bathypelagic regions (1000–4500 m). Lakes are probable more variable in their residence times (Pace and Prairie, Chapter 6) and thus more difficult to generalize upon. Thus, in a general way we might expect estuaries and rivers to export a greater proportion of their organic budget than the oceanic systems.

From the data compiled in Table 14.3 it would appear that this expectation is nowhere near met. The residence time is clearly not a major determinant of the export from the system, for whereas the water residence times vary by a factor of 10 000–100 000, the export ratios vary only by an order of 10, with no obvious relation to the water residence time. This would suggest that the first factor—the resistance to decay is the prime determinant, presumably the residence time, even in actively flushed systems, is sufficient to allow essentially the labile organic material to the decomposed.

14.4.2 The contribution of the major ecosystems to global aquatic respiration

Table 14.1 shows the estimated global surface area covered by each of the major aquatic ecosystems of

Table 14.3 Comparison of export ratios and water residence times. Export ratio is calculated as the scale of the annual organic cycle divided by the scale of export

Ecosystem	Scale of organic cycle (Tmol C a^{-1})	Scale of export (Tmol C a^{-1})	Export ratio rounded off	Residence time (years)
Rivers	33	34	1	0.1–0.01
Estuaries	100	34	3	0.1–0.01
Coastal ocean	1500	150	10	1–10
Epipelagic zone	10 000	1500	7	100
Mesopelagic zone	1000	350	3	100
Bathypelagic zone	250	150	1.6	1500

the world. The numbers that appear in the table are based on those used by the authors of the different sections of the book (Chapters 6–11) to estimate the global respiration for each ecosystem. There is still considerable uncertainty in all these estimates of global coverage. For example, there is a fairly wide range of published estimates of coastal and open ocean surface area (Ducklow and McCallister in press). Roehm, in Chapter 6, discusses the uncertainty of current estimates of freshwater wetlands, particularly those in tropical regions. Likewise, Middelburg *et al.* (Chapter 11) point out the difficulties of estimating the actual surface area covered by different major benthic communities, such as coral reefs or macroalgal beds. It is clear that the uncertainty in the estimates of global coverage contributes to the uncertainty of global estimates of respiration. As global estimates derive from the product of area estimates and mean or median rates, we have a degree of uncertainty even at the primary level.

Table 14.1 and Fig. 14.6 (and also Fig. 14.7) show the estimates of the global respiration of each of the major aquatic ecosystems. All these estimates are based on the extrapolation of direct measurements of respiration to the global ecosystem, but the exact form of this extrapolation varies among the chapters, because authors have chosen the approach most appropriate to the type of data and system. In most cases, the authors have provided some central value that represents the most likely global respiration for each ecosystem, and that takes into consideration the constraints of the data and other factors.

The oceans, because of their massive area, dominate the estimate of global aquatic respiration followed, interestingly by freshwater wetlands. The

oceans give rise to the greatest uncertainty in the overall budget, not only in absolute terms but also in its relative error. The oceanic global budget comprises six separate elements: estimates for the coastal ocean water column and sediments, and estimates for three pelagic layers, and the open-ocean sediments. These six components have varying degrees of certainty over their accuracy. The estimate for the sediments are probably quite secure: they give values of 0.18 Pmol C a^{-1} for the oceanic sediments, 0.166 Pmol C a^{-1} for the coastal sediments, and a further 0.455 Pmol C a^{-1} for the macrophytes and corals. Aristegui *et al.* (Chapter 10) find agreement of values for the deep ocean and especially for the mesopelagic zone, where a number of different approaches converge on a figure of 1.36 Pmol C a^{-1}. Thus, despite the extreme undersampling of this zone, the estimate is probably quite robust. The value of 0.16 Pmol C a^{-1} for the bathypelagic zone is probably less robust, but it is a small contributor to the overall budget. Similarly, although no estimate is made for the deep-sea trenches, these again are likely to be a very minor contributor to the budget. The major problem lies with the estimate for the epipelagic (euphotic) zone of the ocean, which is by far the major component (ca. 90%) of the overall oceanic budget, because there are strong temporal and spatial sampling biases in the available data. Extrapolation of mean values of epipelagic ocean respiration results in a global estimate of 13.4 Pmol O$_2$ a^{-1} (12 Pmol C a^{-1})[2], which Williams and Robinson (Chapter 9) regard to be more likely an overestimate rather than an underestimate. An

[2] This assumes a *R.Q.* = 0.9, see Table 14.1.

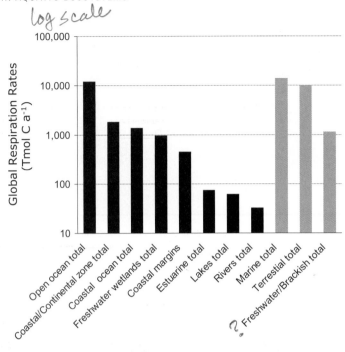

log scale

Figure 14.6 The global annual respiration estimated for each of the major aquatic ecosystems of the biosphere. Data ranked by the rates. Source of data, Table 14.1.

alternative estimate for surface oceanic respiration can be derived from extrapolating the annual respiration at the oligotrophic HOTS station to the entire surface ocean (Robinson and Williams, Chapter 9). This approach yields an estimate of 10 $PmolO_2 a^{-1}$ (9 $PmolCa^{-1}$). The correct figure for the global epipelagic ocean respiration lies within this range.

The total respiration in the world's oceans can be derived from the sum of the contribution of coastal pelagic (1.2 $PmolCa^{-1}$), coastal benthic (0.63 $PmolCa^{-1}$), open ocean surface (9–12 $PmolCa^{-1}$), mesopelagic (1.36 $PmolCa^{-1}$), bathypelagic (0.33 $PmolCa^{-1}$) and deep open ocean sediment (0.18) respirations, this gives a total oceanic respiration in the order of 12.5–15.5 $PmolCa^{-1}$ (Table 14.1).

marine total —

Freshwater wetlands emerge as an important contributor to global aquatic respiration (approximately 1 $PmolCa^{-1}$), almost of the same magnitude as the coastal ocean. Lakes, rivers, and estuaries contribute modestly to the global aquatic respiration, but these systems respire large amounts of organic matter exported from terrestrial ecosystems so their contribution is important in terms of the

fate of net terrestrial ecosystem production. The sum of the respiration of the major aquatic ecosystems yields a range of global aquatic respiration values of 12–19 $PmolCa^{-1}$, with a probable range of 14–17 $PmolCa^{-1}$.

14.4.3 The contribution of anaerobic pathways to global aquatic respiration

The importance of aquatic respiration should not be viewed simply as a loss of organic matter, and suboxic respiration in the water column of oceans and some lakes, as well as anaerobic processes that dominate many benthic ecosystems, play a role that is much more critical than is implied by their direct contribution to the carbon flux. For example, based on the denitrification and N_2 production rates in Table 12.1 (Codispoti *et al.*, Chapter 12) and various plausible assumptions over stoichiometry, one can infer a total suboxic water column respiration would be equivalent to the production of approximately 5.6 $TmolCO_2 a^{-1}$. Thus, the direct contribution of suboxic respiration to carbon dioxide production in

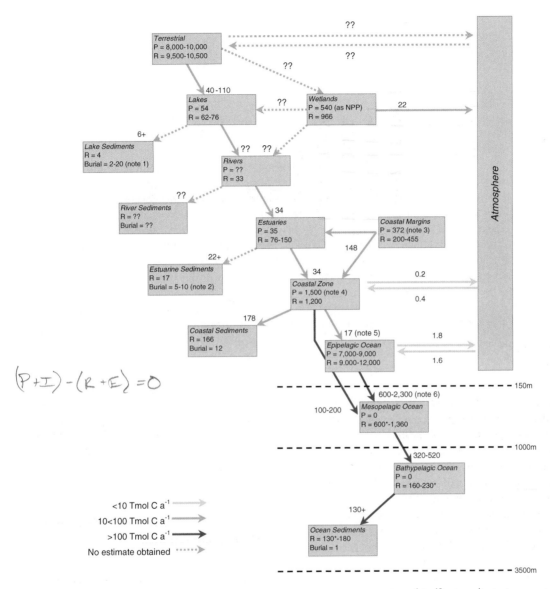

$$(P + I) - (R + E) = 0$$

Figure 14.7 Organic fluxes. A flow chart linking the major aquatic ecosystems. Rates as T mol C a^{-1} (10^{12} mol C a^{-1}). The data are primarily taken from Table 14.1, additional data for the mesopelagic and bathypelagic zones and the deep ocean sediments from Andersson *et al.*, (2004) are marked with an asterisk; the data for the mesopelagic has been scaled for a depth of 150m, rather than 200m, as is used in their paper. Note no attempt has been made in the figure to correct the data to achieve mass balance, as there is no certainty, in a many cases, which estimate is the more reliable.

Notes:

1. Estimates made from sediment accumulation rates and from in sediment respiration rates.
2. As note 1.
3. Coastal margins include estimates for salt marshes, macroalgae, seagrasses and coral reefs. Data from Duarte and Cebrián (1996). Note the productivity estimate is for NPP.
4. The coastal zone excludes estimates for salt marshes, macroalgae, seagrasses and coral reefs.
5. Assuming only the riverine DOM passes to the epipelagic open ocean, the remainder of the export from the coastal ocean passes down the shelf slope.
6. Data from Robinson and Williams (Chapter 10) Box 9.4.

the ocean is negligible relative to that of oxic respiration, but as Codispoti *et al.* point out, this process nevertheless plays a key role in the functioning of the oceans, because it is a major control on the fixed-N pool in the ocean. Suboxic respiration may impact global ocean C flux not so much through its direct production of carbon dioxide, but rather through its indirect influence over global ocean productivity.

suboxic

Likewise, anaerobic processes dominate metabolism in many marine coastal and lake benthic communities, and their importance may be less as a source of carbon dioxide or as a sink of organic matter, and more as a control of the biogeochemical cycling of key elements such as iron and sulfur (see King, Chapter 2; Middelburg *et al.* Chapter 11).

anoxic

Methanogenesis, in particular, is of global importance, because although the flux of methane is much smaller than the flux of carbon dioxide, the impact of the former as a greenhouse gas is much larger. Methanogenesis is very important in many freshwater wetlands, and as Roehm (Chapter 6) suggests, the relevance of wetlands to global greenhouse gas budgets lies not only by their role in carbon dioxide storage or emission, but also as potentially one of the largest sources of methane to the atmosphere. Many freshwater wetland ecosystems, particularly at high latitudes act as small, but nevertheless significant, net sinks of organic matter and carbon dioxide, but at the same time, act as net sources of methane. In terms of carbon equivalents, the amount of methane produced is much smaller than the carbon dioxide stored, but whether the amount of carbon sequestration is capable of offsetting the greenhouse gas effect of the methane produced is a matter of further study.

14.4.4 Global aquatic respiration in relation to aquatic primary production

A fundamental question in ecology is the balance between photosynthesis and respiration: to what extent is the system balanced, that is, are the internal production and respiration matched, or to what extent are there imbalances such that the system is reliant upon external subsidies from adjacent ecosystems or anthropogenic discharges. In large ecosystems, such as the oceans, the balance within

major geographical areas is a further pertinent and pressing question. The subject of the balance of photosynthesis and respiration has been a matter of lively debate (del Giorgio *et al.* 1997, Duarte and Agusti 1998; Williams 1998; Duarte *et al.* 1999; Williams and Bowers 1999) particularly in relation to the oligotrophic areas of the oceans. There is of course no absolute requirement that the budget should balance, the point was made earlier that all systems contain pools of organic material, which can and do sustain imbalances—diurnal and seasonal imbalances have been long acknowledged as part of normal functioning of ecosystems. Secular changes in the upper ocean DOC have been observed for instance at Station ALOHA in the North pacific Subtropical gyre (Church *et al.* 2002). This would imply a change in balance photosynthesis to respiration ratio. How long such imbalances may be sustained will be related to the size of the pools and the level of carbon cycling—the rate of increase at Station ALOHA was 0.83 mmols m^{-2} a^{-1}, approximately 1% of the annual productivity and would have not been discernable from field studies of the rates of photosynthesis and respiration. In the case of the upper oceans, carbon storage is small, approximately equal to that of the annual organic cycle, so there is limited capacity for that system to run out of equilibrium for extended period by "mining" its own reserves.

The data compiled by the contributors to this book have been assembled in Table 14.1 and Fig. 14.7. As the various authors point out, a number of major aquatic ecosystems notably, the global ocean, come out net heterotrophic, although, subcomponents, for example, the coastal ocean and marine macrobenthic systems, however, may be net autotrophic (see Boxes 9.2 and 9.3). Current estimates of open-ocean NPP lie in range 3 and 4 PmolC a^{-1} (Longhurst *et al.* 1995, Falkowski *et al.* 2000; Fasham *et al.* 2001), whereas Robinson and Williams (Chapter 9) estimate respiration to be in the region of 9–12 PmolC a^{-1}. The difference between the measured rates of respiration and production is so large that it begs further analysis.

The mass-balance approach is valuable in this context in the case of the oceans as, although it cannot provide an independent estimate of both photosynthesis and respiration, it can establish the difference between the two with relatively high precision. The sensitivity of this approach is determined by the precision with which we can estimate the input and output terms. Mass balance calculations (see Chapter 9, Box 9.3) give an excess of respiration over photosynthesis (net heterotrophy) of $20 \text{ Tmol O}_2 \text{ a}^{-1}$ (0.02 Pmol a^{-1}). Of the estimates used in the mass-balance for the ocean as a whole, the greatest source of uncertainty lies with the estimate of the major term—river input. It can probably be prescribed within $\pm 0.01 \text{ Pmol C a}^{-1}$, thus our estimate of $R-P$ is $0.02 \pm 0.01 \text{ Pmol C a}^{-1}$, that is, an uncertainty of less than 0.3% of the individual values for photosynthesis and respiration rates. Thus, annual photosynthetic and respiration rates in the global ocean must be numerically very close—less than half a percent of one another. This leads to the realization that there must be substantial systematic errors in our present estimations of the rates of either or both production or respiration.

In principle the measurement of respiration is simpler and more reliable than photosynthesis as it does not suffer from the uncertainties associated with the cycling and fate of the added isotope, nor does it bring the need to simulate a light field of varying quality and fluctuating quantity. For the most part the principal errors associated with the ^{14}C technique result in underestimates and so taking account of these would tend toward a resolution of the discrepancy observed for the oceans. A prevailing, although not exclusive, view (see Williams *et al.* 1996; Marra 2002) is that the ^{14}C technique provides a measurement of net primary, rather than gross production. This being the case, for comparison with whole community respiration measurements, the ^{14}C-derived rates must be corrected upward for phytoplankton respiration to allow estimates of gross production. Robinson and Williams' analysis of the distribution of respiration within the community suggests a central figure for this correction in the region of a 40%. Given current estimates of open-ocean NPD in range 3 and 4 Pmol C a^{-1} (see above), correcting for autotrophic respiration would give gross production rates in the region of $5–7 \text{ Pmol C a}^{-1}$. Contemporary ^{14}C uptake rates are

mainly derived from uptake into the particulate fraction, thus the loss as extracellular production gives rise to a further underestimate of the true rate of production (del Giorgio and Duarte 2002; Karl *et al.* 2003). It is hard to estimate this error with any certainty but it may add 15% onto production rates, giving a further upward correction for global production to $6–8 \text{ Pmol C a}^{-1}$. The need to present the photoautotrophs with the appropriate light quality and quantity has long been recognized as a problem. Much of the productivity work has been done with deck incubators and even when in experienced hands (see e.g. Hiscock *et al.* 2003) it is clear that the approach incurs problems. One might infer that the conditions presented are more likely to be suboptimal than supraoptimal and would tend to give rise to underestimates. It is not possible to put a number to the frequency or scale of this error. We need to recognize the error is there and that it probably results in an underestimate of production. Even when the above suite of problems are avoided by using *in situ* incubations and oxygen determined rates (the latter circumventing the problems associated with tracers) discrepancies are still found between the measured rates of respiration and photosynthesis in oligotrophic ocean areas (Arístegui and Harrison 2002; Williams *et al.* 2004). It has been proposed that this is due to an intermittency in photosynthesis not seen in respiration (Arístegui and Harrison 2002; Karl *et al.* 2003).

Finally, there appear to be major problems with the models used to produce global maps and estimates of production in that they provide very poor results for productivity in low latitude, warm waters (Ducklow 2003). Characteristically they underestimate productivity, predicting for example rates in the range $50–100 \text{ gC m}^{-2} \text{ a}^{-1}$, whereas contemporary field observations at station ALOHA at the HOT site in the subtropical Pacific gyre give values in the range $180–240 \text{ gC m}^{-2} \text{ a}^{-1}$ (Karl 1999; Williams *et al.* 2004), that is, some $150 \text{ gC m}^{-2} \text{ a}^{-1}$ higher than the model outputs. The point is commonly overlooked than on an annual basis these waters are as, if not more, productive than the nutrient rich waters of the North Atlantic. The apparently oligotrophic warm waters occupy about one-third the areas of the oceans and were this underestimation typical of these type of waters it would give rise to an underestimation of global productivity by some 25% (Ondrusek *et al.* 2001) equivalent to 1.5–2 Pmol a^{-1} in the case of gross production estimates. The cumulative effect of these various errors associated with the estimation of global oceanic production would mean that a gross production rate of 8–11 Pmol C a^{-1}, may be a more appropriate one to compare against the 9–12 Pmol C a^{-1} range that we obtain for open-ocean respiration. Thus, if we are prepared to allow the above corrections to the current estimates of oceanic productivity then were are able to bring the estimates sufficiently close that the remaining discrepancy fall within details of the appropriate *RQ*, areas of ecosystems, and small external carbon subsidies.

Other lines of evidence would in fact point to the upper limit of the above range of primary production. Arístegui *et al.* (Chapter 10) show that different approaches converge to a global rate of dark ocean respiration of around 1.7 Pmol C a^{-1}, and as these authors point out, this range of dark ocean respiration is substantially higher than current estimates of organic matter export from surface layers (i.e. Falkowski *et al.* 2000; Fasham *et al.* 2001; Ducklow and McCallister 2004). In this regard, a global value of open-ocean gross primary production of around 11 Pmol C a^{-1} would accommodate the range of open-ocean NPP required (approximately 6–7 Pmol C a^{-1}, at 40% respiration this would require gross production rates of 8–10 C a^{-1}) to support the estimated dark ocean respiration without having to invoke export ratios substantially above the present accepted ranges (<35%, Falkowski *et al.* 2000).

The calculations by Pace and Prairie (Chapter 7) suggest that lakes worldwide may respire in the order of 0.07–0.02 Pmol C a^{-1} of organic matter that is derived from the terrestrial ecosystem, although as the authors point out, these estimates of excess lake respiration are still very uncertain because neither global lake respiration or photosynthesis are well constrained. Nevertheless, considering that the global net terrestrial production is in the order of 0.25 Pmol C a^{-1} per year, this apparent loss of terrestrial organic matter through lake respiration becomes globally important. Not

surprisingly, most lakes are supersaturated in carbon dioxide (Cole *et al.* 1994), and although there are multiple factors that lead to this supersaturation (discussed by Pace and Prairie, Chapter 7), respiration is undoubtedly a major factor (Jonsson *et al.* 2003). This pattern in carbon dioxide supersaturation and emission is not unique to natural lakes, and extends to human-made reservoirs, which have been known for years to be important sources of carbon dioxide to the atmosphere (Duchemin *et al.* 1995). There is still debate as to the origin of the carbon dioxide that is emitted from reservoirs, but it is now clear that at least a portion of this carbon dioxide derives from the biological oxidation of terrestrial organic matter within the reservoirs (Duchemin *et al.* 1995). Although the respiration in human-made reservoirs was not explicitly considered in the calculations made by Pace and Prairie in Chapter 7, it is an issue that should be considered in future studies. For example, the province of Québec (Canada) holds vast freshwater resources, in the form of over 1 million lakes of different size and characteristics. In Québec, the development of hydroelectric power has resulted in the creation of enormous reservoirs in the north, which have further expanded the already vast lake resource of the province. It is interesting to note that out of the 10 largest lakes in Québec, 4 are now hydroelectric reservoirs built in the past three decades. Similar situations are found worldwide, and the creation of these vast reservoirs highlights the need to understand the role of these newly created ecosystems in terms of carbon cycling and regional greenhouse gas budgets (Prairie *et al.* sumbitted).

The data from Table 14.1, together with other estimates assembles as a flow chart in Fig. 14.7. To our knowledge it is the first attempt to compile such a dataset. No attempt has been made to force the dataset to audit, as we feel it is important that discrepancies are revealed as they represent uncertainties in our knowledge and direst attention where greater understanding is needed. The limitations and uncertainties of the data in Table 14.1 have been discussed above. Of the additional data, the rates of net sedimentation (burial) for lakes and estuaries contain major uncertainties. In part this derives from different approaches to determining the rates. In the figure we have separated out the budget associated with the coastal margin macrophytes (salt marshes, benthic macroalgae, mangroves, and sea grasses). In biogeochemical budgets these ecosystems are often incorporated in the coastal zone estimates. We have chosen to separate them out, in part, as they are ecologically and physiologically very different from the planktonic community. Also presenting them separately reveals the striking figure for the export of organic material from these ecosystems. The figure (148 Tmol C a^{-1}), as the estimates given for production and respiration, was derived from the analysis of Duarte and Cebrián (1996). This source to the ocean far exceeds that from rivers (34 Tmol C a^{-1}). The estimate is dominated by inputs from macroalgae (92 Tmol C a^{-1}) and mangroves (27 Tmol C a^{-1}), both known to be major producers of exportable organic material. The estimate is unlikely to be substantially adrift, the main uncertainty lies not with the areal rates themselves but the appropriate area attributable to the individual ecosystems (Middelburg, personnel communication).

14.4.5 Aquatic versus terrestrial respiration

As was discussed in previous sections, although a portion of the global aquatic respiration is supported by organic matter derived from terrestrial ecosystems, the overwhelming majority of the organic matter respired derives from authochthonous primary production. Given the estimates compiled in the present work, this line of evidence suggests that the global gross aquatic primary production must also be in the order of 14–17 Pmol C a^{-1}. This range is not only higher than any current estimate of aquatic primary production, but as we discuss below, is also higher than the best current estimates of gross terrestrial primary production.

Current estimates place net terrestrial primary production in the order of 4–5 Pmol C a^{-1} (Field et al. 1998; Woodwell et al. 1998; Canadell et al. 2000). The contribution of autotrophs to respiration is estimated at roughly 45–50% of gross terrestrial primary production (Field et al. 1998; Waring et al. 1998; Randerson et al. 2002), so that global terrestrial gross primary production should be in the order of

8–10 Pmol C a^{-1} (Canadell et al. 2000; Randerson et al. 2002). Global soil respiration is fairly well constrained in the range of 6.6–6.8 Pmol C a^{-1} (Raich et al. 2002), of which approximately 1.9 Pmol C a^{-1} is due to root respiration, leaving approximately 4.7 Pmol C a^{-1} to respiration by heterotrophs in terrestrial ecosystems globally. Root respiration represents on average around 40% of total autotrophic respiration (Waring et al. 1998), so the sum of below and above-ground plant respiration is in the order of 4.8 Pmol C a^{-1}, about equal to that of heterotrophs, giving an overall figure for terrestrial respiration in the order of 9.5–10.5 Pmol C a^{-1}, which agrees well with the estimate for global terrestrial gross production that we provided above. Our synthesis suggests that global aquatic respiration must be in the order of 14–17 Pmol C a^{-1}. There is still a large degree of uncertainty in these estimates, but the ranges in aquatic and terrestrial respiration are sufficiently apart to suggest that the current paradigm that primary production in the biosphere is roughly equally split between land and water (Field et al. 1998) probably does not hold. Global aquatic respiration, and thus aquatic gross primary production, very likely exceed global terrestrial respiration and production, possibly by a significant amount.

It would appear that in terrestrial systems, there is a convergence in the values of the contribution of plants to total respiration of around 45–50% (Waring et al. 1998; Randerson et al. 2002), even in ecosystems that are very different structurally. Robinson and Williams in Chapter 9 conclude that algal respiration in the ocean represents 40% of total community respiration on average, but there is a large degree of variance around these estimates and the bias in the observations toward times of active production will have inflated this estimate. The contribution of autotrophs to system respiration seems to be highly variable among aquatic ecosystems, and even within a given type of ecosystem along gradients of productivity (Biddanda et al. 2001; Smith and Kemp 2003). This suggests that one of the main functional differences between aquatic and terrestrial ecosystems is that heterotrophs, particularly heterotrophic bacteria and protists, often contribute a larger fraction of respiration in aquatic ecosystems, at least in the pelagic communities. Another functional difference between aquatic and terrestrial ecosystems is that fungal respiration plays a much greater role in the latter.

terr.

Net terrestrial ecosystem production represents the potential accumulation of carbon by ecosystems. This has most frequently been estimated from the difference between gross primary production and terrestrial ecosystem respiration. This ignores all non-respiratory losses of carbon from the terrestrial ecosystem, such as the carbon dioxide and monoxide production by abiotic processes (i.e. fire), the flux of organic volatile compounds, the export of dissolved and organic matter from the terrestrial ecosystem, and the export of various forms of inorganic carbon, including carbon dioxide originating from soil respiration, to lakes and rivers (Randerson *et al.* 2002). The last three categories play important roles in the functional linkage between terrestrial and aquatic ecosystems. The DOC flux from terrestrial ecosystems has been estimated in the range of 0.016–0.075 $PmolCa^{-1}$, with an additional 0.025 $PmolCa^{-1}$ in the form of particulate organic carbon (Randerson *et al.* 2002), giving a total of 0.04–0.11 $PmolCa^{-1}$. The sum of the potential excess respiration from lakes (0.02 $PmolCa^{-1}$, Pace and Prairie, Chapter 7), rivers (0.02 $PmolCa^{-1}$, Caraco and Cole 1999), and estuaries (0.03 $PmolCa^{-1}$) is in the range of 0.06–0.07 $PmolCa^{-1}$. Added to this, it is estimated that approximately 0.035 $PmolCa^{-1}$ are delivered by the rivers to the ocean (Cauwet 2002), giving a total of 0.09–0.11 $PmolCa^{-1}$, similar to the calculated total export of organic matter from land (Randerson *et al.* 2002), so there is some internal consistency in all these estimates.

oceanic

The net carbon dioxide uptake by the ocean (the ocean carbon "sink") is in the order of 0.16 $Pmol\,a^{-1}$, but most of this net uptake is due to physical processes mediated by the solubility pump (Woodwell *et al.* 1998; Ducklow and MacCallister in press). The biological component is responsible for a large flux of organic matter but little net storage of carbon. The biological pump transfers approximately 1.7 $Pmol\,a^{-1}$ of organic carbon produced in the surface layers of the open oceans to the deep layers, but as Arístegui *et al.* (Chapter 9) point out, most of this organic matter (1.4 $PmolCa^{-1}$) is mineralized

in the mesopelagic layer of the ocean, and is thus returned to the surface waters within the time frame of vertical mixing, which is in the order of decades for the top 1000 m of the oceans. The remainder of the carbon exported from the surface continues the downward flux, where approximately an additional 0.3 $Pmol\,a^{-1}$ are respired. The burial rate of organic matter in the ocean is fairly well constrained: it is estimated at 12 and 2 $TmolCa^{-1}$ for coastal and deep ocean sediments, respectively, and this burial represents mostly the contribution of marine biological processes to long-term carbon sequestration.

The terrestrial ecosystem is also thought to be a major sink for carbon, but there is no equivalent of the solubility pump in terrestrial ecosystems, and most of the net atmospheric CO_2 uptake is due to biological processes. The net carbon uptake by terrestrial ecosystems is estimated to be in the order of 0.15 $Pmol\,a^{-1}$ (similar in magnitude to the oceanic C sink), but as opposed to the oceans, the carbon sequestered from the atmosphere is thought to be stored in living or dead plant biomass, although the mechanisms that lead to this storage are still under debate (Houghton *et al.* 1998; Woodwell *et al.* 1998). If global aquatic gross primary production and respiration are substantially higher than those of terrestrial ecosystems, as we find here, then the biological component of terrestrial ecosystems in general appears to function as a more effective short-term carbon removal mechanism than the oceanic biological pump, at least relative to their respective levels of primary production and total C flux. This would in turn suggest that changes in carbon metabolism in terrestrial ecosystems might have proportionally greater influence on the global carbon budget than shifts in oceanic metabolism, although the latter appears to be a greater contributor to global carbon flux. Aquatic ecosystems, and particularly the world's oceans, appear to be more effective than terrestrial ecosystems in terms of mineralizing the organic matter produced by autotrophs, suggesting that the overall degradation of organic matter may be higher in water than in soils. On the other hand, much of the long-term carbon storage in the biosphere occurs as marine kerogens (Woodwell *et al.* 1998), so the oceans appear to be more effective

source? mechanisms + time scale

sink in the very long-term than the terrestrial ecosystem.

In terms of the role of the biological component in the net fluxes of carbon dioxide between the atmosphere and ecosystems, one of the key questions is whether these are mostly driven by the variability in respiration or in gross primary production. Respiration, and not gross primary production, seems to be the main determinant of the overall carbon balance and storage in many terrestrial ecosystems, particularly those at high latitudes (Valentini *et al.* 2001; Schaefer *et al.* 2002). In a scenario of global warming, temperature-induced increase in NPP may be offset by even larger increase in ecosystem respiration (Vukicevic *et al.* 2001). On the other hand, soil respiration appears to be relatively constant, with an interannual variability of around 0.1–0.2 $Pmol\,C\,a^{-1}$ (Raich *et al.* 2002). It would appear that soils buffer fluctuations in atmospheric carbon dioxide concentrations by providing relatively constant, year round supply of carbon dioxide that modulates the effects of more extreme intra and interannual variations in primary production (Schlesinger and Andrews 2000; Raich *et al.* 2002). Potter and Klooster (1998) have suggested that heterotrophic respiration dampens carbon dioxide fluxes from the terrestrial ecosystem to the atmosphere, with about a lag period of 2 years relative to NPP. This type of information, which is critical to predict the changes in global aquatic respiration and net ecosystem production following long-term changes in aquatic primary production, is almost completely lacking for most aquatic ecosystems.

In contrast, it would appear that in most aquatic ecosystems it is the magnitude and fluctuation in primary production that determines the balance between primary production and respiration (del Giorgio *et al.* 1999; Arístegui and Harrison 2002), driven it is argued by the influx of inorganic nitrogen and phosphorus or other limiting nutrients. In other words, an increase in the P/R ratio in lakes or oceans is most likely to be the result of an increase in primary production than to a decrease in respiration, and vice versa, low P/R ratios usually correspond to situations of low primary production rather than higher than normal respiration. Karl *et al.* (2003)

have further speculated that there may be vigorous but short-lived pulses of primary production in the open ocean, that are difficult to detect using our conventional sampling strategies, and that these pulses play a major role in the metabolism of open oceans. It remains to be tested whether such pulsed pattern of primary production does indeed exist in the open ocean. Their thesis and the weight of evidence (see Robinson and Williams, Chapter 9) suggest that respiration does not follow a similar pattern, and most studies suggest that respiration is considerably less variable than photosynthesis when compared at similar spatial and temporal scales. This is also true for lakes (Pace and Prairie, Chapter 7), estuaries (Hopkinson and Smith, Chapter 8), even coastal benthic communities (Middelburg *et al.*, Chapter 11).

14.5 The biogeochemical and ecological roles of respiration in aquatic ecosystems

There is no question that the role of respiration as the counterpart of production is a key ecological one, and this was recognized early on by aquatic ecologists (Odum 1956; Hutchinson 1957). In aquatic ecosystems, as in the biosphere in general, the rate of respiration generally reflects the rate of supply of organic matter that is available to the biota within the constraints imposed by temperature and other physical and chemical factors. Respiration in this context provides the best estimate of the total flux of organic matter through the biota (Jahnke and Craven 1995), although it does not provide any information as to the sources of the organic matter consumed.

In most papers that have explored the relationship between production and respiration for different ecosystems, production almost invariably appears in the *x*-axis and respiration in the *y*-axis, and the underlying assumption is that production is the independent factor whereas respiration is the dependent factor. To what extent are we justified in placing respiration always as the dependent variable? Or, put into other words: are there feedbacks between respiration and production such that at least some components of respiration might influence photosynthesis and not vice versa? The discussion in the preceding sections suggest that such feedbacks might exist, although we can only speculate at this point how they would function. Respiration is not just a sink of organic matter, but rather plays a major part in structuring the ecosystem that goes well beyond its role as the final step in the mineralization of organic matter (Wetzel 1992). This also implies, for example, that organic matter subsidies to aquatic ecosystems, even if they are extremely small, may also assume a role that goes beyond providing just a small additional amount of substrate to fuel bacterial metabolism. The baseline respiration may determine the capacity of the system to respond to environmental fluctuation and heterogeneity. In the same way a starving bacterial or protozoan cell will be able to respond faster to a sudden input of substrate if they have been able to maintain a minimal level of energy flux within the cell (this in part was the argument of Williams 1990), perhaps whole ecosystems may be able to track environmental change more effectively if it is able to maintain a higher energetic flux.

The corollary of this scenario is that in a generally heterogeneous and fluctuating environment, communities that can maintain a higher level of baseline metabolism may also be able to more efficiently track this environmental heterogeneity and more efficiently utilize the resources. There are various ways by which this could happen. For example, mixotrophy is widespread in all aquatic ecosystems, and although it regarded as an important trophic variant (Stickney et al. 1999), its role as a structuring factor has received less attention. Mixotrophy may be one of the processes that can link the baseline metabolism with the response of the ecosystem to environmental fluctuation. Mixotrophs may complement or even survive on an heterotrophic diet while conditions for autotrophy are not optimal, but this requires a minimum level of heterotrophic activity, and in turn, results in further respiration that is not necessarily linked to the *in situ* level of primary production. However, it is this minimum level of heterotrophic activity that may allow a plankton community that includes mixotrophs, to respond to nutrient inputs and maximize the amount of primary production within a fluctuating environment.

Two ecosystems that have similar average levels of nutrients, light and physical forcing, may conceivably differ in their capacity to respond to fluctuations depending on the minimum level of energy flow that they can maintain. Likewise, the relative magnitude of the baseline respiration may influence interactions between autotrophs and heterotrophs, or between bacteria and other heterotrophs. For example, Jansson et al. (2003) have recently suggested that in lakes receiving and processing relatively large amounts of allochthonous organic matter, primary production may be dampened because the external organic matter enhances the capacity of bacteria to take up and store phosphorous, further limiting primary production. This is an example of how some of the components of respiration are not only relatively independent of primary production, but in fact may play a role in the actual regulation of primary production. Understanding this type of feedback should be one of the priorities of our future research efforts.

14.6 Aquatic respiration and global environmental change

How might global climate change affect global aquatic respiration? We are far from having an answer to this question. As in terrestrial ecosystems, the effect of climate change on aquatic respiration is likely to be complex and ecosystem-specific. There will be a direct physiological effect of an increase in temperature on respiration. Most determinations of Q_{10} for aquatic organisms fall within the range of 2–3 (Williams and del Giorgio, Chapter 1; Hopkinson and Smith, Chapter 8; Robinson and Williams, Chapter 9; Arístegui et al, Chapter 10, Middelburg et al., Chapter 11), except perhaps for methanogenesis, where higher Q_{10} have been measured (Roehm, Chapter 6). The range of Q_{10} values measured in terrestrial and aquatic ecosystems seems to be similar, but the physiological response to temperature changes is not simple in any ecosystem and the assumption of a constant Q_{10} is not always warranted. For example, Pace and Prairie (Chapter 7) suggest that the Q_{10} of lake sediment respiration is a function of the temperature range, and similar patterns have been reported for terrestrial vegetation (Tjoelker *et al.* 2001). The Q_{10} tends to be higher at lower temperatures, so the effect of increasing temperature on aquatic respiration may be proportionately much greater in colder regions.

It is unclear to what extent this physiological response to temperature at the individual level translates into an overall shift in respiration at the ecosystem level, because at this level, increases in individual rates driven by temperature may be offset by changes in the rate of supply or the nature of the organic matter, nutrients, and light. For example, it is very likely that respiration in northern peat bogs, wetlands and lakes will increase with increasing average temperature, because of increased mobilization of organic matter stored in permafrost layers (see Roehm, Chapter 6). There is speculation (i.e. Tranvik and Jansson 2002) that changing weather patterns, particularly in northern latitudes, may result in shifts in organic matter loading to lakes and estuaries from the drainage basins and this will no doubt influence lake and estuarine respiration, independently of the direct effect of a slight increases in temperature of the water (see Pace and Prairie, Chapter 7). For example, Algesten *et al.* (2003) have suggested that variations in hydrology can be more important in determining carbon mineralization and carbon dioxide emissions from lakes than temperature fluctuations. Likewise, heating of the surface layers of the oceans may indeed result in a slight increase in the rate of organic matter cycling and respiration, but again, it is likely that the indirect effects of global climate change may be much more important. For example, sea level changes may induce changes in coastal metabolism (Miller *et al.*, 2001), and could result in organic matter remobilization that could potentially influence coastal and even open-ocean metabolism (del Giorgio and Duarte 2002).

Although global climate change has the potential of affecting respiration in all aquatic systems, it is unlikely to affect respiration in the same way and at similar timescales everywhere. In this context, it is important to identify which are the ecosystems that are the most susceptible to global change, because these may represent effective early warning systems. Freshwater wetlands appear to be particularly sensitive to climate change, and as Roehm

shows in Chapter 6, their global metabolism is very significant, so any shift in net metabolism, resulting from either changes in respiration, photosynthesis or both, may have large consequences on the global C budget. As Roehm (Chapter 6) and Woodwell *et al.* (1998) point out, the possible direction of change in wetlands is difficult to predict, because global warming could result in both a lowering or an increase in the water table, with disparate effects on wetlands respiration and methane production.

The patterns of respiration in relation to temperature may differ between terrestrial and aquatic ecosystems. In terrestrial ecosystems, it appears that although temperature does play a role in determining seasonal changes in respiration within a given system, annual ecosystem respiration is poorly correlated with either mean annual temperature, or mean growing season temperature (Law *et al.* 2002). The surface waters are probably less buffered against changes in temperature than soils, and the direct effect of temperature on respiration may be larger (Rivkin and Legendre 2001, and see Roehm, Chapter 6; Pace and Prairie, Chapter 7; and Hopkinson and Smith, Chapter 8). For example, Hopkinson and Smith (Chapter 8) show that temperature alone explains around 30% of the variability in respiration across estuaries. The climatic effect on terrestrial respiration is extremely complex (Woodwell *et al.* 1998), but there is general consensus that increases in average air temperature will result in increasing soil respiration. For example, Raich *et al.* (2002) predict an increase in around 0.27 Pmol C year °C increase. Temperature is not the only climatic factor that affects soil respiration, and for example, the response of soils to simultaneous increases in temperature but decreases in moisture are difficult to model and therefore to predict.

Likewise, changes in patterns of ocean circulation, inputs of nitrogen and consequently primary production (Falkowski *et al.* 2000) may have a much greater impact on ocean respiration that the direct physiological effect due to the heating of the surface waters. For example, warm water dissolves less oxygen than cold, and as Codispoti *et al.* (Chapter 12) suggest, only minor changes in the dissolved oxygen distribution could lead to large changes in the suboxic respiratory pathways that remove globally significant quantities of fixed-N, that, in turn, could impact primary production and respiration. In this regard, the magnitude and the time lag of the response of respiration to changes in primary production at the ecosystem level is not well known, and should be one of the main focus of future studies. For example, during the large-scale iron addition experiments that were carried out in several oceanic areas a large number of variables were measured, including primary production, carbon sedimentation, and carbon dioxide exchange (i.e. Boyd *et al.* 2000), but it is not clear how the rates of respiration responded to the induced algal blooms. Most scenarios of global change assume that even if there are major changes in aquatic primary production, the relationship between production and respiration at the ecosystem level will remain relatively constant. The notion that the current balance between production, respiration, and carbon export can be extrapolated to a scenario of substantially different food web structure and primary production, has little empirical or theoretical basis, and must be thoroughly tested before any solid conclusion can be reached.

14.7   Conclusions and future prospects

The studies in the book have enabled us to prepare an assessment of respiration in all major aquatic ecosystems—from lakes or marshes to the deep ocean waters and sediments, in many cases along with the associated rates of photosynthesis (see Table 14.1 and Fig. 14.7). Whereas there are uncertainties over the exactness of the estimates—notably for the upper ocean where the coverage is small in comparison with the size of the ocean and its contribution to the global estimates—we are confident that our broad rank order is correct. In addition to pointing out the uncertainty in some of the estimates involved in this global sum, the synthesis we present here highlights some major discrepancies that should perhaps be the focus of future research efforts.

In spite of tremendous variation in system productivity, community structure, and physical forcing, average respiration rates vary by less than fivefold from open-ocean surface waters to lakes

and estuaries (3–16 mmol C m^{-3} d^{-1}). The average areal rates of respiration in the surface water column vary even less and are within a factor of three (50–120 mmol C m^{-2} d^{-1}) among widely different ecosystems (excluding freshwater wetlands that have an order of magnitude higher rates). Likewise, average sediment respiration has a total range of 20-fold, from bathypelagic ocean sediments to estuarine sediments (1.6 and 34 mmol C m^{-2} d^{-1}, respectively). Exceptions to this broad generalization are the waters of the deep ocean layers, where both the volumetric and areal rates are orders of magnitude lower, and freshwater wetlands where extremely high rates are observed. The relative narrow range of variation of respiration in the surface layers is a fundamental characteristic of this process that has profound consequences on the functioning of aquatic ecosystems. It is presumed that the underlying mechanisms for this relative constancy are related to the fact that respiration integrates both the amount and nature of the inputs of organic matter, including local primary production as well as external inputs, over a range of temporal and spatial scales, thus smoothing out short-term variations in any of these parameters.

Heterotrophic bacteria are most likely the single largest contributors to total respiration in most aquatic ecosystems, including sediment and plankton communities. Exceptions are some coastal and estuarine benthic communities where metazoans and macroalgae are the main contributors, or highly productive planktonic systems where total respiration is dominated by the respiration of algae. Mesozooplankton appear to contribute between 5 and over 10% of total pelagic respiration in the oceans, but this component is seldom considered in global calculations of respiration. Particularly crucial is the need to devise improved approaches to separating out the algal component, as this is needed to determine NPP that is, how much of primary production is available to remainder of the community.

At the ecosystem level total organic matter input (the sum of authochthonous and allochthonous inputs) seems to be the primary control of respiration in aquatic ecosystems, but the authors in the different chapters have identified a number of factors, in addition to temperature, that may modulate the relationship between the rate of loading of organic matter and the respiration rate at the ecosystem level. These factors one might expect to include the water (and material) residence in the ecosystem (but see Section 14.4.1), the composition of the organic matter, inorganic nutrient availability (Obernosterer *et al.* 2003), and the structure and composition of the food web (Pace and Cole 1999). This may explain why in most ecosystems respiration is poorly correlated to any single factor, including primary production, carbon loading, chlorophyll or nutrient concentrations, or biomass of organisms. The only exception may be the epilimnion of lakes, where respiration appears to be fairly strongly correlated to total phosphorous and thus to overall system productivity. As we have emphasized throughout this chapter, respiration at the ecosystem level is not a single, uniform process, but the combination of several processes, each of which probably controlled by a different set of factors, and it is thus unlikely that respiration can be predicted or modeled on the basis of a single factor. In particular, we need to gain a better understanding of how the interactions within the food web may feedback on processes that determine respiration.

The respiration in the surface layers of the open oceans represents the largest, and also the most uncertain component of global aquatic respiration budget (Table 14.1 and Figures 14.6 and 14.7). There are also extremely few direct measurements of respiration in the dark layers of the oceans, but the existing data tend to converge. Most of the dark ocean respiration occurs in the mesopelagic layer. The coastal ocean contributes approximately to 1.2 $PmolCa^{-1}$. Freshwater wetlands emerge as an important contributor to global aquatic respiration (approximately 1 $PmolCa^{-1}$), of the same magnitude as the coastal ocean, and are the major aquatic source of organic material (methane) to the atmosphere. Lakes, rivers, and estuaries contribute modestly to the global aquatic respiration, but these systems respire large amounts of allochthonous organic matter exported, so their contribution is important in terms of the fate of net terrestrial ecosystem production. The sum of the global respiration of the major aquatic ecosystems yields a probable central range of 14–17 $PmolCa^{-1}$. This

estimate is higher than the respiration of the terrestrial ecosystem, which is currently estimated to range between 9.5 and 10.5 $PmolCa^{-1}$ (Candell *et al.* 2000).

The balance between respiration and production has been for decades considered an essential characteristic of all ecosystems, and even today there is still considerable debate regarding apparent imbalances that have been reported in oceans and lakes. In reality, there is essentially no imbalance, because aquatic ecosystems simply consume and respire all but a small amount of the organic matter available (within the constraints of time, temperature, and other factors that limit biological activity). The apparent imbalances are artifacts that result from our largely insufficient understanding of the magnitude and regulation of the total organic matter flux in aquatic ecosystems. Aquatic ecologists and biogeochemists have often relied on measurements of production to quantify this total flux of organic matter, but we argue in this chapter that it is the measurement of respiration that provides the most accurate and integral estimate of the total organic carbon flux in all ecosystems. The irony is that for historical reasons (see Williams and del Giorgio, Chapter 1), measurement strategies of respiration have always been subordinated to those of production, and throughout past decades have had to conform to the perceptions of production that were held at the time. As we have tried to emphasize in this synthesis chapter, while it is true that there is a paucity of direct measurements of respiration for most aquatic ecosystems, those that do exist are remarkably consistent in magnitude and pattern both within and across ecosystems, much more so than measurements of primary production. In most aquatic ecosystems, respiration may provide the most effective index of organic matter production.

Our analysis of global aquatic respiration suggests that gross aquatic primary production must be in the order of 14–17 $PmolCa^{-1}$, and we would anticipate that global net aquatic primary production lie in the range from 7 to 10 $PmolCa^{-1}$. These estimates of production based on the extrapolation of respiration measurements are considerably higher than the range obtained using both direct measurements of production, and also

[handwritten margin note: no storage or burial?]

indirectly using carbon or nutrient mass balances. The global rate of mesopelagic respiration in the oceans that results from this analysis is also larger than current estimates of organic matter export from surface layers. Overall, the analysis of respiration in aquatic ecosystems suggests that current approaches may significantly underestimate primary production and organic matter flow in aquatic ecosystems in general and thus at the biospheric level. Future research should focus on reconciling these various observations, so that these major discrepancies may eventually be bridged.

Respiration has traditionally been modeled as a direct function of production, but as has been emphasized throughout this synthesis chapter and the book in general, that this approach is ineffective in capturing the complexity of the regulation of respiration not only at the cellular and organism level, and even more so at the ecosystem and biosphere levels. We urgently need models that specifically address respiration as a process, and not some simple function of photosynthesis, as all the field evidence is that the truth is a great deal more complex than that. Further, it is important to incorporate respiration into models of ecosystem function not as a single, homogenous process, but as the sum of various processes that are regulated differently and have different consequences at the ecosystem level (see Flynn, Chapter 13). In addition to considering respiration as a multi-component process, it will be important to explore feedbacks between respiration, production, and other functional aspects of the ecosystem, such feedbacks are practically nonexistent in current models of carbon flow in aquatic ecosystems although they may be the key to our capacity to predict local and global changes in carbon cycling. In wetlands in particular, the feedbacks are crucial, as without an understanding of these we cannot anticipate the consequences of climate change and particularly the balance between carbon dioxide and methane as end products. The inclusion of the various components of respiration in ecosystem models will allow us to explore the spatial and temporal scales of coupling between respiration, production, and carbon loading, which is one of the main gaps in our current understanding of carbon cycling in the biosphere.

There are still numerous technical problems to surmount regarding the measurement of respiration, including improving the sensitivity of chemical methods (for oxygen, carbon dioxide), the robustness of enzymatic or biochemical methods such as ETS, and the validations of assumptions underlying many of the current whole-system approaches. In many cases we are working at the limits of volumetric precision and it is unlikely that the technique can be further refined and improved. Ratio mass spectrometry approaches largely avoid the volumetric problems and potentially offer gains in a number of aspects of performance. The more widespread use of mass spectrometry will lead to greater use of estimates of rates derived from *in situ* gas exchange studies. *In situ* measurements give better temporal and spatial averaging. As *in situ* techniques yield rates of net metabolism and do not resolve the vertical distribution of processes or the temporal separation of production and consumption, they are best complemented with *in vitro* methods. The rates in the deep parts of many lakes and ocean waters are beyond the capability of *in vitro* measurement of oxygen concentration change. We have seen a revival of enzymatic techniques and the evidence is that they are returning rates that are consistent with other measures of oxygen flux. We can hope to see some of the earlier prejudices toward these techniques dispelled and a growth in this approach, and the development of new and more effective biochemical approaches (Packard *et al.*, 2004). Deepwater oxygen flux measurements are not complicated by the need to correct for concurrent photosynthetic rates and we are now seeing the development of methods that allow the interpretation of *in situ* observations in water masses in terms of respiration. The oceanographic database available to us for such type of analysis is massive and the potential considerable, but these approaches must be further developed. Regardless of the technique used to determine respiration, there remains also the problem of separating respiration into its main components, that is, separating algal from heterotrophic respiration, and of exploring the magnitude and regulation of baseline metabolism in different aquatic ecosystems.

It is all too common for ecologists to complain that all ecosystems, particularly their own pet ones, are undersampled. Indeed most are. The level of undersampling in the case of respiration is without doubt serious and sets limits to the depth of our understanding of this process in the biosphere. All told we estimate that the data we have available to us on respiration in aquatic ecosystems as a whole is some 3500–4000 observations. This is about one-tenth of the number of data on oceanic planktonic photosynthesis that served as the basis for the map plankton of productivity of the oceans that Koblenz-Miske *et al.*, published as far back as 1968. We may expect that number of oceanic observations has probably increased by a factor 10 or more since then, thus being some 100 to 500-fold larger than the database on respiration. The Indian and the Pacific oceans are particularly poorly sampled for respiration. In these two oceans, there is a continuous area of some 200×10^6 km^2 (greater than the combined area of the continents) in which not a single respiration profile has ever been made. Although most of the oceanic respiration occurs in the surface layers, mesopelagic respiration is still a key component in the carbon flux to the deep ocean, and one of the most understudied in the oceanic carbon cycle. Likewise, we have extremely few data from the large lakes of the world, which comprise over 50% of all lake waters, and those in tropical regions. Hydroelectric reservoirs are widespread in the world, and are increasing in magnitude to the point that some of the largest freshwater bodies in the world are now human-created dams. Although these are well known to act as net carbon dioxide sources, there is almost no data on respiration in large reservoirs. Tropical estuaries and inland wetlands have also been greatly undersampled, and these again may play important roles in global carbon metabolism.

It is thus crucial that we expand the current meager respiration database of respiration if we are to advance our understanding of organic matter flow in aquatic ecosystems, but it is equally important to change the conceptual framework in which we will place these new respiration data, by granting these estimates a greater biogeochemical relevance. Although we can presently measure the rate of respiration (either directly as oxygen or carbon dioxide changes), or indirectly (with enzymes or air–water gas exchange, for example) in many aquatic ecosystems, we still do not know the form and origin of the organic carbon is being respired at any point in time. This is important if we are to understand the magnitude and regulation of the time lags between the synthesis of organic matter by autotrophs and its respiration by heterotrophs, and also the different roles that different sources and organic matter pools might have in terms of overall respiration. In this regard, we have emphasized the need to view community or ecosystem respiration as a sum of processes that may be associated to specific fractions of the available organic matter pool, but we still lack general models that link respiration to organic matter dynamics in aquatic ecosystems.

Respiration is one of the two primary physiological processes in nature, and the difference between it and photosynthesis is important ecologically as it determines the scope for growth, the scale of carbon dioxide drawdown or emission. There seems to be little to be gained maintaining, even worse, furthering the present massive disparity between our understanding of these two processes. At the ecosystem level, respiration integrates so many aspects of the functioning, that long-term shifts in respiration may provide the best warning system for global change. It is imperative that the aquatic research community incorporates direct respiration measurements to the existing long-term monitoring programs designed to assess carbon and nutrient flux in aquatic ecosystems. If, as we would argue, there is a need to enable the study and understanding of respiration in aquatic ecosystems to balance that of photosynthesis and other biogeochemical processes, considerable resources will be needed for the research. As there is a limit to total funding available for science, any increase in research on respiration will need to derive from a shift in resources, and inevitable this can only happen at the expense of research in other areas. Whether the community will be prepared to accept this is an open question. It is our hope that the preparation and publication of this book—which gives an assessment of where we are and where we should going, what we do and do not know—will bring home the need to redress the balance.

Acknowledgments

We wish to thank all the authors of the different chapters of this book, whose contributions were essential to carry out this final synthesis. We also thank Hugh Ducklow, Michael Pace, and Jon Cole for providing constructive criticism and comments on this chapter, Paula Conde for help in editing and Ann Michel for the design of the cover. Paul del Giorgio was supported by funds from the National Research and Engineering Council of Canada.

References

Algesten, G., Sobek, S., Bergström, A. -K., Agren, A., Tranvik, L. J., and Jansson, M. 2003. Role of lakes for organic carbon cycling in the boreal zone. *Glob. Change Biol.*, **10**: 1–7.

Andersson, J. H., Wijsman, J. W. M., Herman, P. M. J., Middelburg, J. J., Soetaert, K., and Heip, C. 2004. *Geophys. Res. Lett.*, **31**: L03304 1–4.

Arístegui, J. and Harrison, W. G. 2002. Decoupling of primary production and community respiration in the ocean: implications for regional carbon studies. *Aquat. Microb. Ecol.*, **29**: 199–209.

Baines, S. B. and Pace, M. L. 1991. The production of dissolved organic matter by phytoplankton and its importance to bacteria: patterns across marine and freshwater systems. *Limnol. Oceanogr.*, **36**: 1078–1090.

Baines, S. B., Pace, M. L., and Karl, D. M. 1994. Why does the relationship between sinking flux and planktonic primary production differ between lakes and oceans? *Limnol. Oceanogr.*, **39**: 213–226.

Benner, R., Pakulski, J. D., McCarthy, M., Hedges, J. I., and Hatcher, P. G. 1992. Bulk chemical characteristics of dissolved organic matter in the ocean. *Science*, **255**: 1561–1564.

Biddanda, B. A., Ogdahl, M., and Cotner, J. B. 2001. Dominance of bacterial metabolism in oligotrophic relative to eutrophic waters. *Limnol. Oceanogr.*, **46**: 730–739.

Bjørnsen, P. K. 1988 Phytoplankton exudation of organic matter: why do healthy cells do it? *Limnol. Oceanogr.*, **33**: 151–154.

Blight, S. P., Bentley, T. L., Lefèvre, D., Robinson, C., Rodrigues, R., Rowlands, J., and Williams P. J. le B. 1995. The phasing of autotrophic and heterotrophic plankton metabolism in a temperate coastal ecosystem. *Mar. Ecol. Prog. Ser.*, **128**: 61–74.

Boyd, P. W. *et al.* 2000. A mesoscale phytoplankton bloom in the polar Southern Ocean stimulated by iron fertilization. *Nature*, **407**: 695–702.

Canadell, J. G. *et al.* 2000. Carbon metabolism of the terrestrial biosphere: a multitechnique approach for improved understanding. *Ecosystems*, **3**: 115–130.

Caraco, N. F. and Cole, J. J. 1999. Regional-scale export of C, N, P and sediment: what river data tell us about key controlling variables. In J. D. Tenhunen and P. Kabat (eds) *Integrating Hydrology Ecosystem Dynamics and Biogeochemistry in Complex Landscapes*. John Wiley, New York, pp. 167–170.

Caraco, N. F., Cole, J. J., Finlay, S. E. G., Fisher, D. T., Lampman, G. G., Pace, M. L., and Strayer, D. L. 2000. Dissolved oxygen declines in the Hudson River associated with the invasion of the zebra mussel (Dreissena polymorpha). *Environ. Sci. Technol.*, **34**: 1204–1210.

Carlson, C. A., Bates, N. R., Ducklow, H. W., and Hansell, D. A. 1999. Estimation of bacterial respiration and growth efficiency in the Ross Sea, Antarctica. *Aquat. Microb. Ecol.*, **19**: 229–244.

Cauwet, G., 2002 DOM in the Coastal Zone. In D. A. Hansell and Craig A. Carlson (eds) *Biogeochemistry of Marine Dissolved Matter*. Academic Press, San Diego, pp. 579–609.

Cherrier., J., Bauer, J. E., Druffel, E. R. M., Coffin, R. B., and Chanton, J. P. 1999. Radiocarbon in marine bacteria: evidence for the ages of assimilated carbon. *Limnol. Oceanogr.*, **44**: 730–736.

Church, M. J., Ducklow, H. W., and Karl, D. M. 2002. Multi-year increases in dissolved organic matter inventories at Station ALOHA in the North Pacific Subtropical Gyre. *Limnol. Oceanogr.*, **47**: 1–10.

Cole, J. J. and Caraco, N. F. 2001. Carbon in catchments: connecting terrestrial carbon losses with aquatic metabolism. *Mar. Freshwater Res.*, **52**: 101–110.

Cole, J. J., Caraco, N. F., Kling, G. W., and Kratz, T. K. 1994. Carbon dioxide supersaturation in the surface waters of lakes. *Science*, **265**: 1568–1570.

del Giorgio, P. A. and Davis, J. 2002. Large-scale patterns in DOM lability across aquatic ecosystems. In S. Findlay and R. Sinsbaugh (eds) *DOM in Aquatic Systems*. Academic Press, pp. 400–425.

del Giorgio, P.A. and Duarte, C. M. 2002. Respiration in the open ocean. *Nature*, **420**: 379–384.

del Giorgio, P. A., Cole, J. J., and Cimbleris, A. 1997. Respiration rates in bacteria exceed plankton production in unproductive aquatic systems. *Nature*, **385**: 148–151.

del Giorgio, P. A., Cole, J. J., Caraco, N. F., and Peters, R. H. 1999. Linking planktonic biomass distribution to plankton metabolism and net gas fluxes in northern temperate lakes. *Ecology*, 80: 1422–1431.

Duarte, C.M. and Agustí, S. 1998. The CO_2 balance of unproductive aquatic ecosystems. *Science*, 281: 234–236.

Duarte, C. M. and Cebrián J. 1996. The fate of marine autotrophic production. *Limnol. Oceanogr.*, 41: 1758–1766.

Duarte, C. M., Agustí, S., del Giorgio, P. A., and Cole, J. J. 1999. Regional carbon imbalances in the oceans. *Science*, 284: 1735b.

Duchemin, E., Lucotte, M., Canuel, R., and Chamberland, A. 1995. Production of green house gases CH_4 and CO_2 by hydroelectric reservoirs of the boreal region. *Glob. Biochem. Cyc.*, 9: 529–540.

Ducklow, H. W. 2003. Chapter 1. Biogeochemical provinces: towards a JGOFS synthesis. In M. J. R. Fasham (ed) *Ocean Biogeochemistry: A New Paradigm*. Springer-Verlag, New York, pp. 3–18.

Ducklow, H. W. and Mc Callister, S. L. in press. The biogeochemistry of carbon dioxide in the coastal oceans. In A. R. Robinson and K. Brink (eds) *The Sea*, Vol. 13. Wiley and Sons, New York.

Falkowski, P. J. *et al.* 2000. The global carbon cycle: a test of our knowledge of earth as a system. *Science*, 290: 291–296.

Fasham, M. J. R., Baliño, B. M., and Bowles, M. C. (eds). 2001. A new vision of ocean biogeochemistry after a decade of the Joint Global Ocean Flux Study (JGOFS). *Ambio Spec. Rep.*, 10: 4–30.

Field, C. B., Beherenfield, M. J., Randerson, J. T., and Falkowski, P. 1998. Primary production in the biosphere: integrating terrestrial and oceanic components. *Science*, 281: 237–240.

Hansell, D. A., Bates, N. R. and Gundersen, K. 1995. Mineralization of dissolved organic carbon in the Sargasso Sea. *Mar. Chem.*, 51: 201–212.

Hansell, D. A., Ducklow, H. W., MacDonald, A. M., and O'Neil Baringer, M. 2004. Small net heterotrophy in the North Atlantic Ocean diagnosed from organic matter transports. *Limnol. Oceanogr.*, 49: 1084–1094.

Hiscock, M. R., Marra, J., Smith, J., Walker O., Goericke, R., Measures, C., Vink, S., Olson, R. J., Sosik, H. M., and Barber, R. T. 2003. Primary productivity and its regulation in the Pacific Sector of the Southern Ocean." *Deep-Sea Res. II*, 50: 533–558.

Houghton, R. A., Davidson, E. A., and Woodwell, G. M. 1998. Missing sinks, feedbacks, and understanding the role of terrestrial ecosystems in the global carbon balance. *Glob. Biogeochem. Cycl.*, 12: 25–34.

Hutchinson, G. E. 1957. *A Treatise in Limnology. Vol. 1. Geography, Physics and Chemistry*. John Wiley, NY.

Jackson, J. B. C., *et al.* 2001. Historical overfishing and the recent collapse of coastal ecosystems. *Science*, 293: 629–637.

Jahnke, R. A. and Craven, D. B. 1995. Quantifying the role of heterotrophic bacteria in the carbon cycle: a need for respiration rate measurements. *Limnol. Oceanogr.*, 40: 436–441.

Jansson, M., Karlsson, J., and Blomquist, P. 2003. Allochthonous organic carbon decreases pelagic energy metabolism in lakes. *Limnol. Oceanogr.*, 48: 1711–1716.

Jonsssson, A., Karlsson, J., and Jansson, M. 2003. Sources of carbon dioxide supersaturation in clearwater and humic lakes in northern Sweden. *Ecosystems*, 6: 224–235.

Karl, D. M. 1999. A sea of change: Biogeochemical variability in the North Pacific subtropical gyre. *Ecosystems*, 2: 181–214.

Karl, D. M., Laws, E. A., Morris, P., Williams, P. J. le B., and Emerson, S. 2003. Metabolic balance in the open sea. *Nature*, 426: 32.

Kirchman, D. L., Lancelot, C., Fasham, M. J. R., Legendre, L. Radach, G., and Scott, M. 1993. Dissolved organic matter in biogeochemical models in the ocean. In G. T. Evans and M. J. R. Fasham (eds) *Towards a Model of Ocean Biogeochemical Processes*. Springer-Verlag, Berlin, pp. 209–225.

Koblentz-Mishke, O. J., Volkovinsky, V. V., and Kabanova, J. G. 1968. Noviie dannie o velichine pervichnoi produktsii mirovogo okeana. *Dokl. Akad. Nauk SSSR*, 183: 1186–1192.

Longhurst, A., Sathyendranath, S., Platt T., and Caverhill, C. 1995. An estimate of global primary production in the ocean from satellite radiometer data. *J. Plankton Res.*, 17: 1245–71.

Marra, J. 2002. Approaches to the measurements of plankton production. In P. J. le B. Williams, D. N. Thomas, and Reynolds (eds) *Phytoplankton Productivity*. Blackwell, London, pp. 78–108.

McCallister, S. L. 2002. *Organic Matter Cycling in the York River Estuary, Virginia: An Analysis of Potential Sources and Sinks*. The College of William and Mary, School of Marine Science, Gloucester Point, VA 217 pp.

Middelburg, K. J., Vlug T., and van der Nat, F. J. W. A. 1993. Organic matter mineralization in marine systems. *Glob. Planet. Change*, 8: 47–58.

Miller, W. D., Neubauer, S. C., and Anderson, I. C. 2001. Effects of sea level induced disturbances on high salt marsh metabolism. *Estuaries*, 24: 357–367.

Mopper, K. *et al.* 1991. Photochemical degradation of dissolved organic carbon and its impact on the ocean carbon cycle. *Nature*, 353: 60–62.

Nixon, S. W. 1988. Physical energy inputs and the comparative ecology of lake and marine ecosystems. *Limnol. Oceanogr.*, **33**: 1005–1025.

Obernosterer, I., Kawasaki, N., and Benner, R. 2003. Phosphate limitation of respiration in the Sargasso Sea and uncoupling of P cycling in size fraction experiments. *Aquat. Microb. Ecol.*, **32**: 229–237.

Odum, H. T. 1956. Primary production in flowing waters. *Limnol. Oceanogr.*, **1**: 102–117.

Ogura, N. 1973. Rate and extent of decomposition of dissolved organic matter in surface seawater. *Mar. Biol.*, **13**: 89–93.

Ondrusek, M. E., Bidigare, R. R. Waters, K., and Karl, D. M. 2001. A predictive model for estimating rates of primary production in the subtropical North Pacific Ocean. *Deep-Sea Res. II*, **48** 1837–1863.

Pace, M. L. and Cole, J. J. 1999. Effects of whole-lake manipulations of nutrient loading and food web structure on planktonic respiration. *Can. J. Fish. Aquat. Sci.*, **57**: 487–496.

Packard, T. T., Blasco, D., and Estrada, M. 2004. Modeling physiological processes in plankton on enzyme kinetic principles. *Sci. Mar.* **68**: 49–56.

Potter, C. S. and Klooster, S. A. 1998. Interannual variability in soil trace gas (CO_2, N_2O, NO) fluxes and analysis of controllers on regional to global scales. *Glob. Biogeochem. Cycl.*, **12**: 681–635.

Prairie, Y. T., Caraco, N. F., McDowell, W. H., Tranvik, L. J., Striegl, R. G., Duarte, C. M., Kortelainen, P., Cole, J. J., and Downing, J. A. Plumbing the carbon cycle: integrating inland waters into the terrestrial carbon budget. *Front. Ecol.* (submitted).

Quiñones, R. A. and Platt, T. 1991. The relationship between the f-ratio and the P : R ratio in the pelagic ecosystem. *Limnol. Oceanogr.*, **36**: 211–213.

Raich, J. W., Potter, C. S., and Bhagawati, D. 2002. Interannual variability in global soil respiration, 1980–4. *Glob. Change Biol.*, **8**: 800–812.

Randerson, J. T., Chapin III, F. S., Harden, J. W., Neff, J. C., and Harmon, M. E. 2002. Net ecosystem production: a comprehensive measure of net carbon accumulation by ecosystems. *Ecol. Appl.*, **12**: 937–947.

Reynolds, C. S. 2002. On the interannual variability in phytoplankton production in freshwaters. In P. J. le B. Williams, D. N. Thomas, and C. S. Reynolds (eds) *Phytoplankton Productivity*. Blackwell, London, pp. 187–221.

Rivkin, R. B. and Legendre, L. 2001. Biogenic carbon cycling in the upper ocean: effects of microbial respiration. *Science*, **291**: 2398–2400.

Schaefer, A., Denning, A. S., Suits, N., Kaduk, J., Baker, I., Los, S., and Prihodko, L. 2002. Effect of climate on interannual variability of terrestrial CO_2 fluxes. *Glob. Biogeochem. Cyc.*, **16**: 1102.

Schindler, D. E., Carpenter, S. R., Cole, J. J., Kitchell, J. F., and Pace, M. L.. 1997. Influence of food web structure on carbon exchange between lakes and the atmosphere. *Science*, **277**: 248–251.

Schlesinger, W. H. and Andrews, J. A. 2000. Soil respiration and the global carbon cycle. *Biogeochemistry*, **48**: 7–20.

Schrödinger, E. 1944. *What is Life.* Cambridge University Press, Cambridge

Serret, P., Robinson, C., Fernandez, E., Teira, E., and Tilstone, G. 2001. Latitudinal variation of the balance between plankton photosynthesis and respiration in the E. Atlantic Ocean. *Limnol. Oceanogr.*, **46**: 1642–1652.

Smith, E. M. and Kemp, W. M. 2001. Size structure and the production/respiration balance in a coastal plankton community. *Limnol. Oceanogr.*, **46**: 473–485.

Smith, E. M. and Kemp, M. W. 2003. Planktonic and bacterial respiration along an estuarine gradient: responses to carbon and nutrient enrichment. *Aquat. Microb. Ecol.*, **30**: 251–261.

Stickney, H.L., Hood, R.R., and D.K. Stoecker. 1999. The impact of mixotrophy on planktonic trophic dynamics in marine ecosystems. *Ecol. Model.*, **125**: 203–230.

Tjoelker, M. G., Oleksyn, J., and Reich, P. B. 2001. Modelling respiration of vegetation: evidence for a general temperature-dependent Q_{10}. *Glob. Change Biol.*, **7**: 223–230.

Tranvik, L. J. and Jansson, M. 2002. Terrestrial export of organic carbon. *Nature*, **415**: 861–862.

Valentini, R. *et al.* 2001. Respiration as the main determinant of carbon balance in European forests. *Nature*, **404**: 861–865.

Vukicevic, T., Braswell, B. H. and Schimel, D. 2001. A diagnostic study of temperature controls on global terrestrial carbon exchange. *Tellus*, **53B**: 150–170.

Waring, R. H., Landsberg, J. J., and Williams, M. 1998. Net primary production of forests: a constant fraction of gross primary production? *Tree Physiol.*, **18**: 129–134.

Welch, H. E. and Bergmann, M. A. 1984. Winter respiration of lakes at Saqvaqjuac, N. W. T. *Can. J. Fish. Aquat. Sci.*, **42**: 521–528.

Wetzel, R. G. 1992. Gradient-dominated ecosystems: sources and regulatory functions of organic matter in freshwater ecosystems. *Hydrobiologia*, **229**: 181–198.

Williams, P. J. le B. 1990. The importance of losses during microbial growth: commentary on the measurement, physiology and ecology of the release of dissolved organic material. *Mar. Microb. Food Webs*, **4**: 175–206.

Williams, P. J. le B. 1993. Chemical and tracer methods of measuring plankton production. *ICES Mar. Sci. Symp.,* **197**: 20–36.

Williams, P. J. le B. 1998. The balance of plankton respiration and photosynthesis in the open oceans. *Nature,* **394**: 55–57.

Williams, P. J. le B. 2000. Micro-heterotrophs and the dynamics of dissolved organic material. In D. J. Kirchman (ed) *Microbial Ecology of the Oceans.* Wiley-Liss, New York.

Williams, P. J. le B. and Bowers, D. G. 1999. Regional carbon imbalances in the oceans. *Science,* **28**: 173–174.

Williams, P. J. le B. and Lefèvre, D. 1996. Algal ^{14}C and total carbon metabolisms. 1. Models to account for respiration and recycling. *J. Plankton Res.,* **18**: 1941–1959.

Williams, P. J. le B., Robinson, C., Sondergaard, M., Jespersen, A-M., Bentley, T. L., Lefèvre, D., Richardson, K., and Riemann, B. 1996. Algal ^{14}C and total carbon metabolisms. 2. Experimental observations with the diatom *Skeletonema costatum. J. Plankton Res.,* **18**: 1961–1974.

Woodwell. G. M., Mackenzie, F. T., Houghton, R. A., Apps, M., Gorham, E., and Davidson, E. 1998. Biotic feedbacks in the warming of the earth. *Clim. Change,* **40**: 495–518.

Index

Acanthamoeba,
  oxygen uptake and cell size, 51
  sulfide oxidation in, 48
acetate pathway in wetlands, 90
acetoclastic methanogens, 90
active flux in zooplankton, 70
ADP, 40, 41, 43
aerobic purple bacteria, 36, 37
aerobic respiration, *see under*
  respiration
Alcaligenes, denitrification
  processes in, 230
alternative electron acceptors,
  in coastal benthic communities,
  207
  in ocean, 225
  vertical profiles of, 233
alternative oxidase, 37, 38, 39, 44
alternative oxidase respiration, 105
ammonia oxidizers, 24
ammonifying nitrate/nitrite
  respiration, 24
ammonifying nitrate/nitrite
  reduction, 26
anaerobic ammonia oxidation, 232
anaerobic mineralization, 207
anaerobic respiration,
  cf. aerobic, 19
  contribution to global
    respiration, 286
  end products of, 24
  in protists, 47, 48
  in wetlands, 84
  modes of, 24
  nitrous oxide respiration, 24
  sites of, 23, 24
  vertical zonation in marine
    sediments, 19
anammox reaction, 232
anoxia,
  in lakes, 8

suboxic electron acceptors as
  barrier to, 239
apparent oxygen utilization,
  age of water masses and, 192
  derivation of respiration rates
    from, 193
  distribution in dark ocean, 186,
    192
Arabian Sea, suboxia in, 226
Archaea, 24
  denitrification processes in, 25,
    230
  sulfate respiration in, 29
arsenate reduction, 31
ascorbate, 41
ascorbate peroxidase, 40, 41
ascorbic acid, 41
ATP, 38, 39, 40, 41, 42, 43
ATP synthetase, 39, 40, 42
attenuation and respiration, 158
autochthonous photosynthesis, 163
autotrophic respiration,
  cf. heterotrophic, 269
  description, 270

bacteria,
  aerobic purple, 37
  carbon flux in dark ocean, 190
  contribution to pelagic
    respiration in estuaries, 138
  cyanobacteria, 37
  dark respiration in, 38
  denitrification processes in
    suboxic oceanic waters, 230
  derivation of respiration from
    bacterial production, 169
  early Russian studies of
    importance, 7
  ectosymbiotic sulfate-reducing
    on protists, 49
  endosymbiotic methanogens in
    protists, 49

endosymbiotic purple non-sulfur
    bacteria in protists, 49
  energy yields from respiration in,
    22
  glycolysis in, 38
  green sulfur, 37
  growth efficiency, 169, 191
  halo-respiring anaerobes, 31
  humic acid reduction by, 27
  identification of microbes
    involved in denitrification,
    237
  increased numbers in bottle
    experiments, 151
  inorganic vs organic nutrient use,
    256
  iron-reducing, 26
    interactions with plant roots,
      23
  metal reduction in, 26
  model of growth and respiration,
    257
  models of respiration in, 256
  N-oxide respiration in, 24
    ammonifying nitrate/nitrite
      reduction, 26
    denitrification, 25
  Photobacterium as an oxygen
    sensor, 50
  respiration in, 18–22
    and abundance, 158
    in photolithotrophs, 36–45
    role in respiration as a whole,
      14
    role in sediments, 108
  sulfate respiration in, 27
bacterial respiration,
  chromium, 2, 19
  CO_2 reduction, 30–31
  humic acids, 2, 19
  iron, 2, 19
  manganese, 2, 19

metal reduction, 26–27
nitrate, 2, 19
nitrite, 2, 19
N-oxide respiration, 24–26
perchlorate, 2, 19
selinate, 2, 19
structure, 20
sulfate, 2, 19
sulphate respiration, 27–30
uranium, 2, 19
baseline respiration, description,
    270
bathypelagic waters
    export ratios from, 285
    respiration in, 181–99
    see also dark ocean
    zooplankton respiration in, 70
behavior and respiration rate, 252
benthic chambers, 109
benthic respiration,
    cf. pelagic, 281
    effect of organic matter loading
        on, 111
    effect of primary production
        on, 111
    effect of seasonal variations on,
        130, 131
    effect of temperature on, 110, 131
    effect of trophic status on, 111
    effect of turbulence on, 129
    effect of water depth on, 129, 130
    estimates of, 280
    in coastal waters, 206–21
    in dark ocean, 195
    in estuaries,
        average annual rate, 126
        benthic animals and, 132
        carbon dioxide vs oxygen
            measures of, 127
        cf. open-water whole system
            measures, 140
        controls on, 128
        fine grain sediments cf. coarse
            grain, 132
        measurement of, 124
        organic input and, 129, 130
        oxygen consumption cf.
            carbon dioxide production
            in, 128
        seasonal variation, 130, 131
        temperature and, 131
        turbulence and, 129
        water column depth and, 129,
            130
    models of, 110, 111
    sample depth and, 112

biological pump, 10
biomass,
    derivation of respiration
        distribution from, 169
    derivation of respiration rates
        from, 153
    global assessment of in
        zooplankton, 75
    theoretical calculation of and
        respiration size distribution,
        169
biosphere, linking aquatic
    ecosystems in, 283
Black Sea, chemical profiles in, 233
body mass, effect on respiration
    rate in zooplankton, 64
bogs, see under peatlands
brown algae, 44
burial efficiencies, 195

C isotopes, 44
[14]C studies, 8
    determination of net primary
        production, 170
    oceanic, 8
Calanus euximus,
    oxygen uptake by, 63
    respiration rate in, 65
Calanus finmarchicus, seasonal
    changes in respiration rate,
    66
Calanus hyperboreus, seasonal
    changes in respiration rate,
    66
carbon balance,
    in dark ocean, 199
    in lakes, 103
    in wetlands, 94
carbon concentrating mechanisms
    (CCM), 42
carbon cycle,
    biological pump in, 10
    role of aquatic ecosystems in
        global cycle, 267–99
    role of respiration in, 1
    role of wetlands in, 83, 90
carbon dioxide,
    cf. oxygen measures in benthic
        estuarine respiration, 127
    gas flux measurements, 85
    measurement of flux by dark
        bottle method, 148
    partial pressure as a measure of
        lake respiration, 115
    partial pressure changes in lakes,
        105

production,
    dynamics of in wetlands, 89
    global respiration estimates for
        wetlands, 94
    in peatlands, 86
    in swamps and marshes, 87, 88
    in wetlands, 84
    reduction (hydrogenotrophic
        methanogenesis), 30
catabolism, 1
Caulerpa spp., community
    respiration measures in, 217
cell cycle and oxygen uptake, 51
CEPEX study, 166
CF_0-CF_1 ATP synthetase, 41
chamber methods of gas flux
    measurement, 85
Chaos, oxygen uptake and cell
    volume, 52
chemoautotrophs, 208
chemolithotrophs, 208
chloroachniophytes, 37
chemoorganotroph, 36, 37
Chlamydomonas, 43
chlororespiration, 38, 40, 44
chlorophyll,
    effect on benthic respiration, 112
    effect on pelagic respiration in
        estuaries, 136
    plankton respiration and, 107
    respiration and, 158, 159
chlororespiration, 40
    diagram, 41
    relative importance of, 44
chromium in suboxic waters, 229
ciliate respiratory rates, 54
citrate synthase measures in
    zooplankton, 59
climate change,
    aquatic respiration and, 294
    suboxia and, 241
coastal waters,
    benthic communities, 206–21
        see also sediments, coastal
    cf. open waters, 192
    classification of sediments, 209
    community respiration
        measures,
        in coral reefs, 212
        in mangroves, 213
        in salt marshes, 213
        in submerged macrophytes,
            214
    estimates of respiration rates
        in, 277
    export ratios from, 285

coastal waters (*cont.*)
   inputs to dark ocean from, 187
   measurements of respiration in
      benthos, 207, 208
   primary production constraints
      in, 220
   respiration in sediments of, 208
community respiration,
   bacterial contribution, 167
   latitudinal distribution of, 164
      and community production,
         165
      and gross production, 165
   phytoplankton contribution, 167
   protist contribution, 167
   respiration rates, 154
      algal community size structure
         and, 157
      global database, 155
      latitudinal distribution of
         observations, 155, 156
      seasonal distribution of
         observations, 155, 156, 157
   zooplankton contribution, 167
community structure, link with
   respiration, 281
containment errors, 150
coral reefs, 206
   community respiration measures
      in, 212
   respiration rate in, 207
cryptophyte algae, 37
cryptophytes, 37
cyanide, 38
cyanobacteria, 36, 37, 38, 42, 43, 44
cyclic electron flow, 42
cyclonic eddies, 73
Cymodocea nodosa, community
   respiration measures in, 216
cytochrome oxidase, 37, 38, 39, 44
cytochromes, 39, 41
cytochromes, 20, 39, 41
   architecture in bacteria, 21

Daphnia magna, respiration rate in
   and feeding, 65
dark bottle method, 8, 104, 148
   disruption of diel rhythms
      by, 150
   global respiration rates, 171
dark ocean,
   active transport of organic matter
      by plankton, 187
   annual organic carbon flux
      in, 197

apparent oxygen utilization
   in, 186, 192, 193
bacterial carbon flux in, 190
benthic respiration in, 195
carbon budget for, 199
coastal inputs to, 187
deep-water mass flow patterns
   in, 193
delivery of dissolved organic
   carbon to, 185
delivery to particulate organic
   carbon to, 184
depth and respiration rate in, 190
in vitro oxygen consumption
   measures in, 188
mesozooplankton respiration
   rates, 192
microplankton
   electron transport system
      activity in, 188
   respiration in, 189
organic matter supply, 183
pelagic respiration in, 188
relative importance of, 44
respiration budget for, 196
respiration in deep-sea
   sediments, 208
respiration rates in, 195
sediment respiration in, 194
decomposition,
   definition, 11
   in peatlands, 86
   processes, 1
deep-water mass flow patterns in
   dark ocean, 193
degradation, definition, 11
dehydroascorbate, 41
denitrification, 24, 25, 208
   alternative pathways for, 231
   canonical, 229
   constraints on in sediments, 240
   contribution to global
      respiration, 286
   controls on fixed nitrogen pool in
      oceanic waters, 240
   definition, 229
   enzyme systems involved in, 237
   identification of microbes
      involved in, 237
   in Arabian Sea, 226
   in presence of oxygen, 230
   incubation techniques, 238
   organisms involved in, 230
   Redfield ratios, 230
   role in climate change, 241
Desulfobulbus, 29

Desulfonema, 29
Desulfosporosinus sp., 29
Desulfotomaculum sp., 29
diapause in copepods, 66
Diaptomus kenia, 65
diatoms, 39
diel rhythms,
   disruption by dark-bottle
      methods, 150
   in respiration rate of
      zooplankton, 65
   vertical distribution of
      zooplankton and, 70
dinitrogen measurement, 236
dinophyte algae, 37, 44
dissimilatory nitrate reduction, 24
dissolved inorganic carbon,
   changes in and carbon dioxide
      measures, 105
   flux in estuarine benthos, 127
   production in coastal water
      benthos, 208
dissolved organic carbon,
   active transport of by plankton,
      187
   delivery to dark ocean, 185
   mineralization
      and apparent oxygen
         utilization, 192, 193
      and carbon flux, 194
      and sediment oxygen
         consumption, 194
   plankton respiration and, 107
   supply to dark ocean, 183
dissolved organic matter,
   export from photic zone, 183
   supply to dark ocean, 183

ecosystems,
   models of, 263
   organic matter movements
      through, 15
   respiration and function of, 1
   respiration in a whole ecosystem
      context, 14
eddies, 72
   cyclonic, 73
eddy-covariance technique of gas
   flux measurement, 86
electrical potentials of redox pairs,
   22
electron acceptors,
   alternatives
      in coastal benthic
         communities, 207

in ocean, 225
  vertical profiles of, 233
concentrations of in suboxic
    waters, 228
examples of, 19, 31
electron donors, 4
electron flux through ecosystem, 13
electron transport systems, 20, 21
  activity in dark ocean, 188
  assay of, 152
  in denitrification, 237
  measures in zooplankton, 59
  techniques, 10
embryophytes, 37
endosymbiosis, 36
energy dissipation reactions,
    Mehler reaction, 3, 4
  photorespiration, 3, 4
energy provision, reactions
    involved in, 3
energy yields, electrical potentials
    of redox pairs, 22
Enhalus acoroides, community
    respiration measures in, 216
Enterobacteriaceae, 24
environmental change and aquatic
    respiration, 294
enzymatic methods, 58
epipelagic waters,
  export ratios from, 285
  zooplankton respiration in, 67
estuaries,
  definition, 122
  estimates of respiration rates
    in, 277
  export ratios from, 285
estuarine respiration, 122–42
  benthic,
    average annual rate, 126
    benthic animals and, 132
    carbon dioxide vs oxygen
      measures of, 127
    cf. open-water whole system
      measures, 140
    controls on, 128
    fine grain sediments cf. coarse
      grain, 132
    measurement of, 124
    organic input and, 129, 130
    oxygen consumption cf.
      carbon dioxide production
      in, 128
    seasonal variation, 130, 131
    temperature and, 131
    turbulence and, 129

water column depth and, 129,
    130
  measurement of, 123
  open-water whole system,
    cf. component-derived
      measures, 140
    gross production and, 138
    measures of, 138
  open-water whole system
    respiration, 125
  pelagic,
    cf. open-water whole system
      measures, 140
    chlorophyll and, 136
    communities contributing
      to, 137
    effects of substrate supply
      on, 134
    effects of temperature on, 134,
      135
    links to production, 136
    measurement of, 125
    rates of, 132, 133
ETS technique,
  introduction, 10
  measurements, 58–59, 68, 69, 70,
    71
  principle, 152–153
Euglena, enolases in, 40
euglenoid algae, 37
eukaryotes,
  dark respiration in, 39
  glycolysis in, 38
Euphausia mucronata, oxygen
    uptake by, 63
Euphausia superba, seasonal changes
    in respiration rate, 67
Euplotes, oxygen uptake and cell
    size, 51

facultative anaerobes, 49
feeding and respiration rate in
    zooplankton, 65
fens, *see under* peatlands
fermentation,
  energy yield, 22
  in protists, 48, 49
  microaerobic conditions in
    protists, 49
  pathways, 19
  role of acetate and hydrogen in
    links with methanogens, 21
  systems of, 22, 23
ferredoxin, 38, 42
filaments from upwelling, 72, 74
Firmicutes, 24

flow-through method, 58
free water respiration in lakes, 113
fresh waters,
  direct respiration measurements
    in, 8
  estimates of respiration rates in,
    277
  export ratios from, 285
  global estimate of respiration in
    lakes, 115
  lakes, 103–18, 277
  respiration in zooplankton, 57–73
  respiration studies in, 10
  see also lakes, rivers, wetlands
  wetland ecosystems, 83–96
frontal systems, 72
fungi, denitrification processes in,
    230

gas flux measurements, 84
  carbon dioxide, 85
  chamber techniques, 85
  eddy-covariance technique, 86
  methane, 85
  seasonal changes in wetlands, 91,
    92
Geobacter metallireducens, 27
Geobacter sulfurreducens, 27
Geobacteriaceae, 27
Geothrix sp., 27
global environmental change, 294
global respiration,
  contribution of aquatic
    ecosystems to, 284
  in coastal benthic communities,
    207, 218
  in dark ocean, 196, 198
  primary productivity and, 288
  significance of, 267–99
glutamate dehydrogenase
    measures in zooplankton, 59
glutathione, 41
glycerate, 43, 44
glycine, 43
glycolate, 37, 38, 42, 43, 44
glycolate dehydrogenase, 43
glycolate oxidase, 38, 43, 44
glycolysis, 5, 37, 38, 39
  in photolithotrophs, 37
  stoichiometry of metabolism of a
    three-carbon compound, 4
glyoxylate, 38
Gnathophausia ingens, oxygen
    uptake by, 63
green algae, 37, 40, 44
green sulfur bacteria, 36, 37

growth,
efficiency, 153
bacterial, 191
derivation of respiration
distribution from, 169
food access and oxygen uptake,
52
model of growth and respiration
in bacteria, 257
rate, vs respiration rate, 251

H_2O_2, 40, 43
Halimeda spp., community
respiration measures in, 217
halo-respiring anaerobes, 31
Halodule uninervis, community
respiration measures in, 216
Halodule wrightii, community
respiration measures in, 216
haptophyte algae, 37, 44
haptophytes, 37
heterokont algae, 37
heterokonts, 37
heterotrophic respiration, 24
cf. autotrophic, 269
description, 270
history of respiration measurement
in aquatic ecosystems, 6
humic acid reduction, 27
hydrogen peroxide and errors in
oxygen measures, 151
hydrogenosomes, 49
hydrogenotrophic methanogenesis,
30
hydrogenotrophs, 24
hydrostatic pressure, effect on
respiration rate, in
zooplankton, 62
hydroxypyruvate, 43

interspecies hydrogen transfer, 23
iodate concentrations in suboxic
waters, 228
iodide concentrations in suboxic
waters, 228
iron,
concentrations of in suboxic
waters, 229
stresses on phytoplankton, 255
iron reduction, 26, 208
iron-reducing bacteria, 26, 27
interactions with plant roots, 23
isocitrate dehydrogenase, plankton
respiration measurement
and, 154

kelp, community respiration
measures in, 217
Klebsiella aerogenes, 251
Krebs cycle, *see under* TCA cycle
krill, 67

lactate dehydrogenase measures in
zooplankton, 59
Lake Baikal, 116
Lake Memphremagog, 115, 116
Lake Superior, 118
lakes,
benthos
effect of organic matter loading
on benthic respiration, 111
effect of primary production
on benthic respiration, 111
effect of temperature on
benthic respiration, 110
effect of trophic status on
benthic respiration, 111
respiration in, 108, 118
depth and respiration in, 107
depth and sediment respiration,
113
diel oxygen dynamics in, 113
and measurement of
respiration in, 113, 114
estimates of respiration rates in,
277
food web and respiration in, 107
free water respiration in, 113
cf. dark bottle estimates of
respiration in, 114
global estimate of primary
production in, 116
global estimate of respiration in,
115
models of plankton respiration
in, 106
net carbon balance in, 103
organic matter loading and
respiration rates in, 106
planktonic respiration in, 104
respiration in, 103–18
in deep waters, 118
in sediments of, 108
spatial distribution of respiration
in, 114, 115
temperature and respiration
rates in, 106
trophic condition and respiration
rates in, 106
leaky pipe model, 235
leucine incorporation, 191

light, effect on respiration rate, in
zooplankton, 62
Loxodes, anaerobic respiration in, 48

macroalgae,
community respiration measures
in, 214
respiration rate in, 207
macrophytes,
in benthos of coastal waters, 206
respiration measures in
estuarine, 123
submerged, 214
malate dehydrogenase measures in
zooplankton, 59
manganese,
concentrations of in suboxic
waters, 228, 229
oxidation by nitrate, 232, 233
reduction, 26, 208
mangroves, 206
community respiration measures
in, 213
respiration rate in, 207
marine waters,
bathypelagic ocean respiration,
181–99
^{14}C studies in, 8
control on oceanic primary
productivity, 241
controls on fixed nitrogen pool
in, 240
dark ocean processes, 181–99
early studies of, 182
estimates of respiration rates in,
277
mesopelagic ocean respiration,
181–99
ocean circulation models and
organic matter
mineralization, 195
respiration,
depth distribution, 156
distribution within plankton
groups, 166–168, 169, 171
mass balances, 172–175
regional distribution of
measurements, 155–156
respiration relationship with
other parameters, 158–159
size distribution, 168–169, 171
temporal distribution of
observations, 157
respiration in zooplankton, 57–73
suboxic respiration in, 225–43

surface, mass-balance
  calculations for, 173
surface marine waters, 147–76
time development of respiration
  studies, 9
marshes,
  biogeochemistry and ecology of,
    87
  estimates of respiration rates in,
    277
  see also salt marshes
mass-balance calculations, 173
  for coastal water column, 174
  for open-ocean water column,
    174
  for open-ocean water epipelagic
    zone, 175
  for whole oceanic water column,
    173
measurements,
  anomalous observations, 152
  at ecosystem level, 275
  benthic chambers, 109
  by modelling, 123, 124
  chamber methods, 85
  dark bottle method, 104
  derivation of respiration rate
    from biomass, 153
  derivation of respiration rate
    from growth efficiency, 153
  eddy-covariance technique, 86
  enzymatic methods, 58
  errors of containment, 150
  errors of interpretation, 151
  estuarine respiration, 123
  flow-through method, 58
  free water measures cf. dark
    bottle estimates of
    respiration, 114
  free water respiration, 113
  gas flux, 84
  global respiration rate in dark
    ocean, 196, 198
  in anoxic environments, 14
  in coastal benthic communities,
    207
  in oxic environments, 12
  in suboxic waters, 236
  in surface marine waters, 147–76
  inaccuracies between dark and
    light measures, 104
  inverse analysis, 153
  limits of detection, 148
  of benthic respiration in
    estuaries, 124

of open-water whole system
  respiration in estuaries, 125
of pelagic respiration in
  estuaries, 125
of planktonic respiration lakes,
  104
of respiration in sediments, 108
of protist respiration, 49
of zooplankton respiration, 58
open-water whole system
  approach, 123
potential respiration, 152
procedural errors, 149
random and systematic errors,
  149
sealed-chamber method, 58
sediment core incubations, 109
techniques for, 49, 50
vertical profiles of
  decomposition products, 109
Meganyctiphanes norvegica, oxygen
  uptake by, 63
Mehler reaction, 3, 37
  cf. dark measures of oxygen
    consumption, 104
  diagram, 41
  in oxic environments, 12
  in photolithotrophs, 40
  mechanism, 105
  relative importance of, 44
  risk of error in respiration
    measures and, 149
Mehler-peroxidase reaction, 36, 37,
  38, 40, 42, 44
membrane inlet mass spectrometry,
  50
membrane-associated electron
  transport proteins, 19
mesopelagic waters, 181–99
  export ratios from, 285
  see also dark ocean
  zooplankton respiration in, 68
mesozooplankton,
  models of respiration, 262
  respiration rates, 192
metabolic power index in
  zooplankton, 60
metabolic scope in zooplankton, 60
metal reduction, 26
metazooplankton biomass, 57
methane,
  gas flux measurements, 85
methane production,
  dynamics of in wetlands, 89
  emissions in wetlands, 92

global respiration estimates for
  wetlands, 94
in peatlands, 86
in rice paddies, 88, 89
in swamps and marshes, 87, 88
latitude of wetland and, 91
methanogenesis,
  endosymbiotic bacteria in
    protists, 49
  hydrogenotrophic, 30
  latitude of wetland and, 91
  wetlands and, 84
methanogens,
  acetoclastic, 90
  role of acetate and hydrogen in
    links with fermentation, 21
methylotrophs, 24
Metridia longa, seasonal changes in
  respiration rate, 66
microaerophiles, 49
microbial biomass
  underestimation, 7
microelectrodes, 50
micronekton, biomass, 76
microplankton respiration in dark
  ocean, 189
microzooplankton, models of
  respiration, 261
migration of zooplankton, 71
migratory flux, 71
mineralization,
  anaerobic in coastal benthic
    communities, 207
  definition, 11
  from oxygen fields, 192
  in dark ocean, 182
  in wetlands, 91
  ocean circulation models and,
    195
  of organic matter in dark ocean,
    192, 194
Mitochondria, 39, 40, 43
models,
  autotrophic vs heterotrophic, 253
  basal respiration, 250
  incorporation of plankton
    respiration in ecosystem
    function models, 248–64
  metabolic respiration, 250
  of bacterial respiration, 256
    constants for, 258
    equations for, 259
  of ecosystems, 263
  of growth and respiration in
    bacteria, 257

models (*cont.*)
    of phytoplankton respiration, 254
    of zooplankton respiration, 261
    platforms for, 250
molybdenum in suboxic waters,
        229
monodehydroascorbate, 41
mortality and respiration rate, 252
mussel reefs, 123

NAD+, 38, 39
NADH, 38, 39, 43
NADP+, 38, 39, 41
NADPH, 3, 39, 40, 41, 42
N-oxide respiration, 24
    ammonifying nitrate/nitrite
        reduction, 26
    denitrification, 25
nekton biomass, 57
net community production, 164
    concept of, 273
    latitudinal distribution and
        community respiration, 165
net ecosystem exchange, 89
net ecosystem production, of
        estuaries, 123
net primary production,
    concept of, 273
    estimates of respiration from
        14-C determination, 170
nitrate,
    concentrations of in suboxic
        waters, 228
    reduction *in Loxodes*, 48
nitrification at low oxygen tensions,
        234
nitrite,
    concentrations of in suboxic
        waters, 228
    maxima in suboxic waters, 227
nitrogen,
    controls on fixed nitrogen pool in
        oceanic waters, 240
    input to photic zone of ocean, 183
nitrogen fixers, 24
nitrous oxide,
    cycling in suboxic waters, 234
    enhancing the source, 241
    gradients in suboxic waters, 227,
        228
nutrient loading in estuaries, 122

O isotopes, 37, 38, 44
ocean circulation models, 195
oceans, *see under* marine waters

Ochromonas,
    oxygen uptake and cell volume,
        52
    oxygen uptake and starvation, 52
organic fluxes, 287
organic matter,
    effect on benthic respiration, 111
    links to respiration, 273
    loading in lakes and respiration
        rate, 106
    measurements in polluted
        waters, 7
    movement of through
        ecosystems, 15
    respiration as a measure of flow,
        15
oxidases, terminal in bacteria, 21
oxidation,
    definition, 11
    electrical potentials of redox
        pairs, 22
    half-reaction, 21
oxidative pentose phosphate
        pathway, 37, 40
oxidative phosphorylation, 37, 38,
        39
oxygen,
    apparent oxygen utilization in
        dark ocean, 186
    cf. carbon dioxide measures in
        benthic estuarine
    respiration, 127
    concentrations for suboxia, 225
    consumption,
        chemical oxidation and, 104
        hypolimnetic, 108
        in dark cf. in light, 104
        in dark bottles as measure of
            planktonic respiration, 104
        in sediments of dark ocean, 196
        in vitro measures of in dark
            ocean, 188
        measurement of in lake
            plankton, 104
        rates in mesopelagic zone, 193
    cycling, reactions involved in, 3
    dynamics in lakes, 113
    effect on respiration rate, in
        zooplankton, 63
    flux, anomalous observations *in
        vitro*, 152
    measurements
        accurate, 151
        of consumption, 50
        of dissolved, 236

    of flux by dark bottle method,
        148
    precision limits on, 149
    suboxia, 225–43
    tension, 50
    uptake,
        by sediments vs water depth,
            211
        cell cycle and, 51
        cell size and, 51
        factors affecting, 50
        food access and growth, 52
        measures in protists, 47
        oxygen tension and, 51
        protist volume and, 54
        starvation and, 52, 53
        temperature and, 51
oyster reefs, 123

Paramecium, oxygen uptake and
        cell volume, 52
particulate organic carbon,
    active transport of by plankton,
        187
    delivery to dark ocean, 184
    measures of, 184
    respiration and, 158
    sediment traps for, 184, 185
    supply to dark ocean, 184
particulate organic matter, supply
        to dark ocean, 183
peatlands,
    biogeochemistry and ecology of,
        86
    estimates of respiration rates in,
        277
    primary productivity in, 87
pelagic respiration,
    in dark ocean, 188
    in estuaries,
        cf. open-water whole system
            measures, 140
        chlorophyll and, 136
        communities contributing to,
            137
        effects of substrate supply on,
            134
        effects of temperature on, 134,
            135
        links to production, 136
        measurement of, 125
        rates of, 132, 133
pelagic waters,
    cf. benthos, 281
    estimates of respiration rates in,
        278

permafrost, thawing of wetlands on, 95
peroxisome, 43
PGA, 43
P-glycolate, 43
pH,
    effect on respiration rate, in zooplankton, 63
    of marshes, 87
    of peatlands, 86
phosphorus,
    plankton respiration and, 107
    sediment respiration rate and, 112
photoautotrophs, reactions in, 3
Photobacterium as an oxygen sensor, 50
photolithotrophs, 36, 38
    alternative oxidases in, 38
    chloroachniophytes, 37
    chlororespiration in, 38, 40
    cryptophytes, 37
    cytochrome oxidase in, 38
    dark respiration in, 37
    diversity of, 36, 37
    embryophytes, 37
    glycolate oxidase in, 38
    glycolysis in, 37, 38
    green algae, 37
    green algae peculiarities, 40
    haptophytes, 37
    heterokonts, 37
    Mehler reaction in, 38, 40
    red algae, 37
    relative importance of respiration processes in, 44
    respiration in, 36–45
    RUBISCO reaction in, 38, 42
photooxidation, 104
photoreduction, 255
photorespiration, 40–3
    cf. dark measures of oxygen consumption, 104
    mechanism, 105
    risk of error in respiration measures and, 149
photosynthesis,
    autochthonous, 163
    depth-integrated, 163
    diagram, 13
    estimation from ocean color, 171
    frequency distribution of rate, 161
    history of study in aquatic ecosystems, 6
    measurements of, 1

relationship with respiration, 11
relationship with respiration rate, 158
time series with respiration, 160
volumetric respiration and, 162
photosynthesis, consequences of introduction of 14C technique, 8, 9
photosynthetically active radiation, gas effluxes in wetlands and, 92
phytoplankton,
    dark nutrient assimilation and vertical migration, 255
    iron stress and respiration, 255
    models of respiration, 254
    photoreduction and respiration, 255
    respiration in photolithotrophs, 36–45
plankton respiration,
    autotrophic vs heterotrophic models, 253
    chlorophyll and, 107
    community rates, 154
    dissolved organic carbon and, 107
    distribution within community, 166
        calculation from biomass, 166
        derivation from growth efficiencies, 169
        derivation from rates of production, 169
        estimate from ^{14}C determination of net primary production, 170
        inverse analysis, 170
        size distribution, 168
    global database, 155
    global estimate in lakes, 117
    in pelagic estuarine waters, 125
    in situ measures, 148
    in surface marine waters, 147–76
    in the dark ocean, 182
    incorporation in ecosystem function models, 248–64
    lake depth and, 107
    limit of detection, 148
    methods of measurement, 147
        product flux, 148
        reactant flux, 148
    models of, 106, 248, 249
    phytoplankton models, 254
    total phosphorus and, 107

plastids, 39, 40, 43
    chlororespiration in, 40
    glycolysis in, 40
plastoquinone (PQ), 39, 43
pollution, organic matter measurements in, 7
Posidonia oceanica, community respiration measures in, 216
predation,
    lack of in containment experiments, 150
    loss of in containment methods, 150
pressure, *see under* hydrostatic pressure
primary production,
    control on oceanic, 241
    effect on benthic respiration, 111
    estimates of respiration from ^{14}C determination of net, 170
    global aquatic respiration and, 288
    global estimate for lakes, 116
    global estimates for wetlands, 94
    in coastal waters, 220
    in estuaries, 122
    in peatlands, 87
    in swamps and marshes, 87
    link with respiration, 282
    variability of cf. respiration, 272
production processes, 1
production rate, derivation of respiration distribution from, 169
productivity,
    link with respiration, 282
    measurements of, 2
    measures of, 8
α-Proteobacteria, 24
β-Proteobacteria, 24
γ-Proteobacteria, 24
    iron-reduction in, 27
δ-Proteobacteria, 27
    sulfate respiration in, 29
Proteobacteria, denitrification in, 25
protists,
    aerobic metabolism in, 48
    anaerobic metabolism in, 48
    cell size and oxygen uptake, 51
    denitrification processes in, 230
    ectosymbiotic sulfate-reducing bacteria on, 49

protists (*cont.*)
  endosymbiotic methanogens
    in, 49
  endosymbiotic purple non-sulfur
    bacteria in, 49
  energy expenditure in, 47
  facultative anaerobes, 49
  fermentation in, 48, 49
  food access and growth, 52
  measuring respiratory rates in, 49
  microaerophiles, 49
  oxygen uptake by, 50
  respiration in, 47–54
    rate in, 47
    scales and size, 53
  starvation and oxygen uptake in,
    52, 53
proton donors, 4
proton flux through ecosystem, 13
protozooplankton,
  biomass, 57
  physiology, 57
PsI, 42
PsII, 42
Pseudocohnilembus, oxygen uptake
  and starvation, 53
Pseudomonas, denitrification
  processes in, 230
Purple bacteria, 42
pyruvate, 39
pyruvate kinase, measures in
  zooplankton, 59

Q cycle, 40

red algae, 37, 42, 44
Redfield ratios, 209, 248
redox potentials, 22
redox zonation, 207
reduction,
  electrical potentials of redox
    pairs, 22
  half-reaction, 21
remineralization, 11
respiration,
  aerobic cf. anaerobic, 19
  aquatic vs terrestrial, 291
  autotrophic vs heterotrophic
    models, 253
  bacterial, 18–32
  benthic rates, 109
  biochemistry of, 2
  biogeochemical role, 293
  components of aquatic, 269–72
    variation and regulation of, 272
  conceptual basis, 11

contribution of anaerobic to
  global respiration, 286
dark, 37
definition, 1, 11
diagram, 13
ecological role, 293
ecological terminology, 11
function and interactions, 250
global environmental change
  and, 294
history of study in aquatic
  ecosystems, 6
in estuaries, 122–42
in sediments, 108
in zooplankton, 57–77
link with community structure,
  281
link with productivity, 282
links to organic matter pools, 273
measurement of, 12
  in polluted waters, 7
pathways within, 268
processes involved in, 2
relationship with primary
  production, 11
requiring light, 40
role of bacteria in, 14
see also by type
single cell level cf. ecosystem
  level, 268
suboxic, 225–43
time series with photosynthesis,
  160
volumetric and photosynthesis,
  162
whole ecosystem context, 14
respiration rate,
  attenuation and, 158
  bacterial abundance and, 158
  bacterial models, 256
  behavior and, 252
  body mass and, in zooplankton,
    64
  bottle effects, 150–151
  by zooplankton community, 76
  cf. growth rate in *Klebsiella*, 251
  chlorophyll and, 158, 159
  community, 154
  depth distribution, 156
  depth-integrated, 163
  derivation from activity and
    growth efficiencies, 153
  derivation from biomass, 153
  diel rhythms in, in zooplankton,
    65

distribution within plankton
  community, 166
errors of interpretation, 151–152
errors, procedural, 149–150
errors, random and systematic,
  149
feeding and, in zooplankton, 65
frequency distribution of, 161
global,
  assessment of in zooplankton,
    75
  dark bottle studies, 171
  database, 155
  estimation from direct field
    observation, 171
  from correlation analysis, 172
  from ocean organic
    mass-balance calculations,
    172
hydrostatic pressure and, in
  zooplankton, 61
in benthic communities, 207
in coastal macroalgae
  communities, 217
in coastal macrophyte
  communities, 214
in coastal sea grass communities,
  216
in coral reefs, 212
in dark ocean and depth, 190
in major aquatic ecosystems, 276
in mangroves, 213
in salt marshes, 213
latitudinal distribution, 155, 156
light and, in zooplankton, 62
limit of detection, 148–149
measurements of,
  at ecosystem level, 275
  in wetlands, 85
mesozooplankton, 192
models of, 249
mortality and, 252
microplankton in dark ocean, 189
oxygen and, in zooplankton, 63
particulate organic carbon and,
  158
pH and, in zooplankton, 63
relationship with photosynthesis,
  158
salinity and, in zooplankton, 64
seasonal changes in, in
  zooplankton, 66
seasonal distribution, 155
sediment grain size and, 210
size distribution and calculation
  of biomass, 169

spatial variability and, in
    zooplankton, 67
temperature and, 252
    in wetlands, 89
    in zooplankton, 60, 69
turbulence and, in zooplankton,
    62
vertical distribution of, in
    zooplankton, 68
zooplankton models, 261
respiratory quotient,
    anomalous values, 152
    calculation of, 5
    categories of, 6
    definition, 5
    for coastal waters, 208
    in benthic estuarine
        reproduction, 127
    measures in zooplankton, 59
rhizopods, respiratory rates in, 54
rice paddies, biogeochemistry and
    ecology of, 88
rivers,
    estimates of respiration rates in,
        277
RUBISCO, 36, 37, 38, 42, 43, 45
RUBISCO reaction, 3, 37
    diagram, 43
    in oxic environments, 12
    in photolithotrophs, 42
    relative importance of, 44
    risk of error in respiration
        measures and, 149

salicylhydroxamic acids, 38
salinity, effect on respiration rate in
    zooplankton, 64
salt marshes,
    community respiration measures
        in, 213
    respiration rate in, 207
sea grasses, 206
    community respiration measures
        in, 214
    metabolism of communities, 215
    respiration rate in, 207
    seasonal changes in metabolism,
        217
sealed-chamber method, 58
seas, *see under* marine waters
seasons,
    variation in carbon dioxide
        dynamics in wetlands, 91, 92
    variation in methane dynamics
        in wetlands, 91, 92

variation in respiration rate of
    zooplankton with, 66
sediments,
    benthic respiration, 109
        in dark ocean, 195
        in estuaries, 125
    burial efficiencies in, 195
    coastal,
        accumulating cf.
            non-accumulating, 211
        approaches to study of, 207
        bottum-up estimates in, 219
        carbon budget restraints, 219
        classification of, 209
        contributions of communities
            to, 209
        global respiration rate in,
            scaling problems in, 218
        mobile muds, 210
        oxygen uptake vs water depth,
            211
        primary production
            constraints, 220
        respiration measurement in,
            209
        sandy sediments, 210
    constraints on denitrification in,
        240
    core incubations, 109
    effect of temperature on
        respiration in, 110
    estimates of respiration rates in,
        277
    fine grain cf. coarse grain, 132
    grain size and respiration in, 210
    measuring respiration in, 108,
        109
    models of benthic respiration, 110
    oxygen consumption and organic
        carbon flux, 194
    oxygen consumption as a
        function of sediment area, 109
    respiration,
        ambient oxygen
            concentrations and, 112
        global estimate in lakes, 117
        in dark ocean and, 195
        in lakes, 108
        sample depth and, 112
    sediment trap data of particulate
        organic carbon, 184
    sublittoral, 207
    terminal electron acceptors and
        depth, 28
    vertical zonation of anaerobic
        metabolism in marine, 19

selenate reduction, 31
serine, 43
Shewanella putrifaciens, 26
Starvation and oxygen uptake in
    protists, 52, 53
suboxia,
    anthropogenic effects, 242
    as barrier to anoxia, 239
    chemical features of water
        column in, 227
    climate change and, 241
    concentrations of electron
        acceptors in suboxic waters,
        228
    contribution of suboxic
        respiration to global
        respiration, 286
    definition, 225
    denitrification processes, 229
    dinitrogen measurements in, 236
    dissolved oxygen measurements
        in, 236
    in Arabian Sea, 226
    in ocean water column, 225–43
    incubation techniques for
        studying, 238
    methods of measurement, 236
    nitrification and nitrous oxide
        gradients in, 227, 228, 234
    nitrous oxide cycling and, 234
    nutrient determination in, 237
    occurrence of, 227
    sequence of respiration in, 229
    significance of respiration in, 239
    temporal variability, 242
    terminology used, 226
substrate oxidation systems, 20
succinate dehydrogenase measures
    in zooplankton, 58
sufur hexafluoride, 148
sulfate,
    recycling in wetlands, 91
    reduction, 208, 225
        ectosymbiotic sulfate-reducing
            bacteria on protists, 49
        in benthic estuarine
            reproduction, 127
    respiration, 27
        enzymes involved in, 29
        in freshwater sediments, 27
        in marine benthos, 28
        links with methanogenesis, 30
        sequential reduction in, 29
sulfide,
    concentrations in suboxia, 225,
        226

sulfide (*cont.*)
  oxidation, in *Acanthamoeba*, 48
  oxidizers, 24
  production, suboxic electron
    acceptors as barrier to, 239
sulfidogens, 28
superoxide, 40
superoxide dismutase, 40, 41
swamps,
  biogeochemistry and ecology of,
    87
  estimates of respiration rates in,
    277
symbiosis,
  ectosymbiotic sulfate-reducing
    bacteria on protists, 49
  endosymbiotic methanogens on
    protists, 49
  endosymbiotic purple non-sulfur
    bacteria in protists, 49

TCA cycle, 5, 37, 38, 39, 44, 45
temperature,
  carbon dioxide production in
    wetlands and, 90
  effect on benthic respiration, 110
  effect on estuarine benthos, 131
  effect on oxygen uptake, 51
  effect on pelagic respiration in
    estuaries, 134, 135
  effect on respiration rate, in
    zooplankton, 60
  effect on wetland dynamics, 89
  effects on plankton respiration in
    lakes, 106
  methane production in wetlands
    and, 90
  respiration rate and, 252
  in zooplankton, 69
terminal electron acceptor
    reduction, 20
terminal electron acceptors and
    sediment depths, 28
terminal oxidation system, 5
terminology, 11
terrestrial respiration vs aquatic,
    291
Tetrahymena,
  oxygen uptake and cell size, 51
  oxygen uptake and cell volume,
    52
  oxygen uptake and starvation, 52
Thalassia sp., community
    respiration measures in, 216
Thermodesulfobacterium, 29

Thermodesulfovibrio, 29
TCA cycle, 37, 38, 39, 44, 45
thylakoid, 42
total inorganic carbon, precision
    limits on measures of, 149
total organic carbon, from coastal
    waters to dark ocean, 187
triple isotope composition, 154
trophic condition, effects on
    plankton respiration in
    lakes, 106
trophic status, effect on benthic
    respiration, 111
turbulence, effect on respiration
    rate, in zooplankton, 62

ubiquinone (UQ), 39, 43
upwelling, filaments from, 72, 74
uranium in suboxic waters, 229

vertical migration,
  active flux and, 71
  dark nutrient assimilation by
    phytoplankton and, 255
volumetric photosynthesis, 162
volumetric respiration, 162

Warburg technique, 49
water level and gas effluxes in
    wetlands, 92
wetlands,
  biogeochemistry and ecology of,
    86
  carbon cycle in, 90
  carbon dioxide dynamics in, 89
  carbon dioxide respiration in, 88
  carbon production, controls on,
    90
  definition of, 83
  effect of latitude on dynamics of,
    89
  emissions from, 84
  estimates of respiration rates in,
    278
  global area of, 83
  global extent and distribution, 93
  global respiration estimates for,
    93
  growing season cf. non-growing
    season, 95
  methane dynamics in, 89
  methane respiration in, 88

respiration rates in, 85
  seasonal changes in gas effluxes,
    91, 92
  thawing over permafrost, 95
  water level and water level in, 92
Winkler dark bottle measurements,
    10

yeast, oxygen uptake and cell size,
    51

zoobenthos, biomass, 57
zooplankton,
  active flux in, 70
  divisions of, 57
  energy budgets in, 57
  euryhaline species, 64
  global assessment of respiration,
    75
  in bathypelagic zone, 70
  in epipelagic zone, 67
  in mesopelagic zone, 68
  large scale variability and
    respiration, 73
  measuring respiration rates in,
    enzymatic methods, 58
    flow-through method, 58
    sealed-chamber method, 58
  mesoscale variability and
    respiration, 72
  metabolic changes and activity,
    60
  metabolic scope in, 60
  migration of, 71
  respiration,
    models of, 261
    mesozooplankton, 262
    microzooplankton, 261
  respiration rate in, 57–77
    body mass and, 64
    diel rhythms in, 65
    feeding and, 65
    hydrostatic pressure and, 62
    light and, 62
    oxygen and, 63
    pH and, 63
    salinity and, 64
    seasonal changes, 66
    spatial variability and, 67
    temperature and, 60
    turbulence and, 62
  respiratory quotient in, 59
  stenohaline species, 64
  temperature and respiration rate,
    69

vertical distribution,
  of biomass, 68
  of respiration rate, 68
vertical distribution and diel
  rhythms in, 70

zooplankton respiration,
  latitudinal distribution, 76
  mesoscale varibility, 72
  seasonal changes, 66–67
  vertical distribution, 67–72

Zostera marina, community
  respiration measures
  in, 216
Zostera noltii, community
  respiration measures in, 216

Arrhenius eq. → temp. dependence of a chem. rxn. rate (k)

$$k = A e^{\frac{-E_a}{RT}}$$

E_a - activation energy

R - gas constant
    $= 8.314 \, J/K \cdot mole$

A - constant specific
    to a rxn.